Soils and Landscape Restoration

Soils and Landscape Restoration

Edited by

John A. Stanturf
Estonian University of Life Sciences, Tartu, Estonia

Mac A. Callaham, Jr.
USDA Forest Service, Athens, Georgia, United States

ELSEVIER

ACADEMIC PRESS
An imprint of Elsevier

Academic Press is an imprint of Elsevier
125 London Wall, London EC2Y 5AS, United Kingdom
525 B Street, Suite 1650, San Diego, CA 92101, United States
50 Hampshire Street, 5th Floor, Cambridge, MA 02139, United States
The Boulevard, Langford Lane, Kidlington, Oxford OX5 1GB, United Kingdom

British Library Cataloguing-in-Publication Data
A catalogue record for this book is available from the British Library

Library of Congress Cataloging-in-Publication Data
A catalog record for this book is available from the Library of Congress

ISBN: 978-0-12-813193-0

For Information on all Academic Press publications
visit our website at https://www.elsevier.com/books-and-journals

Publisher: Candice Janco
Editorial Project Manager: Emily Thomson
Production Project Manager: Joy Christel Neumarin
                          Honest Thangiah
Cover Designer: Matthew Limbert

Typeset by MPS Limited, Chennai, India

Working together
to grow libraries in
developing countries

www.elsevier.com • www.bookaid.org

Contents

List of Contributors

J.T. Bauer
Department of Biology, Miami University, Oxford, Ohio, United States

Roland Bobbink
B-WARE Research Centre, Radboud University, Nijmegen, The Netherlands

Rachel L. Brockamp
Department of Soil Science, University of Saskatchewan, Saskatoon, SK, Canada

Matt D. Busse
U.S. Department of Agriculture, Forest Service, Pacific Southwest Station, Albany, CA, United States

Loren B. Byrne
Department of Biology, Marine Biology and Environmental Science, Sustainability Studies Program, Roger Williams University, Bristol, RI, United States

Mac A. Callaham Jr.
U.S. Department of Agriculture, Forest Service, Athens, GA, United States

Michael D. Casler
U.S. Department of Agriculture, Agricultural Research Service, U.S. Dairy Forage Research Center, Madison, WI, United States

H.A. Cray
School of Environment, Resources & Sustainability, University of Waterloo, Waterloo, Ontario, Canada

B. Dell
College of Science, Health, Engineering and Education, Murdoch University, Murdoch, WA, Australia

Jan Frouz
Institute for Environmental Studies, Charles University, Prague, Czech Republic; Biology Centre of the Academy of Sciences of the Czech Republic, Institute of Soil Biology & SoWa Research Infrastructure, České Budějovice, Czech Republic

Richard A. Hallett
U.S. Department of Agriculture, Forest Service, Northern Research Station, Durham, NH, United States

R.J. Harper
College of Science, Health, Engineering and Education, Murdoch University, Murdoch, WA, Australia

Jim Harris
School of Energy, Environment and Agrifood, Cranfield University, Bedfordshire, United Kingdom

C.A. Havrilla
Merriam Powell Center for Environmental Research & Center for Ecosystem Science and Society, Northern Arizona University, Flagstaff, Arizona, United States

Eric J. Jokela
School of Forest Resources and Conservation, University of Florida, Gainesville, FL, United States

Martin F. Jurgensen
College of Forest Resources and Environmental Science, Michigan Technological University, Houghton, MI, United States

K. Katzensteiner
Institute of Forest Ecology, University of Natural Resources and Life Sciences (BOKU), Vienna, Austria

J.M. Kranabetter
B.C. Ministry of Forests, Lands, and Natural Resource Operations, Victoria, BC, Canada

Jackson M. Leonard
U.S. Department of Agriculture, Forest Service, Rocky Mountain Research Station, Flagstaff, AZ, United States

Chung-Ho Lin
College of Agriculture, Food, and Natural Resources, University of Missouri, Columbia, MO, United States

Palle Madsen
InNovaSilva, Vejle, Denmark

M.J. McTavish
Faculty of Forestry, University of Toronto, Toronto, Ontario, Canada

S.D. Murphy
School of Environment, Resources & Sustainability, University of Waterloo, Waterloo, Ontario, Canada

Daniel G. Neary
U.S. Department of Agriculture, Forest Service, Rocky Mountain Research Station, Flagstaff, AZ, United States

M. Oelbermann
School of Environment, Resources & Sustainability, University of Waterloo, Waterloo, Ontario, Canada

Deborah S. Page-Dumroese
U.S. Department of Agriculture, Forest Service, Rocky Mountain Research Station, Moscow, ID, United States

J.H. Pedlar
Natural Resources Canada, Canadian Forest Service, Great Lakes Forestry Centre, Sault Ste. Marie, ON, Canada

Andrej Pilipović
Institute of Lowland Forestry and Environment, University of Novi Sad, Novi Sad, Serbia

C.E. Prescott
Faculty of Forestry, University of British Columbia, Vancouver, BC, Canada

J.K. Ruprecht
College of Science, Health, Engineering and Education, Murdoch University, Murdoch, WA, Australia

E.J. Sayer
Lancaster Environment Centre, Lancaster University, Lancaster, United Kingdom

K.R.J. Smettem
College of Science, Health, Engineering and Education, Murdoch University, Murdoch, WA, Australia; School of Civil, Environmental and Mining Engineering, The University of Western Australia, Crawley, WA, Australia

S.J. Sochacki
College of Science, Health, Engineering and Education, Murdoch University, Murdoch, WA, Australia

John A. Stanturf
Institute of Forestry and Rural Engineering, Estonian University of Life Sciences, Tartu, Estonia

Rudy van Diggelen
Ecosystem Management Research Group, University of Antwerp, Antwerp, Belgium

Erik Verbruggen
Plants and Ecosystems, University of Antwerp, Antwerp, Belgium

C. Weston
University of Melbourne, School of Ecosystem and Forest Sciences, Melbourne, Australia

Sharon L. Weyers
U.S. Department of Agriculture, Agricultural Research Service, North Central Soil Conservation Research Laboratory, Morris, MN, United States

R.S. Winder
Natural Resources Canada, Canadian Forest Service, Pacific Forestry Centre, Victoria, BC, Canada

Ronald S. Zalesny, Jr.
U.S. Department of Agriculture, Forest Service, Northern Research Station, Rhinelander, WI, United States

Soils are fundamental to landscape restoration

John A. Stanturf[1], Mac A. Callaham Jr.[2] and Palle Madsen[3]
[1]*Institute of Forestry and Rural Engineering, Estonian University of Life Sciences, Tartu, Estonia*
[2]*U.S. Department of Agriculture, Forest Service, Athens, GA, United States*
[3]*InNovaSilva, Vejle, Denmark*

1.1 Introduction

Soils have long been recognized as integral to the establishment and productivity of agricultural and forestry systems; because of this, soils are frequently manipulated with particular outcomes in mind. However, the role of soils in determining or maintaining the character and integrity of natural systems has been underappreciated. Thus soils are less frequently considered when management goals involve restoration of degraded systems back into more natural states. The prevailing dogma pertaining to soils in restoration schemes has been an attitude of benign neglect. The reasoning seems to be that if the most serious insults are removed, and vegetation reestablished, soil will recover on its own without any need for further intervention. This attitude is most forcefully displayed by the argument for the superiority of passive restoration over more active intervention (Bechara et al., 2016; Chazdon and Uriarte, 2016; Meli et al., 2017). Recently, however, a more nuanced understanding has developed of the critical feedbacks between soil and vegetation (Heneghan et al., 2008; Kardol and Wardle, 2010; van der Bij et al., 2018; Wardle and Van der Putten, 2002) that should raise awareness of the importance of considering soils in restoring natural ecosystems.

A clearer focus has emerged on the complexity of soil ecosystems and on the properties of soils that promote plant and animal diversity, ecosystem functioning, and the provision of ecosystem services. This view has been informed by some of the earliest theories of pedology that refers to the soil formation state factors proposed by Jenny (1941, 1961), wherein soil formation is viewed as a function of interactions among climate, organisms, relief, parent material, and time. More recent thinking recognizes the cumulative impact of long-term management on soils (Richter and Yaalon, 2012; Yaalon, 2007; Yaalon and Yaron, 1966), expressed as soil memory (Baer et al., 2012), and more generally ecosystem legacies (Frelich et al., 2018; Jõgiste et al., 2017; Johnstone et al., 2016). The state factors provide a useful conceptual model for examining soil and landscape restoration. Climate change has a significant impact on soils, including decoupling

Soils and Landscape Restoration. DOI: https://doi.org/10.1016/B978-0-12-813193-0.00001-1

important soil—vegetation linkages. Even though climate manipulation is hardly feasible, the feedback of soils to the atmosphere (the soil—plant—atmosphere continuum) should be recognized (Philip, 1966). There are numerous examples of manipulations of relief, parent material, and organisms resulting in acceleration of recovery processes that normally would require lengthy time intervals.

The accelerated organization and recovery of the ecosystem is one essential goal of restoration science (Gann et al., 2019; Holl and Kappelle, 1999; SERI, 2004). Here we will employ an inclusive definition of restoration that "encompasses any management activity which alters or accelerates the trajectory of recovery to a more naturally functioning system" (Stanturf, 2016; Stanturf et al., 2014a,b). Other chapters in this volume discuss the twin aspects of soils and restoration: (1) specific consideration of methods to accelerate recovery of soil processes in different biomes or (2) pathways that avoid obstacles to recovery that exist due to soil factors. The latter aspect refers to characteristics or substances in soil inimical to plant establishment and growth or due to the lack of soil caused by erosion or removal. In this chapter, we present an overview of roles that soils can play in landscape restoration.

Restoration ecology is a relatively young discipline, and thus debate continues surrounding the theoretical underpinnings and the basic language of restoration, leading to sometimes widely divergent views of what constitutes restoration in practice. For example, the definition of "restored" can be unclear or even contentious, as illustrated by the question: restore to what point in the past? Especially in the Americas and Oceania where human occupancy is recent on an archeological timescale, the reference of historical past is often chosen as the point just prior to European colonization, somewhat minimizing the manipulation of the environment by indigenous populations, in spite of ample evidence that these people exerted tremendous influence on the ecosystems they inhabited (Barlow et al., 2012; Clement et al., 2015; Krech III, 1999; Raymond, 2007). In northern climes, another reference point is immediately after retreat of the glaciers, but this ignores the subsequent movement of species under changing climate and the loss of megaherbivores (Bradshaw and Sykes, 2014). Indeed, the past reality of major changes in climate, human population distribution, and development of ecosystems argues strongly for avoiding too specific a target for restoration.

To illustrate the limited value of historical vegetation as a strict guide for restoration, we offer the following thought experiment: (1) pick any point in time from the past 10,000 years, (2) imagine ourselves transported to that time, and (3) accept the task of maintaining the vegetation in the exact state that we found it for the next 2000 years. What would be our likelihood of success? If we use the pollen record for the Eastern United States as a guide (e.g., Ballard et al., 2017), it would be unlikely that we could produce a period of vegetative stability that persisted for 2000 years, regardless of our best efforts. The wisdom of using historic reference points that assume quasiequilibrium of plant assemblages and stability of climate is questionable, given the likelihood of substantial change in climate and land use over the rest of this century (Stanturf, 2015, 2016).

Nevertheless, restoration of forest and landscape has emerged as a powerful concept that has broad international support (Besseau et al., 2018; Brancalion et al., 2019; Maginnis and Jackson, 2007; Mansourian et al., 2017, 2020). Before moving on to discuss soils and restoration, we will present the international policy context surrounding restoration efforts, responding to the query, why is restoration important?

1.2 Policy context

Interest in soil restoration is embedded within a variety of international policy initiatives with somewhat different emphases but all relevant to restoration. The Changwon Initiative of the United Nations Convention to Combat Desertification (UNCCD) is most directly aimed at soil restoration. The target of net land degradation neutrality (LDN) that developed from the UNCCD aims to maintain or improve the condition of land resources, including restoration of natural and semi-natural ecosystems (Akhtar-Schuster et al., 2017; Cowie et al., 2018; Orr et al., 2017; Safriel, 2017). In the same way a target of the 2010 Strategic Plan of the Convention on Biological Diversity (now being revised) is no net biodiversity loss and net positive impacts on biodiversity (CBD, 2010), including soil biodiversity (Geisen et al., 2019; Nielsen et al., 2011; Wagg et al., 2014). In the climate arena, REDD+ (reducing emissions from deforestation and degradation) attempts to mitigate climate change by encouraging retention of carbon in forested ecosystems through sequestration and avoided deforestation (Alexander et al., 2011; Parrotta et al., 2012; Putz and Nasi, 2009).

Forest restoration has attracted particular attention and enjoys much international support since the inception of the Bonn Challenge in 2011. The Bonn Challenge sets a goal of 150 million ha of the world's deforested and degraded land to be brought into restoration by 2020 and 350 million ha by 2030 (Mansourian et al., 2017a,b). The foundation of the Bonn Challenge is forest landscape restoration (FLR), a "planned process that aims to regain ecological integrity and enhance human wellbeing in deforested or degraded landscapes" (WWF and IUCN, 2000). Although different interpretations of FLR have emerged in the interim (e.g., Lamb et al., 2012; Sabogal et al., 2015), all agree on the fundamental aspects of FLR: its long-term nature, scale, and equal emphasis on ecological and social aspects. As of this writing, 172.35 million ha in 55 countries have committed to the Bonn Challenge (https://www.bonnchallenge.org/about-the-goal#commitments), although the level of achievement has been questioned (Fagan et al., 2020). Nevertheless, restoration efforts have a long history with varied objectives (Stanturf et al., 2014b); later, we briefly describe five historical efforts (Northeastern and Southeastern United States, Puerto Rico, Denmark, and South Korea).

These restoration efforts can potentially contribute to meeting national commitments to the Aichi Biodiversity Targets (CBD, 2010) and the LDN goal

(Akhtar-Schuster et al., 2017). Many countries have included restoration and sustainable land use in their Nationally Determined Contributions under the Paris Climate Agreement (Brancalion and Chazdon, 2017; Woolf et al., 2018). In 2019 the United Nations designated the period of 2021−30 as the "Decade of Ecosystem Restoration." The purpose of the decade is to accelerate progress toward meeting existing global restoration goals and build on regional efforts. Ecosystem restoration is an inclusive concept, aimed at reversing degradation and regaining ecological functionality of a wide array of terrestrial and marine ecosystems. Ecosystem restoration addresses the goals of the Rio Conventions on biodiversity, desertification, and climate change (https://www.unenvironment.org/news-and-stories/press-release/new-un-decade-ecosystem-restoration-offers-unparalleled-opportunity). Whether ecosystem restoration will fully come to pass or not remains to be seen (Cooke et al., 2019).

Along with the concern for climate change, reversing biodiversity loss, and the need to protect natural areas have long been an international focus. Since the Brundtland Report in 1987 (Brundtland, 1987) called for saving 10%−12% of the globe in protected area, proposals have expanded to half of the world's terrestrial and aquatic area ("Nature Needs Half" or a "Global deal for Nature"; Dinerstein et al., 2019; Locke, 2014). While this goal is admirable, the reality is that many already protected areas are degraded (Leverington et al., 2010; Terra et al., 2014) and will require restoration (Cairns et al., 2012; Janishevski et al., 2015; Mappin et al., 2019).

1.3 Nature of soils

Soils are natural bodies, occurring in the upper layer of the Earth's crust. Soils are formed by weathering of geologic parent material through physical, chemical, and biological processes. Components of soils include mineral particles that vary with geologic parent material, organic matter, water, air, and living organisms. The mineral fraction determines soil texture, reflecting the relative amount of sand, silt, and clay particles (Fig. 1.1). These solid particles are acted on by biological, chemical, and physical processes to form aggregates that in turn define soil structure. Soil structure is formed by the size, organization, and shape of soil aggregates. Texture and structure influence porosity and bulk density that determine how gases and fluids move through or remain in soil, defining the role of soils in hydrologic processes and feedbacks to the atmosphere.

Important chemical properties of soils include reaction (pH), redox (oxygenation status), and cation-exchange capacity (CEC). Secondary clay minerals (e.g., smectite, vermiculite, and illite), along with organic matter, are the primary source of CEC. Carbon and nutrients in soils vary by biome and management, and their persistence and cycling are influenced by soil physical and chemical properties (FAO and ITPS, 2015).

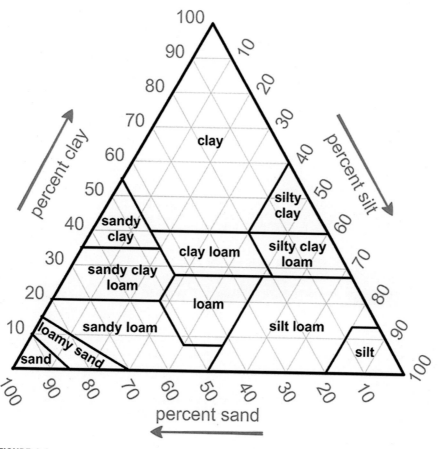

FIGURE 1.1

Soil textural triangle showing the percentages of sand, silt, and clay in 12 basic texture classes. To determine the texture of a soil the components must add to 100%. The texture is where the lightly shaded lines from the three components meet. For example, a loam texture could have 50% sand, 20% clay, and 30% silt-sized particles.

Courtesy of Soil Science Division Staff, 2017. Soil Survey Manual, U.S. Department of Agriculture Handbook 18. Government Printing Office, Washington, DC (Soil Science Division Staff, 2017), Fig. 3.7.

A distinctive feature of soils is the arrangement in horizontal layers that are used to describe soils (Fig. 1.2), giving rise to classifications that group similar soils based on diagnostic soil horizons (Arnold, 2016). Two classifications have global applicability, the FAO World Reference Base for Soil Resources (http://www.fao.org/3/i3794en/I3794en.pdf) and the USDA soil taxonomy (Soil Survey Staff, 1999). Describing soils and understanding how they formed is the province

FIGURE 1.2

Soil profiles illustrative of the diversity of soils. These profiles are of the 12 major orders of soil taxonomy. Top, from left to right: Entisols, Inceptisols, Alfisols, Mollisols, Ultisols, Oxisols. Bottom, from left to right: Aridisols, Andisols, Vertisols, Histosols, Spodosols, Gelisols.

Modified from US Department of Agriculture, Natural Resources Conservation Service; https://www.nrcs.usda. gov/wps/portal/nrcs/detail/soils/survey/class/maps/?cid = nrcs142p2_053589.

of pedology. Concepts have changed over time, but current understanding is highly influenced by the work of Jenny (1941) and others (Bockheim and Gennadiyev, 2010; Jenny, 1961). Jenny (1941) refined the ideas of Russian pedologist Vasily V. Dokuchaev and presented soil formation state factors as

$$\text{Soil} = \int (Cl, O, R, P, T \ldots)$$

where Cl is the climate, O is the organisms, R is the relief, P is the parent material, and T is the time.

These state factors have been modified for digital soil mapping by adding factors for spatial attributes (McBratney et al., 2003) and human action, atmosphere, and water (Grunwald et al., 2011). We use this conceptual framework to present the varied ways that soils should be considered in restoration.

Spatially, soils can be viewed in a hierarchy, from the pedon to the landform. Pedons are natural soil volumes large enough to incorporate all the soil horizons present and their relationships. The pedon is the smallest unit or volume of soil; it

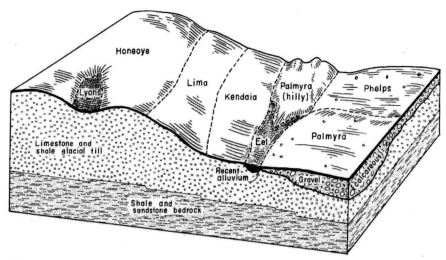

FIGURE 1.3

Soil catena. Block diagram of the Lima—Honeoye association and the Palmyra association in northwestern Tompkins County, New York. This glaciated landscape in the Finger Lakes region of central New York developed in Wisconsin glacial till. Most of the series (i.e., Honeoye, Lima, Palmyra, and Phelps) are classified as Hapludalfs (Alfisols have a clay-enriched subsoil and relatively high fertility.); the Kendaia and Lyons series are Aquepts, wet Inceptisols with incipient horizon development.

Courtesy of USDA Soil Conservation Service, 1965. Soil Survey Tompkins County, New York, Washington, DC.

usually has a surface area of approximately 1 m² and extends from the ground surface down to bedrock. Pedons are the building blocks that make up both soil taxonomic classes and soil mapping units. The polypedon is a parcel of contiguous pedons, all of which have characteristics lying within the defined limits of a single soil series. Comparing soils to a biological entity, pedons are similar to an individual tree, while a polypedon is comparable to a species (Johnson, 1963). Soils in a landscape are arranged across a slope called a catena or toposequence (Fig. 1.3, catena), a group of distinct soils arrayed down a slope. Each soil type (facet or polypedon) differs somewhat from its neighbors, but all occur on the same underlying parent material and developed under the same climate (Huggett, 1975). Zooming out, landforms are terrain features on the Earth's surface. The four main types of landforms are mountains, hills, plateaus, and plains, with many subtypes and variants according to climate and geology (Deng, 2007). Landform and slope comprise the relief factor in Jenny's equation.

Biology is another state factor. Vegetation type and productivity are governed at the macroscale by climate and regional weather patterns and locally by soil type. Aboveground, the litter returned to the soil surface may be incorporated into soil by organisms (e.g., earthworms and other invertebrates; Coleman et al., 2018;

Heneghan et al., 2008). Within soil, plant roots and associated fungi provide carbon and nutrients to other microbes that in turn decompose, adding to soil carbon stocks. An emerging concept in plant and soil sciences is the holobiont, an assemblage of a host and the many other species living in or around it, which together form a discrete ecological unit (Hassani et al., 2018; Vandenkoornhuyse et al., 2015). The microbial diversity in a given soil, and the seeming ability of plants to manipulate their microbiota to adjust to their environment (Vandenkoornhuyse et al., 2015), suggests that this may be an important mechanism for adapting to future climates. The biological and ecological components of soils are discussed in detail in other chapters (Callaham and Stanturf, 2020).

Moisture relations and water movement over and through soils are critical to their functioning. Hydrologic processes affect soil formation, structure, stability, and erosion. Water moving through soil recharges groundwater and lakes, streams, and rivers. Water stored in soil returns to the atmosphere through surface evaporation and plant transpiration and is one determinant of plant growth. Texture and organic matter, as well as soil depth, determine the moisture-holding capacity of soils. At one extreme, deep clayey soils, peat soils, and soils developed on volcanic substrates and have high capacity for moisture storage. At the other extreme, sandy soils and very shallow soils have low capacity. Besides storage capacity, another important characteristic is the tension with which moisture is held in the soil, which affects how much water is available for plant uptake. Loamy soils, with similar amounts of sand, silt, and clay particles, are generally balanced with respect to the moisture stored and how much is available to plants. Drainage plays an important role in oxygen status of soils (poorly drained soils develop reducing conditions), important for adequate root development. Drainage is related to soil type, texture, and slope (Coyle et al., 2016; FAO and ITPS, 2015). For millennia, humankind has manipulated water conditions in soils, with excess water being modified by drainage works and water deficiencies being mitigated by irrigation.

1.4 Scale and complexity

In natural systems, heterogeneity of soil physical, chemical, and biological properties is a product of the interaction of the soil formation state factors (Jenny, 1941, 1961). This heterogeneity is manifested as a mosaic of microhabitats across the landscape and can consist of patches ranging from 1 to 100 m^3 in size. Important factors contributing to this heterogeneity include the effects of large-scale topography (i.e., slope, aspect, shape, and elevation), to the effects of individual plants on underlying soil. For example, the soils on ridgetops are typically shallower and drier than soils in valleys, and at finer resolution, underlying soils will reflect the differences in litter chemistry and soil biota that are characteristic of a particular plant species (e.g., Boettcher and Kalisz, 1990; Fox and

Comerford, 1990; Meinders and Van Breemen, 2005). In sum, the processes that give rise to soil heterogeneity result in a diversity of niche spaces for diverse assemblages of plant, microbe, and animal species (Tsiafouli et al., 2015; Wagg et al., 2014). On lands formerly in agriculture, especially intensive forms of management, past actions sought as much as possible to homogenize soil physical and chemical characteristics in order to optimize niche space for the crop species. To this end, land leveling, plowing, harrowing, and other physical manipulation of soils affect structure and moisture in order to limit competition from other plants (i.e., weeds), and nutrient additions aim to promote growth and yield of the crop species.

1.5 Importance of soil for ecosystem services

Soils provide important ecosystem services, determined by their natural capital, a function of the core soil properties of texture, mineralogy, and organic matter (Palm et al., 2007; Robinson et al., 2013). While texture and mineralogy change slowly over time, soil organic matter content responds quickly to manipulation and disturbance (González-Pérez et al., 2004; Kukuļs et al., 2019; Palm et al., 2007; Schmidt et al., 2011). Different soil types contribute to ecosystem services in several and varied ways (Table 1.1). Plant productivity and diversity are probably the most important services, from a human standpoint, followed by water cycling and increasingly by climate regulation.

The concept of soil quality as a component of environmental quality has been defined as "the capacity of a soil to function within ecosystem and land-use boundaries to sustain biological productivity, maintain environmental quality, and promote plant and animal health" (Doran and Parkin, 1994). Originally focused on plant productivity, particularly food crops, soil quality has been broadened to encompass other ecosystem functions provided by soils (Bünemann et al., 2018) and to other land uses such as forestry (Neary, 2020; Page-Dumroese et al., 2000; Schoenholtz et al., 2000). Because soils are complex, with site-specific properties and linkages between soil functions and soil-based ecosystem services, assessing soil quality to determine management effects requires baseline or reference values (Bünemann et al., 2018). The original focus on identifying soil-quality parameters focused more on physical and chemical properties; recent developments have included greater emphasis on soil biodiversity and ecological functions under the rubric of soil health (Doran et al., 2002).

Soil health is defined as the continued capacity of soil to function as a vital living ecosystem that sustains plants, animals, and humans (Allen et al., 2011; Maikhuri and Rao, 2012). To date, the soil health concept has been applied primarily to agriculture. For example, a soil health database includes 42 soil health indicators and 46 background indicators (e.g., climate, elevation, and soil type). This database focuses on four main conservation management methods: cover crops, no-tillage, agroforestry systems, and organic farming (Jian et al., 2020).

Table 1.1 Contribution of the main soil taxonomy orders to major ecosystem services.

Soil taxonomy orders	Characteristic features	Lower levels of taxonomy	Ecosystem services[a]
Alfisols	Naturally fertile soils with high base saturation and a clay-enriched subsoil horizon		Food security
Alfisols and Ultisols	Illuviation of clay and biological enrichment of base cations	Glossic Great Groups	Biomass
Alfisols, Ultisols, Inceptisols, and Oxisols	Well-drained clayey soils with deep profiles	Different Great Groups	Food security, water
Alfisols, Ultisols, Inceptisols, Entisols, and Mollisols	Poorly drained	Aquic suborders and epiaquic Great Groups	Water storage
Alfisols, Ultisols, and Mollisols	Seasonally water saturated or flooded	Albaqualfs, Albaquults, and Argialbolls	Very few
Andisols	Relatively young soils, mostly of volcanic origin; characterized by unique minerals with poorly organized crystalline structure		Food security, water
Aridisols	Soils of arid regions		Very few
Entisols	Sandy soils that may have diagnostic horizons below 1 m	Psamments	Biomass
Entisols	Soils developed in fluvial, lacustrine, or marine deposits	Fluvents, fluvent subgroups	Food security
Entisols	Very shallow with continuous rock occurring at or near the surface, and soils that are very gravelly	Lithic subgroups	Water runoff
Entisols	Youngest soils with no pedogenic horizons	Orthents	Biomass
Gelisols	Permanently frozen soils with permafrost in the subsoil, development dominated by cryogenic processes		Climate change

(Continued)

Table 1.1 Contribution of the main soil taxonomy orders to major ecosystem services. *Continued*

Soil taxonomy orders	Characteristic features	Lower levels of taxonomy	Ecosystem services[a]
Histosols	Organic materials are deposited in a wetland environment to form peat		Climate change, water
Inceptisols	Young soils just beginning to show profile development		Food security, climate, water, cultural
Inceptisols	Mountainous soils of cool humid climates covered by meadows or sparse forests	Umbric great group in Aquept suborder, Humic subgroups	Water runoff
Mollisols	Soils with a Mollic horizon which occur most frequently in the transitional areas between boreal forests and steppes or temperate tallgrass prairie		Food security, climate
Oxisols	Highly weathered tropical soils with low natural fertility	Some soils have Fe accumulations (plinthic, petroplinthic, or pisoplinthic horizons)	Biomass
Spodosols	Acid soils with low fertility and accumulations of organic matter and iron and aluminum oxides in the subsoil		Biomass
Ultisols	soils with low base status and a clay-enriched subsoil		Biomass, food security
Vertisols	Expansive clayey soils that shrink and swell with changes in moisture content		Food security
Nonsoils	Remnants of natural soils and where soils were radically transformed by different human activities, for example, mine spoils, urban spoils		Infrastructure

[a]All soils contribute to biodiversity, to some extent, under native vegetation depending upon previous disturbances.
Based on FAO and ITPS, 2015. Status of the World's Soil Resources (SWSR)—Main Report. Food and Agriculture Organization of the United Nations and Intergovernmental Technical Panel on Soils, Rome, Italy, p. 650.

1.6 Restoration objectives

Much of the literature on restoration emphasizes vegetation and seeks to recover historic conditions, the same or similar vegetation composition and structure as reference sites or putative natural conditions (Gann et al., 2019; SERI, 2004; Stanturf et al., 2014b). This emphasis has been questioned as it ignores that most landscapes have been manipulated by humans, whether for good or ill (Barlow et al., 2012; Bowman et al., 2011; Dirzo et al., 2014; Goudie, 2018; Lewis et al., 2015). In addition, it presumes relatively stable climatic conditions will continue into the future, contrary to the current understanding of the dynamism of climate systems and the likelihood of altered future climate (Berdugo et al., 2020; Cattau et al., 2020; Colvin et al., 2020; Newman, 2019; Pfeifer et al., 2019). Setting objectives requires knowledge of, and a consensus on, an appropriate baseline for a healthy soil in a landscape, land use, and management context, taking into account the starting (current) conditions (Allen et al., 2011; Fargione et al., 2018; Kibblewhite et al., 2008; Maikhuri and Rao, 2012; Zerga, 2015).

1.6.1 Recovery of function

Objectives for soil restoration aim to recover soil functions by reversing degradation (Bouma and McBratney, 2013; IPBES, 2018; McBratney et al., 2014) and returning soils to healthy conditions (Allen et al., 2011; Doran et al., 2002; Maikhuri and Rao, 2012). The reality is that soils in much of the world have been manipulated to the point of degradation (Hooke et al., 2017; Hudson and Alcántara-Ayala, 2006; Lal et al., 2003; Pandit et al., 2020; Yaalon, 2007), and the widespread human footprint cannot be ignored.

1.6.2 Setting objectives

Understanding the starting point for restoration is the first step in developing objectives and choosing appropriate interventions (Spathelf et al., 2018; Stanturf et al., 2017, 2019). Severely degraded soils pose physical, chemical, and biological barriers to restoration of natural systems; examples of such soils include mined land (Macdonald et al., 2015; Parrotta and Knowles, 2001) and severely eroded landscapes (IPBES, 2018; Zika and Erb, 2009). Under less degraded conditions, soil characteristics will determine recovery potentials, possibly by limiting establishment of native vegetation and certainly in determining which species are suitable for conditions with or without manipulation of soil. One approach to objective setting is to ask two questions (Stanturf et al., 2019): "Do we have it? and Do we want it?". The answers to these questions lead to four possible strategies for preserving or eliminating current conditions, or achieving or avoiding some future conditions (Table 1.2).

Table 1.2 Setting objectives for soil restoration can be aided by answering two questions: "Do we have it? and Do we want it?".

Avoid	Preserve
• Disturbing fragile soils	• Plant cover
• Exposing surfaces to erosion	• Recharge areas
• Compacting surfaces	• Ecosystem legacies (e.g., deadwood, slash, and residues)
• Altering hydroperiod	
Add	Remove
• Organic matter (e.g., biochar)	• Degrading activities
• Mulches	• Overgrazing
• Fertility and moisture amendments	• Invasive species (plants and animals)
• Mycorrhizal inoculum	

Answering these questions leads to four strategies—avoiding and removing degrading activities and conditions, and preserving and adding activities and material that promote soil health. Examples of actions that implement these strategies are shown in each block.

1.6.3 **What to avoid**

Avoiding changes that inhibit plant growth, especially degradation of fragile soils, is imperative regardless of the starting soil conditions. Fragile soils are those vulnerable to degradation due to their inherent properties that make them inhospitable to establishing or maintaining vegetation cover. This includes infertile sandy soils, soils affected by salts (i.e., saline and sodic), or soils with the potential to develop extremely low pH (i.e., acid sulfate potential soils). These soils are all prone to bare surfaces that are subject to erosion (FAO and ITPS, 2015; Fitzpatrick, 2002). Chapters elsewhere in this volume provide further detail on these soils (Harper et al., 2020; Stanturf, 2020).

Other threats to soil functions come from inappropriate resource management (FAO and ITPS, 2015), as detailed in guidelines for maintaining forest quality and health (Allen et al., 2011; Bünemann et al., 2018; Kibblewhite et al., 2008; Maikhuri and Rao, 2012; Page-Dumroese et al., 2000; Schoenholtz et al., 2000). Practices that compact soils include use of agricultural and forestry machinery and overgrazing (Hansson et al., 2019; Mariotti et al., 2020; Vanderburg et al., 2020) and should be avoided or managed to mitigate effects. Many practices disturb soil surfaces at varying frequency in the course of management, subjecting soil organic carbon to accelerated decomposition. Loss of soil organic carbon over time reduces biochemical activity and physical stability of soils, adversely affecting productivity, hydrology, and climate regulation functions (Obalum et al., 2017; Schmidt et al., 2011). Sustainable management and restoration of grassland and forest soils are described in other chapters in this volume (McTavish et al., 2020; Page-Dumroese et al., 2020; Prescott et al., 2020).

1.6.4 **What to preserve**

Soils are protected from erosion by maintaining appropriate levels of plant cover, in terms of species and densities under the prevailing climate. Appropriate levels and plant species must be assessed in the context of landscape, land use, and management. Adapting to future climatic conditions may require altering plant species or densities; for example, one adaption to future drier conditions and droughts is lowering the density or changing the species composition of forest stands in order to reduce competition for soil moisture (D'Amato et al., 2013; Guldin, 2018; Steckel et al., 2020).

Maintaining or improving natural hydrologic conditions and protecting watersheds will become increasingly difficult as urban areas expand into agricultural or wildlands, infrastructure development decreases permeable surfaces, and potable water becomes scarce under warmer and drier climate (Bruijnzeel, 2004; D'Odorico et al., 2018; FAO, 2011; Harper et al., 2019; King and Keim, 2019). Afforestation, in particular, should be cognizant of hydrologic conditions, especially in semiarid regions (Farley et al., 2005; Reisman-Berman et al., 2019; Tarin et al., 2020). Choice of species, including water-use efficiency, planting density, and location within a watershed, are important considerations. For example, in areas where hydrology is important for groundwater recharge, high-density plantings of species with high water-use traits should be avoided (Schwärzel et al., 2020; Stavi, 2019).

Explicit consideration of past land use and the legacies of management manipulations should lead to more effective restoration efforts. Ecosystem legacies consist of biotic or abiotic material or information entities (adaptations to historical disturbance regimes that remain after disturbances); the totality of legacies has been termed ecosystem memory (Frelich et al., 2018; Jõgiste et al., 2017, 2018; Johnstone et al., 2016). Different legacies, with different longevity and strength, result from anthropogenic disturbances. For example, consider the persistent effects due to conversion from native vegetation to row crops or pasture systems compared to natural disturbances such as windthrow (Frelich et al., 2018; Jõgiste et al., 2018). Some legacies are specific to soils (Baer et al., 2012); belowground material legacies are largely soil characteristics as affected by disturbances, occurring at microscale (Stanturf et al., 2020a,b). The most important soil legacy to preserve is soil organic matter (Obalum et al., 2017; Schmidt et al., 2011). In forests, this also means maintaining sufficient amounts and types of deadwood that contributes not only to conservation of organic matter but also to saproxylic organisms that comprise up to one-third of all forest species (Oxbrough and Pinzón, 2020). Deadwood habitat is diverse (e.g., standing snags vs downed stems) and changes over time as decomposition progresses.

1.6.5 **What to add**

Soil amendments are generally used to improve conditions for plant growth; these include organic matter such as biochar (Sohi et al., 2010; Weyers and Brockamp,

2020), mineral and organic fertilizers (Hartemink, 2006), and inoculum of mycorrhizae or other microbes (Aprahamian et al., 2016; Ciadamidaro et al., 2017; Neuenkamp et al., 2019). The efficacy of microbial inoculations has been limited, and the microbial community in the inoculum is usually overwhelmed by the microbes already endemic to a particular site (Winder et al., 2020). On the other hand, Wubs et al. (2016) found that a whole soil inoculum, which included critical soil faunal components, had stronger effects on restoration outcomes in severely degraded former agricultural soils. Generally, bare-root planting stock is sufficiently inoculated by microbes in the nursery soil, but inoculation may be needed when direct seeding or planting species with very specific microbial symbionts.

Under dry conditions, different mulches can improve moisture relations, especially important for seeds and seedlings (Chirwa and Mahamane, 2017; Stanturf et al., 2020a,b; Yirdaw et al., 2017). Water-absorbing gels have been utilized with good results under some conditions (Abedi-Koupai et al., 2008; Chirino et al., 2011; Vallejo et al., 2012).

1.6.6 What to remove (or reduce)

In addition to the soil degrading activities to avoid, there are practices and conditions that should be removed or reduced. Three things that cause soil degradation are overgrazing by wild or domestic animals, invasion by exotic plants and animals, and excess or too little water caused by altered hydrologic regime. Overgrazing by livestock in pastures and wildlands, especially in drier climates, reduces plant cover, accelerating erosion (D'Odorico et al., 2012; Middleton, 2018; Runyan and D'Odorico, 2016). High stocking levels of ungulates in temperate and boreal wildlands have been shown to reduce plant cover and alter species composition, leading to altered nutrient cycling, barriers to forest regeneration, and promotion of invasive plants (Lake and Leishman, 2004; Rooney et al., 2015; Royo et al., 2010). Invasive exotic plants can alter fire regimes (Brooks et al., 2004; D'Antonio and Vitousek, 1992; Gaertner et al., 2017); invasive earthworms alter nutrient cycling and can negatively impact soil biodiversity (Bohlen et al., 2004; Callaham et al., 2006). The removal of invasive species has had variable effectiveness at large scale (Flory and Clay, 2009; Kettenring and Adams, 2011; Krueger-Mangold et al., 2006; Reid et al., 2009; White et al., 2019), but reducing levels of invasives by manipulating vegetation can be effective (Madritch and Lindroth, 2009).

1.7 Historic examples of large-scale restoration

Several historic examples of large-scale restoration illustrate different trajectories following abandonment of agriculture and recovery of forest vegetation. In the sections that follow, we explore and compare passive and active restoration strategies, using passive restoration examples from Northeastern United States and Puerto Rico, and active restoration examples from Southeastern United States, Denmark, and South Korea. These examples show that one critical factor is how degraded the soils are from the previous land use. In Northeastern United States and Puerto Rico, forests recovered after agricultural abandonment largely without intervention. In the Southeastern United States, Denmark, and South Korea, active restoration was required because of the greater degradation caused by previous agriculture, grazing, and armed conflict. These examples provide important insights into the role and importance of soil in the process of afforestation, and the degree to which soil can affect outcomes when forests reestablish on landscapes.

1.7.1 Passive restoration: land clearing and abandonment/ recovery in Northeast and Puerto Rico, the United States

The first example occurred in the New England states following colonial land clearing and subsequent land abandonment beginning in the early 1800s; the second example, from Puerto Rico, occurred a century later.

Historical records suggest that forest clearing at its peak across New England was in the range of 50%–80% of the area (Foster et al., 2003; Foster, 2002; Williams, 1992). Much of this land was cleared using nonmechanized techniques such as crosscut saw and draft animal log skidding. The agricultural uses for these lands included primarily subsistence or local market food plots and pasturage for livestock production of meat and fiber, but there were some industrial-scale food production areas, and wool export was particularly important (Bell, 1989; Donahue, 2007). Wide-scale industrialization in the New England area, along with the availability of Midwestern lands for agricultural development, led to human demographic shifts and farm abandonment, and most of the land experienced natural colonization and forest development. The resultant mixed broadleaf forests (e.g., *Acer, Fagus, Betula*, and *Quercus*) and white pine/eastern hemlock [*Pinus strobus* L./*Tsuga canadensis* (L.) Carr.] came to dominate New England landscapes (Foster, 2002). These forests now face challenges associated with continued dynamics in forest vegetation brought about by invasive insects (e.g., hemlock woolly adelgid and emerald ash borer) that are changing their composition.

Passive restoration in Puerto Rico occurred as land was abandoned from active agriculture, either spontaneously or after government intervention. Beginning in the 1950s, government policies encouraging industrialization led to land abandonment, accompanied by reduced pressure on native forests that allowed deforested areas to recover (Grau et al., 2003). In some cases, farmers were allowed to remain until canopy closure after intercropping food plants with commercial tree species (Robinson et al., 2014). Some new forests resulted from abandoned shade coffee plantations and others developed by natural regeneration on abandoned and degraded

FIGURE 1.4

Puerto Rico was largely deforested by subsistence farming. Following abandonment and some reforestation, lush tropical forests developed. (Left) Farmer clearing a garden patch near his home on the Piza tract of the Luquillo Unit. This man had already worked 8 h on a CCC project on this tract. By living on the parcel under a special use permit, he can use his spare time for garden work and is able to prevent theft of crops. (Middle) Aerial of the road under construction leading to El Verde camp where road 186 crosses Espíritu Santo River. This area is now part of the El Yunque National Forest. (Right) View of the Rio Espíritu Santo corridor, El Yunque National Forest a few months (8 April 2018) before Hurricane Maria. Photo taken from Highway 186 Rio Espíritu Santo bridge in the El Yunque National Forest.

Courtesy of (Left and middle) USDA Forest Service and (right) Jerry Bauer, US Forest Service.

agricultural lands (Fig. 1.4). The resulting novel combinations of species included a mix of native and nonnative tree species. These new forests significantly differed from mature native forests; they had fewer endemic species, fewer large trees, higher soil bulk density, and lower soil carbon and litter stocks. Over time, however, these new forests developed and maintained forest cover and eventually began to facilitate native regeneration of native tree species (Lugo and Helmer, 2004).

1.7.2 Active restoration in Southeastern United States, Denmark, and South Korea

The clearing of primary forest in the Southeastern states in the United States was followed by intensive industrial-scale cotton production in the Piedmont and Atlantic Coastal Plain (Fig. 1.5), with exports of fiber principally to overseas markets. Agricultural practices led to massive degradation of soil resources through erosion and nutrient export in the harvested crop (Trimble, 1974). Cotton agriculture dominated the landscape for many decades, and by the early 20th century, poor soil conditions, insect pests, and the social repercussions of the US Civil War and the end of slavery, all combined to make cotton farming a losing proposition for most landowners.

When we contrast the clearing and use of New England agricultural land with what occurred in the Southeastern United States, it is clear that soil parent material and the intensity of land use significantly influenced the potential for forest

FIGURE 1.5

Eroded gullies in the Piedmont region of the Southeastern United States, caused by intensive cotton farming. Abandoned land was reforested, largely with *Pinus* spp. (Left) Cotton farming in the 1800s depended on manual and equine labor. (Middle) Severe erosion and accelerated gully formation were on sloping land in the Piedmont. (Right) Reforested gully, Calhoun Experimental Forest, Union County, South Carolina. The Calhoun EF was chosen to represent the poorest Piedmont conditions (Metz, 1958); research began there in the 1940s and continues (Richter Jr and Markewitz, 2001).

Courtesy of (Left and Middle Photos), USDA Forest Service; Nelson, D.R., O'Neill, K.P., Coughlan, M.R., Lonneman, M.C., Meyers, Z., and U.S. Department of Agriculture, Forest Service. USDA Forest Service Photographs from the Calhoun Experimental Forest, South Carolina, 1932–1987. Inter-university Consortium for Political and Social Research [distributor], Ann Arbor, MI, 2016-12-07. https://doi.org/10.3886/E100276V3 and (Right Photo) Mac Callaham.

development and the degree of human intervention needed to restore the landscape. The differences between the soils of the northeast and the southeast at the time that farms were abandoned had profound impacts upon the rate and trajectory of forest establishment in the subsequent decades. First, there were inherent differences between soils of the two regions in terms of the age and degree of weathering that the soils had undergone prior to cultivation. The soils of the northeastern states, classified primarily as Inceptisols and Spodosols, were predominantly formed on glacial deposits, thereby younger with greater quantities of primary minerals relative to the ancient and more weathered Ultisols that dominate soils of the south.

Second, the type of agricultural usage was much more intensive both in duration and in technique in cotton farming, and soils in the southeast suffered dramatic losses to erosion as well as losses of organic matter and nutrient capital relative to soils on abandoned farmland in the northeast. These differences in farming led to significant differences in the outcome of forest recovery, with diverse plant communities establishing in New England and species-poor "scrub" dominating successional communities in the southeast. Given the highly degraded condition of Southeastern landscapes, their recovery to productive forest landscapes required coordinated research efforts on the part of the federal and state governments and universities and included massive experimentation in soil stabilization, tree species selection, silvics, and genetic improvement of plant materials. One consequence of this was the afforestation of the Piedmont and Atlantic and Gulf Coastal Plains of the Southern United States following cotton cultivation during the agricultural boom spanning the years 1800–1920; these forests are still dominated by plantation forestry (e.g., Stanturf et al., 2003), and the entire landscape likely remains species poor relative to the original vegetation.

By the 1800s deforestation and shifting agriculture, grazing, and fire created a highly degraded landscape in Jutland, western Denmark. Western and northern parts of Jutland became dominated by heathland. The underlying sandy soils were very acidic (podzols) and became further exposed as the heather was degraded by overgrazing and fire. The exposed sandy soils were eroded by wind, with dunes forming that became mobile and threatened to cover houses and whole villages. The degraded soils made stabilizing the dunes very difficult. Vegetation restoration was needed to improve soil conditions, and restored forests were needed by the local communities to supply wood for construction and energy. Native broadleaf trees could not establish on the degraded heathlands; besides, all seed sources had disappeared centuries ago so that relying on natural regeneration or succession was not an option.

The solution was to introduce nonnative species—particularly conifers (Madsen et al., 2015) and specifically the nonnative mountain pine (*Pinus mugo* Turra), one of the few species that could survive and grow on the degraded sites that were afforested in the 1800s. As these initial forests matured, altered microclimate, and accumulated organic matter, opportunities arose to reintroduce native species as well as other nonnative species with desired functions (Madsen et al., 2015; Stanturf et al., 2018). The reintroduction of native broadleaves such as European beech (*Fagus sylvatica* L.) and Scots pine (*Pinus sylvestris* L.) is occurring along with nonnative Douglas fir [*Pseudotsuga menziesii* (Mirb.) Franco] and Norway spruce [*Picea abies* (L.) Karst], thereby facilitating continuous cover forestry systems (Fig. 1.6). The restored closed canopy forests now share the landscape with other land uses and very distinctive and narrow forest edges or borders occur between land uses.

Ironically, there is now societal pressure to restore heathlands in the region, themselves a legacy of historical degradation, by reducing soil fertility (van Diggelen et al., 2020). Remaining portions of the landscape formerly regarded as degraded (heathland, grazed open or semiopen elements) are now seen as valuable habitats that need protection, conservation management, and even expansion. This is facilitated by legislation, regulations, and subsidies. The rationale is that the degradation and clearing created habitats now considered rich in biodiversity but suffering from widespread intensive farming and expansion of urban areas and infrastructure. The pressure to restore the degraded habitats is because they harbor species that are now rare or lost within Denmark.

Multiple causes of deforestation in the Korean Peninsula included overexploitation of forests for firewood and charcoal, increasing populations of impoverished

FIGURE 1.6

Natural regeneration of mixed species forests on restored site Jutland, western Denmark. About 150 years ago the most tolerant species (e.g., mountain pine and Norway spruce) were planted. As site conditions improved, other mainly nonnative conifers and the native Scots pine established. Today, the focus is on reintroduction of native broad-leaved species.

Courtesy of Palle Madsen.

people accompanied by widespread illegal logging and expansion of agricultural lands, and destruction during the Korean War (Park and Youn, 2017). Reforestation was implemented from 1959 to 1999 to improve soil conditions and control erosion, resulting in rehabilitation of about 97% of the deforested areas (Lee et al., 2015). A series of coordinated national economic plans from 1967 to 1987 was primarily for erosion control and conservation. Reforestation was initiated on a large scale in 1959 mainly by planting fast-growing trees to meet the fuelwood demand. A government program initiated massive reforestation with fast growing, mostly nonnative trees such as *Larix kaempferi* (L.) Carr., *Pinus rigida* Mill., and *Populus* species; nurse trees, such as *Robinia pseudoacacia* L. and *Alnus* species, were also used. Planting programs included terracing, gully stabilization, and adding topsoil to planting holes (Fig. 1.7). The citizenry was mobilized to provide much of the labor, supporting the program as a patriotic endeavor.

These examples show that restoration, either passive or active, can be successful (Lee et al., 2015; Madsen et al., 2015; Stanturf et al., 2018, 2019). Natural processes are restoration assets; seed dispersal, colonization, and self-organization

FIGURE 1.7

Hill slope restoration in Korea. Harsh conditions (eroded soils, droughty sites) required restoration using a combination of physical water control structures to control erosion and tree planting. Restoration treatments in Yeongil District, Gyeongbuk Province, South Korea. (Top) 1973, before planting began but some terracing in place; (middle) 1976, 3-year after restoration treatment; and (bottom) 1981, 8-year after treatment.

Reproduced with permission from Korea Forest Service.

can be augmented by thoughtful interventions (Lugo and Helmer, 2004; Stanturf et al., 2014a). Species composition may be constrained by the degree of soil degradation, but protecting and restoring the soil is a critical first step for facilitating restoration, particularly replenishment of soil organic matter. Restoration of soil organic matter, and establishment of ecological processes associated with forest floor function, may eventually provide opportunities to hasten the reintroduction of native species. Natural processes that follow restoration, either by passive or active means, can produce novel forest ecosystems that can function similar to native forests, even if species composition has been altered (e.g., Stanturf et al., 2018). Nudging these processes in adaptive directions, focusing on functioning is one way forward in the Anthropocene (Finegan, 2017; Lugo and Helmer, 2004; Zalasiewicz et al., 2010).

1.8 Soils and climate change

Projections of future climates are uncertain, especially where and when significant changes will occur. The current situation has been called "spatiotemporal chaos"

(Pielke et al., 2013; Wilby and Dessai, 2010) because of the uncertainty of where and when change will occur. The uncertainty stems from four factors: natural ecological variability, model uncertainty, social system responses to drivers of global change, and the resulting level of future greenhouse gas emissions. What seems certain is that the future will significantly depart from current conditions (Rummukainen, 2012; Williams and Jackson, 2007; Williams et al., 2007). The most recent projections are for a warmer 2100 world (Masson-Delmotte et al., 2018). Globally, human-induced warming has already reached about 1°C above pre-industrial levels, on average (Masson-Delmotte et al., 2018), but since the 1970s, most land regions have warmed faster than this global average. In other words, many regions have already exceeded 1.5°C above preindustrial levels (Kharin et al., 2018; Masson-Delmotte et al., 2018). The climate will continue to warm during the 21st century due to the large inertia of the Earth System (Latif, 2013).

Coupled ocean/atmospheric general circulation models have different levels of skill in projecting precipitation patterns and extremes with lower confidence in precipitation projections than for temperature (Becker et al., 2013). Nevertheless, where mean precipitation is projected to decrease along with higher temperatures, the risk of droughts and precipitation deficits increase (Berdugo et al., 2020). Climate warming will increase the risks from heavy precipitation events in several northern hemisphere high-latitude and/or high-elevation regions, eastern Asia, and eastern North America, including heavy precipitation associated with tropical cyclones. Consequently, flood hazards are projected to increase in some areas (Masson-Delmotte et al., 2018).

Changes in climate over the past century are mostly apparent in individual climate variables (e.g., temperature, precipitation, and frost-free days). Yet interacting and covarying changes among several climate variables are more critical for understanding impacts on socio-ecological systems than changes in single variables (Abatzoglou et al., 2020). Globally, departures of multivariate climate variables (annual climatic water deficit, annual evapotranspiration, average minimum temperature of the coldest month, and average maximum temperature of the warmest month) were nearly three times greater than changes in individual climate variables during 1958−2017 relative to baseline years of 1958−87 (Abatzoglou et al., 2020).

The most direct effect of warmer and drier climates on soils will come from the loss of plant cover, exposing soil surfaces to drying and wind and water erosion (Hooke et al., 2017; Nearing et al., 2004) and, at the extreme, to desertification (Berdugo et al., 2020; D'Odorico et al., 2013; Huang et al., 2020). Less drastic, more subtle effects will come through changes in vegetation and possible lowering of net primary productivity with impacts on food security as well as native vegetation (Allen et al., 2011; Berdugo et al., 2020; Castro et al., 2010; Kardol et al., 2010, 2011; Waldrop and Firestone, 2006). Because changes in aboveground vegetation may occur faster than belowground effects become evident, there could be a decoupling of above- and belowground processes, including nutrient cycling (Bardgett and Van Der Putten, 2014; Lance et al., 2020).

Wildfires are projected to increase in frequency and intensity in many locations (Abatzoglou and Williams, 2016; Earles et al., 2014; Holden et al., 2018; Liu et al., 2010), along with a lengthening of fire season in some ecosystems. More intense fire behavior will lead to losses of soil organic matter (Le Page et al., 2017; Schoennagel et al., 2017; Taufik et al., 2017) and other impacts on soils such as increased erosion rates (Neary and Leonard, 2020). In permafrost regions, projected climate change will alter freezing dynamics and cause thinning of permafrost layers, hastened by increased wildfire (Aaltonen et al., 2019; Harden et al., 2006; Schuur et al., 2008; Tarnocai et al., 2009).

Decomposition processes in soils are sensitive to temperature, and altered climate will affect net C-balance, with feedbacks to the atmosphere (Bradford et al., 2016; Davidson and Janssens, 2006). One method suggested for mitigating climate change is to sequester C in soils (Jobbágy and Jackson, 2000; Lal, 2004). Even more critical is to protect organic carbon already in soil by maintaining vegetation cover and minimizing soil disturbance (Delgado-Baquerizo et al., 2017; Fargione et al., 2018; Kapur et al., 2018). One adaptation method widely discussed is assisting plant colonization or directly moving species within historic ranges or beyond (Dumroese et al., 2015; McLachlan et al., 2007; Pedlar et al., 2012; Thomas, 2011). Soil considerations will be critical in identifying the suitability of the receiving ecosystem, and whether the soil biota are suitable or must also be translocated along with plants. These issues are discussed in detail in a subsequent chapter (Winder et al., 2020).

1.9 Final thoughts

The need to consider soils in landscape restoration is compelling; degraded soils need restoring, and resource management systems (i.e., agriculture and forestry) should seek to maintain beneficial ecosystem legacies and avoid actions that reduce soil health. Soils are fundamental to restoring landscape function and contributing to national commitments to international agreements for conserving biodiversity, reducing land degradation, and mitigating and adapting to climate change. Varied objectives may drive efforts to restore landscape functioning, but even so, soil characteristics may constrain feasible restoration activities.

Soils are natural bodies, formed over time by climate and organisms acting on geological material, influenced by slope and terrain characteristics. Viewing soils in this conceptual framework facilitates understanding of the varied role of soils in landscape restoration. Heterogeneity of soil properties, a product of the interaction of these factors, produces a mosaic of habitats for plants, microbes, and animals. These same factors determine the natural capital that produces important ecosystem services, in particular productivity, diversity, water quality, and climate regulation.

Climate change and its effects on temperature maxima and minima, precipitation amounts and seasonal distribution, evapotranspiration, and extreme weather events (e.g., tropical cyclones or multiyear droughts) are likely but difficult to predict. Nevertheless, future conditions are likely to differ from current and recent historical conditions. Warmer and drier climate could reduce plant cover, exposing soils to accelerated wind and water erosion. Plant species may not adapt quickly enough to changed climate to avoid adverse impacts on productivity, altered fire regimes, and thawing permafrost. Feedback mechanisms from soils to the atmosphere will exacerbate climate heating trends. The need today is great for restoring landscapes, with special focus on stabilization and improvement of their underlying soils, and this need likely will become even greater over the next decades.

References

Aaltonen, H., Köst er, K., Köster, E., Berninger, F., Zhou, X., Karhu, K., et al., 2019. Forest fires in Canadian permafrost region: the combined effects of fire and permafrost dynamics on soil organic matter quality. Biogeochemistry 143 (2), 257–274.

Abatzoglou, J.T., Williams, A.P., 2016. Impact of anthropogenic climate change on wildfire across western US forests. Proc. Natl. Acad. Sci. U.S.A. 113 (42), 11770–11775.

Abatzoglou, J.T., Dobrowski, S.Z., Parks, S.A., 2020. Multivariate climate departures have outpaced univariate changes across global lands. Sci. Rep. 10 (1), 1–9.

Abedi-Koupai, J., Sohrab, F., Swarbrick, G., 2008. Evaluation of hydrogel application on soil water retention characteristics. J. Plant. Nutr. 31 (2), 317–331.

Akhtar-Schuster, M., Stringer, L.C., Erlewein, A., Metternicht, G., Minelli, S., Safriel, U., et al., 2017. Unpacking the concept of land degradation neutrality and addressing its operation through the Rio Conventions. J. Environ. Manage. 195, 4–15.

Alexander, S., Nelson, C.R., Aronson, J., Lamb, D., Cliquet, A., Erwin, K.L., et al., 2011. Opportunities and challenges for ecological restoration within REDD + . Restor. Ecol. 19 (6), 683–689.

Allen, D.E., Singh, B.P., Dalal, R.C., 2011. Soil health indicators under climate change: a review of current knowledge. In: Singh, B.P., Cowie, A.L., Yin Chan, K. (Eds.), Soil Health and Climate Change. Springer, Berlin, pp. 25–45.

Aprahamian, A.M., Lulow, M.E., Major, M.R., Balazs, K.R., Treseder, K.K., Maltz, M.R., 2016. Arbuscular mycorrhizal inoculation in coastal sage scrub restoration. Botany 94 (6), 493–499.

Arnold, R.W., 2016. Soil survey and soil classification. In: Grunwald, S. (Ed.), Environmental Soil-Landscape Modeling. CRC Press, Boca Raton, FL, pp. 50–72.

Baer, S.G., Heneghan, L., Eviner, V., 2012. Applying soil ecological knowledge to restore ecosystem services. In: Wall, D.H., Ritz, K., Six, J., Strong, D.R., van der Putten, W.H. (Eds.), Soil Ecology and Ecosystem Services. Oxford University Press, Oxford, UK, pp. 377–393.

Ballard, J.P., Horn, S.P., Li, Z.-H., 2017. A 23,000-year microscopic charcoal record from Anderson Pond, Tennessee, USA. Palynology 41 (2), 216–229.

Bardgett, R.D., Van Der Putten, W.H., 2014. Belowground biodiversity and ecosystem functioning. Nature 515 (7528), 505.

Barlow, J., Gardner, T.A., Lees, A.C., Parry, L., Peres, C.A., 2012. How pristine are tropical forests? An ecological perspective on the pre-Columbian human footprint in Amazonia and implications for contemporary conservation. Biol. Conserv. 151 (1), 45–49.

Bechara, F.C., Dickens, S.J., Farrer, E.C., Larios, L., Spotswood, E.N., Mariotte, P., et al., 2016. Neotropical rainforest restoration: comparing passive, plantation and nucleation approaches. Biodivers. Conserv. 25 (11), 2021–2034.

Becker, E.J., Van Den Dool, H., Peña, M., 2013. Short-term climate extremes: prediction skill and predictability. J. Clim. 26 (2), 512–531.

Bell, M.M., 1989. Did New England go downhill? Geograph. Rev. 450–466.

Berdugo, M., Delgado-Baquerizo, M., Soliveres, S., Hernández-Clemente, R., Zhao, Y., Gaitán, J.J., et al., 2020. Global ecosystem thresholds driven by aridity. Science 367 (6479), 787–790.

Besseau, P., Graham, S., Christopherson, T. (Eds.), 2018. Restoring Forests and Landscapes: The Key to a Sustainable Future. IUFRO and Global Partnership on Forest and Landscape Restoration, Vienna.

Bockheim, J.G., Gennadiyev, A.N., 2010. Soil-factorial models and earth-system science: a review. Geoderma 159 (3), 243–251.

Boettcher, S., Kalisz, P.J., 1990. Single-tree influence on soil properties in the mountains of eastern Kentucky. Ecology 71 (4), 1365–1372.

Bohlen, P.J., Scheu, S., Hale, C.M., McLean, M.A., Migge, S., Groffman, P.M., et al., 2004. Non-native invasive earthworms as agents of change in northern temperate forests. Front. Ecol. Environ. 2 (8), 427–435.

Bouma, J., McBratney, A., 2013. Framing soils as an actor when dealing with wicked environmental problems. Geoderma 200, 130–139.

Bowman, D.M., Balch, J., Artaxo, P., Bond, W.J., Cochrane, M.A., D'Antonio, C.M., et al., 2011. The human dimension of fire regimes on Earth. J. Biogeogr. 38 (12), 2223–2236.

Bradford, M.A., Wieder, W.R., Bonan, G.B., Fierer, N., Raymond, P.A., Crowther, T.W., 2016. Managing uncertainty in soil carbon feedbacks to climate change. Nat. Clim. Change 6 (8), 751–758.

Bradshaw, R.H., Sykes, M.T., 2014. Ecosystem *Dynamics: From the Past to the Future*. John Wiley & Sons, Oxford.

Brancalion, P.H., Chazdon, R.L., 2017. Beyond hectares: four principles to guide reforestation in the context of tropical forest and landscape restoration. Restor. Ecol. 25, 491–496.

Brancalion, P.H., Niamir, A., Broadbent, E., Crouzeilles, R., Barros, F.S., Zambrano, A.M. A., et al., 2019. Global restoration opportunities in tropical rainforest landscapes. Sci. Adv. 5 (7), eaav3223.

Brooks, M.L., D'Antonio, C.M., Richardson, D.M., Grace, J.B., Keeley, J.E., DiTomaso, J.M., et al., 2004. Effects of invasive alien plants on fire regimes. BioScience 54 (7), 677–688.

Bruijnzeel, L.A., 2004. Hydrological functions of tropical forests: not seeing the soil for the trees? Agric. Ecosyst. Environ. 104 (1), 185–228.

Brundtland, G., 1987. Report of the World Commission on Environment and Development. United Nations General Assembly document A/42/427, New York.

Bünemann, E.K., Bongiorno, G., Bai, Z., Creamer, R.E., De Deyn, G., de Goede, R., et al., 2018. Soil quality—a critical review. Soil Biol. Biochem. 120, 105–125.

Cairns, S., Dudley, N., Hall, C., Keenelyside, K., Stolton, S., 2012. Ecological Restoration for Protected Areas: Principles, Guidelines and Best Practices, vol. 18. IUCN, Gland.

Callaham Jr, M.A., Stanturf, J.A., 2020. Soil ecology and restoration. In: Stanturf, J.A., Callaham, M.A. (Eds.), Soils and Landscape Restoration. Academic Press, New York.

Callaham, M.A., Richter, D.D., Coleman, D.C., Hofmockel, M., 2006. Long-term land-use effects on soil invertebrate communities in Southern Piedmont soils, USA. Eur. J. Soil Biol. 42, S150–S156.

Castro, H.F., Classen, A.T., Austin, E.E., Norby, R.J., Schadt, C.W., 2010. Soil microbial community responses to multiple experimental climate change drivers. Appl. Environ. Microbiol. 76 (4), 999–1007.

Cattau, M.E., Wessman, C., Mahood, A., Balch, J.K., Poulter, B., 2020. Anthropogenic and lightning-started fires are becoming larger and more frequent over a longer season length in the USA. Global Ecol. Biogeogr. 29 (4), 2020, 668–681. <https://doi.org/10.1111/geb.13058>.

CBD, 2010. Strategic plan for biodiversity 2011–2020 and the Aichi targets. Retrieved from CBD, Montreal, QC: <https://www.cbd.int/doc/strategic-plan/2011-2020/Aichi-Targets-EN.pdf>.

Chazdon, R.L., Uriarte, M., 2016. Natural regeneration in the context of large-scale forest and landscape restoration in the tropics. Biotropica 48 (6), 709–715.

Chirino, E., Vilagrosa, A., Vallejo, V.R., 2011. Using hydrogel and clay to improve the water status of seedlings for dryland restoration. Plant Soil 344 (1–2), 99–110.

Chirwa, P.W., Mahamane, L., 2017. Overview of restoration and management practices in the degraded landscapes of the Sahelian and dryland forests and woodlands of East and southern Africa. South For. 79 (2), 87–94.

Ciadamidaro, L., Girardclos, O., Bert, V., Zappelini, C., Yung, L., Foulon, J., et al., 2017. Poplar biomass production at phytomanagement sites is significantly enhanced by mycorrhizal inoculation. Environ. Exp. Bot. 139, 48–56.

Clement, C.R., Denevan, W.M., Heckenberger, M.J., Junqueira, A.B., Neves, E.G., Teixeira, W.G., et al., 2015. The domestication of Amazonia before European conquest. Proc. R. Soc. B: Biol. Sci. 282 (1812), 20150813.

Coleman, D.C., Callaham, M.A., Crossley Jr, D., 2018. Fundamentals of Soil Ecology. Academic Press, New York.

Colvin, R., Crimp, S., Lewis, S., Howden, M., 2020. Implications of climate change for future disasters. In: Lukasiewicz, A., Baldwin, C. (Eds.), Natural Hazards and Disaster Justice: Challenges for Australia and Its Neighbours. Springer Singapore, Singapore, pp. 25–48.

Cooke, S.J., Bennett, J.R., Jones, H.P., 2019. We have a long way to go if we want to realize the promise of the "Decade on Ecosystem Restoration". Conserv. Sci. Pract. 1 (12), e129.

Cowie, A.L., Orr, B.J., Sanchez, V.M.C., Chasek, P., Crossman, N.D., Erlewein, A., et al., 2018. Land in balance: the scientific conceptual framework for Land Degradation Neutrality. Environ. Sci. Policy 79, 25–35.

Coyle, C., Creamer, R.E., Schulte, R.P.O., O'Sullivan, L., Jordan, P., 2016. A functional land management conceptual framework under soil drainage and land use scenarios. Environ. Sci. Policy 56, 39–48.

D'Amato, A.W., Bradford, J.B., Fraver, S., Palik, B.J., 2013. Effects of thinning on drought vulnerability and climate response in north temperate forest ecosystems. Ecol. Appl. 23 (8), 1735–1742.

D'Antonio, C.M., Vitousek, P.M., 1992. Biological invasions by exotic grasses, the grass/fire cycle, and global change. Annu. Rev. Ecol. Syst. 23, 63–87.

D'Odorico, P., Okin, G.S., Bestelmeyer, B.T., 2012. A synthetic review of feedbacks and drivers of shrub encroachment in arid grasslands. Ecohydrology 5 (5), 520–530.

D'Odorico, P., Bhattachan, A., Davis, K.F., Ravi, S., Runyan, C.W., 2013. Global desertification: drivers and feedbacks. Adv. Water Resour. 51, 326–344.

D'Odorico, P., Davis, K.F., Rosa, L., Carr, J.A., Chiarelli, D., Dell'Angelo, J., et al., 2018. The global food-energy-water nexus. Rev. Geophys. 56 (3), 456–531.

Davidson, E.A., Janssens, I.A., 2006. Temperature sensitivity of soil carbon decomposition and feedbacks to climate change. Nature 440 (7081), 165–173.

Delgado-Baquerizo, M., Eldridge, D.J., Maestre, F.T., Karunaratne, S.B., Trivedi, P., Reich, P.B., et al., 2017. Climate legacies drive global soil carbon stocks in terrestrial ecosystems. Sci. Adv. 3 (4), e1602008.

Deng, Y., 2007. New trends in digital terrain analysis: landform definition, representation, and classification. Prog. Phys. Geogr.: Earth Environ. 31 (4), 405–419.

Dinerstein, E., Vynne, C., Sala, E., Joshi, A.R., Fernando, S., Lovejoy, T.E., et al., 2019. A global deal for nature: guiding principles, milestones, and targets. Sci. Adv. 5 (4), . Available from: https://doi.org/10.1126/sciadv.aaw2869, eaaw2869.

Dirzo, R., Young, H.S., Galetti, M., Ceballos, G., Isaac, N.J., Collen, B., 2014. Defaunation in the Anthropocene. Science 345 (6195), 401–406.

Donahue, B., 2007. Another look from Sanderson's farm: a perspective on New England environmental history and conservation. Environ. History 12 (1), 9–35.

Doran, J.W., Parkin, T.B., 1994. Defining and assessing soil quality. In: Doran, J.W., Coleman, D.C., Bezdicek, D.F., Stewart, B.A. (Eds.), Defining Soil Quality for a Sustainable Environment, vol. 35. Soil Science Society of America, Ames, IA, pp. 1–21.

Doran, J.W., Stamatiadis, S., Haberern, J., 2002. Soil Health as an Indicator of Sustainable Management. Publications from USDA-ARS/UNL Faculty, p. 180.

Dumroese, R.K., Williams, M.I., Stanturf, J.A., St Clair, J.B., 2015. Considerations for restoring temperate forests of tomorrow: forest restoration, assisted migration, and bioengineering. N. For. 46 (5–6), 947–964.

Earles, J.M., North, M.P., Hurteau, M.D., 2014. Wildfire and drought dynamics destabilize carbon stores of fire-suppressed forests. Ecol. Appl. 24 (4), 732–740.

Fagan, M.E., Reid, J.L., Holland, M.B., Drew, J.G., Zahawi, R.A., How feasible are global forest restoration commitments? Conserv. Lett. 13 (3), 2020, e12700. https://doi.org/10.1111/conl.12700.

FAO, 2011. The State of the World's Land and Water Resources for Food and Agriculture: Managing Systems at Risk. Routledge, London.

FAO and ITPS, 2015. Status of the World's Soil Resources (SWSR)—Main Report. Food and Agriculture Organization of the United Nations and Intergovernmental Technical Panel on Soils, Rome, Italy, p. 650.

Fargione, J.E., Bassett, S., Boucher, T., Bridgham, S.D., Conant, R.T., Cook-Patton, S.C., et al., 2018. Natural climate solutions for the United States. Sci. Adv. 4 (11). Available from: https://doi.org/10.1126/sciadv.aat1869.

Farley, K.A., Jobbágy, E.G., Jackson, R.B., 2005. Effects of afforestation on water yield: a global synthesis with implications for policy. Glob. Change Biol. 11 (10), 1565–1576.

Finegan, B., 2017. Forest and ecosystem service rehabilitation in the Anthropocene: lessons from contrasting Costa Rican landscapes. Input to IUFRO's International Conference on Forest Landscape Restoration under Global Change—A Contribution to the Implementation of the Bonn Challenge" in Puerto Rico, 6–9 June 2017. Available online: <https://www.iufro.org/download/file/27694/6474/03_BF_Finegan_IUFRO_San_Juan_2017_pdf/> (accessed 13.07.18.).

Fitzpatrick, R.W., 2002. Land degradation processes. In: McVicar, T., Li Rui, W., Fitzpatrick, R., Changming, L. (Eds.), Regional Water and Soil Assessment for Managing Agriculture in China and Australia, vol. 84, ACIAR Monograph No. 84, Australian Centre for International Agricultural Research, Canberra, Australia, 119−129.

Flory, S.L., Clay, K., 2009. Invasive plant removal method determines native plant community responses. J. Appl. Ecol. 46 (2), 434−442.

Foster, D.R., 2002. Thoreau's country: a historical−ecological perspective on conservation in the New England landscape. J. Biogeogr. 29 (10−11), 1537−1555.

Foster, D., Swanson, F., Aber, J., Burke, I., Brokaw, N., Tilman, D., et al., 2003. The importance of land-use and its legacies to ecology and environmental management. BioScience 53, 77−88.

Fox, T., Comerford, N., 1990. Low-molecular-weight organic acids in selected forest soils of the southeastern USA. Soil Sci. Soc. Am. J. 54 (4), 1139−1144.

Frelich, L.E., Jõgiste, K., Stanturf, J.A., Parro, K., Baders, E., 2018. Natural disturbances and forest management: interacting patterns on the landscape. In: Perera, A., Peterson, U., Pastur, G., Iverson, L. (Eds.), Ecosystem Services from Forest Landscapes. Springer, Cham, pp. 221−248.

Gaertner, M., Le Maitre, D.C., Esler, K.J., 2017. Alterations of disturbance regimes by plant and animal invaders. In: Vilà, M., Hulme, P.E. (Eds.), Impact of Biological Invasions on Ecosystem Services. Springer International Publishing, Cambridge, pp. 249−259.

Gann, G., McDonald, T., Walder, B., Aronson, J., Nelson, C., Jonson, J., et al., 2019. International principles and standards for the practice of ecological restoration, 2nd ed. Restor. Ecol. 27 (S1), S1−S46.

Geisen, S., Wall, D.H., van der Putten, W.H., 2019. Challenges and opportunities for soil biodiversity in the anthropocene. Curr. Biol. 29 (19), R1036−R1044.

González-Pérez, J.A., González-Vila, F.J., Almendros, G., Knicker, H., 2004. The effect of fire on soil organic matter—a review. Environ. Int. 30 (6), 855−870.

Goudie, A.S., 2018. Human Impact on the Natural Environment, eighth ed Wiley Blackwell, New York.

Grau, H.R., Aide, T.M., Zimmerman, J.K., Thomlinson, J.R., Helmer, E., Zou, X., 2003. The ecological consequences of socioeconomic and land-use changes in postagriculture Puerto Rico. AIBS Bull. 53 (12), 1159−1168.

Grunwald, S., Thompson, J.A., Boettinger, J.L., 2011. Digital soil mapping and modeling at continental scales: finding solutions for global issues. Soil Sci. Soc. Am. J. 75 (4), 1201−1213.

Guldin, J.M., 2018. Silvicultural options in forests of the southern United States under changing climatic conditions. N. For. 50, 71−87.

Hansson, L., Šimůnek, J., Ring, E., Bishop, K., Gärdenäs, A.I., 2019. Soil compaction effects on root-zone hydrology and vegetation in boreal forest clearcuts. Soil Sci. Soc. Am. J. 83 (s1), S105−S115.

Harden, J.W., Manies, K.L., Turetsky, M.R., Neff, J.C., 2006. Effects of wildfire and permafrost on soil organic matter and soil climate in interior Alaska. Glob. Change Biol. 12 (12), 2391−2403.

Harper, R., Smettem, K.R.J., Ruprecht, J.K., Dell, B., Liu, N., 2019. Forest-water interactions in the changing environment of south-western Australia. Ann. For. Sci. 76 (4), 95. Available from: https://doi.org/10.1007/s13595-019-0880-5.

Harper, R.A., Dell, B., Ruprecht, J.K., Sochacki, S.J., Smettem, K.R.J., 2020. Salinity and the reclamation of salinized lands. In: Stanturf, J.A., Callaham, M.A. (Eds.), Soils and Landscape Restoration. Academic Press, New York.

Hartemink, A.E., 2006. Soil fertility decline: definitions and assessment. Encycl. Soil. Sci. 2, 1618–1621.

Hassani, M.A., Durán, P., Hacquard, S., 2018. Microbial interactions within the plant holobiont. Microbiome 6 (1), 58. Available from: https://doi.org/10.1186/s40168-018-0445-0.

Heneghan, L., Miller, S.P., Baer, S., Callaham Jr, M.A., Montgomery, J., Pavao-Zuckerman, M., et al., 2008. Integrating soil ecological knowledge into restoration management. Restor. Ecol. 16 (4), 608–617.

Holden, Z.A., Swanson, A., Luce, C.H., Jolly, W.M., Maneta, M., Oyler, J.W., et al., 2018. Decreasing fire season precipitation increased recent western US forest wildfire activity. Proc. Natl. Acad. Sci. U.S.A. 115 (36), E8349–E8357.

Holl, K., Kappelle, M., 1999. Tropical forest recovery and restoration. Trends Ecol. Evol. 14, 378–379.

Hooke, J., Sandercock, P., Cammeraat, L., Lesschen, J.P., Borselli, L., Torri, D., et al., 2017. Mechanisms of degradation and identification of connectivity and erosion hotspots. In: Hooke, J.M., Sandercock, P. (Eds.), Combating Desertification and Land Degradation. Springer, Cham, Switzerland, pp. 13–37.

Huang, J., Zhang, G., Zhang, Y., Guan, X., Wei, Y., Guo, R., 2020. Global desertification vulnerability to climate change and human activities. Land Degradation Dev. 31 (4), 1380–1391. <https://doi.org/10.1002/ldr.3556>.

Hudson, P.F., Alcántara-Ayala, I., 2006. Ancient and modern perspectives on land degradation. Catena 65 (2), 102–106.

Huggett, R.J., 1975. Soil landscape systems: a model of soil genesis. Geoderma 13 (1), 1–22.

IPBES, 2018. Summary for policymakers of the thematic assessment report on land degradation and restoration of the Intergovernmental Science-Policy Platform on Biodiversity and Ecosystem Services. IPBES, Bonn.

Janishevski, L., Santamaria, C., Gidda, S., Cooper, H., Brancalion, P., 2015. Ecosystem restoration, protected areas and biodiversity conservation. Unasylva 66 (3), 19–27.

Jenny, H., 1941. Factors of Soil Formation: A System of Quantitative Pedology. McGraw Hill, New York.

Jenny, H., 1961. Derivation of state factor equations of soils and ecosystems. Soil Sci. Soc. Am. J. 25 (5), 385–388.

Jian, J., Du, X., Stewart, R.D., 2020. A database for global soil health assessment. Sci. Data 7 (1), 16. Available from: https://doi.org/10.1038/s41597-020-0356-3.

Jobbágy, E.G., Jackson, R.B., 2000. The vertical distribution of soil organic carbon and its relation to climate and vegetation. Ecol. Appl. 10 (2), 423–436.

Jõgiste, K., Korjus, H., Stanturf, J.A., Frelich, L.E., Baders, E., Donis, J., et al., 2017. Hemiboreal forest: natural disturbances and the importance of ecosystem legacies to management. Ecosphere 8 (2), e01706.

Jõgiste, K., Frelich, L.E., Laarmann, D., Vodde, F., Baders, E., Donis, J., et al., 2018. Imprints of management history on hemiboreal forest ecosystems in the Baltic States. Ecosphere 9 (11), e02503.

Johnson, W.M., 1963. The pedon and the polypedon. Soil Sci. Soc. Am. J. 27 (2), 212–215.

Johnstone, J.F., Allen, C.D., Franklin, J.F., Frelich, L.E., Harvey, B.J., Higuera, P.E., et al., 2016. Changing disturbance regimes, ecological memory, and forest resilience. Front. Ecol. Environ. 14 (7), 369–378.

Kapur, S., Aydın, M., Akça, E., Reich, P., 2018. Climate change and soils. In: Kapur, S., Akça, E., Günal, H. (Eds.), The Soils of Turkey. Springer International Publishing, Cham, Switzerland, pp. 45−55.

Kardol, P., Wardle, D.A., 2010. How understanding aboveground−belowground linkages can assist restoration ecology. Trends Ecol. Evol. 25 (11), 670−679.

Kardol, P., Cregger, M.A., Campany, C.E., Classen, A.T., 2010. Soil ecosystem functioning under climate change: plant species and community effects. Ecology 91 (3), 767−781.

Kardol, P., Reynolds, W.N., Norby, R.J., Classen, A.T., 2011. Climate change effects on soil microarthropod abundance and community structure. Appl. Soil Ecol. 47 (1), 37−44.

Kettenring, K.M., Adams, C.R., 2011. Lessons learned from invasive plant control experiments: a systematic review and meta-analysis. J. Appl. Ecol. 48 (4), 970−979.

Kharin, V., Flato, G., Zhang, X., Gillett, N., Zwiers, F., Anderson, K., 2018. Risks from climate extremes change differently from 1.5°C to 2.0°C depending on rarity. Earth's Fut. 6 (5), 704−715.

Kibblewhite, M., Ritz, K., Swift, M., 2008. Soil health in agricultural systems. Philos. Trans. R. Soc. B: Biol. Sci. 363 (1492), 685−701.

King, S.L., Keim, R.F., 2019. Hydrologic modifications challenge bottomland hardwood forest management. J. For. 117, 504−514.

Krech III, S., 1999. The Ecological Indian: Myth and History. WW Norton & Company, London.

Krueger-Mangold, J.M., Sheley, R.L., Svejcar, T.J., 2006. Toward ecologically-based invasive plant management on rangeland. Weed Sci. 54 (3), 597−605.

Kukuļs, I., Kļaviņš, M., Nikodemus, O., Kasparinskis, R., Brūmelis, G., 2019. Changes in soil organic matter and soil humic substances following the afforestation of former agricultural lands in the boreal-nemoral ecotone (Latvia). Geoderma Regional 16, e00213.

Lake, J.C., Leishman, M.R., 2004. Invasion success of exotic plants in natural ecosystems: the role of disturbance, plant attributes and freedom from herbivores. Biol. Conserv. 117 (2), 215−226.

Lal, R., 2004. Soil carbon sequestration impacts on global climate change and food security. Science 304 (5677), 1623−1627.

Lal, R., Iivari, T., Kimble, J.M., 2003. Soil Degradation in the United States: Extent, Severity, and Trends. CRC Press, Boca Raton, FL.

Lamb, D., Stanturf, J., Madsen, P., 2012. What is forest landscape restoration? In: Stanturf, J., Lamb, D., Madsen, P. (Eds.), Forest Landscape Restoration Integrating Natural and Social Sciences. Springer, Dordrecht, pp. 3−23.

Lance, A.C., Carrino-Kyker, S.R., Burke, D.J., Burns, J.H., 2020. Individual plant-soil feedback effects influence tree growth and rhizosphere fungal communities in a temperate forest restoration experiment. Front. Ecol. Evol. 7 (500). Available from: https://doi.org/10.3389/fevo.2019.00500.

Latif, M., 2013. Uncertainty in climate change projections. In: Hasselmann, K., Jaeger, C., Leipold, G., Mangalagiu, D., Tàbara, J.D. (Eds.), Reframing the Problem of Climate Change. Routledge, New York, pp. 31−48.

Le Page, Y., Morton, D., Hartin, C., Bond-Lamberty, B., Pereira, J.M.C., Hurtt, G., et al., 2017. Synergy between land use and climate change increases future fire risk in Amazon forests. Earth Syst. Dyn. 8 (4), 1237−1246.

Lee, D., Park, P., Park, Y., 2015. Forest restoration and rehabilitation in the Republic of Korea. In: Stanturf, J.A. (Ed.), Restoration of Boreal and Temperate Forests, Second Edition. CRC Press, Boca Raton, FL, pp. 217–232.

Leverington, F., Costa, K.L., Pavese, H., Lisle, A., Hockings, M., 2010. A global analysis of protected area management effectiveness. Environ. Manage. 46 (5), 685–698.

Lewis, S.L., Edwards, D.P., Galbraith, D., 2015. Increasing human dominance of tropical forests. Science 349 (6250), 827–832.

Liu, Y., Stanturf, J., Goodrick, S., 2010. Trends in global wildfire potential in a changing climate. For. Ecol. Manage. 259 (4), 685–697.

Locke, H., 2014. Nature Needs Half: a necessary and hopeful new agenda for protected areas in North America and around the World. George Wright Forum 31 (3), 359–371.

Lugo, A.E., Helmer, E., 2004. Emerging forests on abandoned land: Puerto Rico's new forests. For. Ecol. Manage. 190 (2–3), 145–161.

Macdonald, S.E., Landhäusser, S.M., Skousen, J., Franklin, J., Frouz, J., Hall, S., et al., 2015. Forest restoration following surface mining disturbance: challenges and solutions. N. For. 46 (5–6), 703–732.

Madritch, M.D., Lindroth, R.L., 2009. Removal of invasive shrubs reduces exotic earthworm populations. Biol. Invasions 11 (3), 663–671.

Madsen, P., Jensen, F.A., Fodgaard, S., 2015. Afforestation in Denmark. In: Stanturf, J.A. (Ed.), Restoration of Boreal and Temperate Forests, second ed., CRC Press, Dordrecht, pp. 201–216.

Maginnis, S., Jackson, W., 2007. What is FLR and how does it differ from current approaches? In: Rietbergen-McCracken, J., Maginnis, S., Sarre, A. (Eds.), The Forest Landscape Restoration Handbook. Earthscan, London.

Maikhuri, R.K., Rao, K.S., 2012. Soil quality and soil health: a review. Int. J. Ecol. Environ. Sci. 38 (1), 19–37.

Mansourian, S., Dudley, N., Vallauri, D., 2017a. Forest landscape restoration: progress in the last decade and remaining challenges. Ecol. Restor. 35 (4), 281–288.

Mansourian, S., Stanturf, J.A., Derkyi, M.A.A., Engel, V.L., 2017b. Forest Landscape Restoration: increasing the positive impacts of forest restoration or simply the area under tree cover? Restor. Ecol. 25 (2), 178–183.

Mansourian, S., Parrotta, J., Balaji, P., Bellwood-Howard, I., Bhasme, S., Bixler, R.P., et al., 2020. Putting the pieces together: integration for forest landscape restoration implementation. Land Degradation Dev. 31 (4), 419–429. <https://doi.org/10.1002/ldr.3448>.

Mappin, B., Chauvenet, A.L., Adams, V.M., Di Marco, M., Beyer, H.L., Venter, O., et al., 2019. Restoration priorities to achieve the global protected area target. Conserv. Lett. 12, e12646.

Mariotti, B., Hoshika, Y., Cambi, M., Marra, E., Feng, Z., Paoletti, E., et al., 2020. Vehicle-induced compaction of forest soil affects plant morphological and physiological attributes: a meta-analysis. For. Ecol. Manage. 462, 118004.

Masson-Delmotte, V., Zhai, P., Pörtner, H., Roberts, D., Skea, J., Shukla, P., et al., 2018. Global warming of 1.5°C. An IPCC Special Report on the Impacts of Global Warming of 1.5°C Above Pre-Industrial Levels and Related Global Greenhouse Gas Emission Pathways, in the Context of Strengthening the Global Response to the Threat of Climate Change, Sustainable Development, and Efforts to Eradicate Poverty. IPCC.

McBratney, A., Field, D.J., Koch, A., 2014. The dimensions of soil security. Geoderma 213, 203–213.

McBratney, A.B., Mendonça Santos, M.L., Minasny, B., 2003. On digital soil mapping. Geoderma 117 (1), 3−52.

McLachlan, J.S., Hellmann, J.J., Schwartz, M.W., 2007. A framework for debate of assisted migration in an era of climate change. Conserv. Biol. 21 (2), 297−302.

McTavish, M.J., Cray, H.A., Murphy, S.D., Bauer, J.T., Havrilla, C.A., Oelbermann, M., et al., 2020. Sustainable management of grassland soils. In: Stanturf, J., Callaham, M. (Eds.), Soils and Landscape Restoration. Academic Press, New York.

Meinders, M., Van Breemen, N., 2005. Formation of soil-vegetation patterns. In: Lovett, G.M., Jones, C.G., Turner, M.G., Weathers, K.C. (Eds.), Ecosystem Function in Heterogeneous Landscapes. Springer, New York, pp. 207−227.

Meli, P., Holl, K.D., Benayas, J.M.R., Jones, H.P., Jones, P.C., Montoya, D., et al., 2017. A global review of past land use, climate, and active vs. passive restoration effects on forest recovery. PLoS One 12 (2), e0171368.

Metz, L.J., 1958. The Calhoun Experimental Forest. USDA Forest Service Southeastern Forest Experiment Station, Asheville, NC, 24 pp.

Middleton, N., 2018. Rangeland management and climate hazards in drylands: dust storms, desertification and the overgrazing debate. Nat. Hazards 92 (1), 57−70.

Nearing, M., Pruski, F., O'Neal, M., 2004. Expected climate change impacts on soil erosion rates: a review. J. Soil Water Conserv. 59 (1), 43−50.

Neary, D.G., 2020. Advances in nutrient and water management in forestry: monitoring, maintaining, and restoring forest health. In: Stanturf, J.A. (Ed.), Achieving Sustainable Management of Boreal and Temperate Forests. Burleigh Dodds Science Publishing, Cambridge, pp. 413−445.

Neary, D.G., Leonard, J.M., 2020. Restoring fire to forests: contrasting the effects on soils of prescribed fire and wildfire. In: Stanturf, J.A., Callaham Jr., M.A. (Eds.), Soils and Landscape Restoration. Academic Press, New York.

Neuenkamp, L., Prober, S.M., Price, J.N., Zobel, M., Standish, R.J., 2019. Benefits of mycorrhizal inoculation to ecological restoration depend on plant functional type, restoration context and time. Fungal Ecol. 40, 140−149.

Newman, E.A., 2019. Disturbance ecology in the Anthropocene. Front. Ecol. Evol. 7 (147), 147.

Nielsen, U.N., Ayres, E., Wall, D.H., Bardgett, R.D., 2011. Soil biodiversity and carbon cycling: a review and synthesis of studies examining diversity−function relationships. Eur. J. Soil Sci. 62 (1), 105−116.

Obalum, S.E., Chibuike, G.U., Peth, S., Ouyang, Y., 2017. Soil organic matter as sole indicator of soil degradation. Environ. Monit. Assess. 189 (4), 32−50.

Orr, B., Cowie, A., Castillo Sanchez, V., Chasek, P., Crossman, N., Erlewein, A., et al., 2017. Scientific conceptual framework for land degradation neutrality. In: Paper Presented at the A Report of the Science-Policy Interface. United Nations Convention to Combat Desertification (UNCCD), Bonn, Germany.

Oxbrough, A., Pinzón, J., 2020. Advances in understanding forest ecosystem services: conserving biodiversity. In: Stanturf, J.A. (Ed.), Achieving Sustainable Management of Boreal and Temperate Forests. Burleigh Dodds Science Publishing, Cambridge, pp. 211−238.

Page-Dumroese, D., Jurgensen, M., Elliot, W., Rice, T., Nesser, J., Collins, T., et al., 2000. Soil quality standards and guidelines for forest sustainability in northwestern North America. For. Ecol. Manage. 138 (1-3), 445−462.

Page-Dumroese, D.S., Busse, M.D., Jurgensen, M.F., Jokela, E.J., 2020. Sustaining forest soil quality and productivity. In: Stanturf, J., Callaham, M.A. (Eds.), Soils and Landscape Restoration. Academic Press, New York.

Palm, C., Sanchez, P., Ahamed, S., Awiti, A., 2007. Soils: a contemporary perspective. Annu. Rev. Environ. Resour. 32, 99−129.

Pandit, R., Parrotta, J.A., Chaudhary, A.K., Karlen, D.L., Vieira, D.L.M., Anker, Y., et al., 2020. A framework to evaluate land degradation and restoration responses for improved planning and decision-making. Ecosyst. People 16 (1), 1−18.

Park, M.S., Youn, Y.-C., 2017. Reforestation policy integration by the multiple sectors toward forest transition in the Republic of Korea. For. Policy Econ. 76, 45−55.

Parrotta, J.A., Knowles, O.H., 2001. Restoring tropical forests on lands mined for bauxite: examples from the Brazilian Amazon. Ecol. Eng. 17 (2−3), 219−239.

Parrotta, J.A., Wildburger, C., Mansourian, S., 2012. Understanding Relationships between Biodiversity, Carbon, Forests and People: The Key to Achieving REDD + Objectives. International Union of Forest Research Organizations (IUFRO), Vienna.

Pedlar, J.H., McKenney, D.W., Aubin, I., Beardmore, T., Beaulieu, J., Iverson, L., et al., 2012. Placing forestry in the assisted migration debate. BioScience 62 (9), 835−842.

Pfeifer, S., Rechid, D., Reuter, M., Viktor, E., Jacob, D., 2019. 1.5°, 2°, and 3° global warming: visualizing European regions affected by multiple changes. Regional Environ. Change 19 (6), 1777−1786.

Philip, J.R., 1966. Plant water relations: some physical aspects. Annu. Rev. Plant. Physiol. 17 (1), 245−268.

Pielke, R.A., Wilby, R., Niyogi, D., Hossain, F., Dairuku, K., Adegoke, J., et al., 2013. Dealing with complexity and extreme events using a bottom-up, resource-based vulnerability perspective. In: Sharma, A.S., Bunde, A., Dimri, V.P., Baker, D.N. (Eds.), Extreme Events and Natural Hazards: The Complexity Perspective. American Geophysical Union, Washington, DC, pp. 345−359.

Prescott, C., Katzensteiner, K., Weston, C., 2020. Soils and restoration of forested landscapes. In: Stanturf, J., Callaham, M. (Eds.), Soils and Restoration. Academic Press, New York.

Putz, F.E., Nasi, R., 2009. Carbon benefits from avoiding and repairing forest degradation. In: Realising REDD, p. 249.

Raymond, H., 2007. The ecologically noble savage debate. Annu. Rev. Anthropol. 36, 177−190.

Reid, A.M., Morin, L., Downey, P.O., French, K., Virtue, J.G., 2009. Does invasive plant management aid the restoration of natural ecosystems? Biol. Conserv. 142 (10), 2342−2349.

Reisman-Berman, O., Keasar, T., Tel-Zur, N., 2019. Native and non-native species for dryland afforestation: bridging ecosystem integrity and livelihood support. Ann. For. Sci. 76 (4), 114. Available from: https://doi.org/10.1007/s13595-019-0903-2.

Richter Jr, D.D., Markewitz, D., 2001. Understanding Soil Change: Soil Sustainability Over Millennia, Centuries, and Decades. Cambridge University Press, Cambridge.

Richter, D., Yaalon, D.H., 2012. "The changing model of soil" revisited. Soil Sci. Soc. Am. J. 76 (3), 766−778.

Robinson, D., Hockley, N., Cooper, D., Emmett, B., Keith, A., Lebron, I., et al., 2013. Natural capital and ecosystem services, developing an appropriate soils framework as a basis for valuation. Soil Biol. Biochem. 57, 1023−1033.

Robinson, K., Bauer, J., Lugo, A., 2014. Passing the Baton from the Tainos to Tomorrow: Forest Conservation in Puerto Rico; FS-862. US Department of Agriculture Forest Service, International Institute of Tropical Forestry, San Juan, Puerto Rico.

Rooney, T.P., Buttenschøn, R., Madsen, P., Olesen, C.R., Royo, A.A., Stout, S.L., 2015. Integrating ungulate herbivory into forest landscape restoration. In: Stanturf, J.A. (Ed.),

Restoration of Boreal and Temperate Forests, second ed. CRC Press, Boca Raton, FL, pp. 69–84.

Royo, A.A., Collins, R., Adams, M.B., Kirschbaum, C., Carson, W.P., 2010. Pervasive interactions between ungulate browsers and disturbance regimes promote temperate forest herbaceous diversity. Ecology 91 (1), 93–105.

Rummukainen, M., 2012. Changes in climate and weather extremes in the 21st century. Wiley Interdisc. Rev.: Clim. Change 3 (2), 115–129.

Runyan, C., D'Odorico, P., 2016. Global Deforestation. Cambridge University Press, Cambridge.

Sabogal, C., Besacier, C., McGuire, D., 2015. Forest and landscape restoration: concepts, approaches and challenges for implementation. Unasylva 66 (245), 3–10.

Safriel, U., 2017. Land Degradation Neutrality (LDN) in drylands and beyond—where has it come from and where does it go. Silva Fennica 51 (1B), 1650.

Schmidt, M.W., Torn, M.S., Abiven, S., Dittmar, T., Guggenberger, G., Janssens, I.A., et al., 2011. Persistence of soil organic matter as an ecosystem property. Nature 478 (7367), 49–56.

Schoenholtz, S.H., Van Miegroet, H., Burger, J.A., 2000. A review of chemical and physical properties as indicators of forest soil quality: challenges and opportunities. For. Ecol. Manage. 138 (1–3), 335–356.

Schoennagel, T., Balch, J.K., Brenkert-Smith, H., Dennison, P.E., Harvey, B.J., Krawchuk, M.A., et al., 2017. Adapt to more wildfire in western North American forests as climate changes. Proc. Natl. Acad. Sci. U.S.A. 114 (18), 4582–4590.

Schuur, E.A., Bockheim, J., Canadell, J.G., Euskirchen, E., Field, C.B., Goryachkin, S.V., et al., 2008. Vulnerability of permafrost carbon to climate change: implications for the global carbon cycle. AIBS Bull. 58 (8), 701–714.

Schwärzel, K., Zhang, L., Montanarella, L., Wang, Y., Sun, G., 2020. How afforestation affects the water cycle in drylands: a process-based comparative analysis. Glob. Change Biol. 26 (2), 944–959.

SERI, 2004. The SER international primer on ecological restoration. Retrieved from <http://www.ser.org/resources/resources-detail-view/ser-international-primer-on-ecological-restoration>.

Sohi, S., Krull, E., Lopez-Capel, E., Bol, R., 2010. A review of biochar and its use and function in soil. Adv. Agron. 105, 47–82.

Soil Science Division Staff, 2017. Soil Survey Manual, U.S. Department of Agriculture Handbook 18. Government Printing Office, Washington, DC.

Soil Survey Staff, 1999. Soil Taxonomy: A Basic System of Soil Classification for Making and Interpreting Soil Surveys U.S. Department of Agriculture Handbook 436, second ed. Natural Resources Conservation Service, Washington, DC.

Spathelf, P., Stanturf, J., Kleine, M., Jandl, R., Chiatante, D., Bolte, A., 2018. Adaptive measures: integrating adaptive forest management and forest landscape restoration. Ann. For. Sci. 75 (2), 55. Available from: https://doi.org/10.1007/s13595-018-0736-4.

Stanturf, J.A., 2015. Future landscapes: opportunities and challenges. N. For. 46 (5–6), 615–644.

Stanturf, J.A., 2016. What is forest restoration? In: Stanturf, J.A. (Ed.), Restoration of Boreal and Temperate Forests, second ed. CRC Press, Boca Raton, FL, pp. 1–16.

Stanturf, J.A., 2020. Landscape degradation and restoration. In: Stanturf, J.A., Callaham, M.A. (Eds.), Soils and Landscape Restoration. Academic Press, New York.

Stanturf, J.A., Kellison, R.C., Broerman, F.S., Jones, S.B., 2003. Productivity of southern pine plantations: where are we and how did we get here? J. For. 101 (3), 26–31.

Stanturf, J., Palik, B., Dumroese, R.K., 2014a. Contemporary forest restoration: a review emphasizing function. For. Ecol. Manage. 331, 292–323.

Stanturf, J.A., Palik, B.J., Williams, M.I., Dumroese, R.K., Madsen, P., 2014b. Forest restoration paradigms. J. Sustain. For. 33, S161–S194.

Stanturf, J., Mansourian, S., Kleine, M., 2017. Implementing Forest Landscape Restoration, A Practitioner's Guide. International Union of Forest Research Organizations, Vienna, Austria.

Stanturf, J.A., Madsen, P., Sagheb-Talebi, K., Hansen, O.K., 2018. Transformational restoration: novel ecosystems in Denmark. Plant. Biosyst. 152 (3), 536–546.

Stanturf, J.A., Kleine, M., Mansourian, S., Parrotta, J., Madsen, P., Kant, P., et al., 2019. Implementing forest landscape restoration under the Bonn Challenge: a systematic approach. Ann. For. Sci. 76 (2). Available from: https://doi.org/10.1007/s13595-019-0833-z.

Stanturf, J.A., Botman, E., Kalachev, A., Borissova, Y., Kleine, M., Rajapbaev, M., et al., 2020a. Dryland forest restoration under a changing climate in central Asia and Mongolia. Mongolian J. Biol. Sci. 18 (1), 3–18.

Stanturf, J.A., Frelich, L., Donoso, P.J., Kuuluvainen, T., 2020b. Advances in managing and monitoring natural hazards and forest disturbances. In: Stanturf, J. (Ed.), Achieving Sustainable Management of Boreal and Temperate Forests. Burleigh Dodds Science Publishing, Cambridge, pp. 627–716.

Stavi, I., 2019. Seeking environmental sustainability in dryland forestry. Forests 10 (9), 737.

Steckel, M., del Río, M., Heym, M., Aldea, J., Bielak, K., Brazaitis, G., et al., 2020. Species mixing reduces drought susceptibility of Scots pine (Pinus sylvestris L.) and oak (Quercus robur L., Quercus petraea (Matt.) Liebl.)—site water supply and fertility modify the mixing effect. For. Ecol. Manage. 461, 117908.

Tarin, T., Nolan, R.H., Medlyn, B.E., Cleverly, J., Eamus, D., 2020. Water-use efficiency in a semi-arid woodland with high rainfall variability. Glob. Change Biol. 26 (2), 496–508.

Tarnocai, C., Canadell, J.G., Schuur, E.A.G., Kuhry, P., Mazhitova, G., Zimov, S., 2009. Soil organic carbon pools in the northern circumpolar permafrost region. Glob. Biogeochem. Cycles 23 (2). Available from: https://doi.org/10.1029/2008GB003327.

Taufik, M., Torfs, P.J., Uijlenhoet, R., Jones, P.D., Murdiyarso, D., Van Lanen, H.A., 2017. Amplification of wildfire area burnt by hydrological drought in the humid tropics. Nat. Clim. Change 7 (6), 428–431.

Terra, T.N., dos Santos, R.F., Costa, D.C., 2014. Land use changes in protected areas and their future: the legal effectiveness of landscape protection. Land. Use Policy 38, 378–387.

Thomas, C.D., 2011. Translocation of species, climate change, and the end of trying to recreate past ecological communities. Trends Ecol. Evolution 26 (5), 216–221.

Trimble, S.W., 1974. Man-Induced Soil Erosion on the Southern Piedmont. Ankeny, IA.

Tsiafouli, M.A., Thébault, E., Sgardelis, S.P., De Ruiter, P.C., Van Der Putten, W.H., Birkhofer, K., et al., 2015. Intensive agriculture reduces soil biodiversity across Europe. Glob. Change Biol. 21 (2), 973–985.

Vallejo, V.R., Smanis, A., Chirino, E., Fuentes, D., Valdecantos, A., Vilagrosa, A., 2012. Perspectives in dryland restoration: approaches for climate change adaptation. N. For. 43 (5–6), 561–579.

van der Bij, A.U., Weijters, M.J., Bobbink, R., Harris, J.A., Pawlett, M., Ritz, K., et al., 2018. Facilitating ecosystem assembly: plant-soil interactions as a restoration tool. Biol. Conserv. 220, 272–279.

van Diggelen, R., Bobbink, R., Frouz, J., Harris, J.A., Verbruggen, E., 2020. Converting agricultural lands into heathlands: the relevance of soil processes. In: Stanturf, J.A., Callaham, M.A. (Eds.), Soils and Landscape Restoration. Academic Press, New York.

Vandenkoornhuyse, P., Quaiser, A., Duhamel, M., Le Van, A., Dufresne, A., 2015. The importance of the microbiome of the plant holobiont. N. Phytol. 206 (4), 1196–1206.

Vanderburg, K.L., Steffens, T.J., Lust, D.G., Rhoades, M.B., Blaser, B.C., Peters, K., et al., 2020. Trampling and cover effects on soil compaction and seedling establishment in reseeded pasturelands over time. Rangel. Ecol. Manage. 73 (3), 452–461. Available from: https://doi.org/10.1016/j.rama.2020.01.001.

Wagg, C., Bender, S.F., Widmer, F., van der Heijden, M.G., 2014. Soil biodiversity and soil community composition determine ecosystem multifunctionality. Proc. Natl. Acad. Sci. U.S.A. 111 (14), 5266–5270.

Waldrop, M.P., Firestone, M.K., 2006. Response of microbial community composition and function to soil climate change. Microb. Ecol. 52 (4), 716–724.

Wardle, D.A., Van der Putten, W.H., 2002. Biodiversity, ecosystem functioning and above-ground-below-ground linkages. In: Loreau, M., Naeem, S., Inchausti, P. (Eds.), Biodiversity and Ecosystem Functioning: Synthesis and Perspectives. Oxford University Press, Oxford, pp. 155–168.

Weyers, S.L., Brockamp, R.L., 2020. Biochar for the restoration and rehabilitation of managed systems. In: Stanturf, J.A., Callaham Jr, M.A. (Eds.), Soils and Landscape Restoration. Academic Press, New York.

White, E.R., Cox, K., Melbourne, B.A., Hastings, A., 2019. Success and failure of ecological management is highly variable in an experimental test. Proc. Natl. Acad. Sci. U.S.A. 116 (46), 23169–23173.

Wilby, R.L., Dessai, S., 2010. Robust adaptation to climate change. Weather 65 (7), 180–185.

Williams, J.W., Jackson, S.T., 2007. Novel climates, no-analog communities, and ecological surprises. Front. Ecol. Environ. 5 (9), 475–482.

Williams, J.W., Jackson, S.T., Kutzbach, J.E., 2007. Projected distributions of novel and disappearing climates by 2100 AD. Proc. Natl. Acad. Sci. U.S.A. 104 (14), 5738–5742.

Williams, M., 1992. Americans and Their Forests: A Historical Geography. Cambridge University Press, Cambridge.

Winder, R.S., Kranabetter, J.M., Pedlar, J., 2020. Adaptive management of landscapes for climate change: how soils influence the assisted migration of plants. In: Stanturf, J.A., Callaham, M.A. (Eds.), Soils and Landscape Restoration. Academic Press, New York.

Woolf, D., Solomon, D., Lehmann, J., 2018. Land restoration in food security programmes: synergies with climate change mitigation. Clim. Policy 18 (10), 1260–1270.

Wubs, E.J., Van der Putten, W.H., Bosch, M., Bezemer, T.M., 2016. Soil inoculation steers restoration of terrestrial ecosystems. Nat. Plants 2 (8), 1–5.

WWF and IUCN, 2000. Minutes of the Forests Reborn Workshop. WWF and IUCN, Segovia.

Yaalon, D.H., 2007. Human-induced ecosystem and landscape processes always involve soil change. BioScience 57 (11), 918–919.

Yaalon, D.H., Yaron, B., 1966. Framework for man-made soil changes-an outline of metapedogenesis. Soil Sci. 102 (4), 272–277.

Yirdaw, E., Tigabu, M., Monge, A.A., 2017. Rehabilitation of degraded dryland ecosystems—review. Silva Fennica 51, . Available from: https://doi.org/10.14214/sf.1673, article id 1673.

Zalasiewicz, J., Williams, M., Steffen, W., Crutzen, P., 2010. The new world of the Anthropocene. Environ. Sci. Technol. 44 (7), 2228–2231.

Zerga, B., 2015. Rangeland degradation and restoration: a global perspective. Point J. Agric. Biotechnol. Res. 1, 37–54.

Zika, M., Erb, K.-H., 2009. The global loss of net primary production resulting from human-induced soil degradation in drylands. Ecol. Econ. 69 (2), 310–318.

Soil ecology and restoration science

2

Mac A. Callaham Jr.[1] **and John A. Stanturf**[2]

[1]*U.S. Department of Agriculture, Forest Service, Athens, GA, United States*
[2]*Institute of Forestry and Rural Engineering, Estonian University of Life Sciences, Tartu, Estonia*

2.1 Introduction

Restoration ecology is a relatively young discipline and is recognized as the inevitable, applied outgrowth from the broader maturing discipline of Ecology (Hobbs et al., 2011). From its outset some 30–40 years ago (the Society for Ecological Restoration was founded in 1988), there has been recognition that soils are an important ecosystem component and that they must be considered explicitly when restoration plans are made (e.g., Bradshaw, 1983). In spite of this, and partly perhaps because soils are an opaque medium and difficult to study, there has still been relatively little work that explicitly (or experimentally) examines the role that soils play with respect to restoration outcomes. This is particularly true when goals are complex, for example, the restoration of a diverse plant community, where soil processes and functions are necessarily biological/ecological ones (Heneghan et al., 2008; Baer et al., 2012; Van der Bij et al., 2018).

The intimate relationships between soils and the plant communities that develop under natural conditions are at the heart of ecosystems (and indeed biomes), and arise from interactions and feedbacks to such a degree that it can be difficult to separate one from another conceptually (i.e., to consider a particular soil apart from its vegetative associates). Examples of this are found in the association of mollisols with grass-dominated vegetation in the steppes and prairies of North America, or the association of Oxisols with tropical rainforest vegetation across the globe. To a certain extent, these associations of soils with vegetation are taken for granted, and assumptions are sometimes made that simply adding appropriate vegetation to a site will eventually result in proper soil functions.

Across the globe, there is a spectrum of types and degrees of ecosystem degradation, and each of these comes with its own relative level of disturbance to soil (Chapter 5: Landscape Degradation and Restoration). These include, on one end of the spectrum, the total removal and/or destruction of the soil surface layers to a depth of several meters (e.g., mountaintop removal, and open pit mining operations, landfill, or urban development) and range to much more diffuse or difficult to detect

Soils and Landscape Restoration. DOI: https://doi.org/10.1016/B978-0-12-813193-0.00002-3

disturbances such as climate change or atmospheric deposition of nutrients or toxics, which may or may not perceptibly influence soil ecological processes at all. There is a temporal component to ecosystem disturbance as well, and the duration of a particular disturbance (e.g., intensive agriculture that may persist for decades or centuries vs a one-time clearing of native vegetation which is then allowed to regrow) will inform the legacy of such disturbances and will dictate the amount of effort and investment needed to achieve restoration. Finally, as detailed by Stanturf et al. (Chapter 1: Soils are Fundamental to Landscape Restoration), the actual physical, chemical, and geomorphic make-up of soils can influence the responses of a particular land area to disturbance, and some soils are far more susceptible to perturbation than others and thus must be considered more carefully when disturbances occur.

The language and terminology of restoration ecology have been a matter of considerable debate and have been beset with questions about what constitutes an appropriate target for restoration of ecosystems. In fact, The Society for Ecological Restoration engages in a continual process of updating and revising its principles and standards, with much progress in recent years (Gann et al., 2019). These issues have been addressed by Stanturf et al. (Chapter 1: Soils are Fundamental to Landscape Restoration), but we briefly mention this again here to establish the context for our attitudes and approaches to restoration in this chapter specifically, and for this whole volume in general. We have consciously adopted the most liberal view of what constitutes restoration, and done so in consideration of the fact that "success" will have very different criteria for people with differing problems. To examine this point, consider the following hypothetical situations: for a land manager restoring an area impacted by acid mine tailings, success might simply be the establishment of a single nonnative grass species at the site (perhaps decreasing erosion and immobilizing heavy metals), whereas for another land manager, success might only be declared if a reproducing population of a rare wildflower was restored to the understory of an otherwise speciose forest. For the mine-land restoration practitioner, soil pH and metal toxicity are the central problems, and these must be mitigated through practical manipulations such as addition of lime and careful selection of a plant species that can tolerate the existing soil conditions. For the forest restorationist the situation may require a more nuanced approach and could even require experimentation, as there could be a specific and critical soil fungal symbiont missing from local soils, or some specific nutrient or hydrologic requirement that must be met before the desired plants will thrive. Notably, both of the abovementioned scenarios involve increasing the local plant diversity by one species, but the outcomes for the respective ecosystems are quite different. Also notable is that in spite of the dramatic differences in these two restoration scenarios, knowledge of soil function and soil ecological relationships would likely improve the probability of success.

Thus, for our purposes, restoration is an appropriate term to use for any improvement in any condition or any functional attribute of an ecosystem. This approach fully embraces the last (no. 8) of the International Society of Restoration Ecology's "Principles of Restoration," which holds that "ecological restoration is part of a continuum of restorative activities" (Gann et al., 2019).

Importantly, this continuum allows the definition to extend to "bioremediation" and "sustainable management" of ecosystems and explicitly includes sites contaminated through industrial activities and agricultural ecosystems (row-crop or animal production) and applied forestry ecosystems (biomass, fiber, lumber). In our view, this is an appropriate extension of the restoration concept so long as bioremediation or sustainable management is defined by an objective to improve conditions (such as productivity, decreased need for external inputs, or increased resilience) over the long term. Further, we subscribe to the point of view that the best measure of restoration success is achievement of objectives and that the most progress is to be made in the arena of deciding what constitutes a legitimate set of objectives which will ultimately lead to management promoting the development of ecosystems that are more stable, more diverse, and more resilient to disturbances (Hallett et al., 2013; Wagner et al., 2016).

2.2 Soil ecology

Soil ecology is its own discipline with all the same subdivisions as the broader discipline of ecology sensu lato (Coleman et al., 2018). In other words, soils can be examined from the standpoint of biotic populations, communities, and ecosystems; and further, soil can be examined with respect to ecological processes such as energy flow, nutrient cycles, and biogeochemical stoichiometry. As such, the study of soil ecology is necessarily the study of soil *function*, which is the basis of whole ecosystem productivity.

Ecology is broadly defined as the study of organisms interacting with their environment, so it follows then that soil ecology is the study of organisms living in soil interacting with their environment. This environment is structured by physical and chemical interplay between minerals, gases, water, and the organisms themselves. As previously mentioned, the relationship between soil and its respective vegetation is an intimate one, and the basic characteristics of the soil in terms of its texture, water-holding capacity, and inherent fertility (dictated by age, weathering rates, parent material, clay mineralogy, etc.), all have bearing on the type and amount of plant life that a soil can support. Of course, the plants exert tremendous reciprocal influence on the soils they inhabit, and the combination of seasonality, temperature, precipitation, and increasingly (as more of Earth's surface is modified by human activity), the vegetation management regime employed (or the legacy thereof) all combine to determine certain attributes of the underlying soils (e.g., Schmidt et al., 2011; Mayer et al., 2020). Among these attributes are the type, quality, and amount of organic matter (OM); diversity and community structure of the nonplant biota; and even the physical properties of porosity, aggregate stability, and bulk density (Foster et al., 2003; Baer et al., 2012; Jõgiste et al., 2017).

OM derived from photosynthesis is the source of energy for all soil ecosystems and is thus the engine that drives most soil ecological interactions (the few exceptions involve chemoautotrophic microbes in specific habitats and conditions). Not surprisingly then, the production of OM, its subsequent cycling through food-webs, and its biochemical transformation into stable, long-term storage pools of soil OM (SOM) have been topics of intense focus from the earliest days of soil ecology until the present (Lehmann and Kleber, 2015). A second focal point for soil ecological research has been describing the sheer abundance and diversity of the microbes and fauna that inhabit soil ecosystems. Soil ecosystems have been frequently referred to as the final frontier in biodiversity studies (Orgiazzi et al., 2016), and although there is debate around the actual figures, all agree that organisms in soil account for a substantial fraction of the total global biotic diversity and that most of the biological diversity of soils remains to be discovered and documented, particularly among soil bacteria, archaea, and fungi (e.g., Fierer and Jackson, 2006; Bates et al., 2011; and Tedersoo et al., 2014, respectively), but this is also indicated for soil animals such as nematodes and microarthropods (e.g., Wu et al., 2011), as well as for macroinvertebrates such as earthworms (Phillips et al., 2019). An understanding of interplay between SOM, all the various biota, and the process of OM stabilization in soils, is at the heart of employing soil ecological knowledge to restoration practice, and we suggest that optimization of these processes, under different sets of conditions, should become an explicit goal of restoration practitioners.

2.3 Soil ecology and ecosystem restoration

Given the proposition that knowledge of soil ecological relationships should help to improve outcomes in restoration scenarios, it next seems reasonable to ask what aspects of soil ecology are best to consider, and what outcomes might be reasonable to expect. In the following section, we will briefly discuss various aspects of soil science and soil ecology with specific reference to implications and applications in ecosystem restoration. As a matter of practicality, we will present information for each aspect more-or-less individually and then conclude with a section dealing with interactions and the resulting processes.

2.3.1 Soil physical properties

Soil physical properties are largely derived from parent material (e.g., weathered bedrock, Aeolian silt, marine or lacustrine sediment, etc.), and other geomorphic attributes such as slope, elevation, and aspect. These factors ultimately interact with climate, weathering, and time to give each soil its own distinctive combination of sand, silt, and clay-sized particles, which may or may not include larger fragments of rock; in sum, the soil's texture. The importance of soil texture is discussed in more detail in Stanturf et al. (Chapter 1: Soils are Fundamental to Landscape Restoration, see Fig. 1.1). From a soil ecology standpoint, texture largely determines the water-holding capacity of a particular soil, and the amount of water is strongly related to the types and numbers of soil organisms that will inhabit and influence that soil. In general, sandy soils retain less water, clayey soils retain more, and silty soils occupy the middle ground. Inasmuch as water availability influences plant productivity, these different textured soils then also have general characteristics of SOM storage, with sandy soils typically being relatively low on the SOM spectrum and clayey soils on the high end.

Soil structure is another physical attribute that can have a strong influence upon soil properties and functions, most notably water and gas infiltration (as influenced through porosity), and SOM storage. Soil structure comes about through various processes both physical and ecological and is the result of soil particles forming into larger bodies called aggregates. Although aggregates can form through strictly physical processes (such as freeze/thaw cycles or shrinking and swelling of clays with wetting and drying cycles), in many soils the process of aggregate formation is known to be dominated by biological interactions. At the smallest scale, microaggregates are held together by charges associated with the particles themselves, but at each successive increase in the size of aggregates, the roles of soil biota become increasingly important. Microaggregates are formed into aggregates held together by microbial products (polysaccharides), and larger aggregates are held together by fungal hyphae and plant roots and their exudates, and so on (Tisdall and Oades, 1982; Hu et al., 1995). Macroaggregates (those >2000 μm in diameter) are often associated with soil invertebrate activity, including earthworms (Snyder et al., 2009; Hallam and Hodson, 2020). Closely related to aggregate structure, porosity is another important soil attribute and is essentially a measure of the space in soil that is not occupied by solid particles. Pore space is available for the movement and storage of water and gases, and it also provides living space for the microfauna in soils (protozoa, nematodes, mites, and other microarthropods) (Fig. 2.1). Porosity is itself influenced by the activity of the larger soil fauna, and earthworms, ants, cicadas, and many other macroarthropods produce macropores that are involved in water and gas movements (Dean, 1992; Jackson et al., 2003; Capowiez et al., 2014). Absence of porosity is one major symptom of soil compaction or other disturbances and is typically associated with dramatically lower populations of soil organisms. Compaction can contribute to poor water infiltration, and thus to runoff and soil erosion, often resulting in decreased plant growth.

In the landscape restoration context, it would seem that there would be very little to do in the way of altering the soil physical attributes just described, and for most practical purposes, this is probably true except for localized site preparation that creates mounds or rows of soil (e.g., Löf et al., 2012; Frouz, Chapter 6: Soil Recovery and Reclamation of Mined Lands) or recreation of microtopography in wetland restoration of former agricultural land that was leveled for irrigation (Bruland and Richardson, 2005). However, there have been experimental attempts to modify biophysical attributes of degraded ecosystems. For example, in grazed grassy woodlands of Australia that have experienced soil degradation

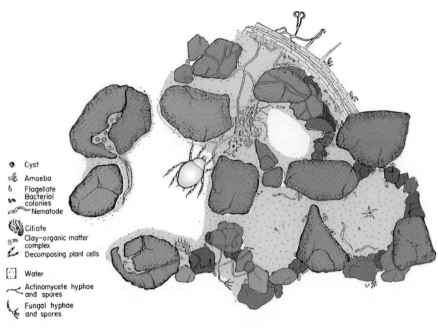

Cyst
Amoeba
Flagellate
Bacterial colonies
Nematode
Ciliate
Clay-organic matter complex
Decomposing plant cells
Water
Actinomycete hyphae and spores
Fungal hyphae and spores

FIGURE 2.1

Soil porosity provides living space for many of the biota inhabiting mineral horizons of soil.

Used by permission. Redrawn from Paul, E.A., Clark, F.E., 1989. Soil Microbiology and Biochemistry. Academic Press, San Diego. Original art by S. Rose and T. Elliott.

and compaction, mechanical aeration followed by biochar amendments to soils, or application of green eucalypt mulch, both proved to decrease soil bulk density and improve water-holding capacity. These improvements were associated with improved performance of a restored native understory plant and are expected to improve resilience of these soils and ecosystems to drought (Prober et al., 2014). In cases of extreme degradation, where surface soils may have been entirely destroyed, or even actively removed from a site, managers have more freedom in terms of reconstructing soils with desirable textural attributes. Frouz (Chapter 6: Soil Recovery and Reclamation of Mined Lands) and Byrne (Chapter 14: Socioecological Soil Restoration in Urban Cultural Landscapes) discuss this in some detail, with particular reference to wholesale construction of "Technosols" in mine reclamation, and urban settings, respectively.

2.3.2 **Soil chemical properties**

Of all the soil chemical properties the SOM content is arguably the most important. This has been alluded to previously in this chapter and was given a detailed treatment in Stanturf et al. (Chapter 1: Soils are Fundamental to Landscape Restoration). Suffice it to say here that the amount and type of SOM influences practically every physical, chemical, and biological interaction that happens in soil (e.g., Mayer et al., 2020). Many restoration approaches initially involve simple addition of OM to soil, often with notable effects such as dramatically improved establishment of tree seedlings and tree growth (e.g., Pinno and Errington, 2015; Somerville et al., 2020). There have been literally hundreds of demonstrations of this basic principle of soil restoration. The overall objective is to boost SOM from exogenous sources, as waiting for this SOM to form through normal processes from vegetation planted on-site (which may or may not successfully establish) may take too long.

Another important, principally chemical, consideration for restoration of soils through remediation, reconstruction, and rehabilitation (terminology of Stanturf et al., 2014) is the concentration of potentially toxic substances, usually metals such as cadmium, arsenic, lead, nickel, mercury, zinc, or other classes of chemicals such as salts, polyaromatic hydrocarbons, or polychlorinated biphenyls. These contaminants are generally produced by various industrial processes and often result from exploration and extraction processes (mining, drilling, pumping, and disposal). Clearly, if concentrations of these chemicals are high enough to be toxic to other biota, then the ecological processes performed by biota will be lost. Such highly contaminated sites may require many years of simple phytoremediation (wherein plants sequester the toxic compounds) before any soil fauna will be able to inhabit the site (see Zalesny et al., Chapter 9: Bioremediation and Soils, for more detail).

Many restoration projects are undertaken on lands that have been taken out of agricultural production, and these sites present a suite of problems to managers attempting restoration. As a general rule, row-cropped agricultural systems are intentionally made to be as homogeneous as possible in terms of soil attributes for the specific benefit of a single species of crop plant (Fig. 2.2). This homogenizing of crop field habitats includes plowing, and often grading of soils to uniform depth and minimal slope, management of pests, and application of nutrients at levels to maximize uptake and growth (often in excess of demand). From an ecological perspective, these activities may be interpreted as optimizing niche space for the crop plant, but this necessarily comes with the collapse of other niches in the ecosystem that would be occupied by other plants (and nontarget organisms affected by pesticide applications). One lasting chemical legacy of agricultural activity in soils is elevated nutrient concentrations, usually of nitrogen or phosphorus. This homogeneous and high availability of nutrients often run counter to the objectives of restoration practitioners who seek to establish heterogeneous habitats with plants and other organisms that exploit different niches in terms of nutrient availability or other limiting resources (e.g., restoring heathlands; see van Diggelen et al., Chapter 13: Converting Agricultural Lands into Heathlands: The Relevance of Soil Processes). One common approach to dealing with high nutrient concentrations has been the practice of "defertilization"

FIGURE 2.2

Agricultural practices such as plowing, harrowing, fertilizer, and pesticide use are intended to homogenize soil resources to provide the optimal niche for the crop plant. This can present challenges to managers interested in restoring such soils to a diverse species assemblage.

Photo by Dwight Burdette—Own work, CC BY 3.0, <https://commons.wikimedia.org/w/index.php? curid = 19574144>.

wherein a substrate with high carbon-to-nitrogen ratio is added to the soil with the intent of immobilizing surplus nutrients (e.g., Baer et al., 2003; Morris and de Barse, 2012), or nutrients are exported by harvesting (Schelfhout et al., 2017) or burning vegetation (Jones et al., 2015), but it is important to note that these treatments sometimes do not have the desired effects on nutrient levels in the time-frame available for study (as in both Schelfhout et al., 2017; Jones et al., 2015). In one of the few studies to actually impose experimental heterogeneity of multiple resources on former agricultural land, Baer et al. (2020) found that resource heterogeneity did positively influence diversity of the restored grassland vegetation, but that this took 20 + years to detect in appreciable amounts.

2.3.3 Soil biological properties

Soil biological attributes are perhaps the most complex and difficult to address aspect of soil ecology in the restoration context, and yet, perhaps the most critical in terms of actual functional restoration of soils and ecosystems. In general, it is true that the soil biota (microbes and fauna) have most often been relegated to the role of "indicators" of restoration success, that is, soil biota are often evaluated in terms of how they respond to different restoration treatments, but not often used as agents of restoration themselves. One major exception to this is that soil microbial inoculums have been used in a fairly large number of restoration projects, but microbes too have been used to indicate progress or success of restoration treatments (e.g., Strickland et al., 2017). For the nonmicrobial biota, there are only a few studies that have actively manipulated various components of the soil biota, and we will summarize findings, applicability, and feasibility of these approaches in the next sections.

2.3.3.1 Soil microbial ecology in restorations
2.3.3.1.1 Whole microbial community

Many studies have manipulated soil microbial communities in restoration projects, generally in the form of whole community microbial inoculums which are produced from soils that represent the desired condition or target for the restored site. An example of this is the work of Hamman and Hawkes (2013) in Florida scrub ecosystems, where the authors collected the surface layers of nearby undegraded soils and applied this material to degraded soils in order to establish populations of crust forming microbes. The results of this study indicated that the application of microbial inoculum did improve the germination of native plant seeds in plots where invasive grasses had been removed, but this positive effect was short-lived and persisted for only one year. This study, and all the others like it are a test of the question, first prominently articulated by Harris (2009) as to whether soil microbial communities are best viewed as "facilitators" or as "followers" of restoration. The answer to this question, as evaluated by a subsequent 10 + years of research, is unclear, with some demonstrations indicating that microbial inoculums help to achieve restoration goals in some settings, whereas they fail to produce results in others (e.g., Grove et al., 2019). However, results from a recent metaanalysis demonstrated that it is probably best to attempt some form of inoculation with microbes, as there is a general trend for improved plant growth and species richness with such inoculations (Neuenkamp et al., 2019).

Whether soil microbial communities are examined as indicators or as agents of restoration, one frequently encountered theme in restoration studies is the importance of soil disturbance legacies, and the difficulty of overcoming certain legacies (Baer et al., 2012; Stanturf et al., 2020). Legacy effects derive from past agricultural uses such as tillage, nutrient enrichment, or pesticide use which encompass many of the physical and chemical soil factors mentioned above. In addition, soil disturbance legacies may originate from biotic factors such as invasion by exotic plants which may involve plant–soil feedbacks that are difficult to dislodge once established (Eviner et al., 2010).

2.3.3.1.2 Mycorrhizae

Mycorrhizae are important determinants of plant community composition in most vegetation types and have been studied extensively with regard to their applicability in restoration ecology for decades (e.g., Miller and Jastrow, 1992; Allen et al., 1995). A complete discussion of research findings with regard to the success or failure of mycorrhizal inoculations in restoration projects is beyond the scope of this chapter, but one emerging theme is the context dependency of success for such treatments. In summary, mycorrhizal inoculations are not always successful and depend upon a number of factors including, but not limited to, the origin of the inoculum (whether commercially or locally acquired), nutrient availability at the restored site, and whether invasive plant species inhabit the site (e.g., Chaudhary et al., 2019). The concept of soil disturbance legacies introduced in the previous section applies to plant–fungus symbioses as well and is illustrated by the work of Lankau et al. (2014), who found that even after an invasive plant (garlic mustard, *Alliaria petiolata*) had been removed from a site, its allelopathic legacy persisted for years. In this case, one of the allelopathic mechanisms was suppression of the arbuscular mycorrhizal (AM) fungus community, and this was detectable for 6 years following removal of garlic mustard, resulting in differences in plant community composition in the plots (Lankau et al., 2014). A final example of mycorrhizal inoculation that emphasizes another consideration for restoration practitioners is that the source of inoculum can influence the resulting diversity of the microbial communities in the receiving soil. Phillips et al. (2020) found that inoculum originating under native chaparral vegetation resulted in greater diversity of AM fungi infecting roots of experimental plants, compared to an inoculum originating under a nonnative plant. However, these authors also found a greater diversity of non-AM fungi in the soils of their experimental plants when the native soil inoculum was used. This result has significant implications for restoration practitioners if total soil microbial diversity is as important as expected in terms of maintaining soil functions under changing conditions (Delgado-Baquerizo et al., 2017).

Many plant species do not provision their seeds with resources for seedlings to use upon germination, and these seedlings are often entirely dependent upon mycorrhizal infection (Van der Heijden et al., 2015). Further, as with many orchids, for example, the degree of specificity between plant and fungal species is quite high (Jacquemyn et al., 2015), and this presents obvious challenges to restoration interests. In any case, it is clear that many plants do poorly without their fungal symbionts, and it behooves the restoration practitioner to have some fairly nuanced understanding of this situation. In short, nowhere is the context dependency of restoration practice more evident than in situations requiring the correct pairing of plant species with appropriate mycorrhizal symbionts. This will be most important when the topsoil has been removed and/or severely degraded, or when the plant introduced is not native to the soil which may lack suitable symbionts (e.g., Hatch, 1936; Miller and Jastrow, 1992).

2.3.3.2 Soil faunal ecology in restoration

2.3.3.2.1 Microinvertebrates/mesofauna

Protozoa, microarthropods, and nematodes make up the bulk of the micro/meso-fauna of soils, and these groups have often been used as indicators of restoration progress and success. This application is primarily due to the incredible diversity documented for all these groups of organisms (Wu et al., 2011). One example of these fauna being used to evaluate restoration techniques is the use of oribatid mites to examine recovery of soils following coal extraction in the Athabasca oil sands region of Alberta, Canada. Oribatid mites were most responsive, in terms of total abundance, species richness, and diversity to the depth of the organic horizon present in the forest floor of reclaimed sites, and this response was independent of the time since reclamation (McAdams et al., 2018). In another study, in this case restoration of urban ecosystems, Schaefer and Hocking (2015) suggested that mite abundance and functional group composition might be used to indicate the presence of certain ecosystem processes, and as a way to delineate whether systems were functional, or merely ornamental.

Although there have been few studies that have specifically manipulated the mesofauna for the purpose of accelerating or promoting restoration, it is clear that these organisms are performing critical ecological functions in soils and have been suggested to indirectly influence the resilience and recovery of soil processes to disturbances. For example, Maraun et al. (1998) conducted a laboratory experiment where soils were subjected to severe freezing and heating treatments and then inoculated half of their microcosms with oribatid mites. The microcosms that received mites exhibited greater respiration rates, and greater retention of nutrients, indicating that the presence and activity of the mites likely transported propagules of various components of the microbial community into litter substrates, stimulated microbial growth, and accelerated the recovery of nutrient and energy cycling processes in these artificial systems.

Nematodes constitute another major component of the soil mesofauna and are known to have major impacts on soil processes, but this information is probably best developed with respect to their status as agricultural pests. Again, the principal use of nematodes has been as indicators of restoration success, in some cases with fewer nematodes being interpreted as success, for example, with restored hydrologic cycles in peatlands in Slovakia (Bobuľská et al., 2019). Nevertheless, there have been a few manipulations involving soil inoculations which include and exclude nematodes, and the results of these indicate an important role of nematodes (and of general diversity of the smaller members of the soil community) in terms of overall soil functionality. Strong evidence is accumulating that soil inoculums that included nematodes were functionally the most complete relative to those that excluded them (Wagg et al., 2014), and this included functions such as support of increased plant diversity, carbon sequestration, and nutrient cycling and retention—all of which are clearly critical to successful ecosystem restoration. Further compelling evidence involving the influence that nematodes may have on restoration trajectories is given in Wubs et al. (2019) who found that the initial introduction of soil nematodes in a whole soil inoculum (along with seeds) had detectable influences on the composition of the plant community 20 years later relative to plots that received only seeds, and importantly, plant community and soil nematode community became increasingly correlated with one another over time. This work illustrates the importance of plant and soil biota interactions and suggests that the composition of inoculums used in restorations should be very carefully considered indeed (and clearly should not consider only the microbial components).

2.3.3.2.2 Macroinvertebrates (ecosystem engineers)

Several groups of soil macroinvertebrates are considered to be ecosystem engineers due to their abilities to physically modify soil habitats. The burrowing, casting, and nest and mound-building activities of earthworms, ants, and termites are among the "engineering" activities that invertebrates perform, and the resulting changes in soil structure can influence biodiversity, hydrologic processes, and even soil formation over the long term (Fig.2.3). These activities have barely been examined in the restoration context, as revealed in the review by Jouquet et al. (2014), and rigorous field testing of several desirable ecosystem functions is needed and has been suggested for more than a decade (Snyder and Hendrix, 2008). In general, the engineering invertebrates have been considered as response variables to vegetative restoration, and in the context of whether their community assembly can be accelerated through restoration of the desired plant community (e.g., Wodika and Baer, 2015).

In one striking example of exploiting the engineering and restorative capacities of termites, Issoufou et al. (2019) employed a traditional soil enrichment practice (termed *zai*) in heavily degraded soils in southwestern Niger, which involved green mulching of marginal agricultural soils. These authors then monitored the soils for termite abundance and then tested various measures of fertility. The overall abundance and diversity of termites were enhanced by the application of zai techniques, and there was a concurrent improvement in soil organic carbon and nitrogen concentrations. Boulton and Amberman (2006) experimentally demonstrated that the activities of harvester ants were responsible for stimulating greater abundances of several groups of soil biota (including nematodes and microarthropods), as well as increased richness within the microbial community. These findings have clear implications for the roles that nest-building ants might have for promoting diversity and function in soil communities, and with applications in restoration planning. In spite of this, we could find no studies where ants have intentionally been manipulated with the goal of improving restoration outcomes.

On the other hand, in projects where ecosystems are constructed on heavily degraded or manipulated sites such as in mine land reclamation, ecosystem engineers, and, in particular, earthworms have been seriously considered and evaluated with regard to their capacity to act as agents of restoration (e.g., Butt, 2008;

FIGURE 2.3

Examples of invertebrate engineers, and the structures they produce.

Top left: These large ant mounds in forests of Czech Republic have significant impacts on surface soil turnover, and organic matter accumulation dynamics.

Bottom left: Organisms such as this earthworm construct burrows and produce castings that influence soil physical attributes such as water and gas movement, and they provide habitat for other organisms.

Right: Biogenic structures such as this termite mound in Tanzania are important features of landscapes and can contribute to soil bioturbation rates and pedogenic processes over long time frames.

(Top left) Photo by Jan Kameníček—Own work, CC BY-SA 3.0, <https://commons.wikimedia.org/w/index.php? curid = 27763799>. (Bottom left) Photo by Rob Hille—Own work, CC BY-SA 3.0, <https://commons. wikimedia.org/w/index.php?curid = 14799364>. (Right) Photo credit: Vierka Maráková, photo in public domain.

Boyer and Wratten, 2010; Frouz et al., 2006). The chapter by Frouz in this volume (Chapter 6: Soil Recovery and Reclamation of Mined Lands) details the findings of a decades-long research program that has given particular attention to the roles of soil macrofauna in the reclamation of mined land in Eastern Europe. In agricultural landscapes which have large inputs of pesticides and fertilizers, hydrologically active biopores (i.e., burrows of earthworms and other organisms) were shown to be "hot spots" for pesticide mineralization (Badawi et al., 2013), and earthworms have also been associated with stimulating denitrification (Depkat-Jakob et al., 2013; Majeed et al., 2013), and this may have particularly interesting implications for restoration of former agricultural lands that have excess nitrogen levels in soils. Earthworms have also been tested for their capacity to improve bioremediation outcomes in toxic materials such as bauxite residues (Di Carlo et al., 2020) and petroleum contaminated soils (Callaham et al., 2002; Schaefer et al., 2005), with generally favorable results, but these have been conducted in the laboratory and require field testing.

It should definitely be mentioned here that not all soil invertebrate ecosystem engineers are considered desirable, and in fact when nonnative soil engineers are introduced, this can cause significant degradations of soil properties and biodiversity. There are many examples of invasive earthworms having negative effects on ecosystems (e.g., Snyder et al., 2011; Ferlian et al., 2018). Likewise, there are cosmopolitan species of ants and termites that cause significant damage to ecological processes and biodiversity (not to mention human health impacts) in the systems where they invade (Evans et al., 2013). For example, in the southeastern United States, the red imported fire ant had dramatic impacts on native ants during the early phases of invasion, but native ant species did recover over time (Morrison, 2002). In spite of the negative effects that nonnative soil engineers may have, the question of whether a nonnative earthworm, for example, is better than no earthworms at all in a particular soil is one that gives pause, as even nonnative earthworms can have positive influences in degraded soils and may improve or accelerate recovery of important soil processes and functions. The philosophical implications and ethical propriety of using known exotic species to achieve restoration objectives are outside the purview of the current chapter!

2.3.3.2.3 Vertebrates

The final group of soil-dwelling animals we will consider is the charismatic megafauna of the soil ecological world—the fossorial vertebrates, consisting principally of mammals, birds, and reptiles. Perhaps not surprisingly, these organisms have received more attention than most of the invertebrate groups with respect to restoration projects, and this is possibly due to their relative tractability as study organisms. Their larger size, reasonable abundances, and manageable diversity relative to their meso- and macrofaunal counterparts are all attributes that make them more attractive to examine (not to mention their fur, feathers, and scales). It is also true that most fossorial vertebrates are involved in some kind of ecological engineering activities and move large amounts of soil around in their burrowing activities. This soil movement has consequences for all levels of community organization from microbes to plants and even to other vertebrate taxa. For example, gopher tortoise burrows in the southeastern United States are known to provide habitat for several mammals, other reptiles, as well as invertebrates, some of which construct auxiliary burrows in the walls of tortoise burrows (Kinlaw and Grasmueck, 2012). Just as with the invertebrate fauna, fossorial vertebrates have been used to evaluate the success of restoration techniques, and Lawer et al. (2019) conducted a metaanalysis of several studies to determine whether passive versus active restoration of mined sites influenced the recovery of fossorial mammals. Their analysis revealed that fossorial and ground-dwelling small mammals exhibited immediate recovery following mine abandonment, but that these organisms decreased in abundance over time postmining.

Other evaluations of soil-dwelling vertebrates have documented the actual influence of these animals on the trajectory of restoration efforts. On overburden dumps associated with diamond mining in South Africa, the only places where vegetation became established were in proximity to whistling rat burrows and mounds (Desmet and Cowling, 1999). The soil properties of the excavated soil associated with these mounds were such that plants could become established in overburden material that would otherwise be toxic to plants, and this was partially attributable to OM inputs deriving from the rodents foraging behavior and excretions. Perhaps the most work on burrowing mammals has been conducted in Australia, where the catastrophic declines in native fossorial mammal populations have been implicated in soil degradation (Martin, 2003) and where reintroductions of these native species have been tested in restoration trials. James and Eldridge (2007) documented that reintroduction of native fossorial marsupials into nature reserves in the arid landscapes of central Australia and noted that pits excavated by these animals resulted in altered and improved soil characteristics—essentially the formation of fertile patches in the landscape. The altered microtopography generated by the digging activity of the animals resulted in materials being intercepted as they moved across the landscape with precious resources such as water, seeds, and other OM collecting in the pits. In another fascinating instance of manipulating fossorial mammals to achieve restoration goals, McCullough Hennessy et al. (2016) experimentally enhanced the population of ground squirrels in California grasslands and found that a combination of squirrel activity and active vegetation management provided significantly more habitat for yet another endangered fossorial animal—the western burrowing owl.

Importantly, like some invasive invertebrates, it should be mentioned that not all ecosystem engineering by fossorial animals is considered desirable, and some ecosystems are susceptible to degradation when populations of burrowing mammals reach high densities. One example of this is in sub-Antarctic islands, where burrowing by introduced mice has destabilized large land areas and contributed to increased erosion rates (in addition to negative effects on nesting seabirds, and invertebrate communities) (Eriksson and Eldridge, 2014). Further, while in some ecosystems the activities of burrowing animals are critical for plant establishment, there are situations where the presence of such animals is detrimental to restoration of vegetation, as was observed in the Sand Fynbos of South Africa where the presence of mole rats was associated with decreased success of the desired plant species (Holmes, 2008).

2.4 Restorative ecological processes in soil

2.4.1 Primary productivity

The astute reader will have noticed that we have thus far not mentioned the role of plants as part of the belowground biota! This is for two reasons, with the first being that we intended to use this chapter to take the focus away from plants in restoration (to the extent possible) and to focus instead on the nonplant components of soil biology. Second, plants do occupy aboveground space in ecosystems, and we intended to keep the focus belowground. However, it is impossible to discuss ecosystem restoration without some acknowledgment of primary productivity belowground, that is, root production. We have presented examples in previous sections demonstrating the importance of SOM, and the difficulty and time involved in bringing concentrations up to a level similar to reference conditions. Clearly, it is not economically or practically feasible to continue to make exogenous inputs of OM over time to build SOM in restored landscapes, so primary productivity, and vigorous root growth is essential. Simply put, the production of roots belowground is the best way to produce appreciable, sustainable amounts of SOM in any restored ecosystem. However, although the storage of C in living roots can recover fairly quickly, the trajectory of restorations indicates that total SOM recovery could take decades or centuries (e.g., Baer et al., 2002, in tallgrass prairie; Matzek et al., 2016, in riparian forests). Nevertheless, the faster roots become established in restored systems, the more microbe-promoting root exudates are available, and the faster the SOM formation process can begin.

2.4.2 Decomposition

Decomposition of OM, both at the soil surface and belowground, is a crucial ecosystem process in functioning ecosystems and is a fundamentally soil-based process. Decomposition integrates relationships between plants, which produce the material to be decomposed, the soil microbes which carry out the decomposition in their metabolism, and soil invertebrates which also metabolize plant detritus but in addition can modulate the activities of microbes and influence soil physical properties. This integrated process is ultimately responsible for the remineralization, and thus recycling of nutrients to plants, as well as forming one of the major pathways for development of SOM (Coleman et al., 2018). Ballantine and Schneider (2009) found that litter decomposition was a critical, but unrealized component of wetland soil restoration. These authors found that even after 55 years of wetland restoration, in spite of similar litter decomposition rates, the restored sites still did not exhibit soil physical properties or SOM concentrations that were consistent with reference wetlands. The lack of development of deep SOM, and persistent high values of bulk density observed in these wetland soils was attributed to the possibility of a missing component of the mixing biota from these soils. Yet litter decomposition can be manipulated for improved and accelerated outcomes in restoration projects as demonstrated by Horodecki et al. (2019) who found that trees planted in certain species combinations produced litter that decomposed more rapidly than litter of several monoculture stands. These authors suggested that judicious selection of species mixes during planting can significantly and positively influence soil development in mine land reclamations.

2.4.3 Bioturbation

Bioturbation is literally the process of the entire biota mixing the surface layers of the soil (Wilkinson et al., 2009). As such, bioturbation is a process that is closely linked to the previous two processes of root production and litter decomposition. Plant roots penetrate and displace, soils wherever they grow, and various forms of macroinvertebrates, and burrowing vertebrates excavate, penetrate, and otherwise riddle the soil with biopores and voids of all shapes and sizes. Some of the soil invertebrates, in particular the ecosystem engineers (ants, termites, and earthworms), have been recognized for their roles in the turnover of surface soils since the time of Charles Darwin, who wrote about the influence of earthworms on the development of "vegetable mold" in a field behind his home in southern England (Butt et al., 2016). Likewise, bioturbation by vertebrates in Australia contributed to the pristine conditions encountered by the early settlers of rangeland in New South Wales who reported the soils to be "soft, spongy and very absorbent" and "like a well tilled field," but deterioration due to grazing (and loss of the native marsupials) resulted in "a hard, clayey smooth surface" (all quotations as reported in Martin, 2003). In the process of casting, nest building, digging, and mound-building, these vertebrate and invertebrate organisms bury OM on the surface, incorporating it into the mineral soil horizons, and contributing over time to SOM stocks. At the same time, these activities influence physical soil properties such as water infiltration and gas movements, and these all in turn influence microbial community development, ultimately culminating in soil profile development.

2.4.4 Soil formation (pedogenesis)

As just alluded to in the previous section, the process of soil formation, and soil profile development (horizonation), is arguably the end goal of any restoration where no soil was previously present, or in situations where soil must be constructed de novo. Functioning pedogenic processes are also critical to restore in situations where soils are present but degraded. This goal, even if not explicitly stated, of restoring the ecological components (primary production, litter decomposition, bioturbation) of soil formation processes, could be viewed as the ultimate test of restoration success, because fully functioning soil necessarily underpins a fully functioning ecosystem.

Frouz et al. (2013) elegantly investigated the interplay between plants, the litter they produce, the relative decomposition rates of that litter, microbial components of soil, and soil fauna relative to soil formation in forest ecosystems restored on coal mining spoils in the Czech Republic. This work is detailed by Frouz (Chapter 6: Soil Recovery and Reclamation of Mined Lands), but to briefly highlight their results here, these authors planted different species of trees in spoils and then followed soil profile development under these plantings for 30 + years in some cases. The results from this work are among the clearest demonstrations of the links between tree species, litter quality, soil invertebrate communities, and soil horizon development patterns, with high-quality litter producing species having a horizon development, and low-quality litter producers displaying O horizon development (see Fig. 7, Chapter. 6: Soil Recovery and Reclamation of Mined Lands). This demonstration is all the more remarkable given the common soil substrate that all these forest stands began with and emphasizes the critical nature of the ecological components of the soil formation process.

2.5 Concluding remarks

Our goal with this chapter was to bring focus to the interconnectedness of soil functions, up to and including the process of soil formation itself. We hope that this treatment of the nuances of soil ecology, and their multiple dependencies upon one another will continue to push restoration science to consider more than single-factor manipulations of soil resources in restoration practice (Callaham et al., 2008). The complexity of relationships in soils is clearly at least partly responsible for the context dependency of restoration approaches and outcomes (Eviner and Hawkes, 2008; Suding, 2011), but this does not mean that generalities and guidelines are impossible to achieve with careful study and thoughtful experimentation. It is our hope that just such guidelines can be produced, with specific regard to enhanced restoration of soil processes in the next decades of research into sustainable management, ecosystem restoration, and climate adaptation.

References

Allen, E.B., Allen, M.F., Helm, D.J., Trappe, J.M., Molina, R., Rincon, E., 1995. Patterns and regulation of mycorrhizal plant and fungal diversity. Plant Soil 170, 47−62.

Badawi, N., Johnsen, A.R., Brandt, K.K., Sørensen, J., Aamand, J., 2013. Hydraulically active biopores stimulate pesticide mineralization in agricultural subsoil. Soil Biol. Biochem. 57, 533−541.

Baer, S.G., Kitchen, D.J., Blair, J.M., Rice, C.W., 2002. Changes in ecosystem structure and function along a chronosequence of restored grasslands. Ecol. Appl. 12, 1688−1701.

Baer, S.G., Blair, J.M., Collins, S.L., Knapp, A.K., 2003. Soil resources regulate productivity and diversity in newly established tallgrass prairie. Ecology 84, 724−735.

Baer, S.G., Heneghan, L., Eviner, V.T., 2012. Applying soil ecological knowledge to restore ecosystem services. In: Wall, D.H., Ritz, K., Six, J., Strong, D.R., Van der Putten, W.H. (Eds.), Soil Ecology and Ecosystem Services. Oxford University Press, Oxford, pp. 377–394.

Baer, S.G., Adams, T., Scott, D.A., Blair, J.M., Collins, S.L., 2020. Soil heterogeneity increases plant diversity after 20 years of manipulation during grassland restoration. Ecol. Appl. 30. Available from: https://doi.org/10.1002/EAP.2014 (accessed 24.04.20.).

Ballantine, K., Schneider, R., 2009. Fifty-five years of soil development in restored freshwater depressional wetlands. Ecol. Appl. 19, 1467–1480.

Bates, S.T., Berg-Lyons, D., Caporaso, J.G., Walters, W.A., Knight, R., Fierer, N., 2011. Examining the global distribution of dominant archaeal populations in soil. ISME J. 5, 908–917.

Bobul̓ská, L., Demková, L., Čerevková, A., Renčo, M., 2019. Impact of peatland restoration on soil microbial activity and nematode communities. Wetlands 39. Available from: https://doi.org/10.1007/s13157-019-01214-2 (accessed 24.04.20.).

Boulton, A.M., Amberman, K.D., 2006. How ant nests increase soil biota richness and abundance: a field experiment. Biodivers. Conserv. 15, 69–82.

Boyer, S., Wratten, S.D., 2010. The potential of earthworms to restore ecosystem services after opencast mining – a review. Basic Appl. Ecol. 11, 196–203.

Bradshaw, A.D., 1983. The reconstruction of ecosystems. J. Appl. Ecol. 20, 1–17.

Bruland, G.L., Richardson, C.J., 2005. Hydrologic, edaphic, and vegetative responses to microtopographic reestablishment in a restored wetland. Restor. Ecol. 13, 515–523.

Butt, K.R., 2008. Earthworms in soil restoration: lessons learned from United Kingdom case studies of land reclamation. Restor. Ecol. 16, 637–641.

Butt, K.R., Callaham Jr., M.A., Loudermilk, E.L., Blaik, R., 2016. Action of earthworms on flint burial – a return to Darwin's estate. Appl. Soil Ecol. 104, 157–162.

Callaham Jr., M.A., Stewart, A.J., Alarcón, C., McMillen, S.J., 2002. Effects of earthworms (*Eisenia fetida*) and wheat (*Triticum aestivum*) straw additions on selected properties of petroleum-contaminated soils. Environ. Toxicol. Chem. 21, 1658–1663.

Callaham Jr., M.A., Rhoades, C.C., Heneghan, L., 2008. A striking profile: soil ecological knowledge in restoration management and science. Restor. Ecol. 16, 604–607.

Capowiez, Y., Sammartino, S., Michel, E., 2014. Burrow systems of endogeic earthworms: effects of earthworm abundance and consequences for soil water infiltration. Pedobiologia 57, 303–309.

Chaudhary, V.B., Akland, K., Johnson, N.C., Bowker, M.A., 2019. Do soil inoculants accelerate dryland restoration? A simultaneous assessment of biocrusts and mycorrhizal fungi. Restor. Ecol. Available from: https://doi.org/10.1111/rec.13088 (accessed 24.04.20.).

Coleman, D.C., Callaham Jr., M.A., Crossley Jr., D.A., 2018. Fundamentals of Soil Ecology, third ed. Academic Press, London.

Dean, W.R.J., 1992. Effects of animal activity on the absorption rate of soils in the southern Karoo, South Africa. J. Grassl. Soc. South. Afr. 9, 178–180.

Delgado-Baquerizo, M., Eldridge, D.J., Ochoa, V., Gozalo, B., Singh, B.K., Maestre, F.T., 2017. Soil microbial communities drive the resistance of ecosystem multifunctionality to global change in drylands across the globe. Ecol. Lett. 20, 1295–1305.

Depkat-Jakob, P.S., Brown, G.G., Tsai, S.M., Horn, M.A., Drake, H.L., 2013. Emission of nitrous oxide and dinitrogen by diverse earthworm families from Brazil and resolution of associated denitrifying and nitrate-dissimilating taxa. FEMS Microbiol. Ecol. 83, 375−391.

Desmet, P.G., Cowling, R.M., 1999. Patch creation by fossorial rodents: a key process in the revegetation of phytotoxic arid soils. J. Arid Environ. 43, 35−45.

Di Carlo, E., Boullemant, A., Poynton, H., Courtney, R., 2020. Exposure of earthworm (*Eisenia fetida*) to bauxite residue: implications for future rehabilitation programmes. Sci. Total Environ. 716.   Available from: https://doi.org/10.1016/j.scitotenv.2020.137126 (accessed 24.04.20.).

Eriksson, B., Eldridge, D.J., 2014. Surface destabilisation by the invasive burrowing engineer *Mus muscula* on a sub-Antarctic island. Geomorphology 223, 61−66.

Evans, T.A., Forschler, B.T., Grace, J.K., 2013. Biology of invasive termites: a worldwide review. Annu. Rev. Entomol. 58, 455−474.

Eviner, V.T., Hawkes, C.V., 2008. Embracing variability in the application of plant-soil interactions to the restoration of communities and ecosystems. Restor. Ecol. 16, 713−729.

Eviner, V.T., Hoskinson, S.A., Hawkes, C.V., 2010. Ecosystem impacts of exotic plants can feed back to increase invasion in western US rangelands. Rangelands 32, 21−31.

Ferlian, O., Eisenhauer, N., Aguirrebengoa, M., Camara, M., Ramirez-Rojas, I., Santos, F., et al., 2018. Invasive earthworms erode soil biodiversity: a meta-analysis. J. Anim. Ecol. 87, 162−172.

Fierer, N., Jackson, R.B., 2006. The diversity and biogeography of soil bacterial communities. Proc. Natl. Acad. Sci. U.S.A. 103, 626−631.

Foster, D., Swanson, F., Aber, J., Burke, I., Brokaw, N., Tilman, D., et al., 2003. The importance of land-use and its legacies to ecology and environmental management. BioScience 53, 77−88.

Frouz, J., Elhottová, D., Kuráž, V., Šourková, M., 2006. Effects of soil macrofauna on other soil biota and soil formation in reclaimed and unreclaimed post mining sites: results of a field microcosm experiment. Appl. Soil Ecol. 33, 308−320.

Frouz, J., Livečková, M., Albrechtová, J., Chroňáková, A., Cajthaml, T., Pižl, V., et al., 2013. Is the effect of trees on soil properties mediated by soil fauna? A case study from post-mining sites. For. Ecol. Manage. 309, 87−95.

Gann, G.D., McDonald, T., Walder, B., Aronson, J., Nelson, C.R., Jonson, J., et al., 2019. International principles and standards for the practice of ecological restoration, 2nd edition. Restor. Ecol. 27, S1−S46.

Grove, S., Saarman, N.P., Gilbert, G.S., Faircloth, B., Haubensak, K.A., Parker, I.M., 2019. Ectomycorrhizas and tree seedling establishment are strongly influenced by forest edge proximity but not soil inoculum. Ecol. Appl. 29.   Available from: https://doi.org/10.1002/eap.1867 (accessed 24.04.20.).

Hallam, J., Hodson, M.E., 2020. Impact of different earthworm ecotypes on water stable aggregates and soil water holding capacity. Biol. Fertil. Soil.   Available from: https://doi.org/10.1007/s00374-020-01432-5 (accessed 24.04.20.).

Hallett, L.M., Diver, S., Eitzel, M.V., Olson, J.J., Ramage, B.S., Sardinas, H., et al., 2013. Do we practice what we preach? Goal setting for ecological restoration. Restor. Ecol. 21, 312−319.

Hamman, S.T., Hawkes, C.V., 2013. Biogeochemical and microbial legacies of non-native grasses can affect restoration success. Restor. Ecol. 21, 58−66.

Harris, J., 2009. Soil microbial communities and restoration ecology: facilitators or followers? Science 325, 573–574.

Hatch, A., 1936. The role of mycorrhizae in afforestation. J. For. 34, 22–29.

Heneghan, L., Miller, S.P., Callaham Jr., M.A., Baer, S.G., Montgomery, J., Rhoades, C.C., et al., 2008. Integrating soil ecological knowledge into restoration management. Restor. Ecol. 16, 608–617.

Hobbs, R.J., Hallett, L.M., Ehrlich, P.R., Mooney, H.A., 2011. Intervention ecology: applying ecological science in the twenty-first century. BioScience 61, 442–450.

Holmes, P.M., 2008. Optimal ground preparation treatments for restoring lowland Sand Fynbos vegetation on old fields. S. Afr. J. Bot. 74, 33–40.

Horodecki, P., Nowiński, M., Jagodziński, A.M., 2019. Advantages of mixed tree stands in restoration of upper soil layers on postmining sites: a five-year leaf litter decomposition experiment. Land Degrad. Dev. 30, 3–13.

Hu, S., Coleman, D.C., Beare, M.H., Hendrix, P.F., 1995. Soil carbohydrates in aggrading and degrading agroecosystems: influences of fungi and aggregates. Agric. Ecosyst. Environ. 54, 77–88.

Issoufou, A.A., Maman, G., Soumana, I., Kaiser, D., Konate, S., Mahamane, S., et al., 2019. Termite footprints in restored versus degraded agrosystems in South-West Niger. Land Degrad. Dev. 31, 500–507.

Jackson, E.C., Krogh, S.N., Whitford, W.G., 2003. Desertification and biopedturbation in the northern Chihuahuan Desert. J. Arid Environ. 53, 1–14.

Jacquemyn, H., Brys, R., Waud, M., Busschaert, P., Lievens, B., 2015. Mycorrhizal networks and coexistence in species-rich orchid communities. N. Phytol. 206, 1127–1134.

James, A.I., Eldridge, D.J., 2007. Reintroduction of fossorial native mammals and potential impacts on ecosystem processes in an Australian desert landscape. Biol. Conserv. 138, 351–359.

Jõgiste, K., Korjus, H., Stanturf, J.A., Frelich, L.E., Baders, E., Donis, J., et al., 2017. Hemiboreal forest: natural disturbances and the importance of ecosystem legacies to management. Ecosphere 8 (2), e01706. 01710.01002/ecs01702.01706.

Jones, R.O., Chambers, J.C., Board, D.I., Johnson, D.W., Blank, R.R., 2015. The role of resource limitation in restoration of sagebrush ecosystems dominated by cheatgrass (*Bromus tectorum*). Ecosphere 6. Available from: https://doi.org/10.1890/ES14-00285.1 (accessed 24.04.2.0).

Jouquet, P., Blanchart, E., Capowiez, Y., 2014. Utilization of earthworm and termites for the restoration of ecosystem functioning. Appl. Soil Ecol. 73, 34–40.

Kinlaw, A., Grasmueck, M., 2012. Evidence for and geomorphologic consequences of a reptilian ecosystem engineer: the burrowing cascade initiated by the Gopher Tortoise. Geomorphology 157–158, 108–121.

Lankau, R.A., Bauer, J.T., Anderson, M.R., Anderson, R.C., 2014. Long-term legacies and partial recovery of mycorrhizal communities after invasive plant removal. Biol. Invasions 16, 1979–1990.

Lawer, E.A., Mupepele, A.-C., Klein, A.-M., 2019. Responses of small mammals to land restoration after mining. Landsc. Ecol. 34, 473–485.

Lehmann, J., Kleber, M., 2015. The contentious nature of soil organic matter. Nature 528, 60–68.

Löf, M., Dey, D.C., Navarro, R.M., Jacobs, D.F., 2012. Mechanical site preparation for forest restoration. N. For. 43, 825–848.

Majeed, M.Z., Miambi, E., Barois, I., Blanchart, E., Brauman, A., 2013. Emissions of nitrous oxide from casts of tropical earthworms belonging to different ecological categories. Pedobiologia 56, 49−58.

Maraun, M., Visser, S., Scheu, S., 1998. Oribatid mites enhance the recovery of the microbial community after a strong disturbance. Appl. Soil Ecol. 9, 175−181.

Martin, G., 2003. The role of small ground-foraging mammals in topsoil health and biodiversity: implications to management and restoration. Ecol. Manage. Restor. 4, 114−119.

Matzek, V., Warren, S., Fisher, C., 2016. Incomplete recovery of ecosystem processes after two decades of riparian forest restoration. Restor. Ecol. 24, 637−645.

Mayer, M., Prescott, C.E., Abaker, W.E.A., Augusto, L., Cécillon, L., Ferreira, G.W.D., et al., 2020. Influence of forest management activities on soil organic carbon stocks: a knowledge synthesis. For. Ecol. Manage. 466, 118127. Available from: https://doi.org/10.1016/j.foreco.2020.118127 (accessed 24.04.20.).

McAdams, B.N., Quideau, S.A., Swallow, M.J.B., Lumley, L.M., 2018. Oribatid mite recovery along a chronosequence of afforested boreal sites following oil sands mining. For. Ecol. Manage. 422, 281−293.

McCullough Hennessy, S., Deutschman, D.H., Shier, D.M., Nordstrom, L.A., Lenihan, C., Montagne, J.-P., et al., 2016. Experimental habitat restoration for conserved species using ecosystem engineers and vegetation management. Anim. Conserv. 19, 506−514.

Miller, R.M., Jastrow, J.D., 1992. The application of VA mycorrhizae to ecosystem restoration and reclamation. In: Allen, M.F. (Ed.), Mycorrhizal Functioning: An Integrative Plant-Fungal Process. Chapman and Hall, New York, pp. 438−467.

Morris, E.C., de Barse, M., 2012. Carbon, fire and seed addition favour native over exotic species in a grassy woodland. Austral Ecol. 38, 413−426.

Morrison, L.W., 2002. Long-term impacts of an arthropod-community invasion by the imported fire ant, *Solenopsis invicta*. Ecology 83, 2337−2345.

Neuenkamp, L., Prober, S.M., Price, J.N., Zobel, M., Standish, R.J., 2019. Benefits of mycorrhizal inoculation to ecological restoration depend on plant functional type, restoration context and time. Fungal Ecol. 40, 140−149.

Orgiazzi, A., Bardgett, R.D., Barrios, E., Behan-Pelletier, V., Briones, M.J.I., Chotte, J.L., et al., 2016. Global Soil Biodiversity Atlas. European Commission, Publications Office of the European Union, Luxembourg.

Phillips, H.R.P., Guerra, C.A., Bartz, M.L.C., Briones, M.J.I., Brown, G.G., Crowther, T.W., et al., 2019. Global distribution of earthworm diversity. Science 366. Available from: https://doi.org/10.1101/587394 (accessed 24.04.20.).

Phillips, M.L., Aronson, E.L., Malz, M.R., Allen, E.B., 2020. Native and invasive inoculation sources modify fungal community assembly and biomass production of a chaparral shrub. Appl. Soil Ecol. 147. Available from: https://doi.org/10.1016/j.apsoil.2019.103370.

Pinno, B.D., Errington, R.C., 2015. Maximizing natural trembling aspen seedling establishment on a reclaimed boreal oil sands site. Ecol. Restor. 33, 43−50.

Prober, S.M., Stol, J., Piper, M., Gupta, V.V.S.R., Cunningham, S.A., 2014. Enhancing soil biophysical condition for climate-resilient restoration in mesic woodlands. Ecol. Eng. 71, 246−255.

Schaefer, V., Hocking, M., 2015. Detecting the threshold between ornamental landscapes and functional ecological communities: soil microarthropods as indicator species. Urban Ecosyst. 18, 1071−1080.

Schaefer, M., Petersen, S.O., Filser, J., 2005. Effects of *Lumbricus terrestris*, *Allolobophora chlorotica* and *Eisenia fetida* on microbial community dynamics in oil-contaminated soil. Soil Biol. Biochem. 37, 2065–2076.

Schelfhout, S., Mertens, J., Perring, M.P., Raman, M., Baeten, L., Demey, A., et al., 2017. P-removal for restoration of *Nardus* grasslands on former agricultural land: cutting traditions. Restor. Ecol. 25, S178–S187.

Schmidt, M.W., Torn, M.S., Abiven, S., Dittmar, T., Guggenberger, G., Janssens, I.A., et al., 2011. Persistence of soil organic matter as an ecosystem property. Nature 478, 49.

Snyder, B.A., Hendrix, P.F., 2008. Current and potential roles of soil macroinvertebrates (earthworms, millipedes and isopods) in ecological restoration. Restor. Ecol. 16, 629–636.

Snyder, B.A., Boots, B., Hendrix, P.F., 2009. Competition between invasive earthworms (*Amynthas corticis*, Megascolecidae) and native North American millipedes (*Pseudopolydesmus erasus*, Polydesmidae): effects on carbon cycling and soil structure. Soil Biol. Biochem. 41, 1442–1449.

Snyder, B.A., Callaham Jr., M.A., Hendrix, P.F., 2011. Spatial variability of an invasive earthworm (*Amynthas agrestis*) population and potential impacts on soil characteristics and millipedes in the Great Smoky Mountains National Park, USA. Biol. Invasions 13, 349–358.

Somerville, P.D., Farrell, C., May, P.B., Livesley, S.J., 2020. Biochar and compost equally improve urban soil physical and biological properties and tree growth, with no added benefit in combination. Sci. Total Environ. 706. Available from: https://doi.org/10.1016/j.scitotenv.2019.135736 (accessed 24.04.20.).

Stanturf, J., Palik, B., Dumroese, R.K., 2014. Contemporary forest restoration: a review emphasizing function. For. Ecol. Manage. 331, 292–323.

Stanturf, J.A., Frelich, L., Donoso, P.J., Kuuluvainen, T., 2020. Advances in managing and monitoring natural hazards and forest disturbances. In: Stanturf, J. (Ed.), Achieving Sustainable Management of Boreal and Temperate Forests (Burleigh Dodds Science Publishing, Cambridge, pp. 627–716.

Strickland, M.S., Callaham Jr., M.A., Gardiner, E.S., Stanturf, J.A., Leff, J., Fierer, N., et al., 2017. Response of soil microbial community composition and function to a bottomland forest restoration intensity gradient. Appl. Soil Ecol. 119, 317–326.

Suding, K.N., 2011. Toward an era of restoration in ecology: successes, failures, and opportunities ahead. Annu. Rev. Ecol. Evol. Syst. 42, 465–487.

Tedersoo, L., Bahram, M., Põlme, S., Kõljalg, U., Yorou, N.S., Wijesundera, R., et al., 2014. Global diversity and geography of soil fungi. Science 346, 1078–1087.

Tisdall, J.M., Oades, J.M., 1982. Organic matter and water-stable aggregates in soils. J. Soil Sci. 33, 141–163.

Van der Bij, A.U., Weijters, M.J., Bobbink, R., Harris, J.A., Pawlett, M., Ritz, K., et al., 2018. Facilitating ecosystem assembly: plant-soil interactions as a restoration tool. Biol. Conserv. 220, 272–279.

Van der Heijden, M.G.A., Martin, F.M., Selousse, M.-A., Sanders, I.R., 2015. Mycorrhizal ecology and evolution: the past, the present, and the future. N. Phytol. 205, 1406–1423.

Wagg, C., Bender, S.F., Widmer, F., van der Heijden, M.G.A., 2014. Soil biodiversity and soil community composition determine ecosystem multifunctionality. Proc. Natl. Acad. Sci. U.S.A. 111, 5266–5270.

Wagner, A.M., Larson, D.L., DalSoglio, J.A., Harris, J.A., Labus, P., Rosi-Marshall, E.J., et al., 2016. A framework for establishing restoration goals for contaminated ecosystems. Integr. Environ. Assess. Manage. 12, 264–272.

Wilkinson, M.T., Richards, P.J., Humphreys, G.S., 2009. Breaking ground: pedological, geological, and ecological implications of soil bioturbation. Earth-Science Rev. 97, 257–272.

Wodika, B.R., Baer, S.G., 2015. If we build it, will they colonize? A test of the field of dreams paradigm with soil macroinvertebrate communities. Appl. Soil Ecol. 91, 80–89.

Wu, T., Ayres, E., Bardgett, R.D., Wall, D.H., Garey, J.R., 2011. Molecular study of worldwide distribution and diversity of soil animals. Proc. Natl. Acad. Sci. U.S.A. 108, 17720–17725.

Wubs, E.R.J., van der Putten, W.H., Mortimer, S.R., Korthals, G.W., Duyts, H., Wagenaar, R., et al., 2019. Single introductions of soil biota and plants generate long-term legacies in soil and plant community assembly. Ecol. Lett. 22, 1145–1151.

Sustaining forest soil quality and productivity

3

Deborah S. Page-Dumroese[1], Matt D. Busse[2], Martin F. Jurgensen[3] and Eric J. Jokela[4]

[1]*U.S. Department of Agriculture, Forest Service, Rocky Mountain Research Station, Moscow, ID, United States*
[2]*U.S. Department of Agriculture, Forest Service, Pacific Southwest Station, Albany, CA, United States*
[3]*College of Forest Resources and Environmental Science, Michigan Technological University, Houghton, MI, United States*
[4]*School of Forest Resources and Conservation, University of Florida, Gainesville, FL, United States*

Forest soil sustainability is not an ideal, fixed, or static goal (Nambiar, 1996; Lal, 2005). Therefore, for this chapter, we define *sustainable soil* as a product of the interaction of soil physical, chemical, and biological properties that function to produce biomass, store carbon (C), bioremediate wastes, regulate water quantity and quality, and promote above- and belowground biodiversity. Ecological and climatic changes, along with pressure from urban growth, all influence soil properties in large and small ways (e.g., land use, soil biodiversity). Because soils are dynamic systems, it is critical to understand the links among soil properties, land management, and forest sustainability (Burger and Kelting, 1999; Page-Dumroese et al., 2009). There are a number of severe forest soil impacts that can occur from management [e.g., coal-mining (Burger et al., 2017), gas pads (Drohan et al., 2012), shale-gas infrastructure (Fink and Drohan, 2015)], but these sites are often spatially limited, or have limited data about restoring and understanding dynamic soil processes (Fink and Drohan, 2015), and are not addressed in this chapter. The focus of this chapter is on if, when, and how forest harvesting and site preparation can affect soil physical, chemical, and biological properties and highlighting several management considerations that help ensure long-term soil productivity (LTSP) and forest sustainability. We also discuss how soil monitoring can provide long-term data to inform adaptive management and contribute to our understanding of healthy forests and soils.

Soils and Landscape Restoration. DOI: https://doi.org/10.1016/B978-0-12-813193-0.00003-5

3.1 Soil health for sustainable forest management

Productive soils are essential to productive forests and are a nonrenewable resource that should be protected, restored, and enhanced to preserve forest production and ecosystem services, such as water quality and quantity, and decreased erosion, and help mitigate climate change through C sequestration (Powers et al., 1998). Recently, there has also been widespread interest in using forest biomass for energy production, fire hazard reduction of overstocked stands, and restoration of forests decimated by drought or insects (He et al., 2014; Miner et al., 2014), which increases the need to understand the relationship of soil properties to site productivity before and after forest operations. Understanding soil impacts requires a dedicated soil monitoring effort. There are numerous international monitoring acts and laws that have been enacted to ensure sustainable practices [e.g., Helsinki Process, 1994, Montreal Process, 1995, Sustainable Forestry Initiative (SFI), Forest Stewardship Council (FSC), American Tree Farm System, state BMP guidelines, and Forest Practices Acts] and which define forest management success as the presence, density, or biomass of plants at a point in time (Ruiz-Jaen and Aide, 2005), but they do not include a metric for soil recovery, sustainability, or improved function (Aronson et al., 2006).

Sustainable forestry at the stand level includes optimizing the harvest unit size, a minimum "green-up" period after harvesting, stocking level, genetic material, wildlife corridors, and the maintenance of water quality (Burger and Kelting, 1999). Similar to international efforts, US guidelines and policy for stand-level forestry also target maintaining soil sustainability. Although many forest criteria address the issue of soil sustainability, there are many complicated factors that make it difficult to determine if soil productivity is affected after harvest operations as discussed later.

3.1.1 Important factors for soil sustainability

Soil loss is irreversible on the human timescale and, therefore, soil conservation is a key to maintaining forest productivity. In 1993 the United Nations convened an international seminar in Montreal, Canada on sustainable development of temperate and boreal forests. This conference led to several multicountry initiatives (e.g., Montreal Process, Helsinki Process) and resulting in criteria and indicators for sustainability which specifically concern soil impacts from forest management. For example, indicator 22 states "area and percent of forest land with significant compaction or change in soil physical properties resulting from human activities." These criteria also include maintaining the C cycle and ecosystem services such as water yield. Soil-based indicators such as erosion, diminished soil organic matter (OM), and compaction are evaluated and are meant to communicate what participating countries want to see as the condition of their forests.

Generally, all ecosystems will recover from disturbance given enough time (Morris et al. 1997, 1983). For example, forest sustainability can be achieved if timber harvesting rotations are longer than the time it takes the soil to recover. If the stress is not too great and the frequency of disturbance is low, the system will eventually return to a predisturbance condition; however, it is critical to understand the starting state of forest conditions (Morris et al., 1997). Successful recovery and future productivity in a changing climate will rely heavily on soil OM and nutrient conservation (Wells and Jorgensen, 1979; Kimmins, 1977).

3.2 **Soil organic matter**

Soil OM is critical for sustaining soil productivity because of its role in nutrient availability, gas exchange, and water supply (Jurgensen et al., 1997; Powers et al., 1998) and is essential for micro- and macrofauna that are active in nutrient cycling, soil aggregation, and disease incidence or prevention (Harvey et al., 1987). Soil C and nutrient pools as well as soil physical properties, such as water-holding capacity and bulk density, are directly correlated with soil OM content (Vance et al., 2018). Forest floor materials also influence vegetation reproduction, erosion, fuel type and fire behavior, and susceptibility to disease, but the exact amount needed to sustain soil productivity varies by site and soil type (Grigal and Vance, 2000).

3.2.1 **Loss of soil organic matter during management**

Many forest management operations can affect soil OM content by altering the inputs and distribution of surface litter, root turnover, and microbial activity. Studies on the effects of harvest operations on soil OM and C formation, however, have produced mixed results. In general, timber harvest operations, especially clearcutting, may alter mineral C pools (Johnson and Curtis, 2001), although the majority of C loss is from the surface organic horizon (Powers et al., 2005; Nave et al., 2010). Some researchers point out that changes in OM are site-specific and often produce a cascade of secondary changes (Clarke et al., 2015). For example, harvest operations that remove woody litter and downed wood may disrupt the input of decomposing C into the mineral soil layer resulting in higher soil temperatures which increase decomposition rates (Covington, 1981), or changes in soil moisture availability (Gordon et al., 1987; Prescott et al., 2000) that could increase or decrease decomposition rates. Harvesting also supports the development of a new aggrading stand that may increase litter C inputs as compared to preharvest levels (Fleming et al., 2006), whereas mixing soil OM into the mineral soil by equipment might also increase decomposition (Jandl et al., 2007). Any reductions in soil OM are likely to persist until the new forest has fully occupied the site

(Powers et al., 2005). On the North American LTSP study sites, OM removal treatments during clearcutting operations (bole only, whole tree, and whole tree + forest floor removal) resulted in little net difference in soil C after 5 years except at the North Carolina study site (Table 3.1). The dramatic increase in soil C in North Carolina was similar to that observed following harvest on the South Carolina Piedmont (Van Lear et al., 1995) and was attributed to root decomposition (Sanchez et al., 2006). However, longer term data from North Carolina indicated that whole tree harvesting + forest removal, resulted in declines in soil C with time were attributed to the loss of the forest floor and its role in supplying C inputs (Powers et al., 2005).

3.3 Soil nutrients

It is well established that soil nutrient limitations constrain plant growth. For example, calcium (Ca) and phosphorus (P) deficiencies limit growth in many tropical areas, whereas nitrogen (N) limitations are fairly common in temperate and boreal systems. Countless research studies on nutrient availability and cycling processes further confirm the importance of site-specific assessments to fully understand the effects of ecosystem disturbance on soil nutrient status (there is no magic or one-size-fits-all solution). In this regard, forest fertilization studies provide a useful mechanism to evaluate site-specific nutrient deficiencies, postharvest changes, and the relationship between soil nutrients and tree growth.

Table 3.1 Mean soil carbon ($\pm$ SE) and changes in bulk density for several long-term soil productivity locations immediately after harvesting and at year 5.

Location	Carbon (Mg ha^{-1})[a]		Bulk density change from preharvest (%)[b]	
	Immediate postharvest	Year 5	Immediate postharvest	Year 5
Idaho	19.7 (2.1)	19.3 (0.7)	22	19
Louisiana	16.4 (0.2)	16.5 (0.3)	2	− 4
North Carolina	12.0 (0.4)	28.2 (1.7)	24	9
British Columbia	29.4 (0.9)	32.5 (1.9)	13	15

[a]Data from Sanchez et al. (2006).
[b]Data from Page-Dumroese et al. (2006).

3.4 Forest fertilization

Soils in the US (Jokela et al., 1991a; Hopmans and Chappell, 1994; Littke et al., 2014), Chile (Donoso et al., 2,009), Brazil (Campoe et al., 2010), Australia (Turner, 1983), and many other locations around the world have benefitted from fertilizer additions for increasing stand growth and value (summarized in Binkley et al., 1995). For example, Albaugh et al. (2007) reported that about 0.567 million ha of pine plantations in the southeastern US have been fertilized annually since 1999. Depending upon the region and soil, N applications alone or in combination with P have been commonly applied at planting or mid-rotation age. On some sites potassium (K) and even micronutrients [i.e., manganese (Mn), copper (Cu)] may limit stand growth after N and P demands have been met (Jokela et al., 1991b; Mandzak and Moore, 1994; Jokela, 2004; Kyle et al., 2005; Vogel and Jokela, 2011; Carlson et al., 2014).

Fertilization resulted in short-term soil changes that minimally altered soil C levels in some ecosystems (Johnson and Curtis, 2001; McFarlane et al., 2009; Nave et al., 2010) but not in others (Harding and Jokela, 1994). This reflects site-specific factors such as background soil fertility and the short-term nature of most studies that do not typically quantify all ecosystem nutrient pools (e.g., soils, vegetation, and forest floor). Furthermore, Vogel et al. (2011) found that fertilization with macro- and micronutrients increased aboveground C relative to controls plots by 40% for loblolly (*Pinus taeda*) pine and 20% for slash pine (*Pinus elliottii*), but only had a marginally significant impact on soil C. Harding and Jokela (1994) reported that, after 25 years, the application of P fertilizers to a P-deficient site increased aboveground biomass and litter accumulation by about 1.5-fold over the untreated control but had no significant effect on soil OM to a depth of 91 cm. McFarlane et al. (2009) examined ponderosa pine (*Pinus ponderosa*) forests of varying fertility and reported that soil C and N increased after fertilization as a function of site quality, with the strongest response on poor sites. Therefore it is important to recognize that fertilization is generally for the benefit of trees, not soil.

3.5 Soil compaction

Soil compaction is often the most prevalent issue that arises from forest harvest operations (Batey, 2009). Compaction from trafficking reduces air-filled pores, thereby increasing soil bulk density (Johnson et al., 2007). Increased bulk density leads to reduced water infiltration, water-holding capacity, and hydraulic conductivity leading to waterlogging or standing water on flat terrain or surface runoff and erosion on steeper slopes (Waggenbrenner et al., 2016; Cambi et al., 2015; Hartemink, 2008). Although all soils are highly susceptible to compaction, forest soils, in particular, with their loose, friable structure and high porosity are particularly at-risk for long-term impacts.

Fine-textured soils and those with low bulk density are more prone to compaction than coarse-textured or higher bulk density soils. It is not uncommon to find compaction in fine-textured soil deep in the mineral soil (50 cm; Hakansson, 1985; Page-Dumroese et al., 2006) and persisting for decades or centuries (Hakansson et al., 1988). Soil OM on the soil surface (slash, forest floor, or mineral soil OM) may reduce the susceptibility of soil to compaction by increasing the resistance to deformation. Ruts are a singular and troubling form of compaction often found in the middle or the sides of skid trails. Typically, the bottom of ruts is sealed, keeping water on the soil surface and creating preferential flow paths that move water and soil downslope.

>Results from the LTSP study showed relatively rapid recovery from harvest-caused compaction in coarse-textured soils, particularly at the soil surface (0–10 cm). Slower recovery was common for fine-textured soils (Table 3.1; Page-Dumroese et al., 2006) and may take as long as four decades to recover to predisturbance levels (Froehlich et al., 1985). Overall, climatic regime; soil texture; and the degree, extent, and severity of compaction determine the rate of recovery (Page-Dumroese et al., 2006). LTSP study sites with coarser-textured soil and warmer temperatures (Louisiana and North Carolina) had much greater recovery of soil bulk density within 5 years as compared with finer-textured soils in cooler climatic regimes, despite some freeze-thaw activity. Interestingly, on 183 clear-cut harvested units on the Kootenai National Forest in northwestern Montana (US) with soil compaction considered detrimental to forest productivity in 1992, 86% of these soils had recovered to predisturbance levels by 2013 (Gier et al., 2018). Soil texture was an important factor in this study, as soils that did not recover were fine-textured.

The consequences of soil compaction can be extreme. Changes in gas exchange, loss of OM, altered root growth, decreased microbial communities, modified greenhouse gas exchange, and ultimately a loss in aboveground growth and production are possible outcomes (Batey, 2009). Further, the time needed for recovery varies depending on the extent of soil physical, chemical, and biological changes induced during the harvest operations and local site factors such as slope, aspect, and climatic conditions. Therefore avoiding or reducing soil impacts during management activities has been recognized as an increasingly important step to maintain forest growth (FAO, 2004).

Degradation of soil properties can be expected during timber harvest operations and subsequent site preparation, although the impact on soil productivity depends on climate regime, soil texture, plant species, and severity of soil changes (Powers et al., 1998). Therefore soil data that focus on (1) baseline soil conditions, (2) the interrelationships of soil type and soil function, and (3) the relationship of soil disturbance severity will help determine how management-induced changes can alter above- and belowground forest productivity. As shown in the following sections, many soil properties are resilient to land management practices while others are more sensitive, with the extent of their response is often dictated at a site-specific level (Burger, 1997).

3.6 **Management impacts on soil properties**
3.6.1 **Harvest operations**

There are few definitive studies that show soil productivity has declined under forest management (Morris and Miller, 1994; Powers, 1990) because site-specific declines in tree growth are often complicated by factors other than—or in addition to—soil disturbance, such as absence of repeated burning (Sheffield et al., 1985) or variable weed competition associated with a redistribution of OM in windrows (Stransky et al., 1985). However, many anecdotal findings from around the world also suggest that management can reduce site productivity through changes in soil properties, especially OM content (Powers, 1990; Walmsley et al., 2009).

3.6.1.1 Precommercial thinning

Precommercial thinning to decrease the risk of wildfire can impact soil physical properties (compaction) or soil fertility (nutrient removal), which may result in reduced residual tree growth or susceptibility to insects or disease attacks. Recent changes in land management methods, particularly in the western US, now favor precommercial thinning to restore fire regimes, watersheds, and wildlife habitat (Jiménez et al., 2011; Churchill et al., 2013). Thinning is often used to restore ecological function in naturally regenerated stands (Edmonds et al., 1989; Brown et al., 2004; Han et al., 2006) and usually generates a significant amount of biomass that must be disposed of to reduce wildfire risk (McKeever and Falk, 2004). Thinning residues can be transported to a bioenergy facility, dispersed across the harvest site by masticating, converted to biochar, or piled and burned (Jones et al., 2010; Creech et al., 2012), each of which can impact soil productivity (Callaham et al., 2012; Scott and Page-Dumroese, 2016). In the US, biomass from precommercial thinnings is usually piled and burned because there are few markets for this material. However, burn piles can be large and may create air quality problems (e.g., smoke and particulates) or degrade soil properties when burned. Local (in- or near-harvest site) production of biochar from nonmerchantable woody residues is an appealing alternative from a soils perspective and has received growing attention in recent years as a method to sequester C, reduce open burning, and avoid soil impacts from pile burning (Busse et al., 2013; Page-Dumroese et al., 2017).

Periodic soil disturbance (e.g., from repeated entries as for thinning) can impact tree growth (Sanchez et al., 2006; Page-Dumroese et al., 2010; Jurgensen et al., 2012) depending on soil type, tree species, and climatic regime (Henderson, 1995; Grigal and Vance, 2000). For example, when various thinning rates were implemented over a 57 year period in a northern Minnesota (US) red pine (*Pinus resinosa*) stand growing on a sandy soil, soil compaction increased with thinning intensity. In contrast, no evidence of soil compaction was found on a rocky, loam soil in a northern Wisconsin (US) sugar maple (*Acer saccharum*) stand that was periodically thinned over a 50 year period (Tarpey et al., 2008). Increased removal of thinned biomass for bioenergy, rather than leaving the residues on the soil surface, can exacerbate short-term nutrient and C losses (Sinclair, 1992), increase soil temperature, or decrease water availability (Harvey et al., 1976; Covington, 1981; Wang et al., 2018). For example, in young balsam fir (*Abies balsamea*) stands in Quebec, Canada, increased soil temperatures after thinning increased cellulose decomposition and nutrient availability (Thibodeau et al., 2000). In a Chinese pine (*Pinus tabulaeformis*) plantation, increased wood decomposition in the mineral soil and on the soil surface was noted after thinning to restore ecosystem functions (Wang et al., 2018). However, Jang et al. (2016) found few impacts on soil properties 50 years after biomass removal and prescribed burning in a western Montana mixed conifer stand. Thinning could also negatively impact soil productivity by providing fuel for uncharacteristically severe wildfires if too much biomass is left on-site (Page-Dumroese et al., 2010).

3.6.1.2 Salvage logging

Salvage logging removes dead and damaged trees left after wind storms, wildfire, drought, disease, or insect attack. This management practice captures economic value and lowers the risk of high-severity ground fires by reducing the amount of coarse wood "jackpiled" on the soil surface (Simon et al., 1994; Passovoy and Fulé, 2006). Such "hot spots" have the potential to kill regeneration (Kemp, 1967) and decrease soil OM and C stocks (Page-Dumroese et al., 2015). Salvage logging can also reduce biodiversity, degrade wildlife habitat, or riparian conditions (Noss et al., 2006). Since these operations are often more intensive and widespread than traditional forms of logging, they exacerbate the amount and severity of soil disturbance following wildfire (McIver and Starr, 2000; Van Nieuwstadt et al., 2001), as is shown in Fig. 3.1. Post wildfire salvage logging can also increase soil erosion, flooding, and impacts on water quality (Lewis et al., 2018), but careful placement of logging debris on skid trails may reduce sediment loss (Spanos et al., 2005).

The season of logging is an important consideration for limiting undesirable soil effects. Soil disturbance considered detrimental to forest growth is typically nominal when conducted on frozen (winter) or dry (summer) soils using equipment matched to the site (Reeves et al., 2012). These two factors (harvest season and equipment used) are relatively easy options to limit the amount, extent, and severity of soil disturbance (McIver and Starr, 2000).

3.6.1.3 Tethered logging

Ground-based harvesting is usually restricted to slopes no greater than 35%, as anything steeper relies on manual felling and raises concerns for human safety (Sessions et al., 2017). In addition, disturbance on steep slopes may increase surface soil displacement, resulting in a greater likelihood of severe erosion and a

FIGURE 3.1

Extensive ruts and compaction after salvage logging a harvest unit killed by beetles.

Photo credit: USDA Forest Service.

decrease in LTSP. Newer logging systems, such as cable-assisted or "tethered" logging operations which uses a wire rope anchored upslope to assist traction (Fig. 3.2), are now being used to harvest steep slopes in mountainous areas of North America, New Zealand, and Europe to reduce soil disturbance (e.g., Sessions et al., 2017; Belart et al., 2019). While equipment slippage may be minimized in tethered logging, compaction (increased ground pressure) or soil displacement can still occur (Visser and Stampfer, 2015). For example, stream and soil disturbance in Oregon and Washington, US conifer stands were greater after tethered logging than after conventional logging operations, although overall detrimental soil impacts were below regulatory thresholds (Chase et al., 2019).

3.7 Harvest operations considerations

- Maintaining organic horizons during and after harvest operations is important and having a known baseline of soil and forest conditions to compare postharvest conditions is essential.

FIGURE 3.2

Tethered logging system in which the back of the harvester is attached to a stump or other equipment.

Photo credit: USDA Forest Service.

- Single-entry thinning operations likely have few impacts on soil sustainability. However, monitoring of soil compaction and OM displacement is encouraged where repeated stand entries are expected.
- Forest residues are important as a long-term soil OM source, whereas forest floor material (litter and duff) is important for short-term turnover of OM. Both types of OM source should be conserved, with the specific amounts dependent on site characteristics and economics.
- Soil compaction is a threat to soil and site sustainability and should be minimized. Selection of the appropriate harvest equipment, season, and slopes will determine the areal extent and depth of compaction and/or ruts.

3.8 Postharvest site preparation

3.8.1 Residue management

Logging residues left following harvests are a variable but important source of soil OM. For example, in the southeastern US approximately 12% of total pine stand volume was left after clear-cut harvesting in Alabama (Schultz, 1997), but in California, US, mixed conifer and broadleaf stand approximately 20%−40% of the stand volume remained as residues (Kizha and Han, 2015). This variability is important to understand and highlights that the amount and distribution of residues at a given site will depend on desired species composition, soil properties, physiographic province, past management practices, harvesting method, understory competition levels, and economics (Nelson et al., 2013). Knowing how much large wood should remain is critical for long-term OM inputs which benefit water-holding capacity and ectomycorrhizal fungi (Jurgensen et al., 1997) and can determine the flora and fauna trajectories of forest and soil restoration activities (Berkowitz, 2013). For example, Harvey et al. (1976) recommended leaving 25−37 Mg ha^{-1} of large woody material on the soil surface in cool-moist ecosystems to build mineral soil OM over time, but the amounts of coarse wood left onsite could range from 10 to 125 Mg ha^{-1} (Jurgensen et al., 1997). Understanding preharvest conditions and the desired stand conditions will dictate the amounts left behind.

The successful establishment of a healthy forest with sustained biodiversity and C sequestration usually requires postharvest management of large-diameter (>7.5 cm) woody residue (Liu et al., 2012; Nave et al., 2019). Depending upon regional preferences and policy, residual logging slash can be left or chipped (masticated) on-site, piled and later burned (Fig. 3.3), or removed for bioenergy production where facilities exist. Residue management generally improves future stand yields and economic

FIGURE 3.3

Slash pile with tree boles up to 6 cm in diameter formed after precommercial thinning a lodgepole pine (*Pinus contorta*) stand in Montana.

Photo credit: USDA Forest Service.

returns, as seedling growth and survival are increased from greater water and nutrient availability and enhanced root growth (Morris and Lowery, 1988; Allen et al., 1990; Lowery and Gjerstad, 1991; Brown et al., 2003). But, caution is advised and the use of appropriate equipment and site preparation methods are encouraged to avoid impacting site productivity by compacting the soil or removing a large amount of organic residues off-site. For example, the placement of nutrient-rich surface litter and mineral soil into adjacent windrows is no longer practiced in the US because of the negative impacts on soil productivity (Morris and Pritchett, 1983).

3.8.1.1 Mechanical site preparation

Intensive forest management practices have increased the use of heavy machinery for improved postharvest seedling establishment and growth (Gent et al., 1984). Mechanical treatments used prior to planting may include disking, roller drum chopping, and subsoiling. These methods are effective at restoring bulk density in the surface 0–15 cm of mineral soil, but equipment pressure may extend compaction below this depth (Reaves and Cooper, 1960) and away from the skid trail (Solgi et al., 2020). Subsoiling may be used on soils having high mechanical resistance (compaction), or with soils having cemented horizons that reduce root penetration, growth, and occupancy within the lower mineral soil profile (Löf et al., 2012). Chopping breaks up existing woody vegetation and crushes the logging slash into smaller pieces to enhance the effectiveness of prescribed burning (Knoepp et al., 2004) and decomposition (Gholz et al., 1985). Similarly, disking may be used to ameliorate surface soil compaction along skid trails, loading ramps, and retired forest roads. As with any surface soil disturbance treatment, an evaluation of the site's erosion potential must be integrated into the prescription process.

3.8.1.2 Pile burning

Slash pile burning is a traditional practice used throughout much of the world for fuel reduction and disposal of harvest residues prior to planting. Nonmerchantable woody residues are moved into piles which can range from small, hand-built piles to imposingly large, tractor-generated piles on log landings, and then burned on-site. This practice is recognized for its effectiveness, low cost, limited risk of fire escape, and simplicity (Rhoades and Fornwalt, 2015). The fuel composition of piles, the amount of soil contained in a pile, and the pile structure and moisture content all influence burn characteristics, and collectively dictate the amount of soil damage that may occur (Massman et al., 2008). Excessive heat from burning large-diameter piles (>15 m), particularly those that contain large-diameter wood, can have large impacts on soil properties (Busse et al., 2013, Smith et al., 2016), as can slow, smoldering fires in poorly cured (wet) piles or in piles containing excessive soil that traps heat and increases the downward heat pulse.

Loss in soil quality is a concern beneath pile burns (Korb et al., 2004; DeSandoli et al., 2016). Soil properties are modified in proportion to fire severity, maximum soil temperatures, and heat duration reached during burning.

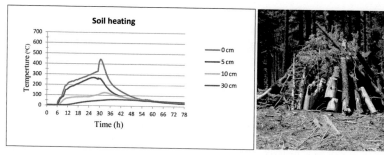

FIGURE 3.4

Temperature and heat duration extremes during burning of a moderate-sized pile containing a mixed composition of conifer fuels.

Photo credit: USDA Forest Service.

FIGURE 3.5

Long-term recovery after severe pile burning on a moist site. Example of a site (A) with many handpiles before burning, (B) 8 months after burning, and (C) 8 years postburning.

Photo credit: USDA Forest Service.

Established thresholds for soil change are typically based on maximum temperature alone, as determined in controlled-setting experiments (Hungerford et al., 1991), whereas the effects of heat duration are poorly quantified. Temperatures that result in the destruction of soil OM, microbial death, or nutrient transformation are often exceeded in the surface mineral soil, even when burning small piles (Busse et al., 2010; 2013). However, since soil is a porous matrix and a poor conductor of heat, extreme heat does not travel much below 5—10 cm in the profile (Fig. 3.4), suggesting that any changes in soil properties will mostly be surficial. Soil moisture content at the time of burning is also a key consideration; moist soils are more effective at quenching the downward heat pulse than dry soils and produce lower maximum temperatures of shorter duration and, consequently, less soil damage. Further, moist sites have been shown to recover vegetation cover fairly rapidly following severe pile burning (Fig. 3.5).

Recent studies have identified a variety of post-fire soil effects, yet they have given few definitive recommendations other than concluding that small-diameter piles are generally benign across landscapes (Tarrant, 1956; Meyer, 2009; Hubbert et al., 2015; Jang et al., 2017). In comparison, soil damage from moderate-sized, tractor-generated piles (5–15 m diameter) is more difficult to generalize and is probably best considered on a site-specific basis that considers the areal extent of pile coverage within a treatment unit (i.e., do the piles occupy a large proportion of the ground surface?). Several studies with moderately large piles showed declines in total soil C and N in the surface 0–15 cm soil layer with concomitant increases in available N (Korb et al., 2004; Esquilín et al., 2007; Rhoades et al., 2015), whereas detrimental effects on soil microbial communities were shown by some (Korb et al., 2004; Esquilín et al., 2007), but not all studies (Smith et al., 2016). York et al. (2009) measured plant growth—a general indicator of sustainable soil health—and found greater 10-year conifer growth within the perimeter of moderate-sized pile scars compared to adjacent unburned sites, which they attributed to fire's positive effect on long-term soil N availability.

3.9 Drainage and bedding

Stand growth and survival are constrained by excessively wet soils, as are access for harvesting and other silvicultural operations. As a result, forest drainage (ditching) became a common mechanical site preparation treatment on both mineral and organic soils in the southeastern US from the 1930s through the mid-1980s (Skaggs et al., 2016). Treated sites were typically very poorly drained (e.g., pocosins), had precipitation amounts that exceeded evapotranspiration rates, and had natural drainage systems that were inadequate for removing excess water. Saturated hydraulic conductivity represents an important soil property influencing drainage, and was used to guide the establishment and spacing between drainage ditches. Growth responses of planted pines were large following drainage, ranging from 80% to almost 1300% (Terry and Hughes, 1975). Yet, despite research conducted in the 1950s–70s that demonstrated the positive impacts of soil drainage on tree growth and site accessibility (Miller and Maki, 1957; Maki, 1960; White and Pritchett, 1970; Terry and Hughes, 1975), the expansion of drainage practices ended in the US in 1990 because of federal wetland protection regulations on jurisdictional wetlands (Fox et al., 2007a; Skaggs et al., 2016). However, forest operations still accrue benefits from existing, "historical" drainage systems through ditch maintenance, replanting, and nutrient management efforts. Today, bedding is now commonly used on such sites to modify the microsite and aid stand establishment and survival.

Raised beds (mounds), commonly made with a bedding plow (Fig. 3.6), are used to improve soil microsites, control competing vegetation, and increase seedling growth (McKee and Shoulders, 1970; McKee and Shoulders, 1974; Crutchfield and Martin, 1983). Single or double-pass bedding operations are

(A)   (B)

FIGURE 3.6

Mechanical site preparation can be done (A) using a subsoiler to prepare a site for planting and (B) a poorly drained, bedded site post-thinning.

Photo credit: Eric Jokela.

conducted in the spring to mid-summer months, which allow the beds to settle before planting, as seedling mortality can occur if roots come in contact with unsettled air pockets. Contour bedding is used on sites with slopes ($>35\%$) to reduce the risks of soil erosion and sediment transport. "Spot" (or bucket) mounding (inverting mounds of soil to create a double organic layer within the mound; Londo and Mroz, 2001) is used rather than continuously mounded rows, thereby reducing runoff while increasing nutrient retention at the planting site (Costantini and Loch, 2002). Bucket mounds are also common in Scandinavia, Canada, and the upper Great Lakes, US region (Takyi and Hillman, 2000; Londo and Mroz, 2001).

Bedding is commonly used to increase tree survival and growth on poorly drained soils characteristic of many lower Coastal Plain sites in the southern US and wetland sites in the northern US (Gent et al., 1984; Gale et al., 1998). High water tables lead to anaerobic soil conditions, which is a problem in young, developing stands that have low leaf area and correspondingly low rates of evapotranspiration. Bedding increases soil aeration by mixing surface organic horizons with mineral soil (Gale et al., 1998). When bedding is combined with burning and harrowing it can also increase nutrient release. For example, this combination of site preparation treatments resulted in a ten-fold increase in soil solution P as compared to burned and chopped site preparation (Gale et al., 1998), and it also resulted in increased K (Burger and Pritchett, 1988). Bedding has also been used in the interior western US and Australia as a method to increase soil water-holding capacity (Cornish, 1993; Page-Dumroese et al., 1997) on drier sites, where it can increase soil OM, cation exchange capacity, N, and decrease soil bulk density. In South Carolina, US soil bedding increased wood decomposition by both microorganisms and termites and this impact on OM content can also alter site sustainability (Jurgensen et al., 2019).

3.10 Herbicide applications

Herbicides are used to control competing vegetation, including unwanted invasive plants, and their application is a common practice in plantation forestry. Over the last several decades, the evolution and use of more intensive silvicultural practices across forested landscapes in the US and Europe (e.g., including the use of genetically improved growing stock, site preparation, fertilizer additions, and understory competition control) have contributed to substantial gains in growth and forest productivity (Fox et al., 2007b; McCarthy et al., 2010). However, the impacts of herbicides on soil and ecosystem processes are variable, but important for understanding their long-term effects on forest development and sustainable soil management practices.

Direct effects: Commonly used forest herbicides are reasonably plant specific and generally act by disrupting or modifying photosynthesis (e.g., hexazinone, atrazine), protein synthesis (glyphosate, imazapyr, sulfometuron methyl), or plant hormone production (triclopyr, 2,4-D). Therefore most of these compounds and have a little lethal effect on soil organisms (Busse et al., 2004). For example, increased soil bacterial growth was found when glyphosate, one of the most widely used compounds around the world (Thompson and Pitt, 2003), was added at high concentrations. This result was because organisms metabolize soil-bound glyphosate as a source of C, N, and P (Fig. 3.7; Ratcliff et al., 2006). Most evidence to date points to the tolerance and resilience of soil organisms to traditional forestry herbicides, although scrutiny is still necessary as new products come on the market.

Indirect effects: Herbicide application can produce lasting changes in plant community structure and tree growth, which could be a factor in the retention of soil nutrients and, therefore, in determining their suitability for sustainable of

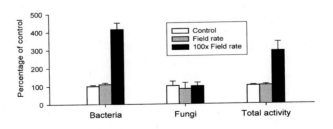

FIGURE 3.7

Effect of glyphosate application on soil bacteria (total number), fungi (hyphal length), and total activity (respiration). Measurements were taken 30 days following glyphosate application to a ponderosa pine plantation soil in northern California.

Based on results from Ratcliff, A.W., Busse, M.D., Shestak, C.J., 2006. Changes in microbial community structure following herbicide (glyphosate) additions to forest soils. Appl. Soil Ecol. 34, 114–124.

forest management systems (Vogel et al., 2011). In southern US pine forests, complete and sustained understory competition control were associated with decreased soil N and C content (Echeverría et al., 2004; Sarkhot et al., 2007; Sartori et al., 2007; Rifai et al., 2010; Vogel et al., 2011). This could potentially impact nutrient retention and long-term stand productivity. Loss of soil C under intensive weed control treatments has been attributed to reductions in woody and herbaceous fine root growth and turnover (Albaugh et al., 1998; Shan et al., 2002). Subedi et al. (2017) examined the inter-rotational effects of sustained weed control (i.e., repeated herbicide treatments during the first 5 years of stand establishment) on the richness, diversity, and composition of understory plants in young loblolly pine stands growing on low-fertility Spodosols in North Central Florida, US. The competition control treatment used in the first rotation effectively reduced the shrubby understory component in the second rotation. The absence of shrubby species such as gallberry (*Ilex glabra*), a dominant manganese (Mn) accumulator (Jokela et al., 1991a; Subedi et al., 2014), may affect nutrient cycling and accentuate subacute deficiencies of this and other nutrients and influence site productivity. However, because the silvicultural treatments used in that study were intensive (sustained weed control and fertilization), the authors noted that soil responses would likely be less pronounced under operational treatments (Subedi et al., 2017).

Post-fire herbicide use, while beneficial to stand establishment and growth, may lead to increased soil erosion and sedimentation rates by reducing plant recruitment and total plant cover (Certini, 2005). Consequently, it is critical to evaluate site erosion potential and factors such as slope, soil cover, anticipated storms, and proximity to rural communities to determine if herbicide applications to remove soil cover will have unintended impacts. Furthermore, the long-term effects of vegetation control on soil sustainability have not been clearly documented, but the LTSP study suggests that there could be a moderate impact of vegetation control on soil nutrients (Busse et al., 1996, 2001; Ponder et al., 2012; Zhang et al., 2013).

Whether herbicides substantially modify forest soil properties or processes depends on the interactions of site, herbicide, and soil characteristics (Nelson and Cantrell, 2002). Soil responses will change depending on herbicide chemistry, soil-herbicide binding potential, herbicide application rate and frequency, vegetation type, soil fertility, and site productivity (Minogue et al., 1991). Clearly, on the one hand, herbicide control of competing vegetation benefits seedling survival and tree growth and may be an important tool for reducing wildfire risk where grasses or shrubs form a dominant understory layer (Ponder et al., 2012; Zhang et al., 2013). On the other, herbicide movement in subsurface water leading to potential contamination of water bodies, increased soil erosion due to loss of vegetation cover, and long-term changes in soil fertility due to a shortfall of understory roots and litter input may lead to unwanted changes to the forest environment.

3.11 Site preparation, bedding, and herbicide considerations

- Determine which site preparation treatments will cause a minimal amount of soil disturbance while still achieving management goals.
- Soil bedding (mounding) can increase planted tree growth by improving soil aeration, concentrating soil OM, and increasing nutrient availability.
- Small burn piles have fewer soil impacts than large or moderately sized piles. Burning when soils are moist will help lessen soil heating and potential damage beneath piles.
- Herbicides reduce competition and increase tree growth, but their indirect effects may alter C pools erosion potential, soil fertility, and water quality.

3.12 Soil sustainability monitoring

As we have pointed out, forest soils are heterogeneous and their responses to harvesting and site preparation are varied. This makes soil monitoring an integral part of forest management so that the data can inform adaptive management by contributing an understanding of the limits of a particular soil to particular management activities and the efficacy of treatments (DeLuca et al., 2010). But, as also previously noted, solid baseline data are needed to link soil type and function changes with management and determine if thresholds indicating a drop in site productivity have been crossed. Monitoring soil sustainability represents a key consideration for both environmental policy and resource management. Soil monitoring refers to the repeated collection of soil data used to describe soil conditions at points in time, requires archiving the data (Page-Dumroese et al., 2009), and can provide information on early changes in soil quality that affect the health or growth of managed forests (Morvan et al., 2008). Within the context of a changing climate, soil monitoring data becomes even more important for proper management of forested landscapes. However, the consistent collection of monitoring data is neither inexpensive to collect nor short-term. Temporal changes associated with changes in soil C, compaction, soil structure, etc. and dynamic processes (e.g., hydrologic function) must be integrated into monitoring efforts to provide detailed soil data to promote continued harvesting or restoration of public and industry lands. Furthermore, soil monitoring can determine unintended consequences of harvest activities and provide data for returning an ecosystem to the intended trajectory (DeLuca et al., 2010).

3.12.1 Descriptive soil quality measures versus functional approaches

Soil monitoring has been made relatively easy with the use of visual indicators (Heninger et al., 1997; Curran et al., 2005; Page-Dumroese et al., 2009) that describe soil displacement, loss of OM cover, and soil mixing, among other visual properties. These visual indicators are a surrogate for soil quality changes associated with land management. However, the change in soil quality must also be related to soil function (support of natural ecosystems, food and fiber production, a base for buildings, etc.; Vegter et al., 1988) and monitoring for these two things (quality and function) can be performed simultaneously. Managers and scientists must understand the limits of individual soils to provide these functions (Nortcliff, 2002) and, as previously noted, these functions will be site-specific (Burger and Kelting, 1999) and require site-specific validation (Page-Dumroese et al., 2012). By using a common soil monitoring protocol on public lands, it is relatively easy to determine where monitoring is needed, enlist citizen scientists in the monitoring effort, get highly detailed and statistically robust data, and conduct analyses of the data. In addition, linking these data to processes such as infiltration or nutrient leaching will provide insight into how soil functions.

3.12.2 Soil functional integrity

Soil functional integrity is the intactness of soil and native vegetation patterns and processes, which indicate whether that soil is either resistant or resilient to disturbances. Healthy soil will maintain inherent functions (e.g., water regulation, plant growth) provided by forested ecosystems (Greenland and Szabolcs, 1994), but they are variable and depend on the interaction of soil physical, chemical, and biological properties (Schoenholtz et al., 2000). Soil OM, function, and resilience are all interconnected. However, the influence of OM on soil properties depends on its spatial distribution (horizontal and vertical), composition (biochemical and functional), and microbial populations (Herrick and Wander, 1997). Therefore properly functioning soil must have intact soil physical, chemical, and biological properties that support: (1) biological activity, (2) water flow, (3) filtering and detoxifying organic and inorganic materials, and (4) storing and cycling nutrients (Seybold et al., 1999).

3.12.3 Ecosystem stability as measured by soil indicators

As we noted previously, most ecosystems will return to predisturbance given enough time and the right climatic, site, and soil conditions. Ecosystem stability (return to an equilibrium after disturbance; Redfearn and Pimm, 1987) is related to the intrinsic properties of OM because both physiochemical and biological properties from the surrounding environment dictate the rate of OM decomposition. Although OM is inherently unstable, it can persist in soils for thousands of years, and it is an ecosystem property that enhances stability (Schmidt et al., 2011). Clearly, land managers influence the amount of OM left on a site, the surrounding environmental properties that govern decomposition rates, vegetative growth, and hydrologic function. Soil monitoring will inform managers about the rates at which soil OM and other soil properties are changing (declining, staying the same, recovering). Since soil OM is often vulnerable to perturbations, an understanding of which ecosystems are resistant to OM losses or resilient (recover OM losses) are key to maintaining both soil and site sustainability.

3.13 Soil monitoring considerations

- Soil monitoring is a crucial activity to determine the long-term success of management activities.
- Repeated monitoring provides an understanding of soil changes over time.
- Policies that encourage monitoring and data archiving in a shared database facilitate data collection and prioritization of ecosystems in need of monitoring.
- Site-specific validation of monitoring data is critical for establishing linkages between soil and site productivity.

3.14 Conclusion

The responses of soil and vegetation to harvest operations and site preparation are complex, making the development of soil sustainability guidelines a balancing act. Environmental values and ecosystem services must be maintained to protect long-term productivity. Adaptive management in forest ecosystems is dependent on soil monitoring to inform the success of treatments and reduce unintended long-term consequences. In addition, long-term field trials, such as the LTSP effort, will provide data on the extent to which a site can tolerate harvesting and site preparation and not lose productivity. Much of our long-term data are based on approximately 30 years of careful measurements which means that as the stands grow we must continue to evaluate whole ecosystem responses to disturbance and repeated stand entry impacts. Of course, changing climates, equipment, and ecosystem services needed will produce varying impacts on soil OM, functions, and sustainability, but careful management can minimize soil physical, chemical, and biological impacts.

References

Albaugh, T.J., Allen, H.L., Dougherty, P.M., Kress, L.W., King, J.S., 1998. Leaf area and above- and belowground growth responses of loblolly pine to nutrient and water additions. For. Sci. 44, 317–328.

Albaugh, T.J., Allen, H.L., Fox, T.R., 2007. Historical patterns of forest fertilization in the southeastern United States from 1969 to 2004. South. J. Appl. For. 31, 129–137.

Allen, H.L., Dougherty, P.M., Campbell, R.G., 1990. Manipulation of water and nutrients — practice and opportunity in southern US pine forests. For. Ecol. Manage. 30, 437–453.

Aronson, C.D., Savage, M., Falk, D.A., Suckling, K.F., Swetnam, T.W., Schulke, T., et al., 2006. Ecological restoration: a new frontier for nature conservation and economics. J. Nat. Conserv. 14, 135–139.

Batey, T., 2009. Soil compaction and soil management—a review. Soil Use Manage. 25, 335–345.

Belart, F., Leshchinsky, B., Sessions, J., Chung, W., Green, P., Wimer, J., et al., 2019. Sliding stability of cable-assisted tracked equipment on steep slopes. For. Sci. 65, 304–311.

Berkowitz, J.F., 2013. Development of restoration trajectory metrics in reforested bottom-land hardwood forests applying a rapid assessment approach. Ecol. Indic. 34, 600–606.

Binkley, D., Carter, R., Allen, H.L., 1995. Nitrogen fertilization practices in forestry. In: Bacon, P.E. (Ed.), Nitrogen Fertilization in the Environment. Marcel Dekker, New York, pp. 421–441.

Brown, J.K., Reinhardt, E.D., Kramer, K.A., Coarse woody debris: managing benefits and fire hazard in the recovering forest, 2003, U.S. Department of Agriculture, Forest Service, Rocky Mountain Research Station, Odgen, UT, Gen. Tech. Rep. RMRS-GTR-105. 16 p.

Brown, R.T., Agee, J.K., Franklin, J.F., 2004. Forest restoration and fire: principles in the context of place. Conserv. Biol. 18, 903–912.

Burger, J.A., 1997. Conceptual framework for monitoring the impacts of intensive forest management on sustainable forestry. In: Hakkila, P., Heino, M., Puranen, E. (Eds.), Forest Management for Bioenergy, 640. The Finnish Forest Research Institute, pp. 147–156. , Research Papers.

Burger, J.A., Kelting, D.L., 1999. Using soil quality indicators to assess forest stand management. For. Ecol. Manage. 122, 155–166.

Burger, J.A., Pritchett, W.L., 1988. Site preparation effects on soil moisture and available nutrients in a pine plantation in the Florida flatwoods. For. Sci. 34, 77–87.

Burger, J., Graves, D., Angel, P., et al., 2017. Chapter 2: The forestry reclamation approach. In: Adams, M.B., (Ed.), *The Forestry Reclamation Approach: Guide to Successful Reforestation of Mined Lands*, 2017, U.S. Department of Agriculture, Forest Service, Northern Research Station, Newtown Square, PA, 2–8, Gen. Tech. Rep. NRS-GTR-169.

Busse, M.D., Cochran, P.H., Barrett, J.W., 1996. Changes in ponderosa pine site productivity following removal of understory vegetation. Soil Sci. Soc. Am. J. 60, 1614–1621.

Busse, M.D., Ratcliff, A.W., Shestak, C.J., Powers, R.F., 2001. Glyphosate toxicity and the effects of long-term vegetation control on soil microbial communities. Soil Biol. Biochem. 33, 1777–1789.

Busse, M.D., Fiddler, G.O., Ratcliff, A.W., 2004. Ectomycorrhizal formation in herbicide-treated soils of differing clay and organic matter content. Water Air Soil Pollut. 152, 23–34.

Busse, M.D., Shestak, C.J., Hubbert, K.R., Knapp, E.E., 2010. Soil physical properties regulate lethal heating during burning of woody residue. Soil Sci. Soc. Am. J. 74, 947–955.

Busse, M.D., Shestak, C.J., Hubbert, K.R., 2013. Soil heating during burning of forest slash piles and wood piles. Int. J. Wildland Fire 22, 786–796.

Callaham, M.A., Scott, D.A., O'Brien, J.J., Stanturf, J.A., 2012. Cumulative effects of fuel management on the soils of eastern US. In: LaFayette, R., Brooks, M.T., Potyondy, J. P., Audin, L., Krieger, S.L., Trettin, C.C. (Eds.), Cumulative Watershed Effects of Fuel Management in the Eastern United States. U.S. Department of Agriculture, Forest Service, Southern Research Station, Asheville, NC, Gen. Tech. Rep. SRS-GTR-161, pp. 202–228.

Cambi, M., Giacomo, C., Neri, F., Marchi, E., 2015. The impact of heavy traffic on forest soils: a review. For. Ecol. Manage. 338, 124–138.

Campoe, O.C., Stape, J.L., Mendes, J.C.T., 2010. Can intensive management accelerate the restoration of Brazil's Atlantic forest? For. Ecol. Manage. 259, 1808–1814.

Carlson, C.A., Fox, T.R., Allen, H.L., Albaugh, T.J., Rubilar, R.A., Stape, J.L., 2014. Growth responses of loblolly pine in the Southeast United States to midrotation applications of nitrogen, phosphorus, potassium, and micronutrients. For. Sci. 60, 157–169.

Certini, G., 2005. Effects of fire on properties of forest soils: a review. Oecologia 143, 1–10.

Chase, C.W., Reiter, M., Homyack, J.A., Jones, J.E., Sucre, E.B., 2019. Soil disturbance and stream-adjacent disturbance from tethered logging in Oregon and Washington. For. Ecol. Manage 454. Available from: https://doi.org/10.1016/j.foreco.2019.117672.

Churchill, D.J., Larson, A.J., Dahlgreen, M.C., Franklin, J.F., Hessburg, P.F., Lutz, J.A., 2013. Restoring forest resilience: from reference spatial patterns to silvicultural prescriptions and monitoring. For. Ecol. Manage. 291, 442–457.

Clarke, N., Gundersen, P., Jönsson-Belyazid, U., Kjønaas, O.J., Persson, T., et al., 2015. Influence of different tree-harvesting intensities on forest soils carbon stocks in boreal and northern temperate forest ecosystems. For. Ecol. Manage. 351, 9–19.

Cornish, P.M., 1993. The effects of logging and forest regeneration on water yields in a moist eucalypt forest in New South Wales. J. Hydrol. 150, 301–322.

Costantini, A., Loch, R.J., 2002. Effects of site preparation on runoff, erosion, and nutrient losses from Pinus plantations established on the coastal lowlands of south-east Queensland, Australia. Aust. J. Soil Res. 40, 1287–1302.

Covington, W.W., 1981. Changes in forest floor organic matter and nutrient content following clear cutting in northern hardwoods. Ecology 62, 41–48.

Creech, M.N., Kirkman, L.K., Morris, L.A., 2012. Alteration and recovery of slash pile burn sites in the restoration of fire-maintained ecosystems. Restor. Ecol. 20, 505–516.

Crutchfield, D.M., Martin, J.P., 1983. Site preparation—coastal plain. Proceedings of the Symposium on the Loblolly Pine Ecosystem (East Region); 1982 December 8–10; Raleigh, NC. North Carolina State University, School of Forest Resources, Raleigh, NC, pp. 49–57.

Curran, M.P., Maynard, D.G., Heninger, R.L., Terry, T.A., et al., 2005. An adaptive management process for forest soil conservation. For. Chron. 81, 717–722.

DeLuca, T.H., Aplet, G.H., Wilmer, B., Burchfield, J., 2010. The unknown trajectory of forest restoration: a call for ecosystem monitoring. J. For. 8, 288–296.

DeSandoli, L., Turkington, R., Fraser, L.H., 2016. Restoration of slash pile burn scars to prevent establishment and propagation of non-native plants. Can. J. For. Res. 46, 1042–1050.

Donoso, P.J., Soto, D.P., Schlatter, J.E., Büchner, C.A., 2009. Effects of early fertilization on the performance of Nothofagus dombeyi planted in the Coastal Range of south-central Chile. Cienc. Invest. Agrar. 36, 475–486.

Drohan, P.J., Finley, J.C., Roth, P., Schuler, T.M., Stout, S.L., Brittingham, M.C., et al., 2012. Perspectives from the field: oil and gas impacts on forest ecosystems: findings gleaned from the 2012 Goddard forum at Penn State University. Environ. Pract. 14, 394–399.

Echeverría, M.E., Markewitz, D., Morris, L.A., Hendrick, R.L., 2004. Soil organic matter fractions under managed pine plantations of the southeastern USA. Soil Sci. Soc. Am. J. 68, 950–958.

Edmonds, R.L., Shaw, D.C., Hsiang, T., Driver, C.H., 1989. Impact of precommercial thinning on development of *Heterobasidion annosum* in western hemlock. In: Otrosina, W.J., Scharpf, R.F. (Tech. Coords.), *Proceedings of the Symposium on Research and Management of Annosus Root Disease (Heterobasidion annosum)*, April 18–21, 1989, Monterey, CA. U.S. Department of Agriculture, Forest Service, Pacific Southwest Forest and Range Experiment Station, Berkeley, CA, Gen. Tech. Rep. PSW-GTR-116. pp. 85–94.

Esquilín, A.E., Stromberger, M.E., Massman, W.J., Frank, J.M., Shepperd, W.D., 2007. Microbial community structure and activity in a Colorado Rocky Mountain forest soil scarred by slash pile burning. Soil Biol. Biochem. 39, 1111–1120.

FAO, 2004. Reduced Impact Logging in Tropical Forests: Literature Synthesis, Analysis, and Prototype Statistical Framework. Forest Harvesting and Engineering Programme, Rome.

Fink, C.M., Drohan, P.J., 2015. Dynamic soil property change in response to reclamation following Northern Appalachian natural gas infrastructure development. Soil Sci. Soc. Am. J. 79, 146–154.

Fleming, R.L., Powers, R.F., Foster, N.W., Kranabetter, J.M., Scott, D.A., et al., 2006. Effects of organic matter removal, soil compaction, and vegetation control on 5-year seedling performance: a regional comparison of long-term soil productivity sites. Can. J. For. Res. 36, 529–550.

Fox, T.R., Jokela, E.J., Allen, H.L., 2007a. The development of pine plantation silviculture in the southern United States. J. For. 10, 337–347.

Fox, T.R., Allen, H.L., Albaugh, T.J., Rubilar, R., Carlson, C.A., 2007b. Tree nutrition and forest fertilization of pine plantations in the southern United States. South. J. Appl. For. 31, 5–11.

Froehlich, H.A., Miles, D.W.R., Robbins, R.W., 1985. Soil bulk density recovery on compacted skid trails in Central Idaho. Soil Sci. Soc. Am. J. 49, 1015–1017.

Gale, M.R., McLaughlin, J.W., Jurgensen, M.F., Trettin, C.C., Soelsepp, T., Lydon, P.O., 1998. Plant community responses to harvesting and post-harvest manipulations in a *Picea-Larix-Pinus* wetland with a mineral substrate. Wetlands 18, 150–159.

Gent Jr., J.A., Ballard, R., Hassan, A.E., Cassel, O.K., 1984. Impact of harvesting and site preparation on physical properties of Piedmont forest soils. Soil Sci. Soc. Am. J. 48, 173–177.

Gholz, H.L., Perry, C.S., Cropper Jr, W.P., Hendry, L.C., 1985. Litterfall, decomposition, and nitrogen and phosphorus dynamics in a chronosequence of slash pine (*Pinus elliottii*) plantations. For. Sci. 31, 463–478.

Gier, J.M., Kindel, K.M., Page-Dumroese D.S., Kuennen, L.J., Soil disturbance recovery on the Kootenai National Forest, 2018. U.S. Department of Agriculture, Forest Service, Rocky Mountain Research Station, Odgen, UT, Gen. Tech. Rep. RMRS-GTR-380. 31 p.

Gordon, A.M., Schlentner, R.E., Van Cleve, K., 1987. Seasonal patterns of soil respiration and CO_2 evolution following harvesting in the white spruce forests of interior Alaska. Can. J. For. Res. 17, 304–310.

Greenland, D.J., Szabolcs, I. (Eds.), 1994. Soil Resilience and Sustainable Land Use. CAB International, Wallingford.

Grigal, D.F., Vance, E.D., 2000. Influence of soil organic matter on forest productivity. N. Z. J. For. Sci. 30, 169–205.

Hakansson, I., 1985. Swedish experiments on subsoil compaction by vehicles with high axle load. Soil Use Manage. 1, 113–116.

Hakansson, I., Voorhees, W.B., Riley, H., 1988. Vehicle and wheel factors influencing soil compaction and crop responses in different traffic regimes. Soil Tillage Res. 11, 239–282.

Han, H.-S., Page-Dumroese, D.S., Han, S.-K., Tirocke, J., 2006. Effect of slash, machine passes, and soil moisture on penetration resistance in a cut-to-length harvesting. Int. J. For. Eng. 17, 11–24.

Harding, R.B., Jokela, E.J., 1994. Long-term effects of forest fertilization on site organic matter and nutrients. Soil Sci. Soc. Am. J. 58, 216–221.

Hartemink, A.E., 2008. Soils are back on the global agenda. Soil Use Manage. 24, 327–330.

Harvey, A.E., Larsen, M.J., Jurgensen, M.F., 1976. Distribution of ectomycorrhizae in a mature Douglas-fir/larch forest soil in western Montana. For. Sci. 22, 393–398.

Harvey, A.E., Jurgensen, M.F., Larsen, M.J., Graham, R.T., 1987. Decaying organic materials and soil quality in the Inland Northwest: a management opportunity. U.S. Department of Agriculture, Forest Service, Intermountain Research Station, Odgen, UT, Gen. Tech. Rep. INT-GTR-225. 20 p.

He, L.X., English, B.C., de la Torre Ugarte, D.G., Hodges, D.G., 2014. Woody biomass potential for energy feedback in the United States. J. For. Econ. 20, 174–191.

Helsinki Process. Proceedings of the Ministerial Conferences and Expert Meetings. Liaison Office of the Ministerial Conference on the Protection of Forests in Europe. Helsinki, Finland. 1994.

Henderson, G.S., 1995. Soil organic matter: a link between forest management and productivity. In: McFee, W.W., et al., (Eds.), Carbon Forms and Functions in Forest Soils. Soil Sci. Soc. Am., Inc., Madison, WI, pp. 419–437.

Heninger, R.L., Terry, T.A., Dobkowski, A., Scott, W., 1997. Managing for sustainable site productivity: Weyerhaeuser's forestry perspective. Biomass Bioenergy 13, 255–267.

Herrick, J.E., Wander, M.M., 1997. Relationships between soil organic carbon and soil quality in cropped and rangeland soils: the importance of distribution, composition, and soil biological activity. In: Lal, R., Kimble, J.M., Follett, R.F., Stewart, B.A. (Eds.), Soil Processes and the Carbon Cycle. CRC Press, Boca Raton, FL, pp. 405–424.

Hopmans, P., Chappell, H., 1994. Growth response of young, thinned Douglas-fir stands to nitrogen fertilizer in relation to soil properties and tree nutrition. Can. J. For. Res. 24, 1684–1688.

Hubbert, K., Busse, M., Overby, S., Shestak, C., Gerrard, R., 2015. Pile burning effects on soil water repellency, infiltration, and downslope water chemistry in the Lake Tahoe Basin, USA. Fire Ecol. 11, 100–118.

Hungerford R.D., Harrington M.G., Frandsen W.H., Ryan K.C. and Niehoff G.J., Influence of fire on factors that affect site productivity, In: Harvey A.E. and Neuenschwander L.F., (Eds.), *Proceedings – Management and Productivity of Western-Montane Forest Soils*.1991, U.S. Department of Agriculture, Forest Service, Intermountain Research Station, Ogden, UT, Gen. Tech. Rep. INT-GTR-280, pp. 32–50.

Jandl, R., Lindner, M., Vesterdal, B., Bauwens, R., Baritz, F., Hagedorn, D.W., et al., 2007. How strongly can forest management influence soil carbon sequestration? Geoderma 137, 253–268.

Jang, W., Page-Dumroese, D.S., Keyes, C.R., 2016. Long-term soil changes from forest harvesting and residue management in the Northern Rocky Mountains. Soil Sci. Soc. Am. J. . Available from: https://doi.org/10.2136/sssaj2015.11.0413222.

Jang, W., Page-Dumroese, D.S., Han, H.-S., 2017. Comparison of heat transfer and soil impacts of air curtain burner burning and slash pile burning. Forests . Available from: https://doi.org/10.3390/f80802978 p.

Jiménez, E., Vega, J.A., Fernánez, C., Fonturbel, T., 2011. Is pre-commercial thinning compatible with carbon sequestration? A case study in a maritime pine stand in northwestern Spain. Forestry 1−9.

Johnson, D.W., Curtis, P.S., 2001. Effects of forest management on soil C and N storage: meta-analysis. For. Ecol. Manage. 140, 227−238.

Johnson, L.R., Page-Dumroese, D., Han, H.-S., 2007. Effects of machine traffic on the physical properties of ash-cap soils. In: Page-Dumroese, D., Mille, R., Mital, J., McDaniel, P., Miller, D. (Eds.), Volcanic-Ash-Derived Forest Soils of the Inland Northwest: Properties and Implications for Management and Restoration. U.S. Department of Agriculture, Forest Service, Rocky Mountain Research Station, Fort Collins, CO, Proceedings, RMRS-P-44, pp. 69−83.

Johnson L.R., Page-Dumroese D. and Han H.-S., Effects of machine traffic on the physical properties of ash-cap soils, In: Page-Dumroese D., Mille R., Mital J., McDaniel P. and Miller D., (Eds.), Volcanic-Ash-Derived Forest Soils of the Inland Northwest: Properties and Implications for Management and Restoration. 2007, U.S. Department of Agriculture, Forest Service, Rocky Mountain Research Station, Fort Collins, CO, Proceedings, RMRS-P-44, pp., 69–83

Jokela, E.J., Allen, H.L., McFee, W.W., 1991a. Fertilization of southern pines at establishment. In: Duryea, M.L., Dougherty, P.M. (Eds.), Forest Regeneration Manual. Kluwer Academic Publishers, Dordrecht, pp. 263−277.

Jokela, E.J., McFee, W.W., Stone, E.L., 1991b. Micronutrient deficiency in slash pine: response and persistence of added manganese. Soil Sci. Soc. Am. J. 55, 492−496.

Jones, G., Loeffler, D., Calkin, D., Chung, W., 2010. Forest treatment residues for thermal energy compared with disposal by on-site burning: emissions and energy return. Biomass Bioenergy 34, 737−746.

Jurgensen, M.F., Harvey, A.E., Graham, R.T., Page-Dumroese, D.S., et al., 1997. Impacts of timber harvesting on soil organic matter, nitrogen, productivity, and health of Inland Northwest forests. For. Sci. 43, 234−251.

Jurgensen, M.F., Tarpey, R., Pickens, J., Kolka, R., Palik, B., 2012. Long-term effect of silvicultural thinnings on soil carbon and nitrogen pools. Soil Sci. Soc. Am. J. 76, 1418−1425.

Jurgensen, M.F., Miller, C.A., Trettin, C.T., Page-Dumroese, D.S., 2019. Bedding of wetland soil: effects of bed height and termite activity on wood decomposition. Soil Sci. Soc. Am. J. 83, S218−S227.

Kemp, J.L., 1967. Epitaph for the giants: the story of the Tillamook Burn. Touchstone Press, Portland, OR.

Kimmins, J.P., 1977. Evaluation of the consequences for future tree productivity of the loss of nutrients in whole-tree harvesting. For. Ecol. Manage. 1, 169−183.

Kizha, A.R., Han, H.-S., 2015. Forest residues recovered from whole-tree timber harvesting operations. Eur. J. For. Eng. 1, 46−55.

Knoepp, J.D., Vose, J.M., Swank, W.T., 2004. Long-term soil responses to site preparation burning in the southern Appalachians. For. Sci. 50, 540–550.

Korb, J.E., Johnson, N.C., Covington, W.W., 2004. Slash pile burning effects on soil biotic and chemical properties and plant establishment: recommendations for amelioration. Restor. Ecol. 12, 52–62.

Kyle, K.H., Andrews, L.J., Fox, T.R., Aust, W.M., Burger, J.A., Hansen, G.H., 2005. Long-term effects of drainage, bedding, and fertilization on growth of loblolly pine (*Pinus taeda* L.) in the coastal plain of Virginia. South. J. Appl. For. 29, 205–214.

Lal, R., 2005. Forest soils and carbon sequestration. For. Ecol. Manage. 220, 242–258.

Lewis, J., Rhodes, J.J., Bradley, C., 2018. Turbidity response from timber harvesting, wildfire, and post-fire logging in the Battle Creek Watershed, northern California. Environ. Manage. 63, 416–432.

Littke, K.M., Harrison, R.B., Zabowski, D., Briggs, A.R., 2014. Assessing nitrogen fertilizer response of coastal Douglas-fir in the Pacific Northwest using a paired-tree experimental design. For. Ecol. Manage. 330, 137–143.

Liu, Y., Wei, X., Guo, X., Niu, D., Zhange, J., Gong, X., et al., 2012. The long-term effects of reforestation on soil microbial biomass carbon in sub-tropic severe red soil degradation areas. For. Ecol. Manage. 285, 77–84.

Löf, M., Dey, D.C., Navarro, R.M., Jacobs, D.F., 2012. Mechanical site preparation for forest restoration. New For. 43, 825–848.

Londo, A.J., Mroz, G.D., 2001. Bucket mounding as a mechanical site preparation technique in wetlands. North. J. Appl. For. 18, 7–13.

Lowery, R.F., Gjerstad, D.H., 1991. Chemical and mechanical site preparation. Forest Regeneration Manual. Springer, Dordrecht, pp. 251–261.

Maki, T.E., 1960. Improving site quality by wet-land drainage. In: Burns, P.Y. (Ed.), Eighth Annual Forestry Symposium: Southern Forest Soils. Louisiana State University Press, Baton Rouge, LA, pp. 106–114.

Mandzak, J.M., Moore, J.A., 1994. The role of nutrition in the health of Inland Western forests. J. Sustain. For. 2, 191–210.

Massman, W.J., Frank, J.M., Reisch, N.B., 2008. Long-term impacts of prescribed burns on soil thermal conductivity and soil heating at a Colorado Rocky Mountain site: a data/model fusion study. Int. J. Wildland Fire 17, 131–146.

McCarthy, N., Bentsen, N.S., Willoughby, I., Balandier, P., 2010. The state of forest vegetation management in Europe in the 21st century. Eur. J. For. Res. 130, 7–16.

McFarlane, K.J., Schoenholtz, S.H., Powers, R.F., 2009. Plantation management intensity affects belowground carbon and nitrogen storage in northern California. Soil Sci. Soc. Am. J. 73, 1020–1032.

McIver J.D. and Starr, L. Environmental effects of post-fire logging: literature review and annotated bibliography, 2000, U.S. Department of Agriculture, Forest Service, Pacific Northwest Research Station, Portland, OR, Gen. Tech. Rep. PNW-GTR-486, 72 p.

McKee Jr, W.H., Shoulders, E., 1970. Depth of water table and redox potential of soil affect slash pine growth. For. Sci. 16, 399–402.

McKee Jr., W.H., Shoulders, E., 1974. Slash pine biomass response to site preparation and soil properties. Soil Sci. Soc. Am. Proc. 38, 144–148.

McKeever, D.B., Falk, R.H., 2004. Woody residues and solid waste wood available for recovery in the United States. 2002 In: *European COST E31 Conference; Management of Recovered Woody Recycling Bioenergy and Other Options*. Proceeding Apr. 22–24, 2004. University Studio Press, Thessaloniki, pp. 307–316.

Meyer, N.J., 2009. Soil and Plant Response to Slash Pile Burning in a Ponderosa Pine Forest (M.Sc. thesis). Montana State University, Bozeman, MT.

Miller, W.D., Maki, T.E., 1957. Planting pines in pocosins. J. For. 55, 659−663.

Miner, R.A., Abt, R.C., Bowyer, J.L., Buford, M.A., Malmsheimer, R.W., et al., 2014. Forest carbon accounting considerations in US bioenergy policy. J. For. 112, 591−606.

Minogue, P.J., Cantrell, R.L., Griswold, H.C., 1991. Vegetation management after plantation establishment. Forest Regeneration Manual. Springer, Dordrecht, pp. 335−358.

Montreal Process. Criteria and indicators for the conservation and sustainable management of temperate and boreal forests. Catalogue Fo42-238/1995E. Hull, Quebec: Canadian Forest Service. 1995.

Morris, L.A., Lowery, R.F., 1988. Influence of site preparation on soil conditions affecting stand establishment and tree growth. South. J. Appl. For. 12, 170−178.

Morris, L.A., Miller, R.E., 1994. Evidence for long-term productivity change as provided by field trials. Impacts of Forest Harvesting on Long-Term Site Productivity. Springer, Dordrecht, pp. 41−80.

Morris L.A. and Pritchett W.L., Effects of site preparation on Pinus elliottii-P. palustris flatwoods forest soil properties, IUFRO Symposium on Forest Site and Continuous Productivity. 1983, U.S. Department of Agriculture, Forest Service, Pacific Northwest Experiment Station, Portland, OR, Gen. Tech. Rep. PNW-GTR-163, pp. 243−251.

Morris, D.M., Kimmins, J.P., Duckert, D.R., 1997. The use of soil organic matter as a criterion of the relative sustainability of forest management alternatives: a modelling approach using FORECAST. For. Ecol. Manage. 94, 61−78.

Morvan, X., Saby, N.P.A., Arrouays, D., LeBas, C., Jones, R.J.A., Verheijen, F.G.A., et al., 2008. Soil monitoring in Europe: a review of existing systems and requirements of harmonisation. Sci. Total. Environ. 391, 1−12.

Nambiar, E.K.S., 1996. Sustained productivity of forests in a continuing challenge to soil science. Soil Sci. Soc. Am. J. 60, 1629−1642.

Nave, L.E., Vance, E.D., Swanston, C.W., Curtis, P.R., 2010. Harvest impacts on soil carbon storage in temperate forests. For. Ecol. Manage. 259, 857−866.

Nave, L.E., Walters, B.F., Hofmeister, K.L., Perry, C.H., Mishra, U., Domke, G.M., et al., 2019. The role of reforestation in carbon sequestration. N. For. 50, 115−137.

Nelson, L.R., Cantrell, R.L., 2002. Herbicide prescription manual for southern pine management. In: Clemson University Extension Document EC 659. 53 p.

Nelson C.D., Peter G.F., McKeand S.E., Jokela E.J., Rummer R.G., Groom L., et al., Pines. 2013, In: Singh B.P., (Ed), Biofuel Crops Production, Physiology and Genetics 2013, 2013, CAB International, Fort Valley State University, Fort Valley, GA, 427–459, 33 p.

Nortcliff, S., 2002. Standardization of soil quality attributes. Agric. Ecosyst. Environ. 88, 161−168.

Noss, R.J., Franklin, J.F., Baker, W.L., Schoennagel, T., Moyle, P.B., 2006. Managing fire-prone forests in the western United States. Front. Ecol. Environ. 4, 481−487.

Page-Dumroese, D.S., Jurgensen, M.F., Harvey, A.E., Graham, R.T., Tonn, J.R., 1997. Soil changes and tree seedling response associated with site preparation in northern Idaho. West. J. Appl. For. 12, 81−88.

Page-Dumroese, D.S., Jurgensen, M.F., Tiarks, A.E., Ponder Jr, F., et al., 2006. Soil physical changes at the North American long-term soil productivity study sites: 1 and 5 years after compaction. Can. J. For. Res. 36, 5651−5664.

Page-Dumroese, D.S., Abbott, A.M., Rice, T.M., 2009. Forest soil disturbance monitoring protocol. Vol. II: Supplementary methods, statistics, and data collection. U.S. Department of Agriculture, Forest Service, Washington, DC, Gen. Tech. Rep. WO-GTR-82b, 70 p.

Page-Dumroese, D.S., Jurgensen, M.F., Terry, T., 2010. Maintaining soil productivity during forest or biomass-to-energy thinning harvests in the western United States. West. J. Appl. For. 25, 5—11.

Page-Dumroese, D.S., Abbott, A.M., Curran, M.P., Jurgensen, M.F., 2012. Validating visual disturbance types and classes used for forest soil monitoring protocols. U.S. Department of Agriculture, Forest Service, Rocky Mountain Research Station, Fort Collins, CO, Gen. Tech. Rep. RMRS-GTR-267, 17 p.

Page-Dumroese D.S., Jain T.J., Sandquist J.E., Tirocke J.M., Errecart J. and Jurgensen M.F., Reburns and their impact on carbon pools, site productivity, and recovery, In: Potter K.M. and Conkling B.L., (Eds.), *Forest Health Monitoring: National Status, Trends, and Analysis* 2015, U.S. Department of Agriculture Forest Service, Southern Research Station, Asheville, NC, Gen. Tech. Rep. SRS-GTR-209, pp.143–149, [Chapter 13].

Page-Dumroese, D.S., Busse, M.D., Archuleta, J.G., McAvoy, D., Roussel, E., 2017. Methods to reduce forest residue volume after timber harvesting and produce black carbon. Scientifica . Available from: https://doi.org/10.1155/2017/27457648 p.

Passovoy, M.D., Fulé, P.Z., 2006. Snag and woody debris dynamics following severe wildfires in northern Arizona ponderosa pine forests. For. Ecol. Manage. 223, 237—246.

Ponder Jr., F., Fleming, R.L., Berch, S., Busse, M.D., Elioff, J.D., et al., 2012. Effects of organic matter removal, soil compaction, and vegetation control on 10th year biomass and foliar nutrition: LTSP continent-wide comparisons. For. Ecol. Manage. 278, 35—54.

Powers, R.F., 1990. Nitrogen mineralization along an altitudinal gradient: interactions of soil temperature, moisture, and substrate quality. For. Ecol. Manage. 30, 19—29.

Powers, R.F., Tiarks, A.E., Boyle, J.R., 1998. Assessing soil quality: practicable standards for sustainable forest productivity in the United States. In: Davidson, E.A. (Ed.), Criteria and Indicators of Soil Quality for Sustainable Forest Productivity. Special Publication 53 of the Soil Science Society of America, Madison, WA, pp. 53—80.

Powers, R.F., Scott, D.A., Sanchez, F.G., Voldseth, R.A., et al., 2005. The North American long-term soil productivity experiment: findings from the first decade of research. For. Ecol. Manage. 220, 31—50.

Prescott, C.E., Maynard, D.G., Laiho, R., 2000. Humus in northern forests: friend or foe? For. Ecol. Manage. 133, 23—36.

Ratcliff, A.W., Busse, M.D., Shestak, C.J., 2006. Changes in microbial community structure following herbicide (glyphosate) additions to forest soils. Appl. Soil Ecol. 34, 114—124.

Reaves, C.A., Cooper, A.W., 1960. Stress distribution and soil compaction under tractor tires. Agric. Eng. 41, 20—31.

Redfearn, A., Pimm, S.L., 1987. Insect outbreaks and community structure. In: Barbosa, P., Schultz, J.C. (Eds.), Insect Outbreaks. Academic Press, San Diego, CA, pp. 99—133.

Reeves, D.A., Reeves, M.C., Abbott, A.M., Page-Dumroese, D.S., Coleman, M.D., 2012. A detrimental soil disturbance prediction model for ground-based timber harvesting. Can. J. For. Res. 42, 821—830.

Rhoades, C.C., Fornwalt, P.J., 2015. Pile burning creates a fifty-year legacy of openings in regenerating lodgepole pine forests in Colorado. For. Ecol. Manage. 336, 203—209.

Rhoades, C.C., Fornwalt, P.J., Paschke, M.W., Shanklin, A., Jonas, J.L., 2015. Recovery of small pile burn scars in conifer forests of the Colorado Front Range. For. Ecol. Manage. 347, 180–187.

Rifai, S.W., Markewitz, D., Borders, B., 2010. Twenty years of intensive fertilization and competing vegetation suppression in loblolly pine plantations: impacts on soil C, N, and microbial biomass. Soil Biol. Biochem. 42, 713–723.

Ruiz-Jaen, M.C., Aide, T.M., 2005. Restoration success: how is it being measured? Restor. Ecology. 13, 569–577.

Sanchez, F.G., Tiarks, A.E., Kranabetter, J.M., Page-Dumroese, D.S., Powers, R.F., Sanborn, P.T., et al., 2006. Effects of organic matter removal and soil compaction on fifth-year mineral soil carbon and nitrogen contents for sites across the Unites States and Canada. Can. J. For. Res. 36, 565–576.

Sarkhot, D., Comerford, N.B., Jokela, E.J., Reeves III, J.B., 2007. Effects of forest management intensity on carbon and nitrogen content in different soil size fractions of a north Florida Spodosol. Plant Soil 294, 291–303.

Sartori, F., Markewitz, D., Borders, B.E., 2007. Soil carbon storage and nitrogen and phosphorus availability in loblolly pine plantations over 4 to 16 years of herbicide and fertilizer treatments. Biogeochemistry 84, 13–30.

Schmidt, M.W., Torn, M.S., Abiven, S., Dittmar, T., Guggenberger, G., Janssens, et al., 2011. Persistence of soil organic matter as an ecosystem property. Nature 478 (7367), 49.

Schoenholtz, S.H., Van Miegroet, H., Burger, J.A., 2000. A review of chemical and physical properties as indicators of forest soil quality: challenges and opportunities. For. Ecol. Manage. 138, 335–356.

Schultz, R.P., 1997. Loblolly pine: the ecology and culture of loblolly pine (*Pinus taeda* L.), Agriculture Handbook, Vol. 713. U.S. Department of Agriculture, Forest Service, Washington, DC, 493 p.

Scott, D.A., Page-Dumroese, D.S., 2016. Woody bioenergy and soil productivity research. Bioenergy Res. 9, 507–517.

Sessions, J., Leshchinsky, B., Chung, W., Boston, K., Wimer, J., 2017. Theoretical stability and traction of steep slope tethered feller-bunchers. For. Sci. 63, 192–200.

Seybold, C.A., Herrick, J.E., Brejda, J.J., 1999. Soil resilience: a fundamental component of soil quality. Soil Sci. 164, 224–234.

Shan, J., Morris, L.A., Hendrick, R.L., 2002. The effects of management on soil and plant carbon sequestration in slash pine plantations. J. Appl. Ecol. 38, 932–941.

Sheffield R.M., Cost N.D., Bechtold W.A., and McClure J.P. Pine growth reductions in the Southeast. Resource Bulletin SE-RP-83, 1985, U.S. Department of Agriculture, Forest Service, Southeastern Forest Experiment Station, Asheville, NC, pp. 117.

Simon, J., Christy, S., Vessels, J., 1994. Clover-mist fire recovery: a forest management response. J. For. 92, 41–44.

Sinclair, T.R., 1992. Mineral nutrition and plant growth response to climate change. J. Exp. Bot. 43, 1141–1146.

Skaggs, R.W., Tian, S., Chescheir, G.M., Amatya, D.M., Youssef, M.A., 2016. Forest drainage. In: Amatya, et al., (Eds.), Forest Hydrology: Processes, Management and Assessment. CABI Publishers, pp. 124–140.

Smith, P., House, J.I., Bustamante, M., Sobocká, J., et al., 2016. Global change pressures on soils from land use and management. Global Change Biol. 22, 1008–1028.

Solgi A., Najafi A., Page-Dumrose D.S. and Zenner E.K., Assessment of soil disturbance caused by different skidding machine types along the margin of the machine operating trail, Geoderma 367, 2020, 114238. https//doi.org/10.1016/j.geoderma.2020.114238.

Spanos, I., Raftoyannis, Y., Goudelis, G., Xanthopoulou, E., Samara, T., Tsiontsis, A., 2005. Effects of post-fire logging on soil and vegetation recovery in a *Pinus halapensis* Mill. forest in Greece. Plant Soil 278, 171−179.

Stransky, J.J., Roese, J.J., Watterston, K.G., 1985. Soil properties and pine growth affected by site preparation after clearcutting. South. J. Appl. For. 9, 40−53.

Subedi, P., Jokela, E.J., Vogel, J.G., Martin, T.A., 2014. Inter-rotational effects of fertilization and weed control on juvenile loblolly pine productivity and nutrient dynamics. Soil Sci. Soc. Am. J. 78, S152−S167.

Subedi, P., Jokela, E.J., Vogel, J.G., 2017. Inter-rotational effects of fertilizer and herbicide treatments on the understory vegetation community in juvenile loblolly pine (*Pinus taeda* L.) stands. For. Sci. 63, 459−473.

Takyi, S.K., Hillman, G.R., 2000. Growth of coniferous seedlings on a drained and mounded peatland in Central Alberta. North. J. Appl. For. 17, 71−79.

Tarpey, R.A., Jurgensen, M.F., Palik, B.J., Kolka, R.K., 2008. The long-term effects of silvicultural thinning and partial cutting on soil compaction in a red pine (*Pinus resinosa* Ait.) and northern hardwood stands in the northern Great Lakes Region of the United States. Can. J. Soil Sci. 88, 849−857.

Tarrant, R.F., 1956. Effects of slash burning on some soils of the Douglas-fir region. Soil Sci. Soc. Am. J. 20, 408−411.

Terry, T.A., Hughes, J.H., 1975. The effects of intensive management on planted loblolly pine (*Pinus taeda* L.) growth on poorly drained soils of the Atlantic Coastal Plain. In: Bernier, B., Winget, C.H. (Eds.), Forest Soils and Forestland Management. Les Presses de L'Universite Laval, Quebec, pp. 351−377.

Thibodeau, L., Raymond, P., Camiré, C., Munson, A.D., 2000. Impact of precommercial thinning in balsam fir stands on soil nitrogen dynamics, microbial biomass, decomposition, and foliar nutrition. Can. J. For. Res. 30, 229−238.

Thompson, D.G., Pitt, D.G., 2003. A review of Canadian forest vegetation management research and practice. Ann. For. Sci. 60, 559−572.

Turner J., A review of forest fertilization programs in Australia, In: Ballard R. and Gessel S.P., (Eds.), *IUFRO Symposium on Forest Site and Continuous Productivity*. 1983, U.S. Department of Agriculture, Forest Service, Pacific Northwest Research Station, Portland, OR, Gen. Tech. Rep. PNW-GTR-163, pp. 349–357.

Vance, E.D., Prisley, S.P., Schilling, E.B., Tatum, V.L., Wigley, T.B., et al., 2018. Environmental implication of harvesting lower-value biomass in forests. For. Ecol. Manage. 407, 47−56.

Van Lear, D.H., Kapeluck, P.R., Parker, M.M., 1995. Distribution of carbon in a piedmont soil as affected by loblolly pine management. In: McFee, W.W., Kelly, J.M. (Eds.), Carbon Forms and Functions in Forest Soils. Soil Science Soc. Am., Madison, WI, pp. 489−501.

Van Nieuwstadt, M.G., Shiel, D., Kartawinata, K., 2001. The ecological consequences of logging in the burned forests of east Kalimantan, Indonesia. Conserv. Biol. 15, 1183−1186.

Vegter, J.J., Roels, J.M., Bavinck, H.F., 1988. Soil quality standards: science or science fiction? Contaminated Soil 88, 309−316.

Visser, R., Stampfer, K., 2015. Expanding ground-based harvesting onto steep terrain. Croatian J. For. Eng. 36, 133−143.

Vogel, J.G., Jokela, E.J., 2011. Micronutrient limitations in two managed southern pine stands planted on Florida Spodosols. Soil Sci. Soc. Am. J. 75, 1117−1124.

Vogel, J.G., Suau, L.J., Martin, T.A., Jokela, E.J., 2011. Long-term effects of weed control and fertilization on the carbon and nitrogen pools of a slash and loblolly pine forest in north-central Florida. Can. J. For. Res. 41, 552−567.

Waggenbrenner, J.W., Robichaud, P.R., Brown, R.E., 2016. Rill erosion in burned and salvage logged western montane forests: effects of logging equipment type, traffic level, and slash treatment. J. Hydrol. 541, 889−901.

Walmsley, J.D., Jones, D.L., Reynolds, B., Price, M.H., Healey, J.R., 2009. Whole tree harvesting can reduce second rotation forest productivity. For. Ecol. Manage. 257, 1104−1111.

Wang, W.W., Page-Dumroese, D., Jurgensen, M., Tirocke, J., Liu, Y., 2018. Effect of forest thinning and wood quality on the short-term decomposition rate in *Pinus tabuliformis* plantation. J. Plant. Res. Available from: https://doi.org/10.1007/s10265-018-10169y.

Wells, C.G., Jorgensen, J.R., 1979. Effect of intensive harvesting on nutrient supple and sustained productivity. In: Leaf, A.I. (Ed.), Impact of Intensive Harvesting on Forest Nutrient Cycling. Symp. Proc. Syracuse, NY. State University of New York, College of Environmental Science and Forestry, Syracuse, NY, pp. 212−230.

White, E.H., Pritchett, W.L., 1970. Water table control and fertilization for pine production in the Flat-woods. Technical Bulletin No. 743. Agricultural Experiment Station. University of Florida, Gainesville, FL.

York, R.A., Thomas, Z., Restaino, J., 2009. Influence of ash substrate proximity on growth and survival of planted mixed-conifer seedlings. West. J. Appl. For. 24, 117−123.

Zhang, J., Powers, R.F., Oliver, W.W., Young, D.H., 2013. Response of ponderosa pine plantations to competing vegetation control in northern California, USA: a meta-analysis. Forestry. 86, 3−11.

Sustainable management of grassland soils

M.J. McTavish[1], H.A. Cray[2], S.D. Murphy[2], J.T. Bauer[3], C.A. Havrilla[4], M. Oelbermann[2] and E.J. Sayer[5]

[1]*Faculty of Forestry, University of Toronto, Toronto, Ontario, Canada*
[2]*School of Environment, Resources & Sustainability, University of Waterloo, Waterloo, Ontario, Canada*
[3]*Department of Biology, Miami University, Oxford, Ohio, United States*
[4]*Merriam Powell Center for Environmental Research & Center for Ecosystem Science and Society, Northern Arizona University, Flagstaff, Arizona, United States*
[5]*Lancaster Environment Centre, Lancaster University, Lancaster, United Kingdom*

4.1 Overview of grassland soils

Grasslands are found in every region of the globe except for Antarctica and Greenland and are estimated to cover approximately 31%—43% of the Earth's land surface (White et al., 2000). Although there are many classification schemes, grasslands are generally defined as terrestrial ecosystems dominated by nonwoody vegetation (Suttie et al., 2005; White et al., 2000). Grasslands occur when there is sufficient moisture available for grasses or forbs, but extensive tree growth is prevented by climate and disturbances such as grazing and natural or anthropogenic fire (Allaby, 2006). There is considerable diversity in the types of grasslands across the world, including savannas, shrublands, meadows, scrub, and tundra (see Dixon et al., 2014 for a classification scheme and map of global grassland types).

Belowground soil structure and processes in grasslands are ecologically unique because of the periodic removal of large amounts of aboveground biomass by disturbance (Johnson and Matchett, 2001). This constant cycling of plant biomass (particularly root detritus resulting from aboveground shoot removal) feeds invertebrate and microbial decomposer communities, creating organic-enriched soil aggregates with higher carbon-to-nitrogen ratios that tend toward relatively slow turnover (Karlen, 2005). Although there is considerable variation in the types of grassland soils globally, many of the most productive and intensively managed soils can be classified as Mollisols following the USDA Soil Classification System; these soils share a characteristically dark, organic matter—enriched topsoil that is generally rich in base cations, including Ca^{2+}, Mg^{2+}, Na^+, and K^+ (Eswaran and Reich, 2005). The surfaces of these grassland soils are often characterized by colloidal aggregates or crusts that slow the impact of precipitation,

Soils and Landscape Restoration. DOI: https://doi.org/10.1016/B978-0-12-813193-0.00004-7

increase infiltration while minimizing water loss to runoff or high impact erosion, and allow for efficient plant root growth (Cambardella, 2005).

Global grasslands and grassland soils provide many indispensable ecosystem services (Egoh et al., 2011). Of these services, there is currently considerable interest in carbon storage and the climate change mitigation potential of grasslands. Grasslands are estimated to contain approximately 34% of the world's terrestrial carbon stocks (compared to 39% in forests and 17% in agroecosystems) (White et al., 2000) and are generally considered to be carbon sinks because of slow rates of organic matter turnover (Jones and Donnelly, 2004; Wang and Fang, 2009). Unlike forests, grasslands tend to store most of their carbon in soils rather than in vegetation (White et al., 2000). In addition to their carbon sequestration potential, grasslands typically have organic matter and base cation−enriched soils that make them well suited to supporting productive plant growth (Eswaran and Reich, 2005), making grasslands key contributors to global food security and a prospective source of low input, sustainable biofuel feedstock (Porqueddu et al., 2016; Tilman et al., 2006). Grasslands are also globally prominent, ecologically unique systems that support rich biodiversity and endemic species, and have considerable cultural, esthetic, and recreational values (Porqueddu et al., 2016; White et al., 2000; Wu et al., 2011).

Given the importance of grasslands and grasslands soils, there is great interest in understanding how best to manage them in order to sustain the many services they provide into the distant future. In the following sections of this chapter, we provide brief overviews of (1) the major threats to grassland soils and the current management challenges and (2) the types of management used for grassland soils. We supplement these summaries with four research highlights describing relevant contemporary, leading-edge research. We end the chapter with a synthesis of some of the major themes observed throughout the chapter and a brief discussion of priorities for future research.

4.2 Threats to grassland soils and management challenges

Grassland soils are threatened by several, primarily anthropogenic, factors that have contributed to their historical, present day, and potential future degradation and disappearance. We have summarized three of the primary threats to grassland soils:

1. *Land-use change*:
   Among global biomes, temperate grasslands, savannas, and shrublands have the greatest disparity between the amount of land use (c. 46%) and extent of habitat protection (c. 4.6%), and many global grasslands have been heavily modified and converted for human purposes (Hoekstra et al., 2005).
   The primary anthropogenic use of grasslands is food production; grasslands have been exploited throughout human history for hunting and gathering, herding and grazing of livestock, and sedentary agriculture (Ramankutty et al., 2008;

Suttie et al., 2005; White et al., 2000). Grasslands are integral to current global food security and will be under increasing pressure as human populations continue to grow (O'Mara, 2012). In addition to providing food, grasslands are a convenient source of fuel and building materials and are relatively easy to develop. As a result, grasslands have been heavily developed for human settlement; approximately twice as many people are estimated to live on former grasslands compared to forests (White et al., 2000) (Fig. 4.1).

Unfortunately, this preferential exploitation of grasslands contributes to their continuing degradation and disappearance. The development of agriculture or settlement removes whole grassland habitats. Harvesting of plant biomass (e.g., animal forage, biofuel feedstock) or grazing animals (e.g., meat, milk) can negatively impact soil carbon stocks and export other essential nutrients from the landscape. Intensive cultivation through tillage, grazing, or harvesting can also disrupt soil physical and chemical processes, breaking down soil structure and aggregates, exposing previously less accessible carbon species to microbial decomposition, and altering the diversity of soil biotic communities (Frey et al., 1999; Menta et al., 2011).

FIGURE 4.1

Black-faced sheep grazing in the Scottish Highlands. Scotland's upland grasslands are thought to cover less than 1% of the country's land area.

Photo: H. Cray.

2. *Climate change*:

Anthropogenic climate change is a complex environmental challenge that threatens grassland soils through warming temperatures, atmospheric carbon dioxide (CO_2) enrichment, and changes in the intensity and frequency of rainfall and fires (Pachauri and Mayer, 2014).

Warming temperatures generally increase rates of organic matter decomposition and soil respiration leading to higher net losses of soil carbon (Rey et al., 2011; Schuerings et al., 2013; Wang and Fang, 2009), though these effects are not universal (Strong et al., 2017). Atmospheric CO_2 enrichment in grasslands can enhance plant growth but may also stimulate increased heterotrophic denitrification in the underlying soils through higher root turnover, raising emissions of the potent and long-lived greenhouse gas nitrous oxide (N_2O) (Niboyet et al., 2011). Drought or irregular rainfall can reduce aboveground productivity (Felton et al., 2020) but may also increase carbon storage as microbial decomposition is reduced and plants allocate increasing biomass to roots (Martí-Roura et al., 2011), while higher intensity rainfall may alter soil hydraulic properties (Caplan et al., 2019) and variably increase the release or storage of carbon from water-stressed and wetter systems, respectively (Vargas et al., 2012; Wang and Fang, 2009). Higher intensity and more frequent fires could release large amounts of both carbon and nitrogen, but the incorporation of remaining ash and dead organic matter to the soil may also increase plant growth and carbon storage (Martí-Roura et al., 2011).

Overall, these diverse potential climate change impacts interact to produce complex trade-offs between higher rates of both decomposition and net primary production (NPP), leaving researchers in agreement that climate change could have strong impacts on grasslands but uncertain whether it will increase or decrease carbon storage in grassland soils (Jones and Donnelly, 2004). Climate change is also expected to have other poorly understood effects on grassland plant litter quality, soil organisms, soil structure, and nutrient cycling (Barnett and Facey, 2016; Walter et al., 2013).

3. *Woody encroachment and biological invasion*:

Grassland soils and soil biota are also altered by changes in the aboveground plant community (Hedlund et al., 2003; Virágh et al., 2011). Changes in vegetation can create competition with desired plant species for light and soil resources (Hautier et al., 2009; Sperry et al., 2006; Zhang et al., 2014); alter fire regimes (Vasquez et al., 2008); and modify the quality, quantity, and timing of organic inputs through the soil from roots and litter (Reed et al., 2005). The two primary sources of changing vegetation in grasslands are woody encroachment and the invasion of exotic forbs or grasses.

Woody encroachment is one of the main challenges facing global grasslands and is generally a result of changes in disturbance regimes, including anthropogenic fire suppression, loss of grazing herbivores, or even overgrazing that disrupts fuel and fire connectivity (Archer et al., 2017; Case and Staver, 2017; Daskin et al., 2016; Joubert et al., 2012). These changes in disturbance regime allow woody vegetation (native or exotic) from contiguous habitat to establish and spread, generally decreasing grassland plant diversity (Ratajczak et al., 2012). Although woody encroachment may or may not support higher rates of carbon sequestration on an ecosystem scale (Smith and Johnson, 2003), encroachment can also result in increased fluxes of trace gases (e.g., NO_x), greater susceptibility to higher severity wildfires, and loss of grassland ecosystem traits (Liao et al., 2006; Liao and Boutton, 2008; Porqueddu et al., 2016).

Although the study of biological invasions and the implications of exotic species for conservation have been the subjects of extensive and ongoing debate (see Chew, 2015; Crowley et al., 2017; Guerin et al., 2018; Kuebbing and Nuñez, 2018; Russell and Blackburn, 2017), it is clear that the introduction of exotic species can cause ecological changes that are both difficult to predict and frequently undesirable (Jeschke et al., 2014; Mack et al., 2000). Because grassland vegetation community structure is often strongly driven by disturbance and competition for key limiting nutrients (e.g., N), altered disturbance regimes or higher nutrient loading from sources such as fertilization or atmospheric deposition can provide a competitive advantage to exotic species over locally adapted taxa (Sperry et al., 2006; Vasquez et al., 2008). Grasslands worldwide have been invaded by many exotic species such as Cheatgrass (*Bromus tectorum*) in the arid and semiarid rangelands of the western US steppe (Sperry et al., 2006; Vasquez et al., 2008), alligator weed (*Alternanthera philoxeroides*) in Chinese annual grasslands (Zhang et al., 2014), and South African Lovegrass (*Eragrostis plana*) in the Pampas grasslands of southern Brazil, Uruguay, and Argentina (Fonseca et al., 2013). In addition to changing the aboveground vegetation community, these exotic species can alter many belowground properties and processes, including soil carbon storage, nutrient cycling, soil structure, hydrology, and flora and faunal diversity.

4.3 Keys to sustainable management of grassland soils

Given the considerable disparity between grassland exploitation and protection (Hoekstra et al., 2005), relatively intact grassland ecosystems will benefit greatly from increased conservation. For the more human-altered grasslands around the world, the challenge is to manage these systems for different goals and services such as food production, climate change mitigation, and protection of a unique ecosystem. Many of the management strategies that have been proposed for grasslands typify "ecological intensification" (Gos et al., 2016), actions aimed at increasing productivity while simultaneously reducing overall ecological impacts (Griffon, 2009). Researchers differ in their opinions on how feasible such a balance is. Focusing on soil carbon storage, reviews by Conant et al. (2017, 2001) suggest that management designed to increase forage quality and grassland productivity generally lead to concomitant increases in soil carbon. In contrast, a review by Eze et al. (2018) reports a net negative effect of management attributed primarily to negative effects of grazing that are partially mitigated by fertilization and liming. The inconsistent conclusions of these different reviews suggest individual grasslands characterized by unique biotic and abiotic conditions, and feedbacks/interactions may vary widely in their responses to management, requiring case-by-case selection of the most appropriate management approaches.

We provide here a brief overview of three of the most common types of grassland management: (1) fire, (2) grazing and mowing, and (3) soil amendments. Given the importance of grassland soils to global carbon cycles and climate change, much of the existing literature focuses on the impacts of management on grassland soil carbon. We also discuss, however, the implications of these different management approaches for other soil properties and processes, including nutrient cycling, soil structure, and belowground biota.

1. *Fire*:

Fire is a disturbance that plays a key role in maintaining many grassland types by preventing the encroachment of woody vegetation or exotic species. In addition to naturally occurring fires, anthropogenic burning has been practiced in many parts of the world as a tool for managing grasslands by historic populations through to the present day (Ratajczak et al., 2014). While the most obvious effects of fire are on aboveground vegetation, fire itself is a soil-forming factor that can directly alter the physical and chemical properties of burned soils (Certini, 2014) (Fig. 4.2).

The effects of prescribed fire on grassland soils are often most pronounced in the first several months following burning and in the top several centimeters of soil (Girona-García et al., 2018; Pereira et al., 2017; Strong et al., 2017), though it can also have longer term effects that influence deeper parts of the whole soil profile. Like wildfires, prescribed fire can have complex trade-off effects on greenhouse gas production, carbon fluxes, and nutrient cycling. Fire combusts organic matter in the topsoil, producing pulsed releases of large amounts of both carbon (CO_2, CO, CH_4) and nitrogen (NO_x, N_2) (Girona-García et al., 2018; Martí-Roura et al., 2011; Niboyet et al., 2011). Fire can also increase soil respiration by removing surface litter and elevating soil temperature and by increasing the production of fine roots and aboveground biomass (Strong et al., 2017). Although many nutrients released from vegetation by burning are highly plant available and plant communities are generally effective at retaining postfire nutrients (Martí-Roura et al., 2013), high-intensity fires that cause extensive plant mortality can severely limit plant uptake and increase nutrient losses to leaching (Augustine et al., 2014; Martí-Roura et al., 2011; Rapp, 1990).

On the other hand, although many grassland microbial populations appear to be relatively resistant to low-intensity fires (Carson and Zeglin, 2018; Dooley and Treseder, 2012), partial removal of the upper organic horizons and direct mortality from higher intensity fire can temporarily reduce microbial biomass and associated respiration (Coyle et al., 2017; Girona-García et al., 2018). New plant growth in response to open ground,

FIGURE 4.2

Prescribed burning of a tallgrass prairie restoration site in Norfolk County, Ontario, Canada. These prescribed burns are considered low complexity and involve trained staff and volunteers.

Photo: H. Cray.

high availability of soluble ash-derived nutrients, including Ca, Mg, and K, changes in soil pH, or water stress—induced root biomass allocation can also change patterns of plant growth and potentially increase carbon storage in vegetation (Kitchen et al., 2009; Pereira et al., 2017; Strong et al., 2017). Fire also exposes surface soil to wind erosion that helps redistribute soil nutrients into a more heterogeneous distribution that can favor grass regrowth over woody encroachment.

Fire has historically been and continues to be an important tool in sustainably managing grasslands and grassland soils; prescribed burns can be effective in controlling the early stages of woody encroachment (O'Connor et al., 2020; Sankey et al., 2012) while simultaneously increasing grassland productivity and overall carbon storage potential (Conant et al., 2017). However, the effectiveness of fire depends on the frequency and intensity with which it is applied (Coyle et al., 2017). Less frequent burns that release smaller amounts of nutrients may be better suited to preserving desired plant communities rather than undesired taxa that may be more competitive in nutrient-enriched soils (Augustine et al., 2014; Reed et al., 2005). Burns can also be seasonally timed and targeted to reduce emissions of greenhouse gases (O'Mara, 2012). Finally, given the often close association between managed grasslands and human activity (e.g., grazing, nearby settlement), the application of fire as a management tool may come into conflict with other land uses or safety concerns, and prescribed burns must be planned with these considerations in mind (Case and Staver, 2017).

2. *Grazing and mowing*:

Like fire, grazing is a disturbance that is important for maintaining grasslands and grassland soils that can occur both naturally (e.g., by resident herbivores) or anthropogenically (e.g., introduction of grazing animals, use as pasture land). Mowing by hand or by machine is also used to harvest biomass or simulate the effects of grazing (Kitchen et al., 2009; Schrama et al., 2013). Grazing and mowing are usually practiced with the primary intent of producing feed for livestock or otherwise managing aboveground vegetation (Jacquemyn et al., 2011), but they also have implications for grassland soil carbon storage, structure, and nutrient cycling.

Recent reviews suggest that grazing herbivores can have highly inconsistent effects on grassland soil properties (McSherry and Ritchie, 2013), though some suggest that if managed correctly, moderate levels of grazing can be effective at increasing grassland productivity and soil carbon storage (Conant et al., 2017; Wang and Fang, 2009). Grazing animals typically consume up to 50% of the annual aboveground NPP, returning between 60% and 90% of the ingested plant biomass to the soil as waste (Maharning et al., 2009); this can transform relatively recalcitrant plant tissues into more labile soil inputs, increasing soil nutrient pools, resource heterogeneity, plant productivity, and plant diversity (Collins et al., 1998; Johnson and Matchett, 2001; Knapp et al., 1999). Grazing can also stimulate a greater allocation of plant growth to belowground biomass, increasing soil carbon storage in roots and soil organic matter (McSherry and Ritchie, 2013).

If managed poorly, however, grazing—particularly overgrazing by maintained high densities of herbivores— can have many undesirable effects. Most animals graze selectively, which can lead to dominance of lower quality plant litters that are generally higher in secondary metabolites and structural compounds such as lignin; this can alter the availability of nutrients to plants and microbes and alter also ecosystem functioning (Maharning et al., 2009). Overgrazing can also impair plant productivity by reducing photosynthetic tissue (Chang et al., 2015; McSherry and Ritchie, 2013), degrade soil structure and hydrology by trampling (Kaine and Tozer, 2005; Monaghan et al., 2005; Steffens et al., 2008), reduce the effectiveness of fire by disrupting fuel connectivity (Archer et al., 2017), and increase erosion as barren soils are left exposed to wind and rain (Hoffmann et al., 2008). In grasslands that have been heavily degraded by overgrazing, temporary reductions or complete exclusion of grazing animals can help restore vegetation cover (Zhang et al., 2015). Maintaining intermediate grazing levels by controlling the intensity and frequency of grazing by stocking levels, fencing, and rotation grazing can help protect managed grassland soils from soil degradation (Chang et al., 2015) (Fig. 4.3).

FIGURE 4.3

Soil erosion in a grazed pasture, Altos de Campana, Panama. These deforested, rolling grassland landscapes are particularly vulnerable to erosional processes.

Photo: H. Cray.

Many of the same principles apply to mowing of grasslands, which can also be an effective management tool if used responsibly. However, compared to grazing, mowing poses some additional management considerations and challenges. Unlike herbivores, mowing is not selective and will generally have a different effect on vegetation community structure (Wang et al., 2016). In addition, if cut biomass is removed (e.g., for forage or biomass feedstock), a large portion of plant-derived nutrients and carbon are exported from the system rather than returned through animal wastes and residues (Kitchen et al., 2009). Long-term mowing with heavy machinery can also lead to soil compaction and reduce aeration, degrading soil structure and hydrology, increasing erosion potential, and favoring more superficial root growth (Schrama et al., 2013).

3. *Soil amendments*:

The addition of amendments such as fertilizers and lime to grassland soils is a more intensive management approach to "ecological intensification" (Gos et al., 2016) that is more frequently encountered in grasslands managed for forage or animal production. Nevertheless, recent reviews by both Conant et al. (2017) and Eze et al. (2018) suggest that amendments can also have cobenefits for soil conservation and carbon storage.

Common amendments include inorganic or organic fertilizers (e.g., animal manure, and mulch) to increase plant nutrient availability. Liming (e.g., lime and calcareous sand) can also be used to increase the productivity of acid−soil grasslands (Heyburn et al., 2017; Mijangos et al., 2010). There is considerable evidence to support using amendments to increase above- and belowground plant growth in grasslands to improve soil structure and increase carbon storage in soil aggregates and plant and microbial biomass (Ammann et al., 2007; Conant et al., 2017, 2001; Eze et al., 2018; Jones et al., 2006; Pavlů et al., 2016; Ryals et al., 2015, 2014; Wang and Fang, 2009). However, depending on local conditions such as temperature and moisture, soil amendment also has complex trade-off effects such as increased soil respiration and production of other greenhouse gases, including N_2O (Conant et al., 2001; Jones et al., 2006; Ryals et al., 2015; Wang and Fang, 2009), and can unpredictably alter the composition and function of the soil microbial community (Bardgett et al., 1996; Maharning et al., 2009).

As with both fire and grazing or mowing, proper application of amendments requires planning appropriate application rates and timing; excessively high or frequent inputs that are poorly matched with vegetation demand will tend to increase undesirable results such as greenhouse gas emissions and nutrient runoff and leaching (O'Mara, 2012). Furthermore, in grasslands where lower nutrient availability contributes to their invasion resistance, intensive amendment can create more fertile soils that competitively disadvantage locally adapted plant and microbial communities and facilitate biological invasion (Vasquez et al., 2008). These poorly planned applications can be expensive and time-consuming while producing limited conservation gain and can have undesirable off-target effects such as eutrophication of nearby aquatic systems (Zhou et al., 2017). Nevertheless, for more heavily managed grasslands, moderate amendments may be a useful tool for comanaging for human productive needs and soil conservation (Conant et al., 2017; Eze et al., 2018). For more natural grasslands that may be characteristically nutrient poor, soil amendments may be desirable as shorter term corrective measures to remediate specific soil conditions such as uncharacteristic nutrient deficiencies or soil acidification.

4.4 Contemporary research highlights

To complement the preceding general discussion of the leading challenges and solutions in the sustainable management of grassland soils, we highlight four areas of leading-edge, contemporary research concerning the sustainable management of grassland soils. The first two address general plant—soil microbial interactions by discussing an underlying conceptual framework (research highlight 1) and the implications of climate change for these interactions (research highlight 2). The third expands on the often overlooked subject of "biocrusts," with emphasis on the impacts of grassland grazing (research highlight 3). The fourth describes novel crop systems used to help sustainably manage agricultural grassland soils in the Rolling Pampas of Argentina (research highlight 4).

Research highlight 1: Plant—soil feedbacks (PSFs) in sustainable soil management. Jonathan T. Bauer (the University of Miami—Ohio).

The PSF framework has helped to link the complexity of plant × microbial interactions to consequences for plant population dynamics and community composition in a diversity of ecosystems, including grasslands (Casper and Castelli, 2007; Reinhart, 2012). These feedbacks arise as soil microbial communities change in response to different plant species, and these changes affect the fitness of the associated plant and cooccurring plant species (Bever et al., 1997). Links between PSFs and plant species abundance (Klironomos, 2002; Mangan et al., 2010), plant life history (Bauer et al., 2015), and invasion dynamics (Suding et al., 2013) support the importance of PSFs for shaping plant community composition. Additional theoretical and empirical work has linked plant × microbial feedbacks to abiotic conditions in the soil (Revilla et al., 2013; Smith-Ramesh and Reynolds, 2017) and to associated ecosystem services (Maron et al., 2011; Baer et al., 2012). Here, we explore how changes to plants, microbes, and soil nutrients may alter PSFs and inform the sustainable management and/or restoration of plant biodiversity and soil ecosystem services in grasslands.

Displacement of native species by anthropogenic disturbance or nonnative species introductions can lead to shifts in PSFs. When anthropogenic disturbance promotes ruderal nonnative species, the resulting changes to soil microbial communities and soil nutrient cycling may reinforce this change in plant community composition (Kulmatiski et al., 2006). In addition, the effects of invasive species on soil microorganisms may directly cause changes in plant community composition or ecosystem function (Ehrenfeld, 2003; Stinson et al., 2006). It is also possible for reestablishment of native plant species to promote recovery of native soil microorganisms and soil ecosystem functions, including nitrogen and carbon cycling (Baer et al., 2002; Bach et al., 2010; Barber et al., 2017).

Changes to grassland soil microbial communities may also change PSFs. For example, the absence of microbial symbionts limits the establishment of some plant species in disturbed environments and consequently alters plant community productivity, nutrient cycling, and soil carbon storage (Bauer et al., 2012; Keller and Lau, 2018). Reintroduction of microbial mutualists can generate positive feedbacks that stabilize the persistence of late successional plant communities (Koziol and Bever, 2019). Similarly, the absence of appropriate mutualists may limit the establishment of introduced species, but the cointroduction of plants and novel mutualists may lead to dramatic transformation of community and ecosystem processes (Nuñez et al., 2009). However, we also note that PSFs have the potential to be resilient to changes in soil microbial communities caused by past anthropogenic disturbance (Bauer et al., 2015, 2017).

Nutrient enrichment is widespread (Elser and Bennett, 2011) and may change PSFs by shifting microbial symbioses from mutualistic to parasitic (Johnson, 1993). It is not well understood how these shifts in plant−microbial symbioses may translate into net effects of PSFs (Smith-Ramesh and Reynolds, 2017), but changes in PSFs under different light levels suggest that resource availability could have large effects on overall PSFs (Smith and Reynolds, 2015). Consequently, global increases in nitrogen and phosphorus availability are likely to be causing changes in PSFs in addition to their direct effects on grassland plant communities.

The PSF framework has provided useful predictions for how interactions among plants, microbial communities, and soil properties may affect plant community composition and function. These insights may allow us to anticipate how various aspects of global change—changes in plant and microbial communities and disruption of nutrient and carbon cycles—will further alter feedbacks among plants, microorganisms, and soils. By anticipating these dynamics, we may be better able to manage soils to maintain plant community diversity and associated ecosystem services in global grasslands (Fig. 4.4).

Research highlight 2: Climate change impacts on plant−soil microbe communities in sustainable management of grasslands. Emma J. Sayer (Lancaster Environment Centre, Lancaster University, United Kingdom).

Soil microbial communities will play a crucial role in the sustainable management of grasslands under climate change. Bacteria and fungi in soils catalyze many processes that underpin soil health, the maintenance of plant productivity (Van der Heijden et al., 2008; Bardgett et al., 2008), and key ecosystem services such as carbon sequestration (Orwin et al., 2015). However, predicting change in soil microbial processes is highly challenging because microbial community composition, structure, and function are affected both directly by abiotic soil conditions and indirectly via the vegetation (Bardgett et al., 2008). Previous assumptions that soil microbial processes will be largely unaffected by the change have been challenged by the collective results of experiments in grasslands and other ecosystems showing that microbial community composition is generally sensitive to change and not immediately resilient after disturbance (Allison and Martiny, 2008). Severe or long-lasting disturbance can select for microbes that are better adapted to the new environment, which could alter the turnover rates of different resources and affect carbon and nutrient cycling.

Shifts in microbial community composition in response to change arise primarily from differences in the growth rates and resource-use efficiencies of the constituent organisms, as well as their capacity to resist or adapt to new conditions (Schimel et al., 2007). Consequently, changes in the vegetation influence soil

FIGURE 4.4

Traditional grassland management and restoration have focused on aboveground vegetation metrics. Moving forward, new discoveries linked to soil microbial communities are broadening our knowledge base and creating new opportunities for conservation and management.

Photo: H. Cray.

microbial community composition via particular plant traits (Orwin et al., 2015) related to microbial resource requirements (Sayer et al., 2017). In general, soil fungi have slower growth and turnover rates and lower nutrient requirements compared to bacteria; the ability of fungi to degrade recalcitrant lignin-rich plant litter may give them an advantage when resource quality and quantity is low, whereas rapid growth and turnover rates make bacteria better competitors for high-quality resources and labile carbon (Waring et al., 2013). Various experiments have demonstrated that plant diversity may be a dominant control of soil microbial communities under environmental change (e.g., Eisenhauer et al., 2010; Lange et al., 2015; Steinauer et al., 2015). Warming, for example, may favor fast-growing bacteria by enhancing plant growth, increasing root turnover (Fitter et al., 1999), and reducing the lignin content in plant tissues (Henry et al., 2005). By contrast, although fungi are generally better able to withstand abiotic drought stress than bacteria (Schimel et al., 2007), strong shifts in soil fungal communities in response to 17 years of experimental summer drought treatments in the UK grassland were attributed to vegetation changes, which altered the quantity and quality of resources available to soil organisms (Fridley et al., 2016; Sayer et al., 2017).

The strong links between above- and belowground communities could be crucial for the sustainable management of grassland soils under climate change, as many of the observed effects of land management on soil microbial communities also appear to be mediated by plants. Grazing intensity (Zhou et al., 2017) and nutrient addition in particular can cause profound and predictable changes to soil microbial community composition by modifying the quality and quantity of plant inputs to the soil. Nonetheless, the coadaptation of plants and soil microorganisms to change may play an important role in regulating soil functioning in the future. For example, the specialization of the soil biotic community for plant material of the same origin results in faster rates of decomposition (the "home-field advantage"; Ayres et al., 2009). Along a similar vein an experiment measuring soil processes in different combinations of plants and soil microbial communities showed that losses of carbon and nitrogen from grassland soils were lower in "matching" plant and microbial communities from the same origin (De Vries et al., 2015). Increasing grassland plant diversity can also enhance the recovery of plant root networks and microbial biomass in restored grasslands (Klopf et al., 2017), and soil microorganisms associated with mature plant communities tend to have greater resource-use efficiency (Eisenhauer et al., 2010; Bölscher et al., 2016), which can increase soil carbon storage and nutrient retention (Klopf et al., 2017).

In sum, the reciprocal exchange of resources between plants and soil microorganisms underpins not only several important ecosystem processes but also their response to change (Gutknecht et al., 2012). Plant-mediated indirect effects of environmental change play a key role in determining the structure and function of soil microbial communities (Bardgett et al., 2008). Hence, land management to increase plant diversity could maintain soil function by providing a greater variety of resources to sustain different soil microbial groups (Eisenhauer et al., 2010), which in turn is likely to improve the resistance of grasslands to climate change.

Research highlight 3: Biocrusts: an often overlooked component of sustainable management plans in grassland ecosystems. Caroline A. Havrilla, (Department of Ecology and Evolutionary Biology, University of Colorado Boulder)

Biological soil crusts (biocrusts), soil surface—dwelling microbial communities dominated by cyanobacteria, bryophytes, lichens, and fungi coexist in patchy mosaics with vascular plants in arid and semiarid grasslands worldwide. Where they occur, biocrusts enhance biodiversity and mediate a suite of critical ecosystem services. Biocrusts, for example, benefit soil structure and stability (Zhang et al., 2006; Bowker et al., 2008a) and influence surface hydrology (Belnap, 2006). Biocrusts are also key intermediaries of biochemical exchanges between the atmosphere and soils, often enhancing soil fertility through increased carbon (Li et al., 2012) and nitrogen fixation (Barger et al., 2016). Moreover, biocrusts can enhance the availability of other mineral nutrients (Jafari et al., 2004; Guo et al., 2008) and increase the abundance and diversity of soil microfauna (Housman et al., 2007; Darby and Neher, 2016).

Through provision of these important soil services, biocrusts can also strongly influence the vascular plant communities with which they coexist (reviewed in Zhang et al., 2016). A recent global metaanalysis of biocrust—plant interactions by Havrilla et al. (2019) revealed that overall biocrusts increase the growth of C_3 and C_4 grass species with $C_3 > C_4$, suggesting biocrusts may increase forage quality and biomass production for livestock and wildlife where they occur. Evidence additionally suggests biocrusts mediate exotic plant invasibility by serving as an ecological filter to exotic seed germination (Zhang et al., 2016, Havrilla et al., 2019).

While biocrusts support a variety of important soil services, they are vulnerable to physical disturbance by compressional forces (reviewed by Zaady et al., 2016). As such, domestic livestock grazing, vehicular traffic, and industrial mineral extraction all pose significant threats to biocrust functioning in grasslands (Zaady et al., 2016). Conservation prioritization and incorporation of biocrusts into sustainable management plans, however, may mitigate these effects (Bowker et al., 2008b). Incorporating biocrusts into sustainable grazing management practices, for example, demonstrates this potential.

Livestock grazing in global rangelands typically has detrimental impacts on biocrusts (Eldridge et al., 2016), decreasing biocrust cover and richness (Root and McCune, 2012; Thomas, 2012). Losses of biocrusts contribute to reductions in carbon and nitrogen fixation (Aranibar et al., 2008), carbon sequestration (Thomas, 2012), and soil stability; changes in soil hydrology (Eldridge et al., 2016); and an increase in exotic plant invasibility. These biocrust losses due to intense grazing can be further exacerbated by increasing aridity associated with global climate change (Mallen-Cooper et al., 2017).

While the effects of grazing on biocrusts are unequivocally negative, the magnitude of these effects can be mitigated by incorporating considerations for biocrusts into sustainable grazing management plans. Selecting sustainable grazing densities and stocking rates, for example, may play an important role in mitigating grazing effects on biocrusts (Zaady et al., 2016). The capacity of lower grazing intensity to help preserve biocrust community intactness and services relative to higher grazing intensity has been evidenced by multiple studies (e.g., Read et al., 2008; Liu et al., 2009). In addition, while no studies have directly compared the influences of different rotational strategies or stocking rates on biocrusts, lower stocking rates, which spatially dilute livestock impacts on soils, may also be more compatible with biocrusts than higher stocking rates. Grazing impacts on biocrusts may also be season dependent (Zaady et al., 2016). In grasslands that receive cool season precipitation, for example, grazing in cooler, wet seasons has been shown to have more moderate effects on biocrusts than summer grazing (Memmott et al., 1998; Wang et al., 2009).

Regardless of the driver of biocrust degradation, it is clear that biocrust considerations should be incorporated into sustainable management plans to mitigate losses of biocrust cover and diversity, particularly in drier grasslands most susceptible to these losses, and to preserve biocrust contributions to grassland functioning.

> *Research highlight 4*: Using novel crop systems in Argentina's Rolling Pampa enhances carbon sequestration and mitigates greenhouse gas emissions. Maren Oelbermann (School of Environment, Resources and Sustainability, University of Waterloo, Waterloo, ON, Canada)

In temperate South America the Rio de la Plata Grasslands (RPG) comprises a land area of 700,000 km^2, including parts of Argentina, Brazil, and Uruguay. The RPG is composed of 19 subregions, one of which is the Rolling Pampa, based on vegetation communities, soil types, and topography (Modernel et al., 2016). Although the Rolling Pampa makes up only 9% of the entire RPG region, it has 1600 native plant species composed of a mixture of C$_3$ and C$_4$ grasses and legumes, and its soils store 5400 Tg carbon to a 30 cm depth (Modernel et al., 2016). By 1910 more than 60% of the natural grasslands in the Rolling Pampa were converted to agriculture (Viglizzo et al., 2001).

Today, the Rolling Pampa is one of the world's largest maize and soybean producing regions and plays a key role in international food security. But intensive agriculture in the Rolling Pampa caused soil erosion, with levels of soil organic carbon (SOC) decreased by 30% (Alvarez, 2001), with attendant effects on soil biodiversity and fertility. Since the 1990s conservation practices have been widely adopted to combat soil degradation. Conservation practices, however, are not sufficient to address long-term soil security, climate change resilience, and greenhouse gas mitigation. Novel maize–soybean intercrop systems, however, help to maintain or increase SOC and provide other agronomic and environmental benefits.

Intercropping is the simultaneous growth of more than one species in the same field, where one crop species is a legume (Vandermeer, 1992). In the Rolling Pampa, maize–soybean intercrops contribute to the long-term immobilization of nitrogen, reducing nitrification rates and thereby abating N$_2$O and CO$_2$ emissions (Dyer et al., 2012; Regehr et al., 2015). The mixed crop arrangement captures resources from different parts of the soil and/or uses resources at different times and/or in different forms, increases microbial diversity and activity, and nutrient translocation (Echarte et al., 2011).

The Rolling Pampa is one of the global regions with a high potential to sequester carbon (Hutchinson et al., 2006). Based on extensive fieldwork with maize–soybean intercrops in the Rolling Pampa (Oelbermann et al., 2011; 2015), Oelbermann et al. (2017) used the Century model to predict changes in SOC stocks in intercrops and sole crops. After 100 years of maize–soybean intercropping, SOC increased by 47% but increased only by 21% in the simulations where maize was the sole crop and by 2% in the soybean only simulations.

Encouragingly, the model predicted SOC stocks in intercrops and sole crops within c. 7% of measured field values. Although crop residue input was 20% lower in the intercrop compared to the maize sole crop, input from mixed residue sources caused complex interactions at the soil—residue interface that influenced carbon transformations differently than in the sole crops, allowing for the accumulation of SOC. This research demonstrates that cereal—legume intercropping is a more sustainable land management practice in the Rolling Pampa, with respect to soil carbon and nitrogen transformations and soil microbial activity that leads to reduced CO_2 and N_2O emissions and enhances the long-term accumulation of SOC compared to sole cropping.

4.5 Synthesis and priorities for future research

From the information provided above in both the overview sections and the research highlights, we have identified four recurring themes pertinent to sustainably managing grassland soils:

Sustainable management of grassland soils requires consideration of interactions between different ecosystem components. In recent decades, scientists have increasingly recognized the importance of interactions between above- and belowground components of ecological systems (Wardle et al., 2004). Aboveground vegetation is a visible, and defining element of grassland ecosystems and many established management efforts such as fire, grazing, and soil amendment are targeted primarily at vegetation. To achieve sustainable management goals, however, requires an understanding of how both the vegetation and management actions are inextricably linked with different biotic and abiotic ecosystem components (Heneghan et al., 2008), including grazing animals, soil organisms, plant roots, soil structure, soil chemistry, and climate (Maharning et al., 2009; Sayer et al., 2013; Wang et al., 2016). Grasslands are often managed using multiple approaches, and the interaction of different tools can have unexpected outcomes (Kitchen et al., 2009). These kinds of complex interactions and the importance of soil and soil processes were an explicit and integral focus of all four of the above-mentioned research highlights. Future work that continues to consider and explore the complex, interconnected impacts of different management approaches will improve our ability to make informed, sustainable management decisions.

Sustainable management of grassland soils involves complex trade-offs. The most common trade-off to navigate in the management of grassland soils is the balance between decomposition and productivity, and the result impacts on soil carbon storage. There are many examples in which application of fire (Strong et al., 2017), grazing and mowing (McSherry and Ritchie, 2013), or soil amendments (Jones et al., 2006) may simultaneously increase or decrease both NPP and decomposition. The magnitude of change for each of these processes will influence the overall rate of carbon cycling and whether a grassland soil in a given context will act as a carbon sink or source. Given our global dependence on grasslands for ecosystem services such as food, fuel, and climate change mitigation, the sustainable management of grasslands often represents another more fundamental trade-offs between balancing the exploitation of these systems for human uses and conserving them for their intrinsic value as unique ecosystems.

While it may not always be possible to fully reconcile these different uses (Eze et al., 2018), literature reviews (Conant et al., 2017, 2001) and research into novel crop systems (research highlight 4) suggest that it is possible to achieve more sustainable practices with multiple cobenefits. Intermediate intensities and frequencies of management interventions (e.g., prescribed burns and grazing) may be most successful at achieving a balance of desirable outcomes (Ward et al., 2016).

Best practices of the sustainable management of grassland soils are highly contextual. Given the considerable diversity in global grasslands in terms of their ecological characteristics, cultural and economic importance to human societies, and the threats facing them (Dixon et al., 2014; Porqueddu et al., 2016), it should not be surprising that the best management practices for grassland soils are not served by a "one-size-fits-all" approach (Eviner and Hawkes, 2008). Although it would be helpful to describe an isolated, general effect of a given intervention, the common message from ongoing research is that effects of factors such as grazing (Eze et al., 2018; McSherry and Ritchie, 2013; Wang and Fang, 2009), rainfall (Barnett and Facey, 2016; Vargas et al., 2012), and fire (Girona-García et al., 2018) are inconsistent and idiosyncratic. Prevailing influences of local factors such as temperature, precipitation, soil texture, vegetation types, and past management legacies require a strongly contextual and case-by-case approach to sustainable grassland soil management (Fig. 4.5).

The sustainable management of grassland soils shares many goals and methods with grassland restoration. When historical grassland ecosystems have been severely degraded by agricultural conversion, overgrazing, or other stressors, restoration may be required to first recover lost ecological properties and processes. As is common in the restoration of many types of ecosystems, methods of grassland restoration have historically focused on the aboveground plant community and include spontaneous or "old-field" succession, seed sowing, planting, translocation, and grazing and mowing (Török et al., 2011). However, as in the sustainable management of grassland soils, there is a growing appreciation for the importance of integrating soil ecology into restoration and the benefits of manipulating soil physical, chemical, and biological properties singly or in combination to achieve restoration goals (Heneghan et al., 2008). Emerging research is expanding tools for grassland restoration to include inoculation of soil fauna and flora to facilitate desirable successional trajectories (Koziol and Bever, 2017; Middleton and Bever, 2012; Török et al., 2011) and management of high soil nutrient levels through topsoil removal or carbon-rich amendments to limit competition from weedy species (Baer et al., 2020; Tesei et al., 2020; Török et al., 2011). Further integration of grassland restoration and long-term, sustainable management of grassland soils will benefit shared goals, including biodiversity conservation and climate change mitigation (Scott et al., 2019; Török et al., 2011).

As global grasslands continue to face new and continuing pressures from anthropogenic climate change and the needs of a growing human population (O'Mara, 2012; Suttie et al., 2005), learning how to sustainably manage these ecosystems and their precious soils will become an increasingly pressing challenge. In addition to the research topics and themes addressed in this chapter, there are many promising directions for

FIGURE 4.5

Grassland habitats in Ecuador. Differences in temperature, rainfall, and altitude can all affect the optimal management of grasslands, even within the same geographic region.

Photo: H. Cray.

future research. We anticipate that priority areas of research will include the changing role of grassland soils in carbon storage and climate change mitigation under different management regimes, and further investigation of the complex links between different above- and belowground ecosystem components. Interactions between plants, soils, and organisms, including microbes and invertebrates such as earthworms, are particularly poorly understood but also likely of considerable ecological importance (Barnett and Facey, 2016; Carson and Zeglin, 2018; Conant et al., 2017; Curry et al., 2008; Sayer et al., 2013). This continuing research and the resulting best practices will have a key role to play in conserving, restoring, and sustainably managing unique and valuable global grassland ecosystems.

References

Allaby, M., 2006. A Dictionary of Plant Sciences. Oxford University Press.
Allison, S.D., Martiny, J.B.H., 2008. Resistance, resilience, and redundancy in microbial communities. Proc. Natl. Acad. Sci. U.S.A. 105, 11512–11519.

Alvarez, R., 2001. Estimation of carbon losses by cultivation from soils of the Argentina Pampa using the Century model. Soil Use Manage. 17, 62–66.

Ammann, C., Flechard, C.R., Leifeld, J., Neftel, A., Fuhrer, J., 2007. The carbon budget of newly established temperate grassland depends on management intensity. Agric. Ecosyst. Environ. Greenh. Gas. Balance Grassl. Europe 121, 5–20.

Aranibar, J.N., Anderson, I.C., Epstein, H.E., Feral, C.J.W., Swap, R.J., Ramontsho, J., et al., 2008. Nitrogen isotope composition of soils, C3 and C4 plants along land use gradients in southern Africa. J. Arid. Environ. 72 (4), 326–337.

Archer, S.R., Andersen, E.M., Predick, K.I., Schwinning, S., Steidl, R.J., Woods, S.R., 2017. Woody plant encroachment: causes and consequences. In: Briske, D. (Ed.), Rangeland Systems, Springer Series on Environmental Management. Springer, Cham, pp. 25–84.

Augustine, D.J., Brewer, P., Blumenthal, D.M., Derner, J.D., von Fischer, J.C., 2014. Prescribed fire, soil inorganic nitrogen dynamics, and plant responses in a semiarid grassland. J. Arid. Environ. 104, 59–66.

Ayres, E., Steltzer, H., Simmons, B.L., Simpson, R.T., Steinweg, J.M., Wallenstein, M.D., et al., 2009. Home-field advantage accelerates leaf litter decomposition in forests. Soil. Biol. Biochem. 41 (3), 606–610.

Bach, E.M., Baer, S.G., Meyer, C.K., Six, J., 2010. Soil texture affects soil microbial and structural recovery during grassland restoration. Soil Biol. Biochem. 42, 2182–2191.

Baer, S.G., Kitchen, D.J., Blair, J.M., Rice, C.W., 2002. Changes in ecosystem structure and function along a chronosequence of restored grasslands. Ecol. Appl. 12, 1688–1701.

Baer, S.G., Heneghan, L., Eviner, V.T., 2012. Applying Soil Ecological Knowledge to Restore Ecosystem Services. Soil Ecol. Ecosyst. Serv. 377–393.

Baer, S.G., Adams, T., Scott, D.A., Blair, J.M., Collins, S.L., 2020. Soil heterogeneity increases plant diversity after 20 years of manipulation during grassland restoration. Ecol. Appl. 30, e02014. Available from: https://doi.org/10.1002/eap.2014.

Barber, N.A., Chantos-Davidson, K.M., Amel Peralta, R., Sherwood, J.P., Swingley, W.D., 2017. Soil microbial community composition in tallgrass prairie restorations converge with remnants across a 27-year chronosequence. Environ. Microbiol. 19, 3118–3131.

Bardgett, R., Hobbs, P., Frostegård, Å., 1996. Changes in soil fungal: bacterial biomass ratios following reductions in the intensity of management of an upland grassland. Biol. Fertil. Soils 22, 261–264.

Bardgett, R.B., Freeman, C., Ostle, N.J., 2008. Microbial contributions to climate change through carbon cycle feedbacks. ISME J. 2, 805–814.

Barger, N.N., Weber, B., Garcia-Pichel, F., Zaady, E., Belnap, J., 2016. Patterns and controls on nitrogen cycling of biological soil crusts. Biological Soil Crusts: An Organizing Principle in Drylands. Springer, Cham, pp. 257–285.

Barnett, K.L., Facey, S.L., 2016. Grasslands, invertebrates, and precipitation: a review of the effects of climate change. Front. Plant. Sci. 7. Available from: https://doi.org/10.3389/fpls.2016.01196.

Bauer, J.T., Kleczewski, N.M., Bever, J.D., Clay, K., Reynolds, H.L., 2012. Nitrogen-fixing bacteria, arbuscular mycorrhizal fungi, and the productivity and structure of prairie grassland communities. Oecologia 170, 1089–1098.

Bauer, J.T., Mack, K.M.L., Bever, J.D., 2015. Plant-Soil Feedbacks as Drivers of Succession: Evidence From Remnant and Restored Tallgrass Prairies. Ecosphere.

Bauer, J.T., Blumenthal, N., Miller, A.J., Ferguson, J.K., Reynolds, H.L., 2017. Effects of between-site variation in soil microbial communities and plant-soil feedbacks on the productivity and composition of plant communities. J. Appl. Ecol. 54, 1028–1039.

Belnap, J., 2006. The potential roles of biological soil crusts in dryland hydrologic cycles. Hydrological Process. 20 (15), 3159–3178.

Bever, J.D., Westover, K.M., Antonovics, J., 1997. Incorporating the soil community into plant population dynamics: the utility of the feedback approach. J. Ecol. 85, 561–573.

Bölscher, T., Wadsö, L., Börjesson, G., Herrmann, A.M., 2016. Differences in substrate use efficiency: impacts of microbial community composition, land use management, and substrate complexity. Biol. Fertil. Soils 52 (4), 547–559.

Bowker, M.A., Belnap, J., Chaudhary, V.B., Johnson, N.C., 2008a. Revisiting classic water erosion models in drylands: the strong impact of biological soil crusts. Soil Biol. Biochem. 40 (9), 2309–2316.

Bowker, M.A., Miller, M.E., Belnap, J., Sisk, T.D., Johnson, N.C., 2008b. Prioritizing conservation effort through the use of biological soil crusts as ecosystem function indicators in an arid region. Conserv. Biol. 22 (6), 1533–1543.

Cambardella, C.A., 2005. Carbon cycle in soils | Formation and Decomposition. In: Hillel, D. (Ed.), Encyclopedia of Soils in the Environment. Elsevier, Oxford, pp. 170–175.

Caplan, J.S., Giménez, D., Hirmas, D.R., Brunsell, N.A., Blair, J.M., Knapp, A.K., 2019. Decadal-scale shifts in soil hydraulic properties as induced by altered precipitation. Sci. Adv. 5. Available from: https://doi.org/10.1126/sciadv.aau6635.

Carson, C.M., Zeglin, L.H., 2018. Long-term fire management history affects N-fertilization sensitivity, but not seasonality, of grassland soil microbial communities. Soil Biol. Biochem. 121, 231–239.

Case, M.F., Staver, A.C., 2017. Fire prevents woody encroachment only at higher-than-historical frequencies in a South African savanna. J. Appl. Ecol. 54, 955–962.

Casper, B.B., Castelli, J.P., 2007. Evaluating plant—soil feedback together with competition in a serpentine grassland. Ecol. Lett. 10, 394–400. Available from: https://doi.org/10.1111/j.1461-0248.2007.01030.x.

Certini, G., 2014. Fire as a soil-forming factor. AMBIO 43, 191–195.

Chang, X., Bao, X., Wang, S., Wilkes, A., Erdenetsetseg, B., Baival, B., et al., 2015. Simulating effects of grazing on soil organic carbon stocks in Mongolian grasslands. Agric. Ecosyst. Environ. 212, 278–284.

Chew, M.K., 2015. Ecologists, environmentalists, experts, and the invasion of the 'second greatest threat.' Int. Rev. Environ. Hist. 1, 17–40.

Collins, S.L., Knapp, A.K., Briggs, J.M., Blair, J.M., Steinauer, E.M., 1998. Modulation of diversity by grazing and mowing in native tallgrass prairie. Science 280, 745–747. Available from: https://doi.org/10.1126/science.280.5364.745.

Conant, R.T., Paustian, K., Elliott, E.T., 2001. Grassland management and conversion into grassland: effects on soil carbon. Ecol. Appl. 11, 343–355.

Conant, R.T., Cerri, C.E.P., Osborne, B.B., Paustian, K., 2017. Grassland management impacts on soil carbon stocks: a new synthesis. Ecol. Appl. 27, 662–668.

Coyle, D.R., Nagendra, U.J., Taylor, M.K., Campbell, J.H., Cunard, C.E., Joslin, A.H., et al., 2017. Soil fauna responses to natural disturbances, invasive species, and global climate change: current state of the science and a call to action. Soil Biol. Biochem. 110, 116–133.

Crowley, S.L., Hinchliffe, S., Redpath, S.M., McDonald, R.A., 2017. Disagreement about invasive species does not equate to denialism: a response to Russell and Blackburn. Trends Ecol. Evol. 32, 228–229.

Curry, J.P., Doherty, P., Purvis, G., Schmidt, O., 2008. Relationships between earthworm populations and management intensity in cattle-grazed pastures in Ireland. Appl. Soil Ecol. 39, 58–64.

Darby, B.J., Neher, D.A., 2016. Microfauna within biological soil crusts. Biological Soil Crusts: An Organizing Principle in Drylands. Springer, Cham, pp. 139–157.

Daskin, J.H., Stalmans, M., Pringle, R.M., 2016. Ecological legacies of civil war: 35-year increase in savanna tree cover following wholesale large-mammal declines. J. Ecol. 104, 79–89.

De Vries, F.T., Bracht Jørgensen, H., Hedlund, K., Bardgett, R.D., 2015. Disentangling plant and soil microbial controls on carbon and nitrogen loss in grassland mesocosms. J. Ecol. 103 (3), 629–640.

Dixon, A.P., Faber-Langendoen, D., Josse, C., Morrison, J., Loucks, C.J., 2014. Distribution mapping of world grassland types. J. Biogeogr. 41, 2003–2019.

Dooley, S.R., Treseder, K.K., 2012. The effect of fire on microbial biomass: a meta-analysis of field studies. Biogeochemistry 109, 49–61.

Dyer, L., Oelbermann, M., Echarte, L., 2012. Soil carbon dioxide and nitrous oxide emissions during the growing season from temperate maize-soybean intercrops. J. Plant. Nutr. Soil Sci. 175, 394–400.

Echarte, L., Della Maggiora, A., Cerrudo, D., Gonzalez, V.H., Abbate, P., Cerrudo, A., et al., 2011. Yield response to plant density of maize and sunflower intercropped with soybean. Field Crop. Res. 121, 423–429.

Egoh, B.N., Reyers, B., Rouget, M., Richardson, D.M., 2011. Identifying priority areas for ecosystem service management in South African grasslands. J. Environ. Manage. 92, 1642–1650.

Ehrenfeld, J.G., 2003. Effects of exotic plant invasions on soil nutrient cycling processes. Ecosystems 6, 503–523.

Eisenhauer, N., Beßler, H., Engels, C., Gleixner, G., Habekost, M., Milcu, A., et al., 2010. Plant diversity effects on soil microorganisms support the singular hypothesis. Ecology 91 (2), 485–496.

Eldridge, D.J., Poore, A.G., Ruiz-Colmenero, M., Letnic, M., Soliveres, S., 2016. Ecosystem structure, function, and composition in rangelands are negatively affected by livestock grazing. Ecol. Appl. 26 (4), 1273–1283.

Elser, J., Bennett, E., 2011. A broken biogeochemical cycle. Nature 478, 29–31.

Eswaran, H., Reich, P.F., 2005. World soil map. In: Hillel, D. (Ed.), Encyclopedia of Soils in the Environment. Elsevier, Oxford, pp. 352–365.

Eviner, V.T., Hawkes, C.V., 2008. Embracing variability in the application of plant–soil interactions to the restoration of communities and ecosystems. Restor. Ecol. 16, 713–729. Available from: https://doi.org/10.1111/j.1526-100X.2008.00482.x.

Eze, S., Palmer, S.M., Chapman, P.J., 2018. Soil organic carbon stock in grasslands: effects of inorganic fertilizers, liming and grazing in different climate settings. J. Environ. Manage. 223, 74–84.

Felton, A.J., Slette, I.J., Smith, M.D., Knapp, A.K., 2020. Precipitation amount and event size interact to reduce ecosystem functioning during dry years in a mesic grassland. Glob. Change Biol. 26, 658–668. Available from: https://doi.org/10.1111/gcb.14789.

Fitter, A.H., Self, G.K., Brown, T.K., Bogie, D.S., Graves, J.D., Benham, D., et al., 1999. Root production and turnover in an upland grassland subjected to artificial soil warming respond to radiation flux and nutrients, not temperature. Oecologia 120 (4), 575–581.

Fonseca, C.R., Guadagnin, D.L., Emer, C., Masciadri, S., Germain, P., Zalba, S.M., 2013. Invasive alien plants in the Pampas grasslands: a tri-national cooperation challenge. Biol. Invasions 15, 1751–1763.

Frey, S.D., Elliott, E.T., Paustian, K., 1999. Bacterial and fungal abundance and biomass in conventional and no-tillage agroecosystems along two climatic gradients. Soil Biol. Biochem. 31, 573–585.

Fridley, J.D., Lynn, J.S., Grime, J.P., Askew, A.P., 2016. Longer growing seasons shift grassland vegetation toward more productive species. Nat. Clim. Change 6, 865–868.

Girona-García, A., Badía-Villas, D., Martí-Dalmau, C., Ortiz-Perpiñá, O., Mora, J.L., Armas-Herrera, C.M., 2018. Effects of prescribed fire for pasture management on soil organic matter and biological properties: a 1-year study case in the Central Pyrenees. Sci. Total Environ. 618, 1079–1087.

Gos, P., Loucougaray, G., Colace, M.-P., Arnoldi, C., Gaucherand, S., Dumazel, D., et al., 2016. Relative contribution of soil, management and traits to co-variations of multiple ecosystem properties in grasslands. Oecologia 180, 1001–1013.

Griffon, M., 2009. What will be the future of the pastures and forage crops in the next decades? Fourrages 200, 539–546.

Guerin, G.R., Martín-Forés, I., Sparrow, B., Lowe, A.J., 2018. The biodiversity impacts of non-native species should not be extrapolated from biased single-species studies. Biodivers. Conserv. 27, 785–790.

Guo, Y., Zhao, H., Zuo, X., Drake, S., Zhao, X., 2008. Biological soil crust development and its topsoil properties in the process of dune stabilization, Inner Mongolia, China. Environ. Geol. 54 (3), 653–662.

Gutknecht, J.L., Field, C.B., Balser, T.C., 2012. Microbial communities and their responses to simulated global change fluctuate greatly over multiple years. Glob. Change Biol. 18, 2256–2269.

Hautier, Y., Niklaus, P.A., Hector, A., 2009. Competition for light causes plant biodiversity loss after eutrophication. Science 324, 636–638. Available from: https://doi.org/10.1126/science.1169640.

Havrilla, C.A., Chaudhary, V.B., Ferrenberg, S., Antoninka, A.J., Belnap, J., Bowker, M.A., et al., 2019. Towards a predictive framework for biocrust mediation of plant performance: a meta-analysis. J. Ecol. 107, 2789–2807. Available from: https://doi.org/10.1111/1365-2745.13269.

Hedlund, K., Santa Regina, I., Van der Putten, W.H., Lepš, J., Díaz, T., Korthals, G.W., et al., 2003. Plant species diversity, plant biomass and responses of the soil community on abandoned land across Europe: idiosyncracy or above-belowground time lags. Oikos 103, 45–58.

Heneghan, L., Miller, S.P., Baer, S., Callaham, M.A., Montgomery, J., Pavao-Zuckerman, M., et al., 2008. Integrating soil ecological knowledge into restoration management. Restor. Ecol. 16, 608–617. Available from: https://doi.org/10.1111/j.1526-100X.2008.00477.x.

Henry, H.A., Cleland, E.E., Field, C.B., Vitousek, P.M., 2005. Interactive effects of elevated CO_2, N deposition and climate change on plant litter quality in a California annual grassland. Oecologia 142 (3), 465–473.

Heyburn, J., McKenzie, P., Crawley, M.J., Fornara, D.A., 2017. Long-term belowground effects of grassland management: the key role of liming. Ecol. Appl. 27, 2001–2012.

Hoekstra, J.M., Boucher, T.M., Ricketts, T.H., Roberts, C., 2005. Confronting a biome crisis: global disparities of habitat loss and protection. Ecol. Lett. 8, 23–29.

Hoffmann, C., Funk, R., Wieland, R., Li, Y., Sommer, M., 2008. Effects of grazing and topography on dust flux and deposition in the Xilingele grassland, Inner Mongolia. J. Arid. Environ. 72, 792–807.

Housman, D.C., Yeager, C.M., Darby, B.J., Sanford Jr, R.L., Kuske, C.R., Neher, D.A., et al., 2007. Heterogeneity of soil nutrients and subsurface biota in a dryland ecosystem. Soil Biol. Biochem. 39 (8), 2138–2149.

Hutchinson, J.J., Campell, C.A., Desjardins, R.L., 2006. Some perspectives on carbon sequestration in agriculture. Agric. For. Meteorol. 142, 288–302.

Jacquemyn, H., Mechelen, C.V., Brys, R., Honnay, O., 2011. Management effects on the vegetation and soil seed bank of calcareous grasslands: an 11-year experiment. Biol. Conserv. 144, 416–422.

Jafari, M., Tavili, A., Zargham, N., Heshmati, G.A., Chahouki, M.Z., Shirzadian, S., et al., 2004. Comparing some properties of crusted and uncrusted soils in Alagol Region of Iran. Pak. J. Nutr. 3 (5), 273–277.

Jeschke, J.M., Bacher, S., Blackburn, T.M., Dick, J.T.A., Essl, F., Evans, T., et al., 2014. Defining the impact of non-native species. Conserv. Biol. 28, 1188–1194.

Johnson, L.C., Matchett, J.R., 2001. Fire and grazing regulate belowground processes in tallgrass prairie. Ecology 82, 3377–3389.

Johnson, N.C., 1993. Can fertilization of soil select less mutualistic mycorrhizae? Ecol. Appl. 3, 749–757.

Jones, M.B., Donnelly, A., 2004. Carbon sequestration in temperate grassland ecosystems and the influence of management, climate and elevated CO_2. N. Phytol. 164 (423–439).

Jones, S.K., Rees, R.M., Kosmas, D., Ball, B.C., Skiba, U.M., 2006. Carbon sequestration in a temperate grassland; management and climatic controls. Soil. Use Manage 22, 132–142.

Joubert, D.F., Smit, G.N., Hoffman, M.T., 2012. The role of fire in preventing transitions from a grass dominated state to a bush thickened state in arid savannas. J. Arid. Environ. 87, 1–7.

Kaine, G.W., Tozer, P.R., 2005. Stability, resilience and sustainability in pasture-based grazing systems. Agric. Syst. 83, 27–48.

Karlen, D.L., 2005. TILTH. In: Hillel, D. (Ed.), Encyclopedia of Soils in the Environment. Elsevier, Oxford, pp. 168–174.

Keller, K.R., Lau, J.A., 2018. When mutualisms matter: rhizobia effects on plant communities depend on host plant population and soil nitrogen availability. J. Ecol. 106, 1046–1056.

Kitchen, D.J., Blair, J.M., Callaham, M.A., 2009. Annual fire and mowing alter biomass, depth distribution, and C and N content of roots and soil in tallgrass prairie. Plant Soil 323, 235–247.

Klironomos, J.N., 2002. Feedback with soil biota contributes to plant rarity and invasiveness in communities. Nature 417, 67–70.

Klopf, R.P., Baer, S.G., Bach, E.M., Six, J., 2017. Restoration and management for plant diversity enhances the rate of belowground ecosystem recovery. Ecol. Appl. 27 (2), 355–362.

Knapp, A.K., Blair, J.M., Briggs, J.M., Collins, S.L., Hartnett, D.C., Johnson, L.C., et al., 1999. Bison increase habitat heterogeneity and alter a broad array of plant, community, and ecosystem processes. BioScience 49, 39–50. Available from: https://doi.org/10.1525/bisi.1999.49.1.39.

Koziol, L., Bever, J.D., 2017. The missing link in grassland restoration: arbuscular mycorrhizal fungi inoculation increases plant diversity and accelerates succession. J. Appl. Ecol. 54, 1301–1309. Available from: https://doi.org/10.1111/1365-2664.12843.

Koziol, L., Bever, J.D., 2019. Mycorrhizal feedbacks generate positive frequency dependence accelerating grassland succession. J. Ecol. 107, 622–632.

Kuebbing, S.E., Nuñez, M.A., 2018. Current understanding of invasive species impacts cannot be ignored: potential publication biases do not invalidate findings. Biodivers. Conserv. 27, 1545–1548.

Kulmatiski, A., Beard, K.H., Stark, J.M., 2006. Soil history as a primary control on plant invasion in abandoned agricultural fields. J. Appl. Ecol. 43, 868–876.

Lange, M., Eisenhauer, N., Sierra, C.A., Bessler, H., Engels, C., Griffiths, R.I., et al., 2015. Plant diversity increases soil microbial activity and soil carbon storage. Nat. Commun. 6, 6707.

Liao, J.D., Boutton, T.W., 2008. Soil microbial biomass response to woody plant invasion of grassland. Soil Biol. Biochem. 40, 1207–1216.

Liao, J.D., Boutton, T.W., Jastrow, J.D., 2006. Organic matter turnover in soil physical fractions following woody plant invasion of grassland: evidence from natural ^{13}C and ^{15}N. Soil Biol. Biochem. 38, 3197–3210.

Liu, H., Han, X., Li, L., Huang, J., Liu, H., Li, X., 2009. Grazing density effects on cover, species composition, and nitrogen fixation of biological soil crust in an inner Mongolia steppe. Rangel. Ecol. Manage. 62 (4), 321–327.

Li, X.R., Zhang, P., Su, Y.G., Jia, R.L., 2012. Carbon fixation by biological soil crusts following revegetation of sand dunes in arid desert regions of China: a four-year field study. Catena 97, 119–126.

Mack, R.N., Simberloff, D., Lonsdale, W.M., Evans, H., Clout, M., Bazzaz, F.A., 2000. Biotic invasions: causes, epidemiology, global consequences, and control. Ecol. Appl. 10, 689–710.

Maharning, A.R., Mills, A.A.S., Adl, S.M., 2009. Soil community changes during secondary succession to naturalized grasslands. Appl. Soil Ecol. 41, 137–147.

Mallen-Cooper, M., Eldridge, D.J., Delgado-Baquerizo, M., 2017. Livestock grazing and aridity reduce the functional diversity of biocrusts. Plant Soil 429, 175–185.

Mangan, S.A., Schnitzer, S.A., Herre, E.A., Mack, K.M.L., Valencia, M.C., Sanchez, E.I., et al., 2010. Negative plant-soil feedback predicts tree-species relative abundance in a tropical forest. Nature 466, 752. –U10.

Maron, J.L., Marler, M., Klironomos, J.N., Cleveland, C.C., 2011. Soil fungal pathogens and the relationship between plant diversity and productivity. Ecol. Lett. 14, 36–41.

Martí-Roura, M., Casals, P., Romanyà, J., 2011. Temporal changes in soil organic C under Mediterranean shrublands and grasslands: impact of fire and drought. Plant Soil 338, 289–300.

Martí-Roura, M., Casals, P., Romanyà, J., 2013. Long-term retention of post-fire soil mineral nitrogen pools in Mediterranean shrubland and grassland. Plant Soil 371, 521–531.

McSherry, M.E., Ritchie, M.E., 2013. Effects of grazing on grassland soil carbon: a global review. Glob. Change Biol. 19, 1347–1357.

Memmott, K.L., Anderson, V.J., Monsen, S.B., 1998. Seasonal grazing impact on ccrypto-gamic crusts in a cold desert ecosystem. J. Range Manage. 51, 547–550.

Menta, C., Leoni, A., Gardi, C., Conti, F.D., 2011. Are grasslands important habitats for soil microarthropod conservation? Biodivers. Conserv. 20, 1073–1087.

Middleton, E.L., Bever, J.D., 2012. Inoculation with a native soil community advances succession in a grassland restoration. Restor. Ecol. 20, 218–226. Available from: https://doi.org/10.1111/j.1526-100X.2010.00752.x.

Mijangos, I., Albizu, I., Epelde, L., Amezaga, I., Mendarte, S., Garbisu, C., 2010. Effects of liming on soil properties and plant performance of temperate mountainous grasslands. J. Environ. Manage. 91, 2066–2074.

Modernel, P., Rossing, W.A.H., Corbeels, M., Dogliotti, S., Picasso, V., Tittonell, P., 2016. Land use change and ecosystem service provision in Pampas and Campos grasslands of southern South America. Environ. Res. Lett. 11, 113002.

Monaghan, R.M., Paton, R.J., Smith, L.C., Drewry, J.J., Littlejohn, R.P., 2005. The impacts of nitrogen fertilisation and increased stocking rate on pasture yield, soil physical condition and nutrient losses in drainage from a cattle-grazed pasture. N. Zeal. J. Agric. Res. 48, 227–240.

Niboyet, A., Brown, J.R., Dijkstra, P., Blankinship, J.C., Leadley, P.W., Le Roux, X., et al., 2011. Global change could amplify fire effects on soil greenhouse gas emissions. PLoS One 6, e20105.

Nuñez, M.A., Horton, T.R., Simberloff, D., 2009. Lack of belowground mutualisms hinders Pinaceae invasions. Ecology 90, 2352–2359.

O'Connor, R.C., Taylor, J.H., Nippert, J.B., 2020. Browsing and fire decreases dominance of a resprouting shrub in woody encroached grassland. Ecology 101, e02935. Available from: https://doi.org/10.1002/ecy.2935.

Oelbermann, M., Regehr, A., Echarte, L., 2015. Changes in soil characteristics after six seasons of cereal–legume intercropping in the Southern Pampa. Geoderma Reg. 4, 100–107.

Oelbermann, M., Echarte, L., Marroquin, L., Morgan, S., Regehr, A., Vachon, K.E., et al., 2017. Estimating soil carbon dynamics in complex agroecosystems using the Century model. J. Plant Nutr. Soil Sci. 180, 241–251.

O'Mara, F.P., 2012. The role of grasslands in food security and climate change. Ann. Bot. 110, 1263–1270.

Orwin, K.H., Stevenson, B.A., Smaill, S.J., Kirschbaum, M.U., Dickie, I.A., Clothier, B.E., et al., 2015. Effects of climate change on the delivery of soil-mediated ecosystem services within the primary sector in temperate ecosystems: a review and New Zealand case study. Glob. Change Biol. 21 (8), 2844–2860.

Pachauri, R.K., Mayer, L., 2014. Intergovernmental panel on climate change. Climate Change 2014: Synthesis Report. Intergovernmental Panel on Climate Change, Geneva.

Pavlů, L., Gaisler, J., Hejcman, M., Pavlů, V.V., 2016. What is the effect of long-term mulching and traditional cutting regimes on soil and biomass chemical properties, species richness and herbage production in *Dactylis glomerata* grassland? Agric. Ecosyst. Environ. 217, 13–21.

Pereira, P., Cerda, A., Martin, D., Úbeda, X., Depellegrin, D., Novara, A., et al., 2017. Short-term low-severity spring grassland fire impacts on soil extractable elements and soil ratios in Lithuania. Sci. Total. Environ. 578, 469–475.

Porqueddu, C., Ates, S., Louhaichi, M., Kyriazopoulos, A.P., Moreno, G., del Pozo, A., et al., 2016. Grasslands in 'Old World' and 'New World' Mediterranean-climate zones: past trends, current status and future research priorities. Grass Forage Sci. 71, 1–35.

Ramankutty, N., Evan, A.T., Monfreda, C., Foley, J.A., 2008. Farming the planet: 1. Geographic distribution of global agricultural lands in the year 2000. Glob. Biogeochem. Cycles 22. Available from: https://doi.org/10.1029/2007GB002952.

Rapp, M., 1990. Nitrogen status and mineralization in natural and disturbed mediterranean forests and coppices. Plant Soil 128, 21–30.

Ratajczak, Z., Nippert, J.B., Collins, S.L., 2012. Woody encroachment decreases diversity across North American grasslands and savannas. Ecology 93, 697–703. Available from: https://doi.org/10.1890/11-1199.1.

Ratajczak, Z., Nippert, J.B., Briggs, J.M., Blair, J.M., 2014. Fire dynamics distinguish grasslands, shrublands and woodlands as alternative attractors in the Central Great Plains of North America. J. Ecol. 102, 1374–1385.

Read, C.F., Duncan, D.H., Vesk, P.A., Elith, J., 2008. Biological soil crust distribution is related to patterns of fragmentation and land use in a dryland agricultural landscape of southern Australia. Landsc. Ecol. 23, 1093–1105.

Reed, H.E., Seastedt, T.R., Blair, J.M., 2005. Ecological consequences of C4 grass invasion of a C4 grassland: a dilemma for management. Ecol. Appl. 15, 1560–1569.

Regehr, A., Oelbermann, M., Videla, C., Echarte, L., 2015. Gross nitrogen mineralization and immobilization in temperate maizesoybean intercrops. Plant Soil 391, 353–365.

Reinhart, K.O., 2012. The organization of plant communities: negative plant–soil feedbacks and semiarid grasslands. Ecology 93, 2377–2385. Available from: https://doi.org/10.1890/12-0486.1.

Revilla, T.A., (Ciska) Veen, G.F., Eppinga, M.B., Weissing, F.J., 2013. Plant-soil feedbacks and the coexistence of competing plants. Theor. Ecol. 6, 99–113.

Rey, A., Pegoraro, E., Oyonarte, C., Were, A., Escribano, P., Raimundo, J., 2011. Impact of land degradation on soil respiration in a steppe (*Stipa tenacissima* L.) semi-arid ecosystem in the SE of Spain. Soil Biol. Biochem. 43, 393–403.

Root, H.T., McCune, B., 2012. Regional patterns of biological soil crust lichen species composition related to vegetation, soils, and climate in Oregon, USA. J. Arid. Environ. 79, 93–100.

Russell, J.C., Blackburn, T.M., 2017. The rise of invasive species denialism. Trends Ecol. Evolution 32, 3–6.

Ryals, R., Kaiser, M., Torn, M.S., Berhe, A.A., Silver, W.L., 2014. Impacts of organic matter amendments on carbon and nitrogen dynamics in grassland soils. Soil Biol. Biochem. 68, 52–61.

Ryals, R., Hartman, M.D., Parton, W.J., DeLonge, M.S., Silver, W.L., 2015. Long-term climate change mitigation potential with organic matter management on grasslands. Ecol. Appl. 25, 531–545.

Sankey, J.B., Ravi, S., Wallace, C.S.A., Webb, R.H., Huxman, T.E., 2012. Quantifying soil surface change in degraded drylands: Shrub encroachment and effects of fire and vegetation removal in a desert grassland. J. Geophys. Res.: Biogeosci. 117. Available from: https://doi.org/10.1029/2012JG002002.

Sayer, E.J., Wagner, M., Oliver, A.E., Pywell, R.F., James, P., Whiteley, A.S., et al., 2013. Grassland management influences spatial patterns of soil microbial communities. Soil Biol. Biochem. 61, 61–68.

Sayer, E.J., Oliver, A.E., Fridley, J.D., Askew, A.P., Mills, R.T., Grime, J.P., 2017. Links between soil microbial communities and plant traits in a species-rich grassland under long-term climate change. Ecol. Evol. 7 (3), 855–862.

Schimel, J., Balser, T.C., Wallenstein, M., 2007. Microbial stress response physiology and its implications for ecosystem function. Ecology 88, 1386–1394.

Schrama, M., Cordlandwehr, V., Visser, E., Elzenga, T., Vries, Y., Bakker, J., 2013. Grassland cutting regimes affect soil properties, and consequently vegetation composition and belowground plant traits. Plant Soil 366, 401–413.

Schuerings, J., Beierkuhnlein, C., Grant, K., Jentsch, A., Malyshev, A., Peñuelas, J., et al., 2013. Absence of soil frost affects plant-soil interactions in temperate grasslands. Plant Soil 371, 559–572.

Scott, D.A., Rosenzweig, S.T., Baer, S.G., Blair, J.M., 2019. Changes in potential nitrous oxide efflux during grassland restoration. J. Environ. Qual. 48, 1913–1917. Available from: https://doi.org/10.2134/jeq2019.05.0187.

Smith-Ramesh, L.M., Reynolds, H.L., 2017. The next frontier of plant–soil feedback research: unraveling context dependence across biotic and abiotic gradients. J. Veg. Sci. 28, 484–494.

Smith, D.L., Johnson, L.C., 2003. Expansion of *Juniperus virginiana* L. in the great plains: changes in soil organic carbon dynamics. Glob. Biogeochem. Cycles 17. Available from: https://doi.org/10.1029/2002GB001990.

Smith, L.M., Reynolds, H.L., 2015. Plant-soil feedbacks shift from negative to positive with decreasing light in forest understory species. Ecology 96, 2523–2532.

Sperry, L.J., Belnap, J., Evans, R.D., 2006. *Bromus tectorum* invasion alters nitrogen dynamics in an undisturbed arid grassland ecosystem. Ecology 87, 603–615.

Steffens, M., Kölbl, A., Totsche, K.U., Kögel-Knabner, I., 2008. Grazing effects on soil chemical and physical properties in a semiarid steppe of Inner Mongolia (P.R. China). Geoderma 143, 63–72.

Steinauer, K., Tilman, D., Wragg, P.D., Cesarz, S., Cowles, J.M., Pritsch, K., et al., 2015. Plant diversity effects on soil microbial functions and enzymes are stronger than warming in a grassland experiment. Ecology 96 (1), 99–112.

Stinson, K.A., Campbell, S.A., Powell, J.R., Wolfe, B.E., Callaway, R.M., Thelen, G.C., et al., 2006. Invasive plant suppresses the growth of native tree seedlings by disrupting belowground mutualisms. PLoS Biol. 4, 727–731.

Strong, A.L., Johnson, T.P., Chiariello, N.R., Field, C.B., 2017. Experimental fire increases soil carbon dioxide efflux in a grassland long-term multifactor global change experiment. Glob. Change Biol. 23, 1975–1987.

Suding, K.N., Stanley Harpole, W., Fukami, T., Kulmatiski, A., Macdougall, A.S., Stein, C., et al., 2013. Consequences of plant-soil feedbacks in invasion. J. Ecol. 101, 298–308.

Suttie, J.M., Reynolds, S.G., Batello, C., 2005. Grasslands of the world (Plant Production and Protection Series No. No. 34). Food and Agriculture Organization of the United Nations, Rome.

Tesei, G., D'Ottavio, P., Toderi, M., Ottaviani, C., Pesaresi, S., Francioni, M., et al., 2020. Restoration strategies for grasslands colonized by Asphodel-dominant communities. Grassl. Sci. 66, 54–63. Available from: https://doi.org/10.1111/grs.12252.

Thomas, A.D., 2012. Impact of grazing intensity on seasonal variations in soil organic carbon and soil CO_2 efflux in two semiarid grasslands in southern Botswana. Philos. Trans. R. Soc. B: Biol. Sci. 367 (1606), 3076–3086.

Tilman, D., Hill, J., Lehman, C., 2006. Carbon-negative biofuels from low-input high-diversity grassland biomass. Science 314, 1598–1600.

Török, P., Vida, E., Deák, B., Lengyel, S., Tóthmérész, B., 2011. Grassland restoration on former croplands in Europe: an assessment of applicability of techniques and costs. Biodivers. Conserv. 20, 2311–2332. Available from: https://doi.org/10.1007/s10531-011-9992-4.

Van der Heijden, M.G.A., Bardgett, R.D., van Straalen, N.M., 2008. The unseen majority: soil microbes as drivers of plant diversity and productivity in terrestrial ecosystems. Ecol. Lett. 11, 296–310.

Vandermeer, J., 1992. The Ecology of Intercropping. Cambridge University Press, Cambridge.

Vargas, R., Collins, S.L., Thomey, M.L., Johnson, J.E., Brown, R.F., Natvig, D.O., et al., 2012. Precipitation variability and fire influence the temporal dynamics of soil CO_2 efflux in an arid grassland. Glob. Change Biol. 18, 1401–1411.

Vasquez, E., Sheley, R., Svejcar, T., 2008. Nitrogen enhances the competitive ability of Cheatgrass (*Bromus tectorum*) relative to native grasses. Invasive Plant. Sci. Manage. 1, 287–295.

Viglizzo, E.F., Lertora, F., Pordomingo, A.J., Bernardos, J.N., Roberto, Z.E., Del Valle, H., 2001. Ecological lessons and applications from one century of low external-input farming in the pampas of Argentina. Agric. Ecosyst. Environ. 83, 65–81.

Virágh, K., Tóth, T., Somodi, I., 2011. Effect of slight vegetation degradation on soil properties in *Brachypodium pinnatum* grasslands. Plant Soil 345, 303–313.

Walter, J., Hein, R., Beierkuhnlein, C., Hammerl, V., Jentsch, A., Schädler, M., et al., 2013. Combined effects of multifactor climate change and land-use on decomposition in temperate grassland. Soil Biol. Biochem. 60, 10–18.

Wang, W., Fang, J., 2009. Soil respiration and human effects on global grasslands. Glob. Planet. Change 67, 20–28.

Wang, X., Zhang, Y., Jiang, J., Yang, W., Guo, H., Hu, Y., 2009. Effects of spring–summer grazing on longitudinal dune surface in southern Gurbantunggut Desert. J. Geograph. Sci. 19 (3), 299–308.

Wang, Z., Ji, L., Hou, X., Schellenberg, M.P., 2016. Soil respiration in semiarid temperate grasslands under various land management. PLoS One 11.

Ward, S.E., Smart, S.M., Quirk, H., Tallowin, J.R.B., Mortimer, S.R., Shiel, R.S., et al., 2016. Legacy effects of grassland management on soil carbon to depth. Glob. Change Biol. 22, 2929–2938.

Wardle, D.A., Bardgett, R.D., Klironomos, J.N., Setälä, H., Van der Putten, W.H., Wall, D. H., 2004. Ecological linkages between aboveground and belowground biota. Science 304, 1629–1633.

Waring, B.G., Averill, C., Hawkes, C.V., 2013. Differences in fungal and bacterial physiology alter soil carbon and nitrogen cycling: insights from meta-analysis and theoretical models. Ecol. Lett. 16, 887–894.

White, R.P., Murray, S., Rohweder, M., 2000. Pilot Analysis of Global Ecosystems: Grassland Ecosystems. World Resources Institute, Washington, DC.

Wu, T., Ayres, E., Bardgett, R.D., Wall, D.H., Garey, J.R., 2011. Molecular study of worldwide distribution and diversity of soil animals. Proc. Natl. Acad. Sci. U.S.A. 108, 17720–17725.

Zaady, E., Eldridge, D.J., Bowker, M.A., 2016. Effects of local-scale disturbance on biocrusts. Biological Soil Crusts: An Organizing Principle in Drylands. Springer, Cham, pp. 429–449.

Zhang, Y.M., Wang, H.L., Wang, X.Q., Yang, W.K., Zhang, D.Y., 2006. The microstructure of microbiotic crust and its influence on wind erosion for a sandy soil surface in the Gurbantunggut Desert of Northwestern China. Geoderma 132 (3–4), 441–449.

Zhang, L., Wang, H., Zou, J., Rogers, W.E., Siemann, E., 2014. Non-native plant litter enhances soil carbon dioxide emissions in an invaded annual grassland. PLoS One 9, e92301.

Zhang, P., Tang, J., Sun, W., Yu, Y., Zhang, W., 2015. Differential effects of conservational management on SOC accumulation in the grasslands of China. PLOS One 10, e0137280.

Zhang, Y., Aradottir, A.L., Serpe, M., Boeken, B., 2016. Interactions of biological soil crusts with vascular plants. Biological Soil Crusts: An Organizing Principle in Drylands. Springer, Cham, pp. 385–406.

Zhou, Q., Daryanto, S., Xin, Z., Liu, Z., Liu, M., Cui, X., et al., 2017. Soil phosphorus budget in global grasslands and implications for management. J. Arid. Environ. 144, 224–235.

Landscape degradation and restoration

5

John A. Stanturf

Institute of Forestry and Rural Engineering, Estonian University of Life Sciences, Tartu, Estonia

5.1 Introduction

An estimated 25% of the world's land area is degraded (FAO, 2011b), threatening global sustainability. More than 50% of the global ice-free land area has been directly modified by humans (Hooke et al., 2012) and 75% of the terrestrial environment has been severely altered by human actions (IPBES, 2019). Agricultural expansion and intensification over the last several millennia has been the obvious culprit (Goldewijk, 2001) with an expansion of settlements and infrastructure increasing in importance since the Industrial Revolution (Kelley and Williamson, 1984; Seto et al., 2012). Deforestation, desertification, soil erosion, loss of productivity potential, biodiversity loss, water shortage, and soil pollution are ongoing processes associated with landscape degradation. Responding to the adverse consequences of these processes requires a two-pronged approach: (1) avoiding or at least reducing degradation and (2) restoring degraded ecosystems.

Policy initiatives, including the Changwon Initiative of the United Nations Convention to Combat Desertification (UNCCD), aim to achieve net land-degradation neutrality (LDN) by 2030. The objective of LDN is to maintain or improve the condition of land resources, including restoration of natural and semi-natural ecosystems (Safriel, 2017; Cowie et al., 2018). Similarly, the 2010 Strategic Plan of the Convention on Biological Diversity set a goal of no net biodiversity loss, and net positive impacts on biodiversity; the Aichi Target 15 specifically calls for countries to restore at least 15% of degraded lands by 2020 (CBD, 2010).

The burgeoning international interest in halting and reversing degradation and restoring landscapes has focused attention on forest landscapes. The Bonn Challenge, resting on a foundation of forest landscape restoration (FLR), is a voluntary effort to realize these international commitments by bringing 150 million ha of the world's deforested and degraded land into restoration by 2020, and 350 million ha by 2030 (Mansourian et al., 2017a). By embedding these efforts into national development objectives to meet Sustainable Development Goals, nations potentially can contribute to meeting their Aichi Biodiversity Targets and the LDN goal. Many countries have included forest restoration and sustainable forest

Soils and Landscape Restoration. DOI: https://doi.org/10.1016/B978-0-12-813193-0.00005-9

management in their Nationally Determined Contributions under the Paris Climate Agreement. One unexplored contribution is to consider FLR as "green infrastructure" under the Sendai Framework on Disaster Risk Reduction, an especially important potential in seismically active mountainous regions (e.g., Thurman, 2011).

Notwithstanding this impressive international interest, there are underlying tensions that impede clear progress on meeting these goals. A lack of consensus on what constitutes degradation makes it difficult to estimate where and how great the need is for restoration (Hudson and Alcántara-Ayala, 2006; Thompson et al., 2013; Zerga, 2015; Obalum et al., 2017). Restoration is a means to attaining sustainable, productive landscapes, not an end unto itself; yet stakeholders often have differing views on goals, for example, between production versus protection functions (e.g., Mansourian, 2018). Perhaps most importantly, an emphasis on restoring to past conditions may render forested and other landscapes maladapted to future climatic conditions. Thus current restoration programs are distinguished by different approaches and goals, depending on a wide range of perceived degradation levels and forest conditions (Stanturf et al., 2014b; Stanturf, 2016).

Degradation and restoration can be visualized as parallel trajectories (Stanturf, 2016) except that they are not necessarily linear, for example, the hysteresis exhibited by many indicators such as soil organic matter (SOM) content; degradation often occurs quickly but recovery, inevitably, is a long process. Degradation and restoration processes have both intrinsic technical (e.g., physical and biological) as well as extrinsic social elements. For example, soil formation, development, and erosion are natural processes but humans modify and frequently accelerate these processes (Blaikie and Brookfield, 1987). An often contentious question in restoration is what is the end point that constitutes successful restoration? Whether the goal is historic fidelity or future adaptation, the answer is clearly a social choice (Stanturf, 2015).

The twin processes of degradation and restoration operates over several levels of spatial scale, confounding any discussion of their nature and extent as well as choosing appropriate responses. Tempering what is measured and deemed significant is the necessity of examining degradation and restoration from a particular spatial perspective, whether of the soil series or land unit, forest stand or habitat type, watershed or ecosystem, or the landscape (Ghazoul and Chazdon, 2017). Temporal scale is important as well, for example, when differentiating between short-term forest cover removal by ongoing forest management and long-term or permanent change by agricultural conversion.

Less severe forms of degradation occur when land use remains stable but resources are unsustainably managed. Sustainability has been defined as meeting the needs of present and future generations while substantially reducing poverty and conserving the planet's life support systems (Bettencourt and Kaur, 2011; Kates, 2011). Arguments about the nature of sustainability—weak versus strong—have focused on substitutability between the economy and the environment or whether natural capital and manufactured capital are interchangeable (Ayres et al., 2001). Applying these concepts to deciding what constitutes sustainable management (or development) of natural resources in a simplistic fashion is to ask how much change in the natural environment is allowed. The more extreme position of *very* strong sustainability holds that any change constitutes degradation. Accepting a weak sustainability criterion suggests that all change is allowable if it provides material benefit. A compromise version of strong sustainability holds that at least some ecosystems and environmental assets are critical because they are essential and irreplaceable (Ayres et al., 2001). The question then becomes of two parts: what is the baseline and how much departure from the baseline is allowed before it becomes degradation.

Sustainable management is usually defined in terms of ecological, social, and economic criteria (Swart et al., 2004). Some argue that sustainable management is a relative concept that must be evaluated within a context of multiple social demands and rapid environmental change (e.g., Loucks, 2000). The multiple impacts and trade-offs resulting from management actions must be identified so that social or political decision-making processes can resolve conflicts (Loucks, 2000; Emborg et al., 2012; Mansourian, 2017). A full accounting of sustainable management is beyond the scope of this chapter; the emphasis will be on biophysical criteria. The primary focus will be on the landscape scale and change over the long term; however landscape changes are usually the cumulative effects of activities undertaken lower in the spatial hierarchy. Different perspectives will be examined, in particular the nature of degradation processes, drivers, and impacts. Avoiding degradation is always preferable to attempting post-degradation restoration and investing in restoration is pointless unless the underlying cause of degradation has been addressed and halted, or at least mitigated. Nevertheless, restoration of the billions of hectares of degraded landscapes is an urgent need now to recover some of the estimated US\$4.3–20.2 trillion of ecosystem services lost each year to use land change and degradation (ELD, 2015).

5.2 **The landscape perspective**

Landscapes are coupled socioecological systems (Liu et al., 2007; Yang et al., 2018) that include physical features (e.g., mountains, streams, forests, and soils) and constructed features (e.g., buildings, roads, drainage ditches, and mines) together in a particular place as well as connections to other places. Our dependence on the natural environment is widely understood (Fischer et al., 2015), as is our increasing ability, for good or ill, to influence it at local to global scales (Steffen et al., 2015) although there may be less agreement on our ethical obligations toward the environment (Stenmark, 2017).

Landscapes as social constructions (Greider and Garkovich, 1994; Buijs et al., 2006; Halofsky et al., 2014; Martín-López et al., 2017) revolve around perceptions of landscape attributes, the meaning given to them, and material relations (how resources are used) (Greider and Garkovich, 1994). Landscape constructs differ among segments of the population, for example, long-term residents versus newcomers (e.g., Abrams et al., 2014). The meaning individuals attribute to a landscape or elements within it has been termed sense of place (Kibler et al., 2018) and the strength of emotional attachment to a place may affect the perception of degradation (Vorkinn and Riese, 2001; Stedman, 2003).

Landscapes are multifunctional; they are not defined just by what is found within a geographical space. Besides internal landscape dynamics, landscapes are also influenced by external factors such as transportation, migration, international agreements, and of course climate heating. A landscape is also defined by outside actors and factors that shape the landscape and most are affected by globalization; choices made by distant consumers may impact a landscape in ways that are not completely apparent (IPBES, 2018).

The physical concept of landscape suggests a relatively large area, possibly on the order of 1000s–10,000s of hectares (Forman and Godron, 1986), but in practice, the size of a landscape varies by the phenomenon under consideration. Especially from the physical perspective, a landscape is a heterogeneous collection of smaller units that are themselves more or less homogeneous. In geomorphology the landscape is defined by the collection of landforms contained within it (Garner, 1974). To an ecologist a landscape is a mosaic of interacting ecosystems (Forman and Godron, 1986) or a spatially heterogeneous area in at least one factor of interest (Turner et al., 2001). Very generally, a landscape comprises land units that are ecologically similar (Zonneveld, 1989).

5.3 What is degradation?

The human footprint dominates most landscapes, whether for good or ill (Kareiva et al., 2007; Bowman et al., 2011; Hooke et al., 2012; Lewis et al., 2015; Steffen et al., 2015). The human footprint is most apparent in the removal or fragmentation of native vegetation (Riitters et al., 2000; Cochrane, 2001), the simplification and reduction in complexity of managed natural systems (Parrott and Meyer, 2012; Fahey et al., 2018), and in the physical manipulation of soils (Yaalon, 2007). Degradation in dryland areas is termed desertification (D'Odorico et al., 2002; UNCCD, 2017) although the process is more pervasive; occurrence is widespread, not just at desert borders (Grainger et al., 2000; Sivakumar, 2007; Middleton, 2018). Degradation and recovery occur across scales and drivers are interlinked (Ghazoul and Chazdon, 2017), overlain by contemporary disturbances (Fig. 5.1), legacies of the past (Jõgiste et al., 2017; Frelich et al., 2018), and restoration efforts, resulting in landscape mosaics of land units in differing states.

Recent attempts to define degradation, for example, forest degradation, have included efforts to map or delineate degraded lands (Cherlet et al., 2013;

FIGURE 5.1

Salvage logging a beetle-killed forest in southern Germany after a heat wave and severe drought.

Photo John Stanturf.

Hansen et al., 2013; Gibbs and Salmon, 2015; Liu et al., 2015). Although large areas of degraded lands clearly exist, the definition of degradation is inconsistent among efforts to map and estimate the extent of degraded areas. In addition to inconsistent definitions the methods used to estimate the area of degraded land have imposed technical constraints on mapping, especially the spatial and temporal resolution of remotely sensed data such as LANDSAT and MODIS (Vrieling, 2006; Lanfredi et al., 2015; Liu et al., 2015; Mitchell et al., 2017). Over time, definitions of degradation have changed depending on context and policy objectives (Hudson and Alcántara-Ayala, 2006). A general definition of degradation is the loss or diminishment of functions, usually provisioning or regulating functions (UNEP, 1992). Embedded in this definition is the relationship of degraded to a previous nondegraded state, that is, degradation is relative to some baseline condition or concept. On a geological timescale, natural processes of erosion and sediment transport are offset by soil formation, volcanism, and tectonic uplift. Induced (i.e., anthropogenic) degradation is a faster process that results from removal and conversion of native vegetation. Both natural and induced degradation can lead to extreme degradation such as desertification or salinization, but natural and induced degradation differ substantially in their rates of decline (Amundson et al., 2015). Inappropriate land use accelerates natural processes or adds phytotoxic materials to soils (Fitzpatrick, 2002). Anthropogenic processes involving soil change are not all negative; however, draining, terracing, and irrigating have improved physical characteristics and fertilization has improved fertility (Yaalon, 2007).

According to the IPBES (2018), land degradation can occur when a land use change causes a loss of biodiversity, ecosystem functions or services, for example, by deforestation; this also applies when a degraded system is converted to another land use. The terminology to describe resource degradation may stipulate anthropogenic causes or not. For example, compare the definitions of degraded forest from FAO and the CBD:

Changes within the forest which negatively affect the structure or function of the stand or site, and thereby lower the capacity to supply products and/or services (FAO, 2011a; Lamb et al., 2012)

A degraded forest is a secondary forest that has lost, through human activities, the structure, function, species composition or productivity normally associated with a natural forest type expected on that site. Hence, a degraded forest delivers a reduced supply of goods and services from the given site and maintains only limited biological diversity. Biological diversity of degraded forests includes many non-tree components, which may dominate in the under-canopy vegetation. (Convention on Biological Diversity [CBD], 2002, p. 154)

Land and soil degradation have been described variously as processes and conditions depending on disciplinary perspective or policy context (Table 5.1). Historically, land degradation has focused on productivity loss by soil degradation (Hudson and Alcántara-Ayala, 2006; Obalum et al., 2017) especially desertification processes in dryland regions (D'Odorico et al., 2002). Human-caused soil change, which often is degrading, has been termed metapedogenesis (Yaalon and Yaron, 1966) or anthropogenesis (Richter and Yaalon, 2012). Landscape degradation, however, encompasses more than soil degradation and using the terms interchangeably can lead to misinterpretations (Lal et al., 2003). Land degradation focuses on unsustainable uses of forest, rangeland, wetland, or cropland (Ghazoul and Chazdon, 2017). Nevertheless, unsustainable uses of other resources involve degradation of soil resources and close linkage renders them hard to differentiate (FAO, 1976; UNEP, 1992).

Degradation is a complex process and although conceptually it occurs in discrete steps, the activities are not isolated in time or space. Assessing degradation begins from a baseline condition; if the baseline is a pristine state, the first step in degradation would be disruption or removal of native vegetation cover. The baseline need not be pristine conditions; draining land to improve productivity of existing farmland or forest may change plant composition and reduce biodiversity but would be preferable in terms of overall functioning to infrastructure

Table 5.1 Terminology used to describe land and soil degradation, recovery, and restoration in various disciplines and international policy processes.

Theme	Term	Definition	Citation
Land	Land degradation	The many human-caused processes that drive the decline or loss in biodiversity, ecosystem functions, or ecosystem services in any terrestrial and associated aquatic ecosystems.	IPBES (2018)
	Degraded land	The state of land which results from the persistent decline or loss in biodiversity and ecosystem functions and services that cannot fully recover unaided within decadal timescales.	IPBES (2018)
	Desertification	Land degradation in arid, semiarid, and dry subhumid areas resulting from various factors, including climatic variations and human activities.	UNEP (1992)
Soil	Soil acidification	A decrease in acid-neutralizing capacity or an increase in base-neutralizing capacity that results in a decrease in soil pH. Soil acidification can be caused by the production of various acids in the soil (e.g., acid sulphate soils), ion uptake by biota, especially when cations are removed in biomass, nitrogen inputs, and transformation (e.g., fertilizer additions), and the addition of strong dissolved acids, for example, acid deposition in precipitation.	Blake (2005)
	Acid sulfate potential	Soils in which sulfuric acid may be produced, is being produced, or has been produced in amounts that have a lasting effect on main soil characteristics.	Pons (1973)
	Salinization	The process by which water-soluble salts accumulate in the soil; excess salts hinder the growth of crops by limiting their ability to take up water.	Amini et al. (2016)
	Sodic and saline soils	Sodic soils have an exchangeable sodium percentage greater than 6; in saline soils the electrical conductivity exceeds 4 dS/m.	Ghassemi et al. (1995)
	Sedimentation	Processes of detachment, transportation, and deposition of sediment by raindrop impact and flowing water.	Foster and Meyer (1977)
	Lowering water table	Excessive groundwater pumping can lower the water table below the depth required by riparian vegetation for survival and increase CO_2 and methane emissions.	Amlin and Rood (2002)

(Continued)

Table 5.1 Terminology used to describe land and soil degradation, recovery, and restoration in various disciplines and international policy processes. *Continued*

Theme	Term	Definition	Citation
	Waterlogging	The natural flooding and over-irrigation that brings water at underground levels to the surface	Hook (1984)
	Salt water intrusion	The induced flow of saltwater into freshwater aquifers primarily caused by groundwater development near the coast.	Barlow and Reichard (2010)
	Soil fertility decline	A decrease in the levels of soil organic matter, pH, CEC, or plant nutrients.	Hartemink (2006)
	Soil pollution	The presence in the soil of a chemical or substance out of place and/or present at a higher than normal concentration that has adverse effects on any nontargeted organism.	Rodróguez-Eugenio et al. (2018)
	Soil health	The continued capacity of the soil to function as a vital living system, within ecosystem and land-use boundaries, to sustain biological productivity; promote the quality of air and water environments; and maintain plant, animal, and human health.	Doran et al. (2002)
	Soil security	The maintenance or improvement of the world's soil resources so that they can provide sufficient food, fiber, and fresh water, contribute to energy sustainability and climate stability, maintain biodiversity, and deliver overall environmental protection and ecosystem services.	McBratney et al. (2014)
Resources	Deforestation	The removal of a forest or stand of trees where the land is thereafter converted to a nonforest use.	Runyan and D'Odorico (2016)
	Forest degradation	Reduction in the capacity of a forest to produce ecosystem services as a result of anthropogenic and environmental changes.	Thompson et al. (2013)
	Rangeland degradation	Indicated by a shift in species composition, loss of biodiversity, reduction in biomass production less plant cover, low small ruminant productivity, and soil erosion.	Zerga (2015)
Restoration	Revegetation	Plant cover without regard for nativeness of species or structural diversity in order to prevent soil erosion or to provide other functions.	Stanturf et al. (2014b)

(Continued)

Table 5.1 Terminology used to describe land and soil degradation, recovery, and restoration in various disciplines and international policy processes. *Continued*

Theme	Term	Definition	Citation
	Forest landscape restoration	A planned process that aims to regain ecological integrity and enhance human well-being in deforested or degraded landscapes.	IISD (2002)
	Forest and landscape restoration	An active process that brings people together to identify, negotiate, and implement practices that restore an agreed optimal balance of the ecological, social, and economic benefits of forests and trees within a broader pattern of land uses.	Sabogal et al. (2015)
	Ecological restoration	The process of assisting the recovery of an *ecosystem* that has been degraded, damaged, or destroyed.	SERI (2004)
	Functional restoration	Emphasizes the restoration of the underlying abiotic and biotic processes; may result in different structure and composition than the historical reference condition.	Stanturf et al. (2014b)
	Wetland restoration	The return of a wetland and its functions to a close approximation of its original condition as it existed prior to disturbance on a former or degraded wetland site.	NRCS (2010)

CEC, *Cation exchange capacity.*

development on the drained land (Holden et al., 2004; Lõhmus et al., 2015). Causes of degradation, however, may not be straightforward. For example, over-grazing in grasslands is often implicated in desertification, along with accelerated soil erosion (Vetter, 2005; Reid et al., 2014); climate variability, however, can play the dominant role by lengthy periods of lower precipitation reducing productivity and carrying capacity (D'Odorico et al., 2012; Middleton, 2018). Deforestation (D'Odorico et al., 2002; Runyan and D'Odorico, 2016) begins by removing the forest cover, but this is also true of even-aged forest management using clear-cutting. However, deforestation involves another step—the conversion of land use to one that prevents re-growth of the forest vegetation. Sustainable forest management also requires another step following canopy removal, that of regeneration (reforestation) of forest vegetation. Failure to secure adequate regeneration of desirable species would constitute degradation. These examples illustrate that degradation is the complex result of an agent or agents, acting on a system to change conditions relative to a baseline; the most severe degradation often involves a change of land use, over large spatial scale.

Degradation that occurs without a change in land use or cover class can threaten the delivery of ecosystem services such as biodiversity, carbon sequestration, and other provisioning or regulating functions. By one estimate, 761 million ha of managed and natural forests are degraded to some degree (Verdone and Seidl, 2017). In managed systems, long-term sustainability is a goal, and degradation may reduce sustainability by decreasing the capacity for resistance or resilience when disturbed (Millar et al., 2007; O'Hara and Ramage, 2013). For example, the conversion of a native forest to a monospecific plantation clearly results in loss of biodiversity and other functions (Miranda et al., 2017; Manoli et al., 2018). Even less disruptive activities can reduce functioning, for example, harvesting only the largest and best trees without regard for regeneration, so-called "high-grading" (Nyland, 1992; Prévost and Charette, 2018).

Land degradation has generally been associated with mismanagement of natural resources but recent assessments include the effects of climate heating and variability (Grainger et al., 2000; Sivakumar, 2007), alone or in combination with social drivers (MEA, 2005). Recognition is growing of the necessity to strengthen social coping mechanisms in efforts to address biophysical degradation (D'Odorico et al., 2002). Multiple social factors can contribute to land degradation (IPBES, 2018), including globalization (Lenzen et al., 2012), insecure tenure, weak governance (Colfer, 2011; Mansourian, 2017; van Oosten et al., 2018), poor access to markets and financial credit, and demographic change that give rise to extreme poverty, resource scarcity, and inequitable access to resources (e.g., Wardell and Lund, 2006; Gaither et al., 2019).

5.3.1 Extent of degradation

The direct human impact on terrestrial ecosystems is extensive, with as much as 75% of global landscapes affected and only the most inaccessible regions free of human influence (Venter et al., 2016). Even areas set aside for nature protection are impacted (Jones et al., 2018). Conversion of land from native ecosystems such as forests, rangelands and wetlands to agriculture, urban areas, and mineral resource extraction are forms of degradation (Gibbs et al., 2010; Ramankutty et al., 2018). Loss of function, especially productivity, results from effects on the soil resource including erosion and declining SOM. Some sensitive soils are especially prone to declining function from mismanagement. Soil degradation is apparent at multiple scales and adversely affects biodiversity and ecosystem services. Land use change and species extinctions set the stage for most forms of land degradation and biodiversity losses (Hooper et al., 2012).

5.3.2 Land use change

By one estimate, between 1765 and 2005 global forest land cover decreased from 45.4 million km^2 to 30 million km^2 (Fig. 5.2), with approximately 21 km^2 of primary forest remaining (Meiyappan and Jain, 2012). Over that same period,

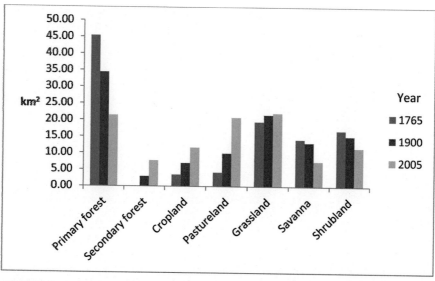

FIGURE 5.2

Change in land use from 1765 to 2005 (data from Meiyappan and Jain, 2012). Greatest change has been in primary forest although other native vegetation types (e.g., savanna and shrubland) have also decreased. Cropland and grazing lands (pasture and grassland) have increased, as has secondary forests.

cropland grew from an estimated $3.5-14.5$ km^2. Of course, there is considerable uncertainty in these estimates, especially in the earlier dates. Conversion to agriculture has been extensive and wetlands in particular have been highly impacted; an estimated 87% of the wetland area has been lost in the last 300 years (Davidson, 2014; Hu et al., 2017; IPBES, 2018). The greatest potential for further conversions to agriculture is in regions with suitable land remaining, including Central and South America, sub-Saharan Africa, and Asia (Lambin and Meyfroidt, 2011; Tscharntke et al., 2012). Increasing human population and settlement trends away from rural to urban and periurban areas also drive land conversions from native vegetation or agriculture (DeFries et al., 2010; Martellozzo et al., 2018; Mughal, 2019).

5.3.3 Erosion

Degradation of agricultural land follows two pathways, change of use by conversion to mined land or urban infrastructure or decline in soil health and productivity (Hartemink, 2006; Daliakopoulos et al., 2016; Montanarella et al., 2016). To a lesser but nontrivial extent, these same pathways apply to managed forests and rangeland (Schoenholtz et al., 2000; Nsengimana et al., 2018). The most

FIGURE 5.3

Intensive cropping for cotton, followed by the a civil war, infestation of the invasive boll weevil pest, and the 1930s Global Economic Depression caused severe land degradation and wide-spread land abandonment in the southern United States. Severe erosion (left photo), including deep gullies (middle photo) resulted. Much of the land was subsequently planted to *Pinus* spp. (right photo). These photos are of the Calhoun Experimental Forest in South Carolina, United States.

Photos USDA Forest Service; Nelson, D.R., O'Neill, K.P., Coughlan, MR., Lonneman, M.C., Meyers, Z., and U.S. Department of Agriculture, Forest Service. USDA Forest Service Photographs from the Calhoun Experimental Forest, South Carolina, 1932–1987. Inter-university Consortium for Political and Social Research (distributor), Ann Arbor, MI, 2016-12-07. https://doi.org/10.3886/E100276V3].

pervasive degradation process is soil erosion, the accelerated removal of topsoil from the land surface through water, wind, or tillage. Erosion mostly occurs when the soil surface is exposed from loss of protective cover and particles are entrained and moved by water or wind flowing over the surface (Fig. 5.3). Tillage erosion is confined to down-slope redistribution of soil within a field. Eventually eroded soil comes to rest and sedimentation occurs. Estimates of global soil erosion by the three mechanisms are $20-30$ Gt year^{-1} by water, 5 Gt year^{-1} by tillage, and 2 Gt year^{-1} by wind erosion (FAO and ITPS, 2015).

Soil erosion by water occurs on all continents where hilly areas are cropped, and this type of erosion is most severe in the tropics due to a combination of high rainfall and population pressure that results in steeper areas being cultivated than in temperate countries (van Oost et al., 2007). Similarly, rangelands and pasturelands in hilly tropical and subtropical areas are more intensively used than in temperate regions and erosion rates approach those of tropical croplands (FAO and ITPS, 2015). Soil erosion rates from forested lands are generally low except following disturbances such as logging or wildfires and then are mostly of concern only on steep slopes and particularly from forest roads (Patric, 1976; Luce and Black, 1999; Ice et al., 2004).

Wind erosion is highly variable and hard to estimate. Satellite studies of dust sources have found that natural sources account for more dust emissions (75%) than anthropogenic sources (25%) (Ginoux et al., 2012). Globally, drylands are the most susceptible to wind erosion (D'Odorico et al., 2002; Middleton, 2018). Many dust sources are relics of a wetter climate during the Quaternary and Holocene that formed alluvial deposits by flooding that are now exposed (Prospero et al., 2002; Goudie, 2018). Anthropogenic sources include the dry lake bed of the Aral Sea in Central Asia where water extracted for irrigation left sediments exposed to severe erosion (Micklin, 2010). In addition, poorly designed irrigation systems in the same region led to agricultural abandonment and a shallow saline groundwater table developed, exposing surface soils to wind erosion (Orlovsky et al., 2001).

5.3.4 Decline in soil organic carbon

SOM and its most important constituent, soil organic carbon (SOC), determine many chemical and physical properties of soil such as pH, nutrient storage, structure, aggregate formation, and stability. Soil texture and SOC content are closely related and together determine moisture relations (holding capacity and release rates) and influence water infiltration rates and resistance to water and wind erosion. Due to its role in the global C-cycle, SOC has received much attention in global change research (Schmidt et al., 2011). Land use change and intensity of management practices greatly influence SOC stocks in most soils, generally in the direction of decline (Table 5.2). Conversion from native vegetation to agriculture and pasture causes SOC declines; the magnitude and rate of loss are higher in tropical soils (as much as 50% reduction by the conversion from forest to agriculture) than soils in temperate regions (FAO and ITPS, 2015). Soil texture also influences SOC loss when land use conversion occurs, with greater SOC losses in coarse-textured than heavy-textured soils (FAO and ITPS, 2015). In addition to disturbances of the soil that increase decomposition rates of SOM, maintenance

Table 5.2 Gains and losses in soil organic carbon after land use change, in percentage.

Gain of SOC (%)				Loss of SOC (%)			
Forest	→	Pasture	+8	Forest	→	Cropland	− 42
Cropland	→	Pasture	+19	Pasture	→	Cropland	− 59
Cropland	→	Plantation	+18	Pasture	→	Plantation	− 10
Cropland	→	Forest	+53	Forest	→	Plantation	− 13

Degradation is relatively fast, recovery and restoration require years to centuries. Magnitude of change is affected by annual precipitation, soil texture, plant species, and soil depth sampled. SOC, Soil organic carbon.
Data from Guo, L.B., Gifford, R., 2002. Soil carbon stocks and land use change: a metaanalysis. Glob. Change Biol. 8, 345–360 (Guo and Gifford, 2002).

and accretion of SOC stocks depends upon the annual rates of litter return; besides removal of the overstory by harvesting or burning, reduced productivity, and decreased litter production will lower SOC over time (Jandl et al., 2007; Lal et al., 2015; Turetsky et al., 2015; Sanderman et al., 2017; Zomer et al., 2017).

5.3.5 Sensitive soils

Some soils are especially vulnerable to further degradation due to their inherent properties that make them inhospitable to vegetation establishment. Besides infertile sandy soils, other soils affected by salts (i.e., saline and sodic) or the potential to develop extremely low pH (i.e., acid sulphate potential soils) are prone to bare surfaces and erosion. Salt-affected soils are widespread, occurring in more than 100 countries and affecting an estimated 1 billion ha (FAO and ITPS, 2015). There are two kinds of salt-affected soils, saline, and sodic. Saline soils have levels of soluble salts (primarily $NaCl$ and Na_2SO_4) and electrical conductivity exceeding 4 dSm^{-1}. Sodic soils have a low salt content but sodium is present in ionic form such that exchangeable sodium percentage (ESP) exceeds 6% (Fitzpatrick, 2002). Soil salinity and sodicity problems occur naturally under low rainfall conditions. Salinity causes dehydration of plants and even low levels of salinity may affect plant growth and species composition (Ross et al., 1994; Bui, 2013; Daliakopoulos et al., 2016). Sodic soils have low permeability and poor structure that affect plant growth in a similar fashion as salinity (Fitzpatrick, 2002).

Degradation of saline soils can be caused by poorly designed irrigation systems (secondary salinity) and by converting deep-rooted native vegetation (Fig. 5.4) to shallow rooted agricultural crops (Fitzpatrick, 2002; Hatton et al., 2003). Irrigation systems that do not provide sufficient water in excess of plant

FIGURE 5.4

Salinized landscapes (left photo) in Western Australia caused by deforestation of the deep-rooted native forests (mostly Eucalyptus) and farming for cereal crops. Salts accumulated in the soil due to lack of flushing and water tables rose (right photo).

Photos Richard Harper, Murdoch University.

demand allow salts to accumulate in the subsoil and eventually affect plant growth (Qadir and Oster, 2004; Wichelns and Qadir, 2015). This is especially a problem in arid and semiarid regions [e.g., Central Asia; Orlovsky et al. (2001)].

Potential acid sulfate soils (PASS) occur naturally, forming under waterlogged conditions in sediments or peats. They contain iron sulfide minerals or their oxidation products. PASS occur mostly in coastal areas and in their natural state, they pose no problems. However, if these minerals are exposed to air by drainage or other lowering of the water table, the sulfides react with oxygen to form sulfuric acid (Dent and Pons, 1995). There are 17–24 million ha of PASS globally, particularly in Africa, eastern Australia, Asia, Latin America, and Europe (Andriesse and van Mensvoort, 2002; Fältmarsch et al., 2008). Acid sulfate soils also form as a result of mining activities from metal sulfides in rocks (Dent, 1986).

Land development projects that drain lowlands and expose the PASS to oxidizing conditions cause extremely low pH and mobilize toxic metal ions inimical to plant growth and aquatic systems (Michael, 2013). Poorly sited and designed rice farms and shrimp ponds in Southeast Asia, often in mangroves underlain by PASS (Fig. 5.5), have been abandoned and the exposed soils and acid drainage are obstacles to rehabilitation of these sites to other uses (Stevenson et al., 1999). Efforts to farm these soils are challenging due to infertility and toxicities as well as effects on human health (Ljung et al., 2009). The key is managing water levels to keeps the PASS from exposure or once exposed, adequately flushing the PASS with freshwater (Burbridge, 2012).

FIGURE 5.5

Exposure of PASS after land conversion from forest to rice farms or shrimp ponds in Indonesia. Oxidation of the PASS resulted in extremely acid conditions leading to poor productivity land abandonment in Southeast Asia. *PASS*, Potential acid sulphate soils.

Photo Peter Burbridge, Newcastle University.

5.3.6 **Contamination**

Soil contamination is a global problem (FAO, 2011b; FAO and ITPS, 2015). Various industrial and agricultural processes contaminate soils directly on local sites through application and movement of agricultural chemicals, industrial waste disposal, mining, military activities, or accidents; contamination can occur indirectly through atmospheric deposition and floods. Indirect or diffuse contaminants affect all ecosystems and runoff and airborne emissions from local sites can contaminate off-site ecosystems. Inorganic and organic compounds that are byproducts of industrial processes include metallic trace-elements, radionuclides, and xenobiotic molecules, including nanoparticles. Organic wastes such as untreated biosolids that are applied to soil are health hazards that can spread infectious diseases. The increasing use of veterinary and human antibiotics and hormones that enter the ecosystem through waste streams adversely impact ecosystems and human health including the incidence of antibiotic-resistant diseases (Jjemba, 2002; Berendonk et al., 2015).

5.3.7 **Biodiversity loss**

The main causes of biodiversity loss are habitat loss and fragmentation, overexploitation of species by humans, pollution, and the impact of invasive species and diseases of wild organisms (IPBES, 2018). Biodiversity loss at landscape scale is driven by many of the other degradation processes already described that cause the loss of habitat, especially of intact ecosystems including areas of uninterrupted forest cover. Of particular concern is the global threat to insect pollinators (Steffan-Dewenter and Westphal, 2008; Potts et al., 2010; Bartomeus and Winfree, 2013) and bird and mammal seed dispersers (Wunderle Jr, 1997; McConkey and O'Farrill, 2016; Rumeu et al., 2017).Tropical forests are some of the most biodiverse ecosystems and are subject to diverse impacts (Bellard et al., 2014; Visconti et al., 2015; IPBES, 2019). For example, large areas of tropical forests in Latin America have been converted to livestock pasture (Grau and Aide, 2008; Gasparri and de Waroux, 2015) and to oil palm plantations in Southeast Asia (Fitzherbert et al., 2008; Richards and Friess, 2015; Manoli et al., 2018). Soil biota (e.g., bacteria, fungi, protozoa, insects, worms, other invertebrates and mammals) are critical to ecosystem functioning, including plant diversity, decomposition, nutrient retention, and nutrient cycling (Holden and Treseder, 2013; Zhou et al., 2018). Soils are a large biodiversity reservoir (Bardgett and Van Der Putten, 2014; Coleman et al., 2018), and losses in soil biodiversity affect multiple ecosystem functions (Wagg et al., 2014; Creamer et al., 2016). Intensified land management, in particular agricultural intensification, may reduce soil biodiversity (Tsiafouli et al., 2015). Although there is considerable functional redundancy in soil communities, some specialized processes may be adversely impacted by intensive management (Nielsen et al., 2011; Veresoglou et al., 2015).

5.3.8 **Desertification**

Desertification is a degradation process that occurs in arid, semiarid, and dry sub-humid areas, not simply at desert margins. Although there are conflicting estimates of the extent of global desertification (largely due to different definitions), an estimated 10%−20% of drylands are degraded and the exposed soil causes dust storms (Fig. 5.6) and air quality problems for people thousands of kilometers away as well as locally (Adeel et al., 2005; MEA, 2005). Desertification has various causes, including climatic variations and human activities, although earlier definitions limited desertification to resource mismanagement (D'Odorico et al., 2002). Loss of vegetative cover and conversion to bare soil begins the desertification process as it exposes the surface to erosion that preferentially removes fine soil particles which are nutrient rich, and simultaneously decreasing soil moisture through compaction and loss of permeability (D'Odorico et al., 2002; Middleton, 2018). In some cases salts accumulate. Poorly managed agriculture and grazing

FIGURE 5.6

Dust storms are commonplace in drylands of Central Asia, such as this one near Elsen tasarkhai, Uvurkhangai province, Mongolia. Exposed surface soils are susceptible to wind erosion and long-range transport.

Photo Ganchudur Tsetsegmaa, Institute of Geography and Geoecology, Mongolian Academy of Sciences.

are the main human culprits that create bare soil conditions, exacerbated by climate variability in drylands such as decadal droughts (Sivakumar, 2007).

5.4 Restoration

The realization that simply preserving extant nondegraded natural ecosystems will be insufficient to address global environmental problems has led to increased reliance on passive and active restoration to counter the effects of landscape degradation (Wiens and Hobbs, 2015; Mappin et al., 2019). Two international voluntary approaches are prominent, LDN (Cowie et al., 2018), and FLR (Besseau et al., 2018). Both approaches are similar in their focus on landscapes, integration of ecological and social values, use of participatory approaches, and desire to reverse environmental degradation. Nevertheless LDN and FLR arise from different disciplinary backgrounds and international political contexts with resulting different kinds of targets (Table 5.3).

5.4.1 Land degradation neutrality

LDN is defined as "a state whereby the amount and quality of land resources necessary to support ecosystem functions and services and enhance food security remain stable or increase within specified temporal and spatial scales and ecosystems" (UNCCD, 2016). It was introduced by the UN Convention to Combat Desertification (UNCCD) in response to the declining health and productivity of land, as a target for Sustainable Development Goal 15 to "Protect, restore and promote sustainable use of terrestrial ecosystems, sustainably manage forests, combat desertification, and halt and reverse land degradation and halt biodiversity loss" (United Nations, 2015; Cowie et al., 2018). The LDN concept is one of no-net-loss of land-based natural capital by reducing, avoiding, and reversing degradation from a baseline year of 2015 with LDN achieved by 2030. Thus the LDN approach recognizes that land will continue to be degraded, but because of other offsets, there will be a zero net loss (Cowie et al., 2018).

The genesis of LDN is the UNCCD; hence, it develops from a background emphasizing desertification and drylands. Nevertheless, LDN can be applied to all land types (Cowie et al., 2018). Commitment to LDN is voluntary, by nation states in a top-down process, to be applied at the level of the catchment or administrative province in spatial planning with national reporting (Cowie et al., 2018). Monitoring and reporting will be on an area basis; that is, whether an area of land experiences significant degradation or improvement in terms of agreed upon indicators; LDN does not consider the magnitude of change (Orr et al., 2017) nor has it been demonstrated to be feasible (Safriel, 2017).

Table 5.3 Forest landscape restoration (FLR) and land-degradation neutrality (LDN) principles compared on the basis of design elements (FLR based on Besseau et al., 2018; LDN based on Cowie et al., 2018; design elements taken from Stanturf et al., 2019).

Design element	FLR principles	FLR description	LDN principles
Scale	Focus on landscapes	FLR takes place within and across entire landscapes, not individual sites, representing mosaics of interacting land uses and management practices under various tenure and governance systems. It is at this scale that ecological, social and economic priorities can be balanced.	Set national LDN targets based on national circumstances.
Governance	Engage stakeholders and support participatory governance	FLR actively engages stakeholders at different scales, including vulnerable groups, in planning and decision-making regarding land use, restoration goals and strategies, implementation methods, benefit sharing, monitoring and review processes.	Reinforce responsible governance: protect human rights, including tenure rights; develop a review mechanism; and ensure accountability and transparency. Protect the rights of vulnerable and marginalized land users.
Visioning	Restore multiple functions for multiple benefits	FLR interventions aim to restore multiple ecological, social and economic functions across a landscape and generate a range of ecosystem goods and services that benefit multiple stakeholder groups.	Seek solutions that provide multiple environmental, economic and social benefits, and minimize trade-offs. Base land use decisions on multivariable assessments, considering land potential, land condition, resilience, social, cultural and economic factors.
Conceptualizing	Maintain and enhance natural ecosystems	FLR does not lead to the conversion or destruction of natural forests or other ecosystems. It	Maintain or enhance land-based natural capital.

(Continued)

Table 5.3 Forest landscape restoration (FLR) and land-degradation neutrality (LDN) principles compared on the basis of design elements (FLR based on Besseau et al., 2018; LDN based on Cowie et al., 2018; design elements taken from Stanturf et al., 2019). *Continued*

Design element	FLR principles	FLR description	LDN principles
	within landscapes	enhances the conservation, recovery, and sustainable management of forests and other ecosystems.	For neutrality the LDN target equals (is the same as) the baseline. Neutrality is the minimum objective: countries may elect to set a more ambitious target. Counterbalance anticipated losses in land-based natural capital with interventions to reverse degradation, to achieve neutrality. Manage counterbalancing at the same scale as land use planning. Counterbalance "like for like" (within the same land type). Apply the response hierarchy in devising interventions for LDN: Avoid > Reduce > Reverse degradation.
Acting	Tailor to the local context using a variety of approaches	FLR uses a variety of approaches that are adapted to the local social, cultural, economic and ecological values, needs, and landscape history. It draws on latest science and best practice, and traditional and indigenous knowledge, and applies that information in the context of local capacities and existing or new governance structures.	Integrate planning and implementation of LDN into existing land use planning processes.

(Continued)

Table 5.3 Forest landscape restoration (FLR) and land-degradation neutrality (LDN) principles compared on the basis of design elements (FLR based on Besseau et al., 2018; LDN based on Cowie et al., 2018; design elements taken from Stanturf et al., 2019). *Continued*

Design element	FLR principles	FLR description	LDN principles
Sustaining	Manage adaptively for long-term resilience	FLR seeks to enhance the resilience of the landscape and its stakeholders over the medium and long term. Restoration approaches should enhance species and genetic diversity and be adjusted over time to reflect changes in climate and other environmental conditions, knowledge, capacities, stakeholder needs, and societal values. As restoration progresses, information from monitoring activities, research, and stakeholder guidance should be integrated into management plans.	Apply a participatory process: include stakeholders, especially land users, in designing, implementing and monitoring interventions to achieve LDN. Monitor using the three UNCCD land-based global indicators: land cover, land productivity (NPP) and carbon stocks (SOC). Use the "one-out, all-out" approach to interpret the result of these three global indicators. Use additional national and subnational indicators to aid interpretation and to fill gaps for ecosystem services not covered by the three global indicators. Apply local knowledge and data to validate and interpret monitoring data. Apply a continuous learning approach: anticipate, plan, track, interpret, review, adjust, and create the next plan.

NPP, *Net primary productivity*; SOC, *soil organic carbon.*

5.4.2 Forest landscape restoration

FLR emerged in 2000 as a response to dissatisfaction with small-scale ecological restoration; FLR was intentional, multidimensional, and large-scale (Stanturf et al., 2014b; Mansourian et al., 2017b). Originally defined as "a planned process that aims to regain ecological integrity and enhance human well-being in deforested or degraded landscapes" (Mansourian, 2005; Lamb et al., 2012; Mansourian et al., 2017b), the FLR concept has changed as it has been elevated to the forefront in the international policy arena under the Bonn Challenge. FLR has been redefined to make it more inclusive by omitting "planned process" and adding "forest *and* landscape restoration" (Mansourian et al., 2017b). FLR does operate at the landscape level and by its emphasis on broad participation it favors a bottom-up approach. Nevertheless, the Bonn Challenge is a voluntary process aimed at national commitments, similar to LDN (although there are several subnational and private sector commitments). The Bonn Challenge is similar to LDN in another way, with its emphasis on the amount of area treated, on hectares rather than functions restored (Mansourian et al., 2017b).

The FLR concept took hold in 2000 but blossomed internationally with the identification of potential areas for different forms of forest restoration that has been mapped with satellite imagery (Hansen et al., 2010). The greatest potential for forest restoration was located in "mosaic" landscapes characterized by mixes of land uses (Stanturf, 2015) where opportunities for forest, agriculture, and agroforestry abounded (Sayer et al., 2013; Zomer et al., 2014; Zomer et al., 2016; Holl, 2017; Zomer et al., 2017). The international response has been the Bonn Challenge for 150 million ha to "be in restoration" by 2030, the New York Declaration on Forests target of 350 million ha by 2050, and regional initiatives mostly in concert with the Bonn Challenge (Stanturf et al., 2019).

5.4.3 Restoration and recovery techniques

Details of specific restoration and recovery methods are beyond the scope of this chapter but can be summarized in general terms. For example, degraded and deforested land in mosaic landscapes have four recovery trajectories: (1) recover to productive, sustainable agriculture use (possibly through intensification); (2) conversion to mixed cropping systems (agroforestry) with woody perennials integrated into crop and livestock systems; (3) restore to actively managed, productive forests; or (4) restore to passively managed, protected forests. Soil recovery is basic to all pathways; this includes reducing soil loss and improving soil quality and soil health, especially maintaining and increasing organic matter (Doran et al., 2002; Wichelns and Qadir, 2015; Amini et al., 2016; Uda et al., 2017; Zomer et al., 2017). Additional benefits will accrue from soil improvement by recovering hydrological functions of infiltration and water retention and possibly flood risk reduction (Anderson et al., 1976; Bruijnzeel, 2004; Farley et al., 2005; Calder and Aylward, 2006; Sidle et al., 2006). Particularly in drylands, range and pasture recovery will be accelerated by managing grazing pressure and improving productivity of pasture and forage crops (Reid et al., 2014; Zerga, 2015; Middleton, 2018); silvopastoral management may be an option (Fig. 5.7).

In all landscapes, recovery of ecological functions should be a goal that includes biotic interactions specific to the landscape; pollinators and seed dispersers are of particular interest. Removing exotic invasive species, restoring fire (Phillips et al., 2012; Brotons et al., 2013; Archibald, 2016; Albar et al., 2018) and natural inundation regimes are needed especially in forests (Hughes and

FIGURE 5.7

Silvopastoral system of livestock management integrated with floodplain *Populus* plantations in Argentina.

Photo John Stanturf.

Rood, 2003; Oswalt and King, 2005; Skiadaresis et al., 2019). Many passive or active forest restoration and sustainable management techniques have been successfully used to restore productivity, conserve biodiversity, and avoid forest degradation at small scale, but scaling up to the landscape remains challenging (Stanturf et al., 2014a; Chazdon, 2015; Dumroese et al., 2015).

Reversing degradation requires time and consistent effort. Even severely degraded ecosystems will recover functions, given sufficient time on the scale of centuries. Past sociopolitical movements have attempted to reverse degradation processes and restore ecosystems from utilitarian, patriotic, or ethical motivations (Bennett, 1935; Scherer, 1995; Stanturf et al., 2019). The current international efforts seemingly with the greatest momentum are FLR, LDN, and the Aichi biodiversity targets (CBD, 2010). Although each focuses on particular aspects of the effects of human utilization of natural systems, mostly to repair past excesses, but implicitly or explicitly, they also argue for more sustainable utilization in the future and the avoidance of further landscape degradation.

References

Abrams, J., Bliss, J., Gosnell, H., 2014. Reflexive gentrification of working lands in the American West: contesting the 'Middle Landscape'. J. Rural. Commun. Dev. 8, 144–158.

Adeel, V., Safriel, U., Niemeijer, D., White, R., 2005. Millennium ecosystem assessment: desertification synthesis. World Resources Institute, Washington, DC.

Albar, I., Amissah, L., Bowman, D., Charlton, V., Cochrane, M., De Groot, B., et al., 2018. Global Fire Challenges in a Warming World-Summary Note of a Global Expert Workshop on Fire and Climate Change. International Union of Forest Research Organizations, Vienna, Austria.

Amlin, N.M., Rood, S.B., 2002. Comparative tolerances of riparian willows and cottonwoods to water-table decline. Wetlands 22, 338–346.

Amini, S., Ghadiri, H., Chen, C., Marschner, P., 2016. Salt-affected soils, reclamation, carbon dynamics, and biochar: a review. J. Soils Sediment. 16, 939–953.

Amundson, R., Berhe, A.A., Hopmans, J.W., Olson, C., Sztein, A.E., Sparks, D.L., 2015. Soil and human security in the 21st century. Science 348, 1261071.

Anderson, H.W., Hoover, M.D., Reinhart, K.G., 1976. Forests and water: effects of forest management on floods, sedimentation, and water supply. General Technical Report PSW-018. US Department of Agriculture, Forest Service, Pacific Southwest Forest and Range Experiment Station, Berkeley, CA, p. 115.

Andriesse, W., van Mensvoort, M., 2002. Distribution and extent of acid sulphate soils. In: Lal, R. (Ed.), Encyclopedia of Soil Science. Marcel Dekker, New York, pp. 1–6.

Archibald, S., 2016. Managing the human component of fire regimes: lessons from Africa. Phil. Trans. R. Soc. B 371, 20150346. Available from: https://doi.org/10.1098/rstb.2015.0346.

Ayres, R., Van den Berrgh, J., Gowdy, J., 2001. Strong versus weak sustainability: economics, natural sciences, and consilience. Environ. Ethics 23, 155–168.

Bardgett, R.D., Van Der Putten, W.H., 2014. Belowground biodiversity and ecosystem functioning. Nature 515, 505–511.

Barlow, P.M., Reichard, E.G., 2010. Saltwater intrusion in coastal regions of North America. Hydrogeol. J. 18, 247–260.

Bartomeus, I., Winfree, R., 2013. Pollinator declines: reconciling scales and implications for ecosystem services. F1000Research 2, 146. Available from: https://doi.org/10.12688/f1000research.2-146.v1.

Bellard, C., Leclerc, C., Leroy, B., Bakkenes, M., Veloz, S., Thuiller, W., et al., 2014. Vulnerability of biodiversity hotspots to global change. Glob. Ecol. Biogeogr. 23, 1376–1386.

Bennett, H.H., 1935. Elements of Soil Conservation, second ed McGraw-Hill Book Company, Inc, London.

Berendonk, T.U., Manaia, C.M., Merlin, C., Fatta-Kassinos, D., Cytryn, E., Walsh, F., et al., 2015. Tackling antibiotic resistance: the environmental framework. Nat. Rev. Microbiology 13, 310–317.

Besseau, P., Graham, S., Christopherson, T. (Eds.), 2018. Restoring Forests and Landscapes: The Key to a Sustainable Future. International Union of Forest Research Organizations and Global Partnership on Forest and Landscape Restoration, Vienna, Austria.

Bettencourt, L.M., Kaur, J., 2011. Evolution and structure of sustainability science. Proc. Natl. Acad. Sci. U.S.A. 108, 19540–19545.

Blaikie, P., Brookfield, H. (Eds.), 1987. Land Degradation and Society. Methuen, London and New York.

Blake, L., 2005. Acid rain and soil acidification. In: Hillel, D. (Ed.), Encyclopedia of Soils in the Environment. Elsevier, New York.

Bowman, D.M., Balch, J., Artaxo, P., Bond, W.J., Cochrane, M.A., D'Antonio, C.M., et al., 2011. The human dimension of fire regimes on Earth. J. Biogeogr. 38, 2223–2236.

Brotons, L., Aquilué, N., De Cáceres, M., Fortin, M.-J., Fall, A., 2013. How fire history, fire suppression practices and climate change affect wildfire regimes in Mediterranean landscapes. PLoS One 8, e62392.

Bruijnzeel, L.A., 2004. Hydrological functions of tropical forests: not seeing the soil for the trees? Agric. Ecosyst. Environ. 104, 185–228.

Bui, E., 2013. Soil salinity: a neglected factor in plant ecology and biogeography. J. Arid. Environ. 92, 14–25.

Buijs, A.E., Pedroli, B., Luginbühl, Y., 2006. From hiking through farmland to farming in a leisure landscape: changing social perceptions of the European landscape. Landsc. Ecol. 21, 375–389.

Burbridge, P.R., 2012. The role of forest landscape restoration in supporting a transition towards more sustainable coastal development. In: Stanturf, J., Lamb, D., Madsen, P. (Eds.), Forest Landscape Restoration. Springer, Dordrecht, pp. 253–273.

Calder, I.R., Aylward, B., 2006. Forest and floods: moving to an evidence-based approach to watershed and integrated flood management. Water Int. 31, 87–99.

CBD, 2010. Strategic Plan for Biodiversity 2011–2020 and the Aichi Targets. United Nations Convention on Biological Diversity. Montreal, Quebec Canada.

Chazdon, R., 2015. Restoring tropical forests: a practical guide. Ecol. Restor. 33, 118–119.

Cherlet, M., Ivits, E., Sommer, S., Tóth, G., Jones, A., Montanarella, L., et al., 2013. Land-productivity dynamics in Europe. Towards Valuation of Land Degradation in the EU. Publications Office of the European Union, Luxembourg.

Cochrane, M.A., 2001. Synergistic interactions between habitat fragmentation and fire in evergreen tropical forests. Conserv. Biol. 15, 1515—1521.

Coleman, D.C., Callaham Jr., M.A., Crossley Jr., D.A., 2018. Fundamentals of Soil Ecology, third ed. Academic Press, London.

Colfer, C.J.P., 2011. Marginalized forest peoples' perceptions of the legitimacy of governance: an exploration. World Dev. 39, 2147—2164.

Cowie, A.L., Orr, B.J., Sanchez, V.M.C., Chasek, P., Crossman, N.D., Erlewein, A., et al., 2018. Land in balance: the scientific conceptual framework for land degradation neutrality. Environ. Sci. Policy 79, 25—35.

Creamer, R., Hannula, S., Van Leeuwen, J., Stone, D., Rutgers, M., Schmelz, R., et al., 2016. Ecological network analysis reveals the inter-connection between soil biodiversity and ecosystem function as affected by land use across Europe. Appl. Soil Ecol. 97, 112—124.

Daliakopoulos, I., Tsanis, I., Koutroulis, A., Kourgialas, N., Varouchakis, A., Karatzas, G., et al., 2016. The threat of soil salinity: a European scale review. Sci. Total Environ. 573, 727—739.

Davidson, N.C., 2014. How much wetland has the world lost? Long-term and recent trends in global wetland area. Mar. Freshw. Res. 65, 934—941.

DeFries, R.S., Rudel, T., Uriarte, M., Hansen, M., 2010. Deforestation driven by urban population growth and agricultural trade in the twenty-first century. Nat. Geosci. 3, 178—181.

Dent, D., 1986. Acid Sulphate Soils: A Baseline for Research and Development ILRI Publication No 39. International Institute of Land Reclamation and Improvement, Wageningen.

Dent, D., Pons, L., 1995. A world perspective on acid sulphate soils. Geoderma 67, 263—276.

D'Odorico, P., Okin, G.S., Bestelmeyer, B.T., 2012. A synthetic review of feedbacks and drivers of shrub encroachment in arid grasslands. Ecohydrology 5, 520—530.

D'Odorico, P., Bhattachan, A., Davis, K.F., Ravi, S., Runyan, C.W., 2013. Global desertification: drivers and feedbacks. Adv. Water Resour. 51, 326—344.

Doran, J.W., Stamatiadis, S., Haberern, J., 2002. Soil health as an indicator of sustainable management. Agric. Ecosyst. Environ. 88, 107—110.

Dumroese, R.K., Williams, M.I., Stanturf, J.A., St Clair, J.B., 2015. Considerations for restoring temperate forests of tomorrow: Forest restoration, assisted migration, and bioengineering. J. For. 46, 947—964.

ELD, 2015. The value of land: prosperous lands and positive rewards through sustainable land management. ELD Initiative, Bonn, Germany.

Emborg, J., Walker, G., Daniels, S., 2012. Forest landscape restoration decision-making and conflict management: applying discourse-based approaches. In: Stanturf, J., Lamb, D., Madsen, P. (Eds.), Forest Landscape Restoration. Springer, Dordrecht, pp. 131—153.

Fahey, R.T., Alveshere, B.C., Burton, J.I., D'Amato, A.W., Dickinson, Y.L., Keeton, W.S., et al., 2018. Shifting conceptions of complexity in forest management and silviculture. For. Ecol. Manage. 421, 59—71.

Fältmarsch, R.M., Åström, M.E., Vuori, K.-M., 2008. Environmental risks of metals mobilised from acid sulphate soils in Finland: a literature review. Boreal. Environ. Res. 13, 444—456.

FAO, 1976. A framework for land evaluation. Food and Agricultural Organization, Soils Bulletin 32, Rome.

FAO, 2011a. Assessing Forest Degradation — Towards the Development of Globally Applicable Guidelines. Food and Agriculture Organization of the United Nations, Rome.

FAO, 2011b. The State of the World's Land and Water Resources for Food And Agriculture (SOLAW)-Managing Systems at Risk. Food and Agriculture Organization of the United Nations, Rome and Earthscan, London.

FAO, ITPS, 2015. Status of the World's Soil Resources (SWSR)—Main Report. Food and Agriculture Organization of the United Nations and Intergovernmental Technical Panel on Soils, Rome, Italy, p. 650.

Farley, K.A., Jobbágy, E.G., Jackson, R.B., 2005. Effects of afforestation on water yield: a global synthesis with implications for policy. Glob. Change Biol. 11, 1565–1576.

Fischer, J., Gardner, T.A., Bennett, E.M., Balvanera, P., Biggs, R., Carpenter, S., et al., 2015. Advancing sustainability through mainstreaming a social–ecological systems perspective. Curr. Opin. Environ. Sustain. 14, 144–149.

Fitzherbert, E.B., Struebig, M.J., Morel, A., Danielsen, F., Brühl, C.A., Donald, P.F., et al., 2008. How will oil palm expansion affect biodiversity? Trends Ecol. Evol. 23, 538–545.

Fitzpatrick, R.W., 2002. Land degradation processes. In: McVicar, T., Li Rui, W., Fitzpatrick, R., Changming, L. (Eds.), Regional Water and Soil Assessment for Managing Agriculture in China and Australia. Australian Centre for International Agricultural Research (ACIAR), Canberra, pp. 119–129.

Forman, R., Godron, M., 1986. Landscape Ecology. Wiley, New York.

Foster, G., Meyer, L., 1977. Soil erosion and sedimentation by water—an overview. In: ASAE Publication No. 4–77. Proceedings of the National Symposium on Soil Erosion and Sediment by Water, Chicago, IL, December 12–13, 1977.

Frelich, L.E., Jõgiste, K., Stanturf, J.A., Parro, K., Baders, E., 2018. Natural disturbances and forest management: interacting patterns on the landscape. In: Perera, A.H., Peterson, U., Pastur, G.M., Iverson, L.R. (Eds.), Ecosystem Services from Forest Landscapes. Springer, New York, pp. 221–248.

Gaither, C.J., Yembilah, R., Samar, S.B., 2019. Tree registration to counter elite capture of forestry benefits in Ghana's Ashanti and Brong Ahafo regions. Land. Use Policy 85, 340–349.

Garner, H., 1974. The Origin of Landscapes. Oxford University Press, New York.

Gasparri, N.I., de Waroux, Yl.P., 2015. The coupling of South American soybean and cattle production frontiers: new challenges for conservation policy and land change science. Conserv. Lett. 8, 290–298.

Ghassemi, F., Jakeman, A.J., Nix, H.A., 1995. Salinisation of Land and Water Resources: Human Causes, Extent, Management and Case Studies. CAB international.

Ghazoul, J., Chazdon, R., 2017. Degradation and recovery in changing forest landscapes: a multiscale conceptual framework. Annu. Rev. Environ. Resour. 42, 161–188.

Gibbs, H.K., Ruesch, A., Achard, F., Clayton, M., Holmgren, P., Ramankutty, N., et al., 2010. Tropical forests were the primary sources of new agricultural land in the 1980s and 1990s. Proc. Natl. Acad. Sci. U.S.A. 107, 16732–16737.

Gibbs, H., Salmon, J., 2015. Mapping the world's degraded lands. Appl. Geogr. 57, 12–21.

Ginoux, P., Prospero, J.M., Gill, T.E., Hsu, N.C., Zhao, M., 2012. Global-scale attribution of anthropogenic and natural dust sources and their emission rates based on MODIS

Deep Blue aerosol products. Rev. Geophys. 50, RG3005. Available from: https://doi. org/10.1029/2012RG000388.

Goldewijk, K.K., 2001. Estimating global land use change over the past 300 years: the HYDE database. Glob. Biogeochem. Cycles 15, 417−433.

Goudie, A., 2018. Dust storms and ephemeral lakes. Desert 23, 153−164.

Grainger, A., Smith, M.S., Squires, V.R., Glenn, E.P., 2000. Desertification, and climate change: the case for greater convergence. Mitig. Adapt. Strateg. Glob. Change 5, 361−377.

Grau, H.R., Aide, M., 2008. Globalization and land-use transitions in Latin America. Ecol. Soc. 13, 16 [online] URL. Available from: http://www.ecologyandsociety.org/vol13/iss2/art16/.

Greider, T., Garkovich, L., 1994. Landscapes: the social construction of nature and the environment. Rural. Sociol. 59, 1−24.

Guo, L.B., Gifford, R., 2002. Soil carbon stocks and land use change: a metaanalysis. Glob. Change Biol. 8, 345−360.

Halofsky, J.E., Creutzburg, M.K., Hemstrom, M.A., 2014. Integrating social, economic, and ecological values across large landscapes. Gen. Tech. Rep. PNW-GTR-896. US Department of Agriculture, Forest Service, Pacific Northwest Research Station, Portland, OR, p. 206.

Hansen, M.C., Stehman, S.V., Potapov, P.V., 2010. Quantification of global gross forest cover loss. Proc. Natl. Acad. Sci. U.S.A. 107, 8650−8655.

Hansen, M.C., Potapov, P.V., Moore, R., Hancher, M., Turubanova, S., Tyukavina, A., et al., 2013. High-resolution global maps of 21st-century forest cover change. Science 342, 850−853.

Hartemink, A.E., 2006. Soil fertility decline: definitions and assessment. Encycl. Soil. Sci. 2, 1618−1621.

Hatton, T., Ruprecht, J., George, R., 2003. Preclearing hydrology of the Western Australia wheatbelt: target for the future. Plant. Soil. 257, 341−356.

Holden, S.R., Treseder, K.K., 2013. A meta-analysis of soil microbial biomass responses to forest disturbances. Front. Microbiol. 4, 163.

Holden, J., Chapman, P., Labadz, J., 2004. Artificial drainage of peatlands: hydrological and hydrochemical process and wetland restoration. Prog. Phys. Geogr. 28, 95−123.

Holl, K.D., 2017. Restoring tropical forests from the bottom up. Science 355, 455−456.

Hook, D.D., 1984. Waterlogging tolerance of lowland tree species of the South. South J. Appl. For. 8, 136−149.

Hooke, R.L., Martín-Duque, J.F., Pedraza, J., 2012. Land transformation by humans: a review. GSA Today 22, 4−10.

Hooper, D.U., Adair, E.C., Cardinale, B.J., Byrnes, J.E., Hungate, B.A., Matulich, K.L., et al., 2012. A global synthesis reveals biodiversity loss as a major driver of ecosystem change. Nature 486, 105−108.

Hudson, P.F., Alcántara-Ayala, I., 2006. Ancient and modern perspectives on land degradation. Catena 65, 102−106.

Hughes, F.M., Rood, S.B., 2003. Allocation of river flows for restoration of floodplain forest ecosystems: a review of approaches and their applicability in Europe. Environ. Manage. 32, 12−33.

Hu, S., Niu, Z., Chen, Y., Li, L., Zhang, H., 2017. Global wetlands: Potential distribution, wetland loss, and status. Sci. Total. Environ. 586, 319−327.

Ice, G.G., Neary, D.G., Adams, P.W., 2004. Effects of wildfire on soils and watershed processes. J. For. 102, 16–20.

IISD, 2002. Summary of the International Expert Meeting on Forest Landscape Restoration 27–28 February 2002. Sustainable Developments, vol. 71, no. 1 (2 March 2002). Available at: <http://www.iisd.ca/crs/sdcfr/sdvol71num1.html>.

IPBES, 2018. In: Scholes, R., Montanarella, L., Brainich, A., Barger, N., Brink, B.T., Cantele, M., Erasmus, B., et al.,Summary for Policymakers of the Thematic Assessment Report on Land Degradation and Restoration of the Intergovernmental Science-Policy Platform on Biodiversity and Ecosystem Services. IPBES Secretariat, Bonn.

IPBES, 2019. Summary for Policymakers of the Global Assessment Report on Biodiversity and Ecosystem Services. Intergovernmental Science-Policy Platform on Biodiversity and Ecosystem Services, Bonn.

Jandl, R., Lindner, M., Vesterdal, L., Bauwens, B., Baritz, R., Hagedorn, F., et al., 2007. How strongly can forest management influence soil carbon sequestration? Geoderma 137, 253–268.

Jjemba, P.K., 2002. The potential impact of veterinary and human therapeutic agents in manure and biosolids on plants grown on arable land: a review. Agric. Ecosyst. Environ. 93, 267–278.

Jõgiste, K., Korjus, H., Stanturf, J.A., Frelich, L.E., Baders, E., Donis, J., et al., 2017. Hemiboreal forest: natural disturbances and the importance of ecosystem legacies to management. Ecosphere 8 (2), e01706. Available from: https://doi.org/10.1002/ecs2.1706.

Jones, K.R., Venter, O., Fuller, R.A., Allan, J.R., Maxwell, S.L., Negret, P.J., et al., 2018. One-third of global protected land is under intense human pressure. Science 360, 788–791.

Kareiva, P., Watts, S., McDonald, R., Boucher, T., 2007. Domesticated nature: shaping landscapes and ecosystems for human welfare. Science 316, 1866–1869.

Kates, R.W., 2011. What kind of a science is sustainability science? Proc. Natl. Acad. Sci. U.S.A. 108, 19449–19450.

Kelley, A.C., Williamson, J.G., 1984. Population growth, industrial revolutions, and the urban transition. Popul. Dev. Rev. 10, 419–441.

Kibler, K., Cook, G., Chambers, L., Donnelly, M., Hawthorne, T., Rivera, F., et al., 2018. Integrating sense of place into ecosystem restoration: a novel approach to achieve synergistic social-ecological impact. Ecol. Soc. 23 (4), 25. Available from: https://doi.org/10.5751/ES-10542-230425.

Lal, R., Iivari, T., Kimble, J.M., 2003. Soil Degradation in the United States: Extent, Severity, and Trends. CRC Press, Boca Raton.

Lal, R., Negassa, W., Lorenz, K., 2015. Carbon sequestration in soil. Curr. Opin. Environ. Sustain. 15, 79–86.

Lambin, E.F., Meyfroidt, P., 2011. Global land use change, economic globalization, and the looming land scarcity. Proc. Natl. Acad. Sci. U.S.A. 108, 3465–3472.

Lamb, D., Stanturf, J., Madsen, P., 2012. What is forest landscape restoration? In: Stanturf, J., Lamb, D., Madsen, P. (Eds.), Forest Landscape Restoration Integrating Natural and Social Sciences. Springer, Dordrecht, pp. 3–23.

Lanfredi, M., Coppola, R., Simoniello, T., Coluzzi, R., Imbrenda, V., Macchiato, M., 2015. Early identification of land degradation hotspots in complex bio-geographic regions. Remote Sens. 7, 8154–8179.

Lenzen, M., Moran, D., Kanemoto, K., Foran, B., Lobefaro, L., Geschke, A., 2012. International trade drives biodiversity threats in developing nations. Nature 486, 109−112.

Lewis, S.L., Edwards, D.P., Galbraith, D., 2015. Increasing human dominance of tropical forests. Science 349, 827−832.

Liu, J., Dietz, T., Carpenter, S.R., Folke, C., Alberti, M., Redman, C.L., et al., 2007. Coupled human and natural systems. AMBIO 36, 639−649.

Liu, Y., Li, Y., Li, S., Motesharrei, S., 2015. Spatial and temporal patterns of global NDVI trends: correlations with climate and human factors. Remote Sens. 7, 13233−13250.

Ljung, K., Maley, F., Cook, A., Weinstein, P., 2009. Acid sulfate soils and human health— a millennium ecosystem assessment. Environ. Int. 35, 1234−1242.

Lõhmus, A., Remm, L., Rannap, R., 2015. Just a ditch in forest? Reconsidering draining in the context of sustainable forest management. Bioscience 65, 1066−1076.

Loucks, D.P., 2000. Sustainable water resources management. Water Int. 25, 3−10.

Luce, C.H., Black, T.A., 1999. Sediment production from forest roads in western Oregon. Water Resour. Res. 35, 2561−2570.

Manoli, G., Meijide, A., Huth, N., Knohl, A., Kosugi, Y., Burlando, P., et al., 2018. Ecohydrological changes after tropical forest conversion to oil palm. Environ. Res. Lett. 13, 064035.

Mansourian, S., 2005. Overview of forest restoration strategies and terms. In: Mansourian, S., Vallauri, D., Dudley, N. (Eds.), Forest Restoration in Landscapes. Springer, pp. 8−13.

Mansourian, S., 2017. Governance and forest landscape restoration: a framework to support decision-making. J. Nat. Conserv. 37, 21−30.

Mansourian, S., 2018. In the eye of the beholder: Reconciling interpretations of forest landscape restoration. Land Degrad. Dev. 29, 2888−2898.

Mansourian, S., Dudley, N., Vallauri, D., 2017a. Forest landscape restoration: progress in the last decade and remaining challenges. Ecol. Restor. 35, 281−288.

Mansourian, S., Stanturf, J.A., Derkyi, M.A.A., Engel, V.L., 2017b. Forest landscape restoration: increasing the positive impacts of forest restoration or simply the area under tree cover? Restor. Ecol. 25, 178−183.

Mappin, B., Chauvenet, A.L., Adams, V.M., Di Marco, M., Beyer, H.L., Venter, O., et al., 2019. Restoration priorities to achieve the global protected area target. Conserv. Lett. e12646. Available from: https://doi.org/10.1111/conl.12646.

Martellozzo, F., Amato, F., Murgante, B., Clarke, K., 2018. Modelling the impact of urban growth on agriculture and natural land in Italy to 2030. Appl. Geogr. 91, 156−167.

Martín-López, B., Palomo, I., García-Llorente, M., Iniesta-Arandia, I., Castro, A.J., Del Amo, D.G., et al., 2017. Delineating boundaries of social-ecological systems for landscape planning: a comprehensive spatial approach. Land Use Policy 66, 90−104.

McBratney, A., Field, D.J., Koch, A., 2014. The dimensions of soil security. Geoderma 213, 203−213.

McConkey, K.R., O'Farrill, G., 2016. Loss of seed dispersal before the loss of seed dispersers. Biol. Conserv. 201, 38−49.

MEA, 2005. Ecosystems and Human Well-being: Synthesis. Island Press, Washington, DC.

Meiyappan, P., Jain, A.K., 2012. Three distinct global estimates of historical land-cover change and land-use conversions for over 200 years. Front. Earth Sci. 6, 122−139.

Michael, P.S., 2013. Ecological impacts and management of acid sulphate soil: a review. Asian J. Water Environ. Pollut. 10, 13−24.

Micklin, P., 2010. The past, present, and future Aral Sea. Lakes Reserv.: Res. Manage. 15, 193−213.

Middleton, N., 2018. Rangeland management and climate hazards in drylands: dust storms, desertification and the overgrazing debate. Nat. Hazards 92, 57−70.

Millar, C.I., Stephenson, N.L., Stephens, S.L., 2007. Climate change and forests of the future: managing in the face of uncertainty. Ecol. Appl. 17, 2145−2151.

Miranda, A., Altamirano, A., Cayuela, L., Lara, A., González, M., 2017. Native forest loss in the Chilean biodiversity hotspot: revealing the evidence. Regional Environ. Change 17, 285−297.

Mitchell, A.L., Rosenqvist, A., Mora, B., 2017. Current remote sensing approaches to monitoring forest degradation in support of countries measurement, reporting and verification (MRV) systems for REDD + . Carbon Balance Manage. 12, 9.

Montanarella, L., Pennock, D.J., McKenzie, N., Badraoui, M., Chude, V., Baptista, I., et al., 2016. World's soils are under threat. Soil 2, 79−82.

Mughal, M.A., 2019. Rural urbanization, land, and agriculture in Pakistan. Asian Geographer 36, 81−91.

Natural Resources Conservation Service Conservation Practice Standard Wetland Restoration, 2010. (ac.) Code 657. <https://www.nrcs.usda.gov/Internet/FSE_DOCUMENTS/nrcs143_026340.pdf>.

Nielsen, U.N., Ayres, E., Wall, D.H., Bardgett, R.D., 2011. Soil biodiversity and carbon cycling: a review and synthesis of studies examining diversity−function relationships. Eur. J. Soil Sci. 62, 105−116.

Nsengimana, V., Kaplin, B.A., Francis, F., Nsabimana, D., 2018. Use of soil and litter arthropods as biological indicators of soil quality in forest plantations and agricultural lands: a review. Entomologie faunistique-Faunistic Entomology 71.

Nyland, R.D., 1992. Exploitation and greed in eastern hardwood forests. J. For. 90, 33−37.

Obalum, S.E., Chibuike, G.U., Peth, S., Ouyang, Y., 2017. Soil organic matter as sole indicator of soil degradation. Environ. Monit. Assess. 189, 32−50.

O'Hara, K.L., Ramage, B.S., 2013. Silviculture in an uncertain world: utilizing multi-aged management systems to integrate disturbance. Forestry 86, 401−410.

Orlovsky, N., Glantz, M., Orlovsky, L., 2001. Irrigation and land degradation in the Aral Sea Basin. In: Breckle, S.-W., Veste, M., Wucherer, W. (Eds.), Sustainable Land Use in Deserts. Springer, Berlin, pp. 115−125.

Orr, B., Cowie, A., Castillo Sanchez, V., Chasek, P., Crossman, N., Erlewein, A., et al., 2017. Scientific conceptual framework for land degradation neutrality. A Report of the Science-Policy Interface. United Nations Convention to Combat Desertification (UNCCD), Bonn.

Oswalt, S.N., King, S.L., 2005. Channelization and floodplain forests: impacts of accelerated sedimentation and valley plug formation on floodplain forests of the Middle Fork Forked Deer River, Tennessee, USA. For. Ecol. Manage. 215, 69−83.

Parrott, L., Meyer, W.S., 2012. Future landscapes: managing within complexity. Front. Ecol. Environ. 10, 382−389.

Patric, J.H., 1976. Soil erosion in the eastern forest. J. For. 74, 671−677.

Phillips, R.J., Waldrop, T.A., Brose, P.H., Wang, G.G., 2012. Restoring fire-adapted forests in eastern North America for biodiversity conservation and hazardous fuels reduction. In: Stanturf, J., Madsen, P., Lamb, D. (Eds.), A Goal-Oriented Approach to Forest Landscape Restoration. Springer, Dordrecht, pp. 187−219.

Pons, LJ Outline of the genesis, characteristics, classification and improvement of acid sulphate soils. In: Proceedings of the 1972 (Wageningen, Netherlands) International Acid Sulphate Soils Symposium, 1973. pp 3–27.

Potts, S.G., Biesmeijer, J.C., Kremen, C., Neumann, P., Schweiger, O., Kunin, W.E., 2010. Global pollinator declines: trends, impacts and drivers. Trends Ecol. Evol. 25, 345–353.

Prévost, M., Charette, L., 2018. Rehabilitation silviculture in a high-graded temperate mixedwood stand in Quebec, Canada. N. For. 1–22.

Prospero, J.M., Ginoux, P., Torres, O., Nicholson, S.E., Gill, T.E., 2002. Environmental characterization of global sources of atmospheric soil dust identified with the Nimbus 7 Total Ozone Mapping Spectrometer (TOMS) absorbing aerosol product. Rev. Geophys. 40, 1002. Available from: https://doi.org/10.1029/2000RG000095.

Qadir, M., Oster, J., 2004. Crop and irrigation management strategies for saline-sodic soils and waters aimed at environmentally sustainable agriculture. Sci. Total Environ. 323, 1–19.

Ramankutty, N., Mehrabi, Z., Waha, K., Jarvis, L., Kremen, C., Herrero, M., et al., 2018. Trends in global agricultural land use: implications for environmental health and food security. Annu. Rev. Plant. Biol. 69, 789–815.

Reid, R.S., Fernández-Giménez, M.E., Galvin, K.A., 2014. Dynamics and resilience of rangelands and pastoral peoples around the globe. Annu. Rev. Environ. Resour. 39, 217–242.

Richards, D.R., Friess, D.A., 2015. Rates and drivers of mangrove deforestation in Southeast Asia, 2000–2012. Proc. Natl. Acad. Sci. U.S.A. 113, 344–349.

Richter, D., Yaalon, D.H., 2012. "The changing model of soil" revisited. Soil Sci. Soc. Am. J. 76, 766–778.

Riitters, K., Wickham, J., O'Neill, R., Jones, B., Smith, E., 2000. Global-scale patterns of forest fragmentation. Conserv. Ecol. 4 (2), 3 [online] URL. Available from: http://www.consecol.org/vol4/iss2/art3/.

Rodríguez-Eugenio, N., McLaughlin, M., Pennock, D., 2018. Soil Pollution: A Hidden Reality. Food and Agriculture Organization of the United Nations, Rome, p. 142.

Ross, M.S., O'Brien, J.J., da Silveira Lobo Sternberg, L., 1994. Sea-level rise and the reduction in pine forests in the Florida Keys. Ecol. Appl. 4, 144–156.

Rumeu, B., Devoto, M., Traveset, A., Olesen, J.M., Vargas, P., Nogales, M., et al., 2017. Predicting the consequences of disperser extinction: richness matters the most when abundance is low. Funct. Ecol. 31, 1910–1920.

Runyan, C., D'Odorico, P., 2016. Global Deforestation. Cambridge University Press, Cambridge.

Sabogal, C., Besacier, C., McGuire, D., 2015. Forest and landscape restoration: concepts, approaches and challenges for implementation. Unasylva 66, 3–10.

Safriel, U., 2017. Land degradation neutrality (LDN) in drylands and beyond—where has it come from and where does it go. Silva Fennica 51 (1650), 19. Available from: https://doi.org/10.14214/sf.1650.

Sanderman, J., Hengl, T., Fiske, G.J., 2017. Soil carbon debt of 12,000 years of human land use. Proc. Natl. Acad. Sci. U.S.A. 114, 9575–9580.

Sayer, J., Sunderland, T., Ghazoul, J., Pfund, J.-L., Sheil, D., Meijaard, E., et al., 2013. Ten principles for a landscape approach to reconciling agriculture, conservation, and other competing land uses. Proc. Natl. Acad. Sci. U.S.A. 110, 8349–8356.

Scherer, D., 1995. Evolution, human living, and the practice of ecological restoration. Environ. Ethics 17, 359–379.

Schmidt, M.W., Torn, M.S., Abiven, S., Dittmar, T., Guggenberger, G., Janssens, I.A., et al., 2011. Persistence of soil organic matter as an ecosystem property. Nature 478, 49–56.

Schoenholtz, S.H., Van Miegroet, H., Burger, J., 2000. A review of chemical and physical properties as indicators of forest soil quality: challenges and opportunities. For. Ecol. Manage. 138, 335–356.

SERI, 2004. The SER International Primer on Ecological Restoration. Society for Ecological Restoration International Science & Policy Working Group, Tucson, AZ, available from http//www.ser.org.

Seto, K.C., Güneralp, B., Hutyra, L.R., 2012. Global forecasts of urban expansion to 2030 and direct impacts on biodiversity and carbon pools. Proc. Natl. Acad. Sci. U.S.A. 109, 16083–16088.

Sidle, R.C., Ziegler, A.D., Negishi, J.N., Nik, A.R., Siew, R., Turkelboom, F., 2006. Erosion processes in steep terrain—truths, myths, and uncertainties related to forest management in Southeast Asia. For. Ecol. Manage. 224, 199–225.

Sivakumar, M., 2007. Interactions between climate and desertification. Agric. For. Meteorol. 142, 143–155.

Skiadaresis, G., Schwarz, J.A., Bauhus, J., 2019. Groundwater extraction in floodplain forests reduces radial growth and increases summer drought sensitivity of pedunculate oak trees (Quercus robur L.). Front. For. Glob. Change 2, 5. Available from: https://doi.org/10.3389/ffgc.2019.00005.

Stanturf, J.A., 2015. Future landscapes: opportunities and challenges. N. For. 46, 615–644.

Stanturf, J.A., 2016. What is forest restoration? In: Stanturf, J.A. (Ed.), Restoration of Boreal and Temperate Forests, second ed. CRC Press, Boca Raton, FL, pp. 1–16.

Stanturf, J., Palik, B., Dumroese, R.K., 2014a. Contemporary forest restoration: a review emphasizing function. For. Ecol. Manage. 331, 292–323.

Stanturf, J.A., Palik, B.J., Williams, M.I., Dumroese, R.K., Madsen, P., 2014b. Forest restoration paradigms. J. Sustain. For. 33, S161–S194.

Stanturf, J.A., Kleine, M., Mansourian, S., Parrotta, J., Madsen, P., Kant, P., et al., 2019. Implementing forest landscape restoration under the Bonn Challenge: A systematic approach. Ann. For. Sci. 76, 50. Available from: https://doi.org/10.1007/s13595-019-0833-z.

Stedman, R.C., 2003. Is it really just a social construction?: The contribution of the physical environment to sense of place. Soc. Nat. Resour. 16, 671–685.

Steffen, W., Richardson, K., Rockström, J., Cornell, S.E., Fetzer, I., Bennett, E.M., et al., 2015. Planetary boundaries: Guiding human development on a changing planet. Science 347, 1259855.

Steffan-Dewenter, I., Westphal, C., 2008. The interplay of pollinator diversity, pollination services and landscape change. J. Appl. Ecol. 45, 737–741.

Stenmark, M., 2017. Environmental Ethics and Policy-Making. Routledge, London.

Stevenson, N., Lewis, R., Burbridge, P., 1999. Disused shrimp ponds and mangrove rehabilitation. In: Streever, W.J. (Ed.), An International Perspective on Wetland Rehabilitation. Kluwer, Dordrecht, pp. 277–297.

Swart, R.J., Raskin, P., Robinson, J., 2004. The problem of the future: sustainability science and scenario analysis. Glob. Environ. Change 14, 137–146.

Thompson, I.D., Guariguata, M.R., Okabe, K., Bahamondez, C., Nasi, R., Heymell, V., et al., 2013. An operational framework for defining and monitoring forest degradation. Ecol. Soc. 18 (2), 20.

Thurman, M., 2011. Natural Disaster Risks in Central Asia: A Synthesis. United Nations Development Programme, Bratislava.

Tscharntke, T., Clough, Y., Wanger, T.C., Jackson, L., Motzke, I., Perfecto, I., et al., 2012. Global food security, biodiversity conservation and the future of agricultural intensification. Biol. Conserv. 151, 53–59.

Tsiafouli, M.A., Thébault, E., Sgardelis, S.P., De Ruiter, P.C., Van Der Putten, W.H., Birkhofer, K., et al., 2015. Intensive agriculture reduces soil biodiversity across Europe. Glob. Change Biol. 21, 973–985.

Turetsky, M.R., Benscoter, B., Page, S., Rein, G., Van Der Werf, G.R., Watts, A., 2015. Global vulnerability of peatlands to fire and carbon loss. Nat. Geosci. 8, 11–14.

Turner, M.G., Gardner, R.H., O'Neill, R.V., 2001. Landscape Ecology in Theory and Practice: Patterns and Process. Springer, New York.

Uda, S.K., Hein, L., Sumarga, E., 2017. Towards sustainable management of Indonesian tropical peatlands. Wetl. Ecol. Manage. 25, 683–701.

UNCCD, 2016. Report on the Conference of Parties on its twelfth session, held in Ankara from 12 to 23 October 2015. Part two: Actions. In: ICCD/COP(12)/20Add.1. UNCCD, Bonn.

UNCCD (Ed.), 2017. Global Land Outlook. UN Convention to Combat Desertification, Bonn, Germany.

UNEP, 1992. The Status of Desertification and Implementation of the United Nations Plan of Action to Combat Desertification. United Nations Environment Program, Nairobi.

United Nations, 2015. Transforming our world: The 2030 agenda for sustainable development. In: A/Res/70/1. United Nations, New York.

van Oost, K., Quine, T., Govers, G., De Gryze, S., Six, J., Harden, J., et al., 2007. The impact of agricultural soil erosion on the global carbon cycle. Science 318, 626–629.

van Oosten, C., Moeliono, M., Wiersum, F., 2018. From product to place—spatializing governance in a commodified landscape. Environ. Manage. 62, 157–169.

Venter, O., Sanderson, E.W., Magrach, A., Allan, J.R., Beher, J., Jones, K.R., et al., 2016. Sixteen years of change in the global terrestrial human footprint and implications for biodiversity conservation. Nat. Commun. 7, 12558.

Verdone, M., Seidl, A., 2017. Time, space, place, and the Bonn Challenge global forest restoration target. Restor. Ecol. 25, 903–911.

Veresoglou, S.D., Halley, J.M., Rillig, M.C., 2015. Extinction risk of soil biota. Nat. Commun. 6, 8862.

Vetter, S., 2005. Rangelands at equilibrium and non-equilibrium: recent developments in the debate. J. Arid. Environ. 62, 321–341.

Visconti, P., Bakkenes, M., Baisero, D., Brooks, T., Butchart, S.H., Joppa, L., et al., 2015. Projecting global biodiversity indicators under future development scenarios. Conserv. Lett. 9, 5–13.

Vorkinn, M., Riese, H., 2001. Environmental concern in a local context: the significance of place attachment. Environ. Behav. 33, 249–263.

Vrieling, A., 2006. Satellite remote sensing for water erosion assessment: a review. Catena 65, 2–18.

Wagg, C., Bender, S.F., Widmer, F., van der Heijden, M.G., 2014. Soil biodiversity and soil community composition determine ecosystem multifunctionality. Proc. Natl. Acad. Sci. U.S.A. 111, 5266–5270.

Wardell, D.A., Lund, C., 2006. Governing access to forests in northern Ghana: micropolitics and the rents of non-enforcement. World Dev. 34, 1887–1906.

Wichelns, D., Qadir, M., 2015. Achieving sustainable irrigation requires effective management of salts, soil salinity, and shallow groundwater. Agric. Water Manage. 157, 31–38.

Wiens, J.A., Hobbs, R.J., 2015. Integrating conservation and restoration in a changing world. BioScience 65, 302–312.

Wunderle Jr, J.M., 1997. The role of animal seed dispersal in accelerating native forest regeneration on degraded tropical lands. For. Ecol. Manage. 99, 223–235.

Yaalon, D.H., 2007. Human-induced ecosystem and landscape processes always involve soil change. Bioscience 57, 918–919.

Yaalon, D.H., Yaron, B., 1966. Framework for man-made soil changes-an outline of metapedogenesis. Soil Sci. 102, 272–277.

Yang, A., Bellwood-Howard, I., Lippe, M., 2018. Social-ecological systems and forest landscape restoration. In: Mansourian, S., Parrotta, J. (Eds.), Forest Landscape Restoration: Integrated Approaches to Support Effective Implementation. Earthscan/Routledge, New York.

Zerga, B., 2015. Rangeland degradation and restoration: a global perspective. Point J. Agric. Biotechnol. Res. 1, 37–54.

Zhou, Z., Wang, C., Luo, Y., 2018. Effects of forest degradation on microbial communities and soil carbon cycling: A global meta-analysis. Glob. Ecol. Biogeogr. 27, 110–124.

Zomer, R.J., Trabucco, A., Coe, R., Place, F., Van Noordwijk, M., Xu, J., 2014. Trees on farms: an update and reanalysis of agroforestry's global extent and socio-ecological characteristics. Working Paper. Center for International Forestry Research, Bogor, Indonesia.

Zomer, R.J., Neufeldt, H., Xu, J., Ahrends, A., Bossio, D., Trabucco, A., et al., 2016. Global tree cover and biomass carbon on agricultural land: the contribution of agroforestry to global and national carbon budgets. Sci. Rep. 6, 29987.

Zomer, R.J., Bossio, D.A., Sommer, R., Verchot, L.V., 2017. Global sequestration potential of increased organic carbon in cropland soils. Sci. Rep. 7, 15554.

Zonneveld, I.S., 1989. The land unit—a fundamental concept in landscape ecology, and its applications. Landsc. Ecol. 3, 67–86.

Soil recovery and reclamation of mined lands

6

Jan Frouz[1,2]

[1]*Biology Centre of the Academy of Sciences of the Czech Republic, Institute of Soil Biology &*
SoWa Research Infrastructure, České Budějovice, Czech Republic
[2]*Institute for Environmental Studies, Charles University, Prague, Czech Republic*

6.1 Introduction

Mining and opencast mining in particular affect all components of ecosystems (Bradshaw, 1997). During opencast mining, material overlying the mined minerals, called overburden, is excavated and deposited in a heap. This heap can be located either inside the mine pit (an internal heap) or outside (an external heap). As a consequence of this, whatever ecosystem that was present at the time prior to mining activity or external heap deposition is basically destroyed, either excavated or buried (Bell and Donnelly, 2006). Additional areas can be impacted or degraded by processing of minerals forming tailings and other deposits. In addition to these immediate effects, the surrounding environment can be affected in many different ways, for example, water pumped into mines can affect local water table depths, resulting in the release of acid mine drainage or other mining water into surface waters. In this chapter, we will focus only on reconstruction of upland ecosystems in overburden heaps. Overburden differs substantially from previous local soils, and some maybe even toxic for plants (Bradshaw, 1997; Frouz et al., 2011a). Overburden "soil" typically is composed of fragmented mixed and dumped parent material, which often weathers intensively after dumping. These soils do not have horizons, and they generally also lack soil structure. In terms of other soil properties, overburden soils differ substantially from native soils in surrounding ecosystems; they typically have extreme pH (either too alkaline or too acidic) and extreme soil texture (Table 6.1). Because sediments that often form overburden materials may be sorted by texture, overburden materials are often characterized by being too sandy or being mixed with a large sand and gravel component or on the other hand having too much clay for proper soil functioning. Finally, and critically, overburden soils lack recent organic matter input, which is the foundation of many important processes such as aggregate formation, sorption of nutrients, and water-holding capacity. Postmining soils also typically have low content of available nutrients, namely, nitrogen (N) (Bradshaw, 1997). This may hamper plant growth, but on the other hand, this is one of the reasons

Soils and Landscape Restoration. DOI: https://doi.org/10.1016/B978-0-12-813193-0.00006-0

Table 6.1 The most common substrate constrains in postmining sites and their ecological consequences.

Substrate property	Mechanism of origin	Constrain	Potential
Low pH	Decrease of pH in overburden is typically connected with weathering sulfur-rich minerals such as pyrite that releases sulfuric acid. Sometimes in tailing as a result of using acid in ore processing.	Low pH can block vegetation development in affected areas. Water that leaches from these areas can be source acid mine drainage, with negatively affect aquatic habitats downstream.	Very small acidic patches (up to 100 m^2) can be beneficial as habitats for specialized insects of fungi, in larger parches negative effects on surrounding aquatic habitats usually prevail
High pH	High pH in overburden is typically connected with carbonates with basic cations and its weathering. In tailing may result from using alkali in ore processing.	Slightly alkaline sites (pH 7.5–8.5) are usually good for vegetation. Very alkaline sites (pH 10) and more hamper vegetation development.	Slightly alkaline substrates (pH 7.5–8.5) may harbor specialized flora
High salt content	Weathering mineral caring basic cations.	Highly salty habitats hamper vegetation development and may affect aquatic life in neighboring small letic waters.	Slightly salty substrates may harbor specialized flora, and promote some aquatic species in neighboring small lentic waters
High sand and gravel content	It is caused by sediment separation in overburden layer during deposition.	Low nutrient content, slow vegetation development in sandy substrate, high potential for erosion.	Low nutrient content can be beneficial for plant species adapted on oligotrophic conditions and some specialized insects that need bare soil
High clay content	It is caused by sediment separation in overburden layer during deposition.	High clay content can cause poor infiltration that may increase runoff and risk of erosion.	Can be beneficial for specialized plant species and some specialized insects that need bare soil
Hydrophobicity	Accumulation of waxy substances and coal dust.	No water infiltration and constrained vegetation development, increased erosion and runoff of sediment in to aquatic habitats.	Very mall parches (up to 100 m^2) can be beneficial as habitat for specialized insects of fungi, in larger parches negative effects associated with erosion and runoff prevail

for high species richness and frequent occurrence of rare and endangered species in postmining land. In most of the developed (industrial) world, ecosystems surrounding mined sites are characterized by having excess N availability (due to deposition) that leads to increasing plant cover but also to a shift in the plant community to less competitive species and consequently to a reduction of species number. In contrast to this general high N availability, in postmining sites where species-rich topsoil has been removed, availability of nutrients is lower, which reduces overall plant biomass but favors less competitive species.

When we restore soil in postmining sites, there is a wide range of possible actions that fall between two extreme scenarios. The first option can be called complete reconstruction, whereas the other extreme is to do nothing and wait for soil formation to occur on its own. Complete reconstruction of soil involves covering overburden with either some suitable material, which does not hamper plant and then layer of A horizon (sometimes A and B horizons) materials transplanted from another location that will be mined in the near future. These are usually transplanted in a way that resemble as much as possible the organization of target soils. The other extreme option, as mentioned previously, is to do nothing and wait for ecological succession to do the work of rebuilding soil. In actual practice, we typically do something that falls in the middle of these extremes, and we prepare some kind of suitable rooting medium and assist in vegetation development that then feeds back to accelerated soil formation.

It is important to note that the goals of postmining restorations may vary considerably depending on socioeconomic conditions in a given location. In some cases, there is more emphasis on restoring the production function of postmining land in terms of agricultural production, while in other cases the restoration goal is to return lands to more natural ecosystems. A final consideration is the type of vegetation to be restored, for example, forest or grassland vegetation, and each of these may require different substrate quality and can benefit from manipulations of the substrate. Clear decisions about the objectives of a particular restoration are essential, because as will be shown later, conditions that may hamper the restoration of highly productive land may actually provide suitable habitat for many rare and endangered species and vice versa.

In this chapter, we will deal with various approaches to restoration of postmining soils, discuss which approaches to choose depending on our target and environmental conditions, and explore how the soil restoration corresponds with whole ecosystem recovery.

6.2 Overburden properties

Overburden materials may vary substantially in their quality and ability to support plant growth (Abakumov and Frouz, 2014; Bradshaw, 1997; Rojik, 2014). Topsoil and subsoil (A and B horizons of recent soils) as well as some quaternary deposits such as loess represent excellent rooting medium that can be used for restoration of grassland and arable soils. Clay materials such as Miocene clays or bentonite represent good rooting medium as well. Suitability of clays increases as the proportion of montmorillonite and illite increases. On the other hand, clay with a high proportion of kaolinite is the worst medium. Rock that weathers easily or that is already partly weathered such as weathered sandstones represent good medium for tree growth. Sand and gravel are difficult to revegetate as they contain low amount of nutrients. However, they can represent valuable oligotrophic substrate for reconstruction of oligotrophic ecosystems, which are typically rare in the contemporary cultural landscape. Sand and gravel may be also used for mixing with media having high content of silt and clay, namely, when rooting medium for forest growth is desired.

Overburden may vary in nutrient content, but in general, substrates rich in silt and clay contain more nutrients than substrates composed mainly of sand. Nutrient availability (specifically nitrogen) in overburden is often poorer than in contemporary cultural landscapes that receive a lot of N from agriculture use and atmospheric deposition. On the other hand, nutrients, which are mostly derived from minerals such as phosphorus or basic cations, can be more abundant in overburden material than in surrounding soil. This is due to phosphorus being depleted in topsoil, particularly in humid environments, but phosphorus is still abundant in the deeper layers brought to the surface. This has been observed in some mines in the Eastern United States (Frouz and Franklin, 2015; Frouz et al., 2013a).

Another important pool of nutrients in undisturbed soils is found in the soil organic matter (SOM). Obviously, the absence of SOM in overburden means that these nutrients are also missing from overburden, and particularly nitrogen, which is the limiting nutrient in most terrestrial ecosystems in temperate climates (Brady and Weil, 1999). Building up SOM stock is thus essential to building the nitrogen pool in restored ecosystems. However, it should be noted that not all overburdens are organic matter poor, but some contain substantial amounts of organic matter in the form of coal as in coal-rich sand or kerogen as in some Miocene clays (Vindušková et al., 2014, 2015). Coal-rich layers often contain pyritic material that may be phytotoxic. However, kerogen in Miocene clays is associated with neutral or slightly alkaline pH and contains large amount of organic carbon (C) and nitrogen (N). Moreover, it has been shown that deep subsurface layers of Miocene clays contain microbes capable of mineralizing this fossil organic matter, which potentially means that N bound in this fossil organic matter can be liberated for plant growth (Frouz et al., 2011b). However, there are few studies exploring role of fossil organic matter in plant nutrition.

Some overburden material may become toxic for plants and soil biota (Frouz et al., 2005, 2011a). The principal causes of toxicity vary. In some cases, toxicity arises due to extremely low pH (or low pH associated with high concentration of heavy metals), whereas in other cases, high salt content may be responsible for toxicity (Bradshaw, 1997). In contrast to other polluted sites where toxicity occurs because of accumulation of some exogenous toxic substances, in postmining sites the toxicity is typically a result of in situ weathering that then produces toxic substances from the overburden. Weathering processes, such as pyrite oxidation, decrease pH and may release some heavy metals. Weathering may also release some other ions that may cause high conductivity of substrate, which may cause toxicity of postmining site soils (Bradshaw, 1997; Frouz et al., 2005, 2011a; Tesnerová et al., 2017). This effect may be enhanced in materials with low sorption capacity, and low content of basic cations such as in sand or kaolinite. Besides toxicity associated with coal accompanied by pyrite, there may be also other mechanisms causing toxicity, for example, sites with high conductivity produced by high contents of sodium—potassium and similar cations that may be toxic to plants and soil fauna due to high osmotic pressure (Frouz et al., 2005). As a consequence, site toxicity may initially increase with material weathering but then decrease as the agent causing toxicity has been leached. However, this natural attenuation of toxicity may take a long time. To illustrate the differences in the time required for recovery of vegetation between toxic and nontoxic overburden, we can refer to examples from the Czech Republic's Sokolov coal mining district. In nontoxic sites, with active reclamation, vegetation recovery can be achieved in 5—10 years, and unmanaged spontaneous succession of vegetation occurs in 10—15 years (Frouz et al., 2014). On the other hand, in sites where pyrite oxidation occurs in overburden material, vegetation may not recover for 50 years (Frouz et al., 2014). The most sensitive indicators of toxicity are biological tests with plants or soil biota (Frouz et al., 2005). Correlation of biological tests with soil chemistry indicated that overburden pH is a very sensitive indicator of potential phytotoxicity (Frouz et al., 2005). If overburden pH is below 3.5, the site is likely to have problems with phytotoxicity. When pH is below 5.5, it is advisable to do some simple bioassays with plants or to conduct chemical analyses of soils to check for content of heavy metals. If the pH level is above 8.5, it is sensible to do simple plant growth bioassays or to check for high soil conductivity. Many of these toxic sites are problematic not only because they hamper restoration of the site itself but they also produce acid mine drainage or other forms of water pollution, which may affect ecosystems downstream. If they are very small (several dozen to a $100 \, m^2$), they may generate patches of alkaline soil, which may not represent a serious problem and may even harbor some unique species. Small sources of acid mine drainage can be treated by construction of wetlands, which can reduce concentrations of iron and other metals 10-fold. However, in larger areas, acid mine drainage is of greater concern because the pollution is not only at the site itself but also extends into surrounding ecosystems. Certainly, the best option is to avoid generation of these sites by burying them in deeper layers of the heap. The sites can be treated with liming or application of other alkaline material, but this is usually effective for only a short time, as subsequent pyrite weathering will release more acids. The only permanent solution is to avoid future weathering by physically isolating any pyritic layers from weathering agents, for example, by a layer of clay (e.g., bentonite) on top of which can be applied nonpyritic topsoil or another suitable rooting medium.

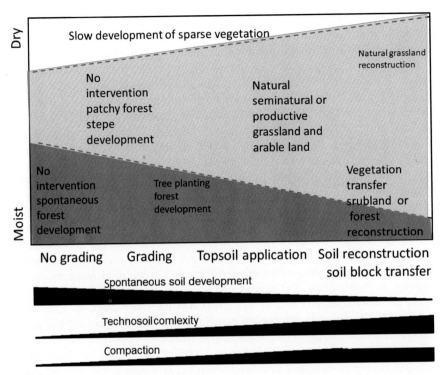

FIGURE 6.1

Scheme of the relationship between complexity in technosoil preparation and suitability of produced substrates for establishment of various types of vegetation in relation to climatic condition. Sites suited best for forest are marked by dark gray and grassland by light gray. Establishment of forest in complex topsoil can be improved by ameliorating compaction (e.g., by chiseling) and reduction of herb and grass cover.

6.3 Overburden preparation and technosoil construction

Overburden sometimes can be used for reclamation work without any further mechanical modification, but in many cases, it is subjected to grading or other modifications involving entries by heavy machinery. Grading increases site accessibility and provides open space for other technical improvements such as planting trees, seeding seeds, and application of fertilizers or topsoil. Each of these operations involves movement of machinery on the site, which directly or indirectly increases compaction. Compacted soils hamper tree establishment, and thus the importance of herbs and grasses increases (Fig. 6.1). Similarly, application of topsoil and improvement of nutrient status may support grasses even more and can make tree establishment more difficult. Nutrient addition can interfere with establishment of many rare and endangered species that require oligotrophic substrates. This is complicated even more by interactions with local climatic conditions. In general, in drier conditions, grassy and shrubby vegetation performs better than forest vegetation (Fig. 6.1). Conducting any given treatment or operation may thus have different effects when applied in different climatic conditions. In moist conditions the compaction caused by topsoil application may deter forest development and should be considered very carefully, whereas in dry conditions, topsoil application leads to rapid recovery of dry grassland—shortgrass prairie (Frouz et al., 2013a) (Fig. 6.2). In this example, we can see that putting more effort into site preparation pays off in shortgrass prairie, as natural soil development processes are very slow and would take a longer time than would be practical to restoration goals. Moreover, as we see later, grasses are less sensitive to compaction. On the other hand, soil compaction is a major problem in forest development, which may be one reason why spontaneous regrowth in loose uncompacted soil is, in some cases, better than planted forest (Fig. 6.2). Based on this, it is clear that time, effort, and resources alone are not enough to guarantee a successful outcome when overburden material is manipulated in reclamation projects. Indeed, techniques that are ideal for improving conditions in one site may reduce site suitability in another and, in many cases (especially forest restoration of for nature conservation purposes), the less intervention, the better.

6.3.1 **Overburden deposition**

The manner and form in which overburden is deposited is of critical importance to how it can be used in reclamation projects. The most common form of overburden is as solid matter but occasionally consists of suspended material in water, where tailings of postprocessed residue ores have been deposited. Deposition of overburden should be undertaken with care to ensure stability of future heaps, because landslides or subsidence may affect previously restored adjacent areas. However, the best methods involved in geotechnical stability are beyond the purview this chapter. Nevertheless, the method by which overburden is deposited does have great potential to affect the future development of postmining sites. First, the basic rule is to bury unsuitable, problematic materials in deep layers and to place more suitable materials near the surface. This usually involves reserving quaternary deposits or other suitable materials (with suitable pH, conductivity, and clay content) and using them to cover geologically older, pyrite rich or other problematic materials. When problematic materials are buried under several meters of other substrates, and oxygen diffusion is thus limited, it is possible to solve the problem of phyotoxicity. This may make deposition of the heap more complex and expensive in the short term, but the price for avoiding generation of acid mine drainage and phytotoxic soils is typically much cheaper and easier than fixing the problem afterward.

In some circumstances, organized or intentional methods of heaping may offer the possibility to improve the textural composition of dumped overburden and consequently soil development by mixing overburden materials. Mixing can be

FIGURE 6.2

Comparison of reclaimed sites with unreclaimed sites developing by spontaneous succession at various locations in the United States, about 15-year-old forest sites in Tennessee (A) show very good performance of succession site, while about 25-year-old sites with shortgrass prairie in Wyoming (B) show unreclaimed site developing by natural succession is much worse relative to reclamation sites.

useful to produce better rooting medium from materials that might not otherwise be particularly good for this purpose. For example, Torbert et al. (1990) tested various mixtures of sandstone and siltstone for tree growth, showing that fast-weathering sandstone represents the best rooting medium. Similar conclusions have been made in correlation studies by Casselman et al. (2006). Based on these correlation studies and mixing experiments, we can summarize the optimal properties of overburden mixture for tree growth as follows. The mixture should have a high proportion of fine soil with a sandy loam texture. Torbert et al. (1990) recommended over 50% of fine soil. This amount of fine soil may not be present at the beginning, but substrates that produce fine soil during relatively rapid initial weathering are also suitable. One benefit of such rapidly weathering substrates is that they have higher macroporosity during early phases of restoration that allows better root penetration and is potentially more resistant to compaction during grading. On the other hand, as pointed by Frouz et al. (2018), suitable rooting medium should also contain a substantial proportion (c. 20%) of unweathered material in the form of medium- and small-sized gravels for the purpose of increasing macroporosity. The presence of gravel is particularly important in overburden with high clay or silt content. Substrates should have low salt content and have slightly acidic pH. The depth of the rooting medium is also important, which should be at least 1.2 m if trees are to be planted (Burger, 2009; Skousen et al., 2011). There has been much less research done regarding characteristics of overburden suitable for herb and grass restorations, but based on results of Frouz et al. (2018), we may expect that these will require more fine soil with higher proportion of clay and silt. In addition, pH is likely to be better at or close to neutral (Table 6.2).

As already mentioned, postmining sites, namely, in Europe were often reported to be valuable habitats for many rare and endangered species (Prach and Hobbs, 2008). In Europe, most of the pristine forests were removed during the Middle Ages, and most of the biodiversity which is of conservation interest is, in fact, composed of species adapted to cultural landscapes, which developed by the end of the Middle Ages in the preindustrial era. The biological communities of these cultural landscapes then suffered from the development of intensive agriculture that increased nutrient load and reduced spatiotemporal heterogeneity of the landscape. Postmining sites have been repeatedly shown to be valuable refuges for species of this preindustrial cultural landscape. These include both plant and animal species, which benefit from the presence of oligotrophic substrates occurring in a highly spatially variable environment. In addition, the presence of material naturally sorted during various deposition techniques such as heaping or flooding generates suitable habitats similar to those created by riverbanks and other rare habitats. Those conditions generate suitable habitats for species that would normally occur in oligotrophic nonforest habitats such as wetlands, marshes, steppe ecosystems, and similar. Preserving or enhancing spatial heterogeneity and original character of loose dumped substrate and allowing spontaneous succession to appear may be among the best options for maintaining and promoting these habitats that are characterized by patches of shrubs, grasses, and bare soil.

Table 6.2 Suggested best practice for restoring various types of landscape in postmining sites.

Target	Approach	Site preparation	Treatment
Support biodiversity in secondary habitats	Succession	Try to maintain as much substrate heterogeneity as possible	Just wait and monitor.
Restoration of natural ecosystem with slow-growing plant in large areas	Ecosystem translocation or inoculum transfer	Try to reconstruct subsoil as much as possible, but keep balance between layering and compaction	If possible, do whole ecosystem translocation of sod transfer, if not transfer topsoil from native habitat, by direct hailing you lay seed or plant species that did not come with topsoil transfer.
Recultivation of productive forest	Forest establishment	Preparing deep porous rooting medium with slightly acidic pH sandy lome structure and minimum compaction	Tree planting. Choosing native tree if possible similar to dominant of local biome. Nitrogen-fixing trees can improve A horizon formation but may reduce success of some late succession trees. Should be used in initial stages of restoration with caution.
Recultivation of productive or seminatural grassland	Grassland establishment	Topsoil application	Seeding grass mixture.

6.3.2 Overburden grading (smooth or rough surface)

After heaping the surface of overburden material is typically rough with many large or small elevations and depressions created during the deposition process. In most reclamation projects, such surfaces are subjected to grading by earth-moving machinery to generate smoother surfaces. Grading increases accessibility of the sites and facilitates subsequent operations such as seeding, planting trees, and application of topsoil. In some cases, grading is simultaneous to the heaping process as it allows easier movement machinery that performs heaping. Comparison of graded and ungraded surfaces shows that grading promotes overburden compaction (Frouz et al., 2018). Compaction is caused not only directly by moving heavy machinery over the surface but may also be a result of secondary processes associated with weathering of clay minerals, as their accumulation can result in the gradual development of a homogeneous compacted and sealed surface layer

(Frouz et al., 2018). In contrast, in ungraded areas where spatial heterogeneity produced by heaping were retained, secondary processes magnify this spatial heterogeneity and lead to creation of small patches of compacted soil, but sites with high macroporosity and presence of stones are also retained, which serve as safe places for tree seedlings to establish. Indeed, a long-term study shows that smooth-graded surfaces over 12 years became gradually covered by an aggressive grass *Calamagrostis epigejos*, while ungraded sites with an undulating surface created by heaping were colonized by woody vegetation dominated by birch (*Betula pendula*), goat willow (*Salix caprea*), and aspen (*Populus tremuloides*) (Fig. 6.3). Thus chronosequence studies focusing on ungraded sites show that these sites move more rapidly toward closed-canopy broadleaf forest at around 20 years of age (Frouz et al., 2008), whereas neighboring smooth-graded areas of the same age are covered by dense grass and a few solitary trees or shrubs. The observation that grading promotes grasses while trees prefer ungraded overburden has also been seen in studies conducted in the Eastern United States (Ashby, 1998; Franklin et al., 2012; Zeleznik and Skousen, 1996) (Fig. 6.4).

The negative effects of grading and compaction have led some practitioners to use earth-moving machinery to actually produce rough and loose surfaces rather than smooth ones. In these microhabitats the tree can be planted in elevated soils where their rooting systems may benefit from loose substrate, but where they will still have access to moisture in depressions (Fig. 6.4). Moreover, depressions will fill in through the erosion of surrounding material. These examples show that loose overburden may be beneficial not only for spontaneous succession but also for planted trees and subsequent development of forest. It should be pointed out that surprisingly little research has been aimed at developing heaping techniques that allow direct tree planting into ungraded overburden in such a way as to allow not only tree growth but also future maintenance and harvesting of the resulting forest. One promising question for future research is whether we can specifically identify a level of spatial heterogeneity that the tree can benefit from at establishment, but which a few decades later would be so reduced by erosion and other natural processes that it would not represent a barrier for tree harvest and other necessary operations associated with forest maintenance.

6.3.3 Topsoil and other soil layer applications

Topsoil is a term used for A (or A and part of B) horizons of forest and meadow soil or the tilled part of arable land that is often salvaged prior to a mining operation and can be used for reclamation (Fig. 6.4). In general, the topsoil can be transferred in different ways that differ in level of soil disturbance. Transferring intact soil monoliths represents a very gentle form of soil transfer; on the other hand, application of topsoil has been stored in stockpiles for quite bit of time, followed by extensive grading results in substantial disturbance of soil, loss of soil aggregates, and SOM (Table 6.2).

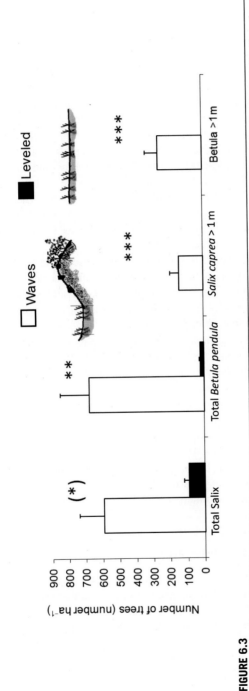

FIGURE 6.3

Number of spontaneously established birch (*Betula pendula*) and willow (*Salix caprea*), and number of birch and willow trees taller than 1 m, in 12-year-old unreclaimed sites that were either leveled or kept in original undulating appearance, significant differences between undulating and leveled surface are marked as follows: (*) $P<.1$, ** $P<.01$, and *** $P<.001$, based on Frouz et al. (2018).

FIGURE 6.4

Grading and topsoil application. Ridge of overburden graded smooth and flat on the top, note that trees planted in the rows perpendicular to the ridge grew well on the sites that remained loose, whereas grasses dominated on smoothed top, Sahara mine, Illinois, United States (A). Example of surface graded to be rough and loose, Kemess Mine, Brithish Columbia, Canada, note trees planted on elevations (B). Heavy machinery spread topsoil, Belle Ayr mine, Wyoming, United States (C).

(B) Photo courtesy of David Poshel.

A general advantage of topsoil application is that it instantly improves soil conditions in the target site (Borůvka et al., 2012). If a natural or seminatural site is used as a source of topsoil, the topsoil can be a valuable source of plant propagules and soil biota, particularly, if the soil is transported by direct hauling without storage in a stockpile prior to application (Boyer et al., 2011). One noteworthy technique for topsoil application is the transfer of whole, intact soil blocks, which are cut from a donor site and reconstructed at the recipient site in as much of a similar shape as possible (this is also sometimes called whole ecosystem transfer) (Bullock, 1998; Good et al., 1999; Waterhouse et al., 2014). In some cases, topsoil may be applied on already reconstructed subsoil. This ecosystem transfer technique has been reported as being a very suitable method of reconstructing valuable natural habitats in postmining soils. The advantage of whole ecosystem transfer is that mature plants, plant propagules, soil biota, etc. are included in the soil blocks. This is particularly advantageous in situations where plant growth and soil development are slow. A disadvantage is that this approach is quite expensive and can only be considered cost-effective if the donor site (which will be mined out) and recipient sites (previously mined) are in reasonably close proximity. Moreover, more research is needed to better understand interactions between the transferred blocks and the underlying overburden. An approach that is similar to the whole ecosystem transfer just discussed is the direct transfer of sod or just direct topsoil transfer, if the sources of plant propagules are natural habitats, it may help to reconstruct these habitats at the donor site. The advantages of this approach are similar to soil block transfers in that soil, plants, and soil biota are transferred at once. The primary difference between the two techniques is that with direct transfer of sod or topsoil, the soil is handled with much less care, which does have the advantage of making this approach much less expensive than whole soil block transfer. To be successful, these direct transfers should be conducted during the optimal season of the year so that transferred soil material stays moist and is not subject to extensive erosion. However, in many cases, topsoil is salvaged and stored in stockpiles for a while, often even several years before being used in reclamation work. Stockpiling of soil often results in compaction and formation of anaerobic conditions inside the pile, which negatively affects seedbank and soil biota (Boyer et al., 2011; Harris et al., 1989). Stockpiles may also become overgrown with weeds, and this ruderalization may increase the seedbank of unwanted species. Due to these issues, stockpiled soils are often assumed to be lower quality sources of plant propagules and soil biota relative to directly transferred soil. Despite this, stockpiled soil still instantly improves soil conditions when applied in reclamation scenarios. For example, Angst et al. (2018) applied stockpiled soils and found that this contained SOM with very similar characteristics to those in the target soils. Compaction is a major problem associated with all forms of topsoil applications, as this activity requires multiple entries of heavy machinery during topsoil transport and spreading. If the applied topsoil is rich in clay, compaction becomes an even worse problem and may inhibit plant growth and be particularly problematic for forest restoration.

Compaction or application of layers that are not naturally compatible may result in the formation of discontinuity layers that can disrupt natural water movement through the soil profile and result in gleification. Such development of water impermeable layers may cause erosion problems and can even result in the formation of detachment layers and landslides. Topsoil application may alter nutrient availability, and this may be a problem if topsoil salvaged from nutrient-rich sites, such as those used for intensive agriculture, has been applied with intention to restore natural or seminatural habitats. Availability of the nutrients together with disturbance caused by salvaging and spreading of the topsoil leads to colonization of the site by ruderal species that outcompetes rare and endangered species specializing on oligotrophic habitats. Similarly, this type of competition may represent a substantial impediment to tree establishment in forest restorations. These problems with topsoil application can be partly solved by chiseling the sites and application of herbicides to control weedy plant competition, but it is always better to evaluate whether topsoil application is necessary at all. Application of topsoil seems to be more beneficial in reconstruction of grassland habitats than in reconstruction of forests and similarly more beneficial in dryer than in moister environments. Frouz et al. (2013a) reported very fast recovery of postmining soil and soil biota during shortgrass prairie restoration on sites where topsoil was spread which was even faster than recovery of forest vegetation on sites with topsoil in postmining sites in the Eastern United States. In some cases, topsoil application can be substituted or supplemented by application of mineral fertilizer and organic amendments. Organic amendments typically contribute little to carbon stock formation but may help in vegetation establishment, particularly important can be amendment with living peat, which can serve as an inoculum for peat layer formation (Abakumov and Frouz, 2014). However, fertilization should be minimalized or avoided when restoring natural or seminatural habitats. If fertilization appears necessary, complex fertilizers with very slow release of nutrients such as rock dust or bone meal or similar should be used.

6.4 **Soil and ecosystem development**

No matter how complex the layering technique of a technosoil may be, the raw surface is always subject to complex natural processes, which take place during soil formation and primary succession. Here we will explore soil and ecosystem development during primary succession in postmining soil. First, because comparing recovery rates during primary succession relative to those in a reclamation effort can show the real added value of reclamation effort (i.e., we can evaluate how much ecosystem development speeds up with reclamation treatments relative to a situation where we did nothing). Second, possessing an understanding of natural processes of ecosystem development is essential to ensure that our reclamation efforts can employ, magnify, or speed up these processes rather than work against them. After preparation of a rooting medium, which was discussed previously, establishment of vegetation is a major step.

Next, we explore the development of key ecosystem parameters in major types of reclaimed vegetation as well as in spontaneously developing ecosystems. We then discuss the underlying mechanisms and practical applications of this knowledge for restoration work. We will present comparisons of spontaneous succession with directly managed establishment of major vegetation types such as grassland, broadleaf forest, and conifers; in cases where information is available, we will pay particular attention to the role of N-fixing plants in the restoration process.

6.4.1 Biodiversity

As previously mentioned, in several mainly European countries, postmining sites were reported to be hot spots for rare and endangered plant species that are rare or absent in surrounding landscapes (Beneš et al., 2003; Prach, 1987; Prach and Hobbs, 2008; Řehounková and Prach, 2006; Tropek et al., 2010). Lack of nutrients, uncompacted base surfaces, and spatial and geochemical heterogeneity are principal causes that support occurrence of these species in postmining sites. Not surprisingly, spontaneous succession sites show higher species diversity compared to reclaimed ones (Mudrák et al., 2010; Tropek et al., 2010). Consequently, allowing spontaneous succession is the recommended management approach to maximize conservation value of these habitats as refuge for rare and endangered species (Gremlica, 2014; Prach and Hobbs, 2008; Tropek et al., 2010). In suitable moisture conditions, ungraded unreclaimed postmining sites develop toward broadleaf forest (Frouz et al., 2008, 2015a,b). Although emerging broadleaf forest may also harbor some species of conservation value, most of the rare and endangered species reported in postmining sites are open-habitat specialists. This situation may necessitate some later management to keep these habitats open. From this point of view, small patches that are difficult to revegetate (hydrophobic, phytotoxic, extremely low nutrients) may be valuable and may remain nonvegetated for decades and maintain the presence of open habitats. Clearly, these areas should not be too big as their conservation value is optimized when they form a mosaic with other habitats, and large bare patches may negatively impact overall landscape benefits. A recent study by Moradi et al. (2018) showed that both ungraded and graded unreclaimed sites may harbor species of conservation value, so partial grading of sites that may then receive no further treatments may also help to produce a mosaic of forested and grassland habitats.

6.4.2 Primary production

Mining sites are usually assumed to be less productive than surrounding unmined landscapes. However, Zipper et al. (2011) showed that harvestable woody biomass on good quality postmining sites can be higher than that in surrounding landscapes. There is not much data available to compare biomass of spontaneously developing sites and technically reclaimed sites where vegetation was established through seeding or planting (Prach and Hobbs, 2008). However, empirical data for grassland vegetation spontaneously developing on graded top-soil was reported to have lower biomass than a cultured meadow reconstruction on applied topsoil (Čížková et al., 2018). Similarly, for forest sites, Pietrzykowski and Krzaklewski (2007), Pietrzykowski (2008), and Gorman et al. (2001) reported lower woody biomass for young spontaneously developing forests in comparison to reclaimed forests in the United States, Poland, and Czech Republic. In contrast, Frouz and Franklin (2015) reported higher woody biomass in 50-year-old unre-claimed sites in Tennessee compared to reclaimed pine plantations. This difference may be attributable to the advanced age of these sites. Frouz et al. (2015a) showed that reclaimed alder plantation had significantly higher biomass than unreclaimed sites at younger ages (younger than 20 years), whereas in older sites (30 years or older), biomass of unreclaimed sites was higher (Fig. 6.5). This is only one case study, and thus cannot be considered a general guide for what to expect during veg-etation development, but in this particular case, there are several reasons why older unreclaimed sites reached higher biomass than reclaimed ones. In particular, a regulator-defined target of reaching closed-canopy as soon as possible required a very high initial tree density (10,000 seedlings per ha). Such a high seedling density is similar to those used when alder is grown in a short rotation (Aosaar et al., 2012) and may cause high intraspecific competition and likely reduced the growth of trees as the plantation aged. Spatial heterogeneity may have been another factor contrib-uting to the observed pattern of lower biomass in older reclaimed sites. Reclaimed sites typically have only one canopy layer, whereas in unreclaimed successional sites a more complex and multilayered canopy can lead to higher light use effi-ciency and biomass production (Hardiman et al., 2011). This is further enhanced by the fact that while the reclaimed sites were graded, unreclaimed sites retained the irregular topography produced by the heaping of overburden. The irregular soil sur-face may be another reason why biomass is larger on older unreclaimed sites than on older reclaimed sites. Wavy surfaces have larger surface area and thus larger surface area of resulting canopy. These wavy canopy surfaces may increase light availability in comparison with a uniform canopy layer. Dense stand structure and reduction in light availability are particularly severe in alder, which as N-fixing trees require high light levels due to their allocation of a large proportion of photo-synthetic assimilates to support the energy-intensive process of N-fixation.

6.4.3 Soil organic matter storage and improvement of other soil properties

Accumulation of SOM is a principal precondition of successful recovery of soil nitrogen pools and nitrogen cycling. Furthermore, SOM also influences other soil properties such as water-holding capacity, sorption capacity, pH, and nutrient availability (Brady and Weil, 1999). As shown in the extensive review of Vindušková and Frouz (2013), postmining sites have excellent ability to sequester

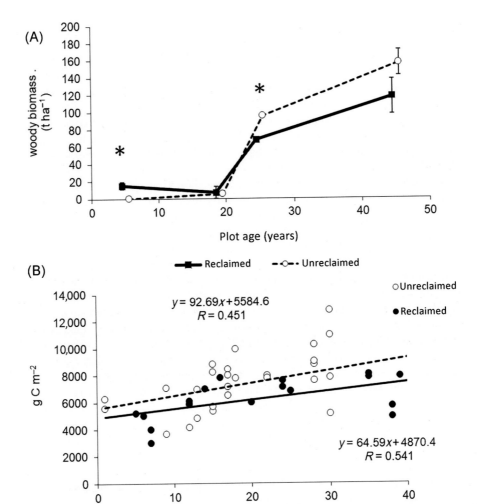

FIGURE 6.5

Woody biomass and soil carbon sequestration in reclaimed and unreclaimed sites. Development of woody biomass, both aboveground and belowground pooled in reclaimed alder plantations and unreclaimed sites overgrown by spontaneous succession in postmining sites near Sokolov (Czech Republic) (A). Soil carbon accumulation in reclaimed alder plantations and unreclaimed sites in the same area (B).

carbon (C). In freshly dumped postmining soil, the rate of C sequestration reached an average of $2.4\,t\,C\,ha^{-1}\,year^{-1}$, which is an order of magnitude greater than rates observed in arable lands converted to meadow or forest (Fig. 6.6). However, this high rate of C sequestration decreased over time and in a ~60-year-old site,

(A)

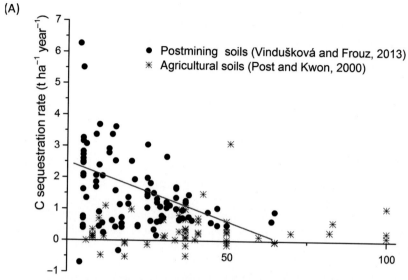

(B)

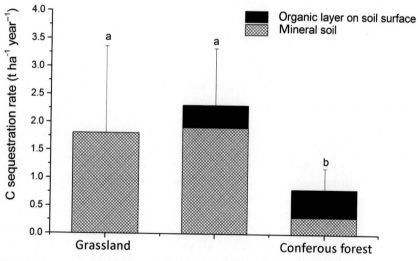

FIGURE 6.6

Carbon sequestration in postmining sites. Comparison of rate of carbon sequestration in postmining sites with carbon sequestration in arable land converted to meadow or forest (A). Comparison of rate of carbon sequestration in mineral soil and organic layer on soil surface in postmining sites reclaimed in to various land uses (B), based on Frouz and Vindušková (2018).

it reached values observed in arable land revegetated by more permanent vegetation cover. The highest rate of C sequestration was observed under broadleaf forest, with lower rates under grassland and coniferous forest (Fig. 6.6). The performance of grasslands and coniferous forests strongly depends on climatic conditions, with grassland performing better in warmer climate while coniferous forest sequestered more C in cold climate (Vindušková and Frouz, 2013). The probable mechanism for faster C sequestration in young postmining soils, as proposed by Frouz (2017), is that these soils are far from saturation and that there are many positive feedback mechanisms by which sequestering some C in one pool may facilitate sequestration of even more C in another pool. When comparing C sequestration rates between actively reclaimed soils and soils under spontaneous ecosystem development (i.e., unmanaged succession), the patterns are similar to those observed for primary production rates. In general, there is not much information available about C sequestration in unreclaimed spontaneously developing sites. Some studies have indicated that unreclaimed sites sequester less C in soil than reclaimed ones, for example, in pine forests in Poland (Pietrzykowski, 2008; Pietrzykowski and Krzaklewski, 2007) or grasslands in Czech Republic (Čížková et al., 2018). In some cases, there was no difference observed in the rate of C sequestration between reclaimed and unreclaimed sites, as shown by comparing data from Šourková et al. (2005) and Frouz and Kalčík (2006) (Fig. 6.5). These data also showed that the rate of C sequestration was much higher in reclaimed sites during early phases of reclamation, but in older sites the rate of C sequestration in unreclaimed sites increased to the point that the total C stored after 50 or 60 years in reclaimed and unreclaimed sites was similar. These patterns of C storage are consistent with observations of primary productivity over decadal timescales for the two different scenarios.

In addition to SOM and nutrient concentrations (e.g., nitrogen), soil C sequestration substantially affects many other soil properties in postmining soils. One of the most apparent is effect of SOM on water-holding capacity (Frouz, 2017; Frouz and Kuráž, 2014). Comparison of ~30-year-old reclaimed and unreclaimed sites showed that reclaimed sites had significantly higher water-holding capacity and, thus, greater soil water storage. However, the reclaimed soil also had higher wilting point that means that plant-available water was not much different between reclaimed and unreclaimed sites. There were also indications that vegetation on reclaimed sites used a bit more water for transpiration. In conclusion, despite significant improvements in water-holding capacity in the alder plantations in the study, it seems that the 30-year-old stands in both reclaimed and unreclaimed sites in Czech Republic had well-developed water regimes broadly comparable with similar sites outside the mining area (Cejpek et al., 2018).

6.4.4 Plant, soil biota, and soil interactions

Plants affect soil both directly and indirectly through their influences on the soil biotic community, which in turn may affect soil formation and plant growth

(Frouz et al., 2008). The interaction between plants and soil biota can be, in principle, divided into two major pathways: the first is associated with roots and the second with litter (Wardle et al., 2004). By interacting with roots, soil biota interact with living part of the plant, which typically immediately affects plant performance. Plants allocate a large proportion of photosynthetic assimilates to their root and rhizosphere symbionts such as mycorrhiza fungi and N-fixers (Smith and Read, 2008). The degree of advantage to the plant derived from these symbionts depends on many factors, including the stage of soil development. For example, in the initial and middle successional stages when N is limiting, N fixation is of advantage, but later, when enough N is accumulated in the ecosystem, costs of N-fixation will outweigh the benefits (Menge et al., 2012; Vitousek and Field, 1999). This may be one of the reasons for poor performance of reclaimed alder plantations described earlier (Section 4.3). Plant performance is affected not only by each associated organism separately but also by their interactions. For example, competition between ectomycorrhizal (EcM) and arbuscular mycorrhizal (AM) fungi may reduce the abundance of AM plants in the understory of a stand of EcM trees, as observed in postmining sites (Knoblochová et al., 2017; Mudrák et al., 2016) but also in elsewhere (Becklin et al., 2012). In contrast, reduction of EcM fungi and increase of AM fungi were observed as an outcome of intensive N input. This may be responsible for lower performance of the seedlings of some EcM trees, namely, oak (*Quercus petraea*) and beech (*Fagus sylvatica*) in reclaimed alder (*Alnus glutinosa*) plantations, in comparison to unreclaimed sites dominated by goat willow (*S. caprea*) (Frouz et al., 2015b). This hypothesis about a negative effect of competition from AM-associated plants on EcM fungi is also supported by lower EcM colonization of *Quercus* seedlings observed when these were planted in alder plantations (Frouz et al., 2015b).

Other influential organisms such as plant pathogens, herbivores, and parasites may also accumulate in the rhizosphere resulting in negative effects on the plants (Cortois et al., 2017; Gibson and Brown, 1991; Kardol et al., 2006; Frouz et al., 2016). Early successional plants typically do not invest in symbionts, and neither do they typically allocate resources to protection from herbivores or pathogens. For example, it has been shown in postmining sites near Sokolov (Czech Republic) that root herbivore pressure may have been a mechanism for rapid replacement of early succession plants by late succession ones (Roubíčková et al., 2012).

Plants growing for a long time (e.g., repeated growing seasons) in a particular volume of soil may cause the accumulation of plant pathogens and/or mutualist symbionts that may affect plants in the next generation: an interaction known as a plant−soil feedback. Plant−soil feedbacks are more often negative during early succession, for example, as conspecific plants grow worse in the next generation when grown repeatedly in the same soil, as the plants do not invest in either symbionts or defenses against pathogens and thus become more sensitive to pathogens. In late succession, positive plant−soil feedbacks are more often observed as plants may benefit from accumulation of beneficial symbionts in soils over time (Kardol et al., 2006).

Another group of interactions is related to plant litter and its effect on nutrient release and formation of the topsoil layer (Laughlin et al., 2015; Ponge, 2013). Litter in the form of dead aboveground plant biomass represents a major source of C and nutrient inputs into the soil. These interactions are essential for soil development and are particularly important in postmining sites as we observe very different soil developments under different plant species planted in the same substrate (Frouz et al., 2013b) (Fig. 6.7). It is generally accepted that fast-growing plants with thin leaves produce nutrient-rich litter that decomposes faster and releases more nutrients relative to plants with more conservative growth strategies (Cornwell et al., 2008). However, besides this direct effect on decomposition and nutrient release, litter quality also affects composition of the soil food web. Soil biota determine the rates of organic matter decomposition and nutrient release. Litter of fast-growing plants promotes bacteria dominated food webs characterized by fast mineralization of organic matter and fast nutrient release. In contrast, litter of conservative plants promotes fungal dominated food webs with slower organic matter turnover and nutrient release (Ponge, 2013). Besides this immediate effect of litter quality on nutrient release, litter quality also plays a principal role in the physical structure of the topsoil either directly or indirectly through impacts on soil food web composition. (Ponge, 2003). Litter of conservative plants generally promotes soil food webs with lower abundance of macrofauna, and consequently, these soils experience little or no mixing of organic matter into the mineral soil. Litter of fast-growing plants promotes the presence of more macrofauna, such as earthworms, which mix organic matter into the mineral soil layers (Frouz, 2018). Consequently, in soils under plants with lower quality litter, organic matter accumulates on the soil surface, but often with intensive leaching of organic matter and nutrients, which may become unavailable to plants. In contrast, in soil profiles developing under plants with higher quality litter, the A horizon becomes deeper as organic matter gets mixed into the soil (Figs. 6.7 and 6.8). Several manipulative experiments and correlation studies have indicated that faunal bioturbation activity is essential to transfer organic matter in the litter layer to soil C storage and A horizon growth (Frouz et al., 2001, 2006, 2007, 2013b). One illustration of this phenomenon can be drawn from different plantations planted in tertiary Miocene clays on a heap formed during lignite mining near Sokolov (Frouz et al., 2013b). In this study, it was clear that depth of A horizon as well as soil C storage closely correlated with earthworm density (Fig. 6.8). In general, soil fauna bioturbation activity increases with decreasing C:N ratio of the litter; however, fauna activity is also affected by soil pH and soil texture and is generally limited in acidic soils and in soil with very high sand content (Frouz, 2018). The level of bioturbation, driven by the interaction of litter quality and soil fauna, affects soil properties such as soil sorption capacity and water-holding capacity (Abakumov and Frouz, 2014; Frouz et al., 2006; Six et al., 2004). In turn, these modifications of the soil environment influence the composition of soil biota and the rates of soil processes (Frouz et al., 2013b,c; Šnajdr et al., 2013). Earthworms shift the composition of the whole soil food web from fungal dominated more to

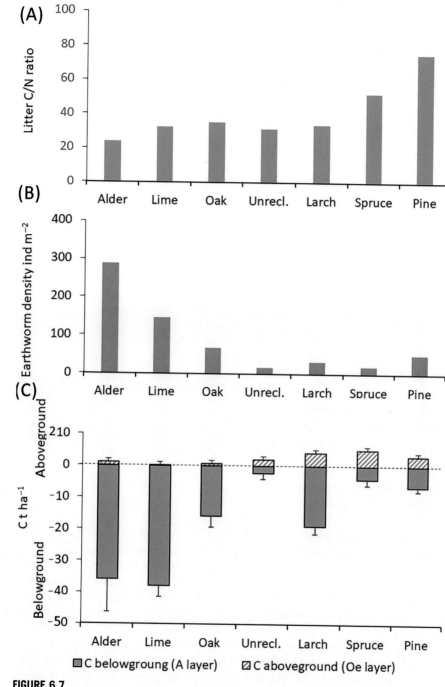

FIGURE 6.7

Litter quality, soil fauna, and soil carbon sequestration in 30-year-old tree stands in postmining sites near Sokolov (Czech Republic), C:N ratio of litter in various tree stands (A), density of earthworms in the same stand (B), and amount of C sequestered in c. 30 years belowground and on the soil surface (C), based on Frouz et al. (2013b).

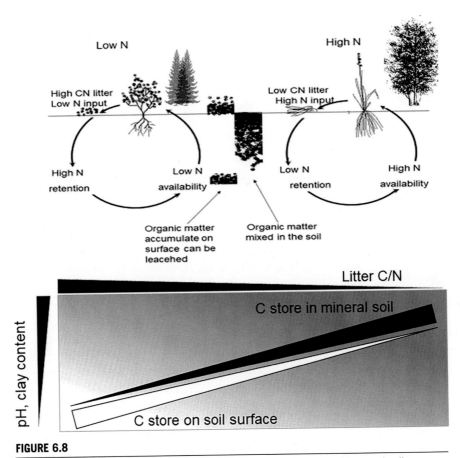

FIGURE 6.8

Scheme of effect of interactions between nitrogen availability, plant traits, and soil formation, which affect the rate of nutrient cycling.

bacterial dominated (Frouz et al., 2013b,c). Earthworm colonization can also reduce mycorrhizal (EcM) colonization in some tree species (Lawrence et al., 2003). These factors contribute to soil formation and may modify soil as a habitat for the plants (Frouz et al., 2007; Ma et al., 2003; Makulec, 2002; Mudrák and Frouz, 2018; Roubíčková et al., 2009; Scheu and Parkinson, 1994; Thompson et al., 1993). These effects of litter quality and bioturbation on soil formation and properties feed back to plant growth and plant community composition (Frouz, 2018; Ponge, 2013; Mudrák and Frouz, 2018). In general, fast-growing plants prefer and promote sites with high levels of bioturbation, fast nutrient turnover, high rate of A horizon formation, and high C storage. In contrast, conservative plants largely rely on EcM fungi that bring nutrients from partly decomposed organic matter and benefit from sites with low bioturbation, high accumulation of organic matter on the soil surface, and slower rate of soil C accumulation, and, similarly as in the previous group of plants, they also promote conditions from which they benefit (Frouz, 2018; Ponge, 2013) (Fig. 6.8).

Root and litter effects working together in reclamation can be illustrated well with the example of N-fixing plants. As already mentioned, an initial planting of N-fixing plants is often recommended to accelerate ecosystem recovery (Mikola et al., 1983; Parkinson, 1978). This makes sense when early objectives include development of communities dominated by fast-growing plants with low C:N ratio litter, and which promote bioturbation, fast C storage in soil, and other soil properties beneficial for these plants. However, when soil conditions reach certain levels of organic matter nitrogen (N) concentrations, then N-fixing plants may no longer be beneficial and should be replaced by other species (Frouz et al., 2015b). Consequently, N-fixing plants should be used either as an initial or short-rotation plantation or in a mixture that will allow replacement of N-fixing plants by other species either spontaneously through interspecific competition or selective thinning. However, if the objective of reclamation is to reconstruct communities with primarily slow-growing species, the use of N-fixing plants does not provide much benefit because the growth of the target species does not respond substantially when provided with more nutrients. In fact, because of the positive plant—soil feedback commonly observed for N-fixers, the advantage they provide for themselves may result in their outcompeting the slower growing target species. One mechanism for this may be that the surplus of nitrogen when soils are influenced by N-fixing plants may negatively affect EcM fungi that may also suppress some of the slower growing plants with conservative growth strategies (Frouz et al., 2015b). Clearly, most of the target plants exist somewhere along a continuum between conservative and fast-growing plants, but it is noteworthy that if the target community is intended to have a large proportion of slow-growing conservative plants, then the more carefully one must consider the degree to which N-fixers should be used during reclamation.

6.5 Conclusion

Many reclamation techniques are focused on facilitation of site improvement and faster initial plant growth and rapid attainment of full vegetative cover. This is achieved by extensive site preparation, application of topsoil, fertilizer, and other amendments and also through manipulation of planting density and using N-fixing plants. In grassland restoration, this approach can be successful although usually resulting in lower plant diversity. In forest restoration, this approach often succeeds in reaching fast canopy closure but at the cost of diminished long-term performance of these stands. High tree density can make the subsequent task of thinning stands more complex and can lead to high levels of competition for light, water, and nutrients and thus reduced growth as age of these stands. Soil compaction and nutrient surplus in the early stages of forest site restoration can reduce root growth and may also disrupt associations of forest trees with their EcM fungi, and this may explain their better performance observed in older unreclaimed sites or stands with less intensive site preparation. In future research, it would be useful to explore techniques that would allow simplified approaches to site preparation, namely, in heap sites with modest levels of spatial heterogeneity, it may be possible to forego grading and other steps of site preparation and thus provide benefits in the form of loose uncompacted rooting medium and some level of site heterogeneity while at the same time allow for access for site management over time. The use of N-fixing plants to speed up ecosystem development can be beneficial, but this must be done carefully in order to avoid the potentially negative consequences listed above. Finally, I suggest that it would be useful in the future to emphasize the potential to restore postmining sites to natural and seminatural habitats. This will require a change of focus, on the part of regulatory bodies, away from the simple short-term objective of rapid recovery of vegetation and more toward restoration of diverse communities and functional soil systems that will possess inherent long-term sustainability and added conservation value.

References

Abakumov, E., Frouz, J., 2014. Humus accumulation and humification during soil development in post-mining soil. In: Frouz, J. (Ed.), Soil Biota and Ecosystem Development in Post Mining Sites. CRC press, Boca Raton, FL.

Angst, G., Mueller, C.W., Angst, Š., Pivokonský, M., Franklin, J., Stahl, P.D., et al., 2018. Fast accrual of C and N in soil organic matter fractions following post-mining reclamation across the USA. J. Environ. Manage. 209, 216–226.

Aosaar, J., Varik, M., Uri, V., 2012. Biomass production potential of grey alder (*Alnus incana* (L.) Moench.) in Scandinavia and Eastern Europe: a review. Biomass Bioenergy 45, 11–26.

Ashby, W.C., 1998. Reclamation with trees pre- and post-SMCRA in Southern Illinois. Int. J. Min. Reclam. Environ. 12, 117–121.

Becklin, K.M., Pallo, M.L., Galen, C., 2012. Willows indirectly reduce arbuscular mycorrhizal fungal colonization in understorey communities. J. Ecol. 100, 343–351.

Bell, F.G., Donnelly, L.J., 2006. Mining and its Impact on the Environment. CRC Press, Boca Raton, FL, p. 536.

Beneš, J., Kepka, P., Konvička, M., 2003. Limestone quarries as refuges for European xerophilous butterflies. Conserv. Biol. 17, 1058–1069.

Borůvka, L., Kozák, J., Mühlhanselová, M., Donátová, H., Nikodem, A., Němeček, K., et al., 2012. Effect of covering with natural topsoil as a reclamation measure on browncoal mining dumpsites. J. Geochem. Explor. 113, 118–123.

Boyer, S., Wratten, S., Pizey, M., Weber, P., 2011. Impact of soil stockpiling and mining rehabilitation on earthworm communities. Pedobiologia 54 (Suppl.), S99–S102.

Bradshaw, A., 1997. Restoration of mined land using natural processes. Ecol. Eng. 8, 255–269.

Brady, N.C., Weil, R.R., 1999. The nature and properties of soils. Prentice Hall, Upper Saddle River, NJ.

Bullock, J.M., 1998. Community translocation in Britain: setting objectives and measuring consequences. Biol. Conserv. 84 (3), 199–214.

Burger, J.A., 2009. Management effects on growth, production and sustainability of managed forest ecosystems: Past trends and future directions. For. Ecol. Manage. 258, 2335–2346.

Casselman, N., Fox, T.R., Burger, J.A., Jones, A.T., Galbraith, J.M., 2006. Effects of silvicultural treatments on survival and growth of trees planted on reclaimed mine lands in the Appalachians. For. Ecol. Manage. 223, 403–414.

Cejpek, J., Kuráž, V., Vindušková, O., Frouz, J., 2018. Water regime of reclaimed and unreclaimed post-mining sites. Ecohydrology 11, e1911.

Čížková, B., Woś, B., Pietrzykowski, M., Frouz, J., 2018. Development of soil chemical and microbial properties in reclaimed and unreclaimed grasslands in heaps after opencast lignite mining. Ecol. Eng. 123, 103–111.

Cornwell, W.K., Cornelissen, J.H.C., Amatangelo, K., Dorrepaal, E., Eviner, V.T., Godoy, O., et al., 2008. Plant species traits are the predominant control on litter decomposition rates within biomes worldwide. Ecol. Lett. 11, 1065–1071.

Cortois, R., Veen, G.F., Duyts, H., Abbas, M., Strecker, T., Kostenko, O., et al., 2017. Possible mechanisms underlying abundance and diversity responses of nematode communities to plant diversity. Ecosphere 8.

Franklin, J.A., Zipper, C.E., Burger, J.A., Skousen, J.G., Jacobs, D.F., 2012. Influence of herbaceous ground cover on forest restoration of eastern US coal surface mines. New For. 43, 905–924.

Frouz, J., 2017. Effects of soil development time and litter quality on soil carbon sequestration: assessing soil carbon saturation with a field transplant experiment along a post-mining chronosequence. Land Degrad. Dev. 28, 664–672.

Frouz, J., 2018. Effects of soil macro- and mesofauna on litter decomposition and soil organic matter stabilization. Geoderma 332, 161–172.

Frouz, J., Franklin, J.A., 2015. Vegetation and soil development in planted pine and naturally regenerated hardwood stands decades after mining. J. Am. Soc. Min. Reclam. 3, 21–40.

Frouz, J., Kalčík, J., 2006. Accumulation of soil organic carbon in relation to other soil characteristic during spontaneous succession in non reclaimed colliery spoil heaps after brown coal mining near Sokolov (the Czech Republic). Ekologia 25, 388–397.

Frouz, J., Kuráž, V., 2014. Soil fauna and soil physical properties. In: Frouz, J. (Ed.), Soil Biota and Ecosystem Development in Post Mining Sites. CRC press, Boca Raton, FL.

Frouz, J., Keplin, B., Pižl, V., Tajovský, K., Starý, J., Lukešová, A., et al., 2001. Soil biota and upper soil layer development in two contrasting post-mining chronosequences. Ecol. Eng. 17, 275–284.

Frouz, J., Krištůfek, V., Bastl, J., Kalčík, J., Vaňková, H., 2005. Determination of toxicity of spoil substrates after brown coal mining using a laboratory reproduction test with *Enchytraeus crypticus* (Oligochaeta). Water Air Soil Pollut. 162, 37–47.

Frouz, J., Elhottová, D., Kuráž, V., Šourková, M., 2006. Effects of soil macrofauna on other soil biota and soil formation in reclaimed and unreclaimed post mining sites: Results of a field microcosm experiment. Appl. Soil Ecol. 33, 308–320.

Frouz, J., Pižl, V., Tajovský, K., 2007. The effect of earthworms and other saprophagous macrofauna on soil microstructure in reclaimed and un-reclaimed post-mining sites in Central Europe. Eur. J. Soil Biol. 43, S184–S189.

Frouz, J., Prach, K., Pižl, V., Háněl, L., Starý, J., Tajovský, K., et al., 2008. Interactions between soil development, vegetation and soil fauna during spontaneous succession in post mining sites. Eur. J. Soil Biol. 44, 109–121.

Frouz, J., Hrčková, K., Lana, J., Krištůfek, V., Mudrák, O., Lukešová, A., et al., 2011a. Can laboratory toxicity tests explain the pattern of field communities of algae, plants, and invertebrates along a toxicity gradient of post-mining sites? Appl. Soil Ecol. 51, 114–121.

Frouz, J., Cajthaml, T., Kříbek, B., Schaeffer, P., Bartuška, M., Galertová, R., et al., 2011b. Deep, subsurface microflora after excavation respiration and biomass and its potential role in degradation of fossil organic matter. Folia Microbiol. 56, 389–396.

Frouz, J., Jílková, V., Cajthaml, T., Pižl, V., Tajovský, K., Háněl, L., et al., 2013a. Soil biota in post-mining sites along a climatic gradient in the USA: simple communities in shortgrass prairie recover faster than complex communities in tallgrass prairie and forest. Soil Biol. Biochem. 67, 212–225.

Frouz, J., Livečková, M., Albrechtová, J., Chroňáková, A., Cajthaml, T., Pižl, V., et al., 2013b. Is the effect of trees on soil properties mediated by soil fauna? A case study from post-mining sites. For. Ecol. Manage. 309, 87–95.

Frouz, J., Thebault, E., Pižl, V., Adl, S., Cajthaml, T., Baldrian, P., et al., 2013c. Soil food web changes during spontaneous succession at post mining sites: a possible ecosystem engineering effect on food web organization? PLoS One 8.

Frouz, J., Špaldoňová, A., Fričová, K., Bartuška, M., 2014. The effect of earthworms (*Lumbricus rubellus*) and simulated tillage on soil organic carbon in a long-term microcosm experiment. Soil Biol. Biochem. 78, 58–64.

Frouz, J., Dvorščík, P., Vávrová, A., Doušová, O., Kadochová, Š., Matějíček, L., 2015a. Development of canopy cover and woody vegetation biomass on reclaimed and unreclaimed post-mining sites. Ecol. Eng. 84, 233–239.

Frouz J., and Vindušková O. Soil organic matter accumulation in postmining sites: potential drivers and mechanisms, in: Munoz M.Á., Zornoza R. (Eds.), Soil management and climate change. effects on organic carbon, nitrogen dynamics and greenhouse gas emissions, Academic Press, 2018.

Frouz, J., Vobořilová, V., Janoušová, I., Kadochová, Š., Matějíček, L., 2015b. Spontaneous establishment of late successional tree species English oak (*Quercus robur*) and European beech (*Fagus sylvatica*) at reclaimed alder plantation and unreclaimed post mining sites. Ecol. Eng. 77, 1–8.

Frouz, J., Mudrák, O., Reitschmiedová, E., Walmsley, A., Vachová, P., Šimáčková, H., et al., 2018. Rough wave-like heaped overburden promotes establishment of woody vegetation while leveling promotes grasses during unassisted post mining site development. J. Environ. Manage. 205, 50–58.

Gibson, C.W.D., Brown, V.K., 1991. The effects of grazing on local colonization and extinction during early succession. J. Veg. Sci. 2, 291–300.

Good, J.E., Wallace, H.L., Stevens, P.A., Radford, G.L., 1999. Translocation of herb-rich grassland from a site in Wales prior to opencast coal extraction. Restor. Ecol. 7 (4), 336–347.

Gorman, J., Skousen, J., Sencindiver, J., et al., 2001. Forest productivity and mine soil development under a white pine plantation versus natural vegetation after 30 years. Land Reclamation, A Different Approach, Albuquerque, NM, June 3−7, 2001. American Society for Surface Mining and Reclamation, Lexington, KY.

Gremlica, T., 2014. Mining land and similar habitats: a barren land or a new wilderness in the cultural landscape? In: Frouz, J. (Ed.), Soil Biota and Ecosystem Development in Post Mining Sites. CRC press, Boca Raton, FL.

Hardiman, B.S., Bohrer, G., Gough, C.M., Vogel, C.S., Curtis, P.S., 2011. The role of canopy structural complexity in wood net primary production of a maturing northern deciduous forest. Ecology 92, 1818−1827.

Harris, J.A., Birch, P., Short, K.C., 1989. Changes in microbial community and physicochemical characteristics of topsoils stockpiled during opencast mining. Soil Use Manage. 5 (4), 161−168.

Kardol, P., Bezemer, T.M., van der Putten, W.H., 2006. Temporal variation in plant-soil feedback controls succession. Ecol. Lett. 9, 1080−1088.

Knoblochová, T., Kohout, P., Püschel, D., Doubková, P., Frouz, J., Cajthaml, T., et al., 2017. Asymmetric response of root-associated fungal communities of an arbuscular mycorrhizal grass and an ectomycorrhizal tree to their coexistence in primary succession. Mycorrhiza 27, 775−789.

Laughlin, D.C., Richardson, S.J., Wright, E.F., Bellingham, P.J., 2015. Environmental filtering and positive plant litter feedback simultaneously explain correlations between leaf traits and soil fertility. Ecosystems 18, 1269−1280.

Lawrence, B., Fisk, M.C., Fahey, T.J., Suárez, E.R., 2003. Influence of nonnative earthworms on mycorrhizal colonization of sugar maple (*Acer saccharum*). New Phytol. 57, 145−153.

Ma, Y., Dickinson, N.M., Wong, M.H., 2003. Interactions between earthworms, trees, soil nutrition and metal mobility in amended Pb/Zn mine tailings from Guangdong, China. Soil Biol. Biochem. 35, 1369−1379.

Makulec, G., 2002. The role of *Lumbricus rubellus* Hoffm. in determining biotic and abiotic properties of peat soils. Pol. J. Ecol. 50, 301−339.

Menge, D.N.L., Hedin, L.O., Pacala, S.W., 2012. Nitrogen and phosphorus limitation over long-term ecosystem development in terrestrial ecosystems. PLoS One 7.

Mikola, P., Uomala, P., Mälkönen, E., 1983. Application of biological nitrogen fixation in European silviculture. Forestry Sci. 9, 279−294.

Moradi, J., Frouz, J., van Diggelen, R., 2018. Facilitating ecosystem assembly: plant-soil interactions as a restoration tool. Biol. Conserv. 220, 272−279.

Mudrák, O., Frouz, J., 2018. Earthworms increase plant biomass more in soil with no earthworm legacy than in earthworm-mediated soil, and favour late successional species in competition. Funct. Ecol. 32, 626−635.

Mudrák, O., Frouz, J., Velichová, V., 2010. Understory vegetation in reclaimed and unreclaimed post-mining forest stands. Ecol. Eng. 36, 783−790.

Mudrák, O., Hermová, M., Tesnerová, C., Rydlová, J., Frouz, J., 2016. Above-ground and below-ground competition between the willow *Salix caprea* and its understorey. J. Veg. Sci. 27, 156−164.

Parkinson, D., 1978. The restoration of soil productivity. In: Holdgate, M.W., Woodman, M.J. (Eds.), The Breakdown and Restoration of Ecosystems. Plenum Publishing Company, New York, pp. 213−229.

Pietrzykowski, M., 2008. Soil and plant communities development and ecological effectiveness of reclamation on a sand mine cast. J. For. Sci. 54, 554−565.

Pietrzykowski, M., Krzaklewski, W., 2007. An assessment of energy efficiency in reclamation to forest. Ecol. Eng. 30, 341−348.

Ponge, J.F., 2003. Humus forms in terrestrial ecosystems: a framework to biodiversity. Soil Biol. Biochem. 35, 935−945.

Ponge, J.F., 2013. Plant-soil feedbacks mediated by humus forms: a review. Soil Biol. Biochem. 57, 1048−1060.

Prach, K., 1987. Succession of vegetation on dumps from strip coal mining, N. W. Bohemia, Czechoslovakia. Folia Geobotanica et. Phytotaxonomica 22, 339−354.

Prach, K., Hobbs, R.J., 2008. Spontaneous succession versus technical reclamation in the restoration of disturbed sites. Restor. Ecol. 16, 363−366.

Řehounková, K., Prach, K., 2006. Spontaneous vegetation succession in disused gravel-sand pits: Role of local site and landscape factors. J. Veg. Sci. 17, 583−590.

Rojik, P., 2014. Geological substrates and heaping process of coal mining operations in the Sokolov Basin, Czech Republic: implications for reclamation and soil development. In: Frouz, J. (Ed.), Soil Biota and Ecosystem Development in Post Mining Sites. CRC press, Boca Raton, FL.

Roubíčková, A., Mudrák, O., Frouz, J., 2009. Effect of earthworm on growth of late succession plant species in postmining sites under laboratory and field conditions. Biol. Fertil. Soils 45, 769−774.

Roubíčková, A., Mudrák, O., Frouz, J., 2012. The effect of belowground herbivory by wireworms (Coleoptera: *Elateridae*) on performance of *Calamagrostis epigejos* (L.) Roth in post-mining sites. Eur. J. Soil Biol. 50, 51−55.

Scheu, S., Parkinson, D., 1994. Effects of invasion of an aspen forest (Canada) by *Dendrobaena octaedra* (Lumbricidae) on plant-growth. Ecology 75, 2348−2361.

Six, J., Bossuyt, H., Degryze, S., Denef, K., 2004. A history of research on the link between (micro)aggregates, soil biota, and soil organic matter dynamics. Soil Tillage Res. 79, 7−31.

Skousen, J., Zipper, C., Burger, J., Barton, C., Angel, P., 2011. Selecting materials for mine soil construction when establishing forest on Appalachian mine sites. For. Reclam. Advis. 8, 1−6.

Smith, S.E., Read, D.J., 2008. Mycorrhizal Symbiosis, Elsevier, third ed.pp. 1−787.

Šnajdr, J., Dobiášová, P., Urbanová, M., Petránková, M., Cajthaml, T., Frouz, J., et al., 2013. Dominant trees affect microbial community composition and activity in post-mining afforested soils. Soil Biol. Biochem. 56, 105−115.

Šourková, M., Frouz, J., Šantrůčková, H., 2005. Accumulation of carbon, nitrogen and phosphorus during soil formation on alder spoil heaps after brown-coal mining, near Sokolov (Czech Republic). Geoderma 124, 203−214.

Tesnerová, C., Zadinová, R., Pikl, M., Zemek, F., Kadochová, Š., Matějíček, L., et al., 2017. Predicting the toxicity of post-mining substrates, a case study based on laboratory tests, substrate chemistry, geographic information systems and remote sensing. Ecol. Eng. 100, 56−62.

Thompson, L., Thomas, C.D., Radley, J.M.A., Williamson, S., Lawton, J.H., 1993. The effect of earthworms and snails in a simple plant community. Oecologia 95, 171−178.

Torbert, J.L., Burger, J.A., Daniels, W.L., 1990. Pine growth variation associated with overburden rock type on a reclaimed surface mine in Virginia. J. Environ. Qual. 19, 88−92.

Tropek, R., Kadlec, T., Karešová, P., Spitzer, L., Kočárek, P., Malenovský, I., et al., 2010. Spontaneous succession in limestone quarries as an effective restoration tool for endangered arthropods and plants. J. Appl. Ecol. 47, 139−147.

Vindušková, O., Frouz, J., 2013. Soil carbon accumulation after open-cast coal and oil shale mining in Northern Hemisphere: a quantitative review. Environ. Earth Sci. 69, 1685−1698.

Vindušková, O., Dvořáček, V., Prohasková, A., Frouz, J., 2014. Distinguishing recent and fossil organic matter − a critical step in evaluation of post-mining soil development − using near infrared spectroscopy. Ecol. Eng. 73, 643−648.

Vindušková, O., Sebag, D., Cailleau, G., Frouz, J., 2015. Methodological comparison for quantitative analysis of fossil and recently derived carbon in mine soils with high content of aliphatic kerogen. Org. Geochem. 89−90, 14−22.

Vitousek, P.M., Field, C.B., 1999. Ecosystem constraints to symbiotic nitrogen fixers: a simple model and its implications. Biogeochemistry 46, 179−202.

Wardle, D.A., Bardgett, R.D., Klironomos, J.N., Setala, H., van der Putten, W.H., Wall, D. H., 2004. Ecological linkages between aboveground and belowground biota. Science 304, 1629−1633.

Waterhouse, B.R., Adair, K.L., Boyer, S., Wratten, S.D., 2014. Advanced mine restoration protocols facilitate early recovery of soil microbial biomass, activity and functional diversity. Basic Appl. Ecol. 15 (7), 599−606.

Zeleznik, J., Skousen, J., 1996. Survival of three tree species on old reclaimed surface mines in Ohio. J. Environ. Qual. 25, 1429−1435.

Zipper, C.E., Burger, J.A., Skousen, J.G., Angel, P.N., Barton, C.D., Davis, V., et al., 2011. Restoring forests and associated ecosystem services on appalachian coal surface mines. Environ. Manage. 47 (5), 751−765.

Salinity and the reclamation of salinized lands

R.J. Harper[1], B. Dell[1], J.K. Ruprecht[1], S.J. Sochacki[1] and K.R.J. Smettem[1,2]
[1]*College of Science, Health, Engineering and Education, Murdoch University,*
Murdoch, WA, Australia
[2]*School of Civil, Environmental and Mining Engineering, The University of Western Australia,*
Crawley, WA, Australia

7.1 Introduction

Soil salinity results from an excess of salts in the soil, with consequent impacts on plant growth. These salts can include the chlorides, carbonates, and sulfates of sodium, calcium, and magnesium (Rhoades, 1993; Rhoades et al., 1992), with sodium chloride being the most common. The salts affect plant function by imposing an osmotic stress that reduces plant water uptake and through toxic concentrations of sodium and chloride. In some cases irrigation waters contain carbonates and this results in increased soil alkalinity, with effects on the availability of nutrients and the stability of soil organic matter (Eaton, 1950). The effects of salinity often manifest themselves in reduced plant growth and crop productivity, with different plant species exhibiting different degrees of salinity tolerance (Rhoades et al., 1992). Salinity also affects soil biological activity (García and Hernández, 1996).

Salinization not only removes arable land from production but in some cases affects water resources, built infrastructure, and remnant biodiversity. This is particularly important where irrigation waters are compromised by salinity as this directly affects food production. Salinity thus intersects with major global concerns, including food security, desertification, and biodiversity protection. Indeed, the salinization of land is considered a form of desertification, under the United Nations Convention to Combat Desertification. In addition, salinization of rivers is considered a global threat to biodiversity and compromises the ecosystem, goods and services of rivers, wetlands, and lakes (Cañedo-Argüelles et al., 2013; Vengosh, 2003). Notable examples of this include the Aral Sea Basin in Central Asia, the Indo-Gangetic Basin in India, Indus Basin in Pakistan, the Yellow River Basin in China, the Euphrates Basin in Syria and Iraq, the Murray-Darling Basin in Australia, and the San Joaquin Valley in the United States (Qadir et al., 2014). Salinization also contributes to the loss of remnant biodiversity, such as in South Western Australia, which

Soils and Landscape Restoration. DOI: https://doi.org/10.1016/B978-0-12-813193-0.00007-2

is a global biodiversity hot spot (Myers et al., 2000), affecting flora and fauna both on the land and in aquatic systems (Halse et al., 2003).

Lands affected by salinity can be considered in terms of the broad framework of primary and secondary salinity (Ghassemi et al., 1995), which can be further modified to account for additional concerns:

- *Primary salinity* is land that is naturally saline.
- *Secondary salinity* occurs where land has been salinized as a result of:
  ○ irrigation;
  ○ dryland salinity that involves changes in landscape- or site-scale water and salt balances;
  ○ inundation from flooding, storm surges, or tsunamis; and
  ○ drying of water bodies, with the exposure of saline soils and sediments.

In an alternative framework, Rengasamy (2006) presents three categories based on salinization processes:

- *Irrigation-associated salinity* where leaching is insufficient the salts brought in by irrigation water can accumulate in the root zone, particularly if drainage is poor. Rising saline groundwater can compound the problem.
- *Groundwater-associated salinity* generally occurs in low-lying discharge areas where the water table is close to the surface and salt is brought to the surface by groundwater movement coupled with high evapotranspiration as a driving force.
- *Nongroundwater-associated salinity* usually associated with landscapes where the water table is deep and drainage is poor. Salts accumulate in the soil from rainfall, weathering, and aeolian accession. If the hydraulic properties of the soil are poor, salt can accumulate in the root zone.

The effects of salinization on arable land are of particular concern in terms of future food security (Qadir et al., 2014; FAO and ITPS, 2015). World population continues to increase toward 11.2 billion people by the year 2100 (United Nations, 2017) and per capita demand is also increasing, leading to pressures to expand food production to meet this need. Much of the change in population is projected to occur in Africa with an increase from 2017 to 2100 of 3.2 billion people, compared to an increase of 3.6 billion people globally. There are suggestions that to meet global food demand by 2050 agricultural production will need to increase by 60%−100% from a 2005/07 baseline (Tillman et al., 2011), although recently Hunter et al. (2017) re-examined this target, with consideration of sustainable intensification and suggested that an increase in production of approximately 25%−75% would be sufficient to meet 2050 food demand. Clearly, some of the expansion of agricultural area to meet this increased demand will come through the use of semisaline land and semisaline irrigation waters as nonsaline sources are fully committed (Ladeiro, 2012). In some areas, groundwater used for irrigation is salinizing as overexploitation results in the reversal of groundwater flows and the intrusion of seawater, while an additional concern lies with climate change, through sea-level rise, associated storm surges, and increased inundation with seawater (Church et al., 2013).

Numerous approaches have been developed to manage salinity and reclaim salinized soils, with varying degrees of success, and these are described in this chapter. We first outline the underlying processes that cause salinity and then address various plant-based (e.g., forestry or agronomic), engineering, and policy management approaches, together with emerging approaches using carbon mitigation investment that might provide funding for broadscale reclamation of salinized land.

7.2 **Global distribution of salinity**

Global estimates of salinity often include both saline and sodic soils (Abrol et al., 1988), with around 1 billion hectares affected (FAO and ITPS, 2015). The FAO uses the definition that salinity occurs when the electrical conductivity in a saturation soil extract is $>4\,\text{dS m}^{-1}$ at 25°C (e.g., Rhoades and Miyamoto, 1990) with different categories of salinity for values less than this (Table 7.1). Sodicity is defined as where the soil has a small salt content, but the soil contains $>6\%$ sodium ions in the exchange complex (FAO and ITPS, 2015). In this chapter we focus on the approximately 412 million hectares of saline soils, which tend to be most prevalent in arid and semiarid regions (Rengasamy, 2006) but do occur in other regions as well (Abrol et al., 1988).

Salinization is a change in the salinity status of a soil with an estimated abandonment of $0.3-1.5$ million hectare year^{-1} of salinized soils (FAO and ITPS, 2015). The area of land that is affected by lower concentrations of salinity, which might have less profound effects on plant growth, is unknown. Rengasamy (2006) termed this "transient salinity" and suggested that in Australia two-thirds of the agricultural area was affected. There are no global estimates of this phenomenon.

Table 7.1 Soil salinity classes and crop growth (Abrol et al., 1988).

Soil salinity class	Conductivity of the soil saturation extract (dS m^{-1})	Effect on crop plants
Nonsaline	0–2	Salinity effects negligible
Slightly saline	2–4	Yields of sensitive crops may be restricted
Moderately saline	4–8	Yields of many crops are restricted
Strongly saline	8–16	Only tolerant crops yield satisfactorily
Very strongly saline	> 16	Only a few very tolerant crops yield satisfactorily

Globally, a major proportion of the land affected by secondary salinity is associated with irrigation. In 2006, out of a global total of 301 million hectares of irrigated land, around 76 million hectares were salinized (FAO and ITPS, 2015) and this was also the fate of historic irrigation developments such as associated with the Euphrates and Tigris Rivers (Hillel, 1991). Contemporary examples of salinity associated with irrigation include the San Joaquin Valley in California, the Amu Darya and Syr Darya Rivers in Uzbekistan (Egamberdiyeva et al., 2007), and Haryana in India (Datta and Jong, 2002).

Dryland salinity is a particular issue across southern Australia where extensive deforestation of deep-rooted perennial plants for agriculture and replacement with shallow-rooted annuals has changed the water balance at regional or landscape scales. This has mobilized salts stored over millennia in the landscape and compromised agricultural soils and remnant biodiversity throughout much of Australia's 100 million hectare wheat—sheep zone, with an estimated 5.7 million hectares affected by or at risk of dryland salinity (National Land and Water Resources Audit, 2001). Affected areas in catchments drain to waterways and as a result rivers have become saline, thus compromising downstream water supplies (Ruprecht and Dogramaci, 2005).

Salinization can also result from flooding or inundation with saline waters, through breaching of dykes, floods, storm surges, or tsunamis or the drying of large inland bodies of water. In the latter case, changes in water balance through changes in climate or diversion of inflow waters can result in previous lake beds being exposed through evaporation. Such is the case with the Aral Sea, where diversion of contributing rivers resulted in a decline of the Sea's area from $68,000 \text{ km}^2$ in 1960 to $14,280 \text{ km}^2$ in 2010 (Micklin, 2007). The level in what is referred to as the "large Aral Sea" declined by more than 29 m and the salinity concentration increased from 10,000 to $130,000 \text{ mg L}^{-1}$ (Gaybullaev et al., 2012). Other lakes where the inflows and surface areas have sharply declined, with the consequent exposure of saline soils, include Lake Urmia in Iran, Lake Chad in Africa, and the Salton Sea in California. The problem is also acute in some large river deltas where multiple drivers are contributing to the salinization of freshwater and soils (Rahman et al., 2019).

7.3 Measurement of salinity and impacts on plant growth

Soil salinity is a continuous variable; with the definition of saline soils varying both across countries and with measurement method. Methods of assessment of salinity vary from the laboratory through to the landscape-scale. The saturation soil extract method involves laboratory analysis through the measurement of electrical conductivity (Rhoades, 1993); however, for rapid field-based assessment, electromagnetic (EM) induction instruments have been used, with data either representing points in the landscape or being used to map the distribution of salinity (Lesch et al., 1992; Williams and Baker, 1982). In some cases these data have been combined with other site information (Lesch et al., 1995). Depending on the EM instrument used, salinity can be mapped in the surface layers or to greater depths (Yao and Yang, 2010; Doolittle and Brevik, 2014). Another approach has been to combine satellite imagery with digital elevation information to predict the distribution and temporal change of dryland salinity (e.g., Furby et al., 2010). Because of the wide range of approaches used to measure salinity, the values presented in different studies are not necessarily directly comparable.

Although salinity is a continuous variable, it can be presented for convenience as a series of categories from nonsaline to very strongly saline (Table 7.1), with associated broad effects on plant species. The physiological tolerance to salinity varies, such that some plant species have no tolerance, whereas others can survive and grow in hypersaline environments. Rhoades et al. (1992) summarize the salinity tolerance of a wide range of crop, forage and horticultural and woody species. Mechanisms of salinity tolerance have been reviewed by Cheeseman (1988) and more recently by Volkov and Beilby (2017). This variation in tolerance of different plant species provides one option for management of saline soils and will be discussed later in this chapter, when describing plant-based management approaches.

As described earlier, a major cause of salinization is through the application of salts in irrigation waters. Salinity can be readily measured in water, via the measurement of electrical conductivity (Rhoades, 1993), because conductivity increases with an increase in ionic concentration. Again, this is a continuous variable, but for convenience the quality of water can be presented as a series of categories (Table 7.2). This approach however does not discriminate between ionic species, with apparent salinity being affected by both cations (Ca^{2+}, Mg^{2+}, Na^+, and K^+) and anions (CO_3^{2-}, SO_4^{2-}, HCO_3^-, and Cl^-) and these can be analyzed and expressed as a mass or milliequivalents per liter and are often expressed as total dissolved salt (TDS) or total soluble salt.

Table 7.2 Classification of saline waters (Rhoades et al., 1992).

Water class	Electrical conductivity (dS m^{-1})	Salt concentration (mg L^{-1})	Type of water
Nonsaline	<0.7	<500	Drinking and irrigation water
Slightly saline	0.7–2	500–1500	Irrigation water
Moderately saline	2–10	1500–7000	Primary drainage water and groundwater
Highly saline	10–25	7000–15,000	Secondary drainage water and groundwater
Very highly saline	25–45	15,000–35,000	Very saline groundwater
Brine	>45	>45,000	Seawater

7.4 Causes of soil salinity

Soil salinization can be considered in terms of the salt and water balance within a soil profile or of a land system; that is the difference in inputs and outputs. Understanding this broad framework provides indications as to useful management approaches.

For example, if more salt is added to a soil than leaves in drainage, there will be salinization. Salt content on a site can increase through inputs from irrigation water, or from inundation with saline waters. The concentration of salt can also increase through evaporation, as seen when a saline water table is close to the surface and contributes to salinity through capillary rise of water. Here the contributing water table is generally <2 m below the surface, with this depth depending on soil properties such as porosity (Ruprecht and Dogramaci, 2005). Evaporative concentration of salt also occurs during drying of inland water bodies.

Salinity can also be related to landscape hydrology, and the interplay of landscape water balance and groundwater systems, as seen with dryland salinity in South Western Australia. Here, the change in landscape water balance has resulted in the pressurization of groundwater systems and consequent movement of water through the deep regolithic landscapes, resulting in the mobilization of stored salt (Schofield, 1992). Salty groundwater can occur close to or discharge on the surface of the landscape from aquifers under pressure and cause salinization. In these situations, drainage can be problematic, and attempts have been made to manipulate the water balance of the whole landscape (George et al., 2012) and thus reduce recharge and saline discharge.

With adequate drainage salts can be leached and the soil's functions recovered. There are numerous examples of this, such as the recovery of soils inundated by seawater following tsunamis in Indonesia (McLeod et al., 2010), the breaching of dykes in Holland (Gerritsen, 2005), and a storm surge in the Mackenzie Delta of Canada that traveled 30 km inland and flooded 30,000 ha (Lantz et al., 2015). Salt can be removed from soils through leaching with nonsaline waters and this requires movement of both salt and water away from the soil surface. The amount of leaching required will depend on the initial salt content of the soil, the quality of the leaching water, and the characteristics of the soil (Rhoades et al., 1992). Conversely, in locations where the water table remains near the surface, such as in discharge areas, or under irrigation without adequate drainage, the salinity problem will remain.

7.5 Managing salinized landscapes: Stabilization or reclamation?

In some cases, it is possible to reverse the effects of salinization; however, a crucial consideration in the reclamation of salinized soils is the desired end point. That is, are the interventions aimed at stabilizing the soils against further change, or reversing the process and restoring it to another state? Apart from the technical issues, the approach taken will also depend on the economic and sociopolitical context of the land in question; in some cases, abandonment of land may be the rational economic response, as the costs of treatment outweigh the economic or social returns, whereas in others the land may be restored to productive agriculture or revegetated with a biodiverse ecosystem.

There are thus several potential approaches:

- *Prevention*: Here the aim is to stop the development of salinity.
- *Stabilization*: Not all situations allow drainage and a reduction of salt concentrations; however, it may be possible to adapt to salinity. One example would be revegetation with salt-adapted plants. This may also be the case in naturally saline lands that have been damaged by overgrazing.
- *Active management*: Here approaches are taken to reverse salinity, such as through drainage and leaching of soils, or the restoration of catchment water balances through recharge control.
- *Land retirement or abandonment*: In some cases the salinized land is simply too saline for plant-based solutions, it may be technically not possible to treat the land or the costs of treatment, such as installation of drainage is uneconomic. In such circumstances abandonment may be the rational approach.

Ruprecht and Dogramaci (2005) describe three major approaches to managing or reversing salinity: these being (1) engineering, (2) plant-based, and (3) policy and legislative changes. In brief, these approaches either aim to prevent, stabilize, or actively reduce salinity, by manipulating a site's salt and water balances.

7.5.1 Engineering approaches

Engineering approaches can be applied to manage both salt and water balances. Examples in irrigation management include controlling the overall salt input through monitoring both the quality and quantity of water applied, as it is often not possible to use alternative sources of water. Irrigation management involves optimizing the quantity of water applied for particular crops, or the recycling of drainage water on more salt-tolerant crops (Levers and Schwabe, 2017; Rhoades et al., 1992). Although desalination technology exists, this is presently too expensive as a source of irrigation water except for intensive very high-value horticultural crops (Burn et al., 2015).

Leaching of salts from salinized soil requires a source of less saline water to leach the salts, drainage, and management of water tables. Excess water that is applied to the surface and moves beneath the root zone is termed the leaching fraction, and the effectiveness of leaching depends on the salinity and volume of the applied water and the infiltration characteristics of the soils (Ruprecht and Dogramaci, 2005; Rhoades et al., 1992). The recovery from inundation can be relatively rapid under the right circumstances.

Water can be carried away via installed pipes and drains or by pumping, with the resultant saline waters either reapplied to other crops, diverted into evaporation ponds (Rhoades et al., 1992; Levers and Schwabe, 2017), or disposed of downstream. This engineering option requires appropriate management of drainage waters and consideration of the likely off-site impacts from salts and possible pesticide or heavy metal contamination (Rhoades et al., 1992; Degens et al., 2012).

A key issue to consider in any restoration program is the properties of the soils that are being treated, and whether they are indeed amenable to drainage and restoration. Consideration is also needed of the subsequent management of reclaimed soils and resulting issues such as dispersal of clay particles and the associated risk of surface crusting (Rhoades et al., 1992), or managing the residual alkalinity caused by the use of carbonate rich waters.

7.5.2 Plant-based approaches

A range of plant-based approaches has been used in the management of salinized soils. These follow three broad strategies and can involve the use of a wide range of plant species, As discussed earlier there are significant differences in the adaptation of different plant species to differences in salinity.

Approaches include:

- on naturally saline land, managing existing vegetation to reduce the risk of damage or replanting native species to rehabilitate ecosystems;
- treating salinized sites with saline-tolerant plant species; and
- managing the site or landscape water balance by lowering the water table through an increase in evapotranspiration.

Less common is phytoremediation or using plants to remove salt from the soil; as generally the amounts of salt stored in saline soils far exceed the amounts that can be removed in plants (Heuperman et al., 2002). However, there are some promising developments in phytoremediation (Imadi et al., 2016). Here the aim of the technique is to improve the soil by using plants with tolerance to high salt contents in order to enhance soil calcium levels and decrease sodium levels, after which the site can be replanted with less salt-tolerant crops. Nikalje et al. (2018), for example, propose harvesting halophytes for industrial use thereby removing salt; however, these techniques have to consider the total salt stored in a soil compared to what will be removed through biomass removal.

7.5.2.1 Managing vegetation on naturally saline land

Naturally saline land is often covered with halophytic species. In the case of shrubs, these may have been damaged by overgrazing and there is consequent interest in restoring these ecosystems. There is an extensive literature, both in terms of grazing management of halophytic shrublands and also in the restoration of these systems (e.g., Barrett-Lennard, 2002; Squires and Ayoub, 1994).

7.5.2.2 Treating salinized soils

A common treatment for salinized land is to revegetate with salt-tolerant species, with this termed "biodrainage" by Heuperman et al. (2002). The selection of plant species depends on factors such the desired end use, climate and degree of soil salinity. If salinity continues to increase, it is possible to change plant species (Rhoades, 1993); however, there is an upper limit of salinity for which this will work. For grazing systems, there is a range of halophytic grazing shrubs (e.g., *Atriplex* spp.) that can be used (Barrett-Lennard, 2002). There is also a range of product options using trees, these including as a bioenergy feedstock (Sochacki et al., 2012) (Fig. 7.1), for carbon sequestration or timber (Marcar and Crawford, 2004; Lambert and Turner, 2000) or using biodiverse species to restore or enhance wildlife habitat (George et al., 2012).

The evidence that such plantings reduce groundwater levels on salinized sites through increased evapotranspiration is mixed (Morris and Collopy, 1999) and success will depend on the local hydrogeological conditions, such as surface water inflows, the transmissivity of near surface water tables, and the discharge from deeper aquifers. For example, in a study in Victoria, Australia, Heuperman (1999) found that after 10 years trees planted in a salinized area decreased the level of an unconfined aquifer by 2−4 m, reversing the hydraulic gradient such that water flowed toward the trees and site salinity increased. Other studies of revegetation with saltbush indicate substantial use of groundwater (e.g., Barrett-Lennard and Malcolm, 2000). An overriding issue with vegetation-based approaches to lower water tables is that if the onsite salt balance has not been

(A)   (B)   (C)

FIGURE 7.1

Salinization is a major issue in South Western Australia affecting around
1 million ha of farmland and nature reserves. Deforestation has pressurized groundwater systems and these have dissolved salt stored in the regolith, with this discharged in lower landscape positions, salinizing both soils and streams. (A) Salinized, formerly productive farmland that has also been affected by water erosion (Wickepin, Western Australia), (B) reforestation of salinized discharge areas with salt-tolerant species *Eucalyptus occidentalis* and *Atriplex nummularia* (Sochacki et al., 2012) (Wickepin, Western Australia), and (C) reforestation of upland recharge areas with belts of mallee *Eucalyptus* (Narrogin, Western Australia) that could be a bioenergy feedstock. Carbon mitigation via bioenergy or sequestration has been investigated as a means of financing the large-scale intervention required to restore the landscape water balance.

managed, then salts will continue to accumulate in the root zone, as the vegetation transpires water (Nosetto et al., 2008; Barrett-Lennard, 2002). This was demonstrated in reforested discharge areas in Western Australia, where although trees decreased the groundwater level, groundwater salinity concentrations increased (Stolte et al., 1997; Archibald et al., 2006).

A range of site management approaches can be used when treating saline land, including the engineering options already described, or soil mounding to overcome localized waterlogging and to encourage leaching of salt. In some cases the lack of plant cover on salinized sites has led to erosion and loss of nutrients, and this requires consideration in site restoration. Wind erosion from salinized areas produces dusts that can represent a human health hazard; Stanturf et al. (2020) review various approaches using vegetation on the salinized, exposed floor of the former Aral Sea. Soil salinity and groundwater levels were critical to successful reestablishment, along with the drought and salinity tolerance of the species used.

A combined engineering and plant-based approach was trialed in the San Joaquin Valley in the United States, where high-value crops are grown on some soils, with drainage induced by pumping. This water was disposed elsewhere in the landscape, in a system termed Integrated Farm Drainage Management, with Levers and Schwabe (2017) demonstrating that the use of a biofuel crop increased the overall profitability of the system and had a lower greenhouse gas footprint than other agricultural crops.

For land that has become alienated from primary production due to salinization, there could be long-term opportunities in introducing engineered salt-tolerant crops. However, despite great effort over recent decades, there has so far been little success in delivering high-yielding, salt-tolerant, and staple food crops to producers. Salt tolerance has been generated in a number of cereal genotypes (e.g., Mujeeb-Kazia et al., 2019; Shahbaz and Ashraf, 2013), but field deployment is limited.

7.5.2.3 Changing landscape-scale water balances

Dryland salinity can increase both stream salinity and the salinity of adjacent soils. As described, this is due to a change in landscape-scale water balance after replacement of deep-rooted perennial species with shallow-rooted species that transpire less. This leads to groundwater rise and mobilization of salts already stored in the landscape. It is thus treated with landscape-scale responses, where the aim is to increase overall evapotranspiration across the landscape, through the protection of existing deep-rooted vegetation or the reintroduction of deep-rooted plants, and thereby reduce groundwater recharge, piezometric pressures, and salt discharge in lower landscape positions.

The landscape approach was successful in controlling saline seeps in the croplands of the United States' Northern Great Plains (Halvorson and Reule, 1980). Here a deep-rooted forage plant (*Medicago sativa*) was planted over 80% of the recharge area and this successfully controlled salinity, whereas 20% coverage didn't. This approach has been recently replicated in the Prairie Provinces of Canada (Wiebe et al., 2010). Here, the area affected by salinity was reduced from 15% in 1981 to 8% by 2011, through changes in the water balance by reducing summer fallows and a 4.8 million hectares increase in the area of permanent cover (FAO and ITPS, 2015).

Both soil and water salinity are major issues in South Western Australia, with around 1 million hectares of soil salinized as a result of deforestation for agriculture (George et al., 1997), and a consequent increase in river salinity in any watershed that has been subjected to land clearing (Halse et al., 2003). Following recognition that disturbing the water balance resulted in the mobilization of salts stored in deep regolith (Peck and Williamson, 1987), subsequent studies indicated that substantial reforestation ($>80\%$ cover) was needed to reverse salinization (George et al., 1999; Bari and Ruprecht, 2003). *Eucalyptus* have been reported to deplete soil water to depths of up to 8 m in 3 years in this region (Harper et al., 2014) with approaches that have resulted in 113,000 ha of reforestation reviewed by Harper et al. (2017). The landscape approach has been successfully demonstrated at scale in the Denmark River watershed in South Western Australia where previous deforestation of 25% of the 56,500 ha watershed resulted in an increase in annual flow-weighted stream salinity from 280 to 1500 mg L^{-1} TDS between 1964 and 1997. Reforestation of 18% of this area with *Eucalyptus globulus* by pulpwood investors resulted in stream salinity reverting to 500 mg L^{-1} TDS by 2017 (Ruprecht et al., 2019).

7.5.3 Policy and legislative approaches

A range of policy and legislative approaches have been used to tackle salinity, these including regulation, research, and education and market mechanisms or economic instruments (Weersink and Wossink, 2005; Ruprecht and Dogramaci, 2005). Examples of regulation include mandating land-use, such as controlling rates of deforestation, the allocation of land for irrigation, managing irrigation schemes through water allocation, or developing rules around the disposal of drainage waters. Education can range from public communications regarding the best practice to formal publications. Economic instruments include procedures such as water pricing and transferable water entitlements, and payment for environmental services.

Successful restoration of salinity at the landscape-scale relies on broadscale land-use change. This is problematic where the most profitable land-use is agriculture, and the replacement by deep-rooted perennials, which have higher evapotranspiration rates, is less attractive. There has therefore been considerable investigation of land-use systems that at least replicate the profitability of the current agricultural system. Thus recent approaches have explored how to make the higher water using farming systems acceptable by making the replacement plants profitable in their own right, such as through the growth of forage plants (Halvorson and Reule, 1980; Wiebe et al., 2010) that can support grazing, the use of plants that can be used to mitigate carbon dioxide through sequestration or bioenergy (Harper et al., 2017; Sochacki et al., 2012; Walden et al., 2017), and timber or payments for environmental services such as water (Townsend et al., 2012).

In some landscapes it may not be possible to achieve the scale of activity needed to manage salinity, either due to the economics of treatment or the nature of the groundwater systems (George et al., 1999; Harper et al., 2017), and here an alternative approach is to follow a containment strategy and only concentrate on the areas that have been salinized (McFarlane et al., 2016). Here plant-based treatments could again be financed through carbon markets (Walden et al., 2017; Harper et al., 2017), in areas where these are in operation.

7.6 Summary and conclusion

Salinity is a major issue in some regions of the world and is predominantly caused by irrigation practice, landscape changes in hydrology, or floods and inundation. It intersects with a range of other issues, including food security, biodiversity protection, and rural livelihoods. The principles of managing salinity can be understood in terms of a soil or site's salt and water balances. There are a range of engineering and plant-based options for reclaiming salinized soils and these have achieved various degrees of success. Combining engineering and plant-based approaches may increase the success of reclamation in some areas. Salinity often occurs over broad areas and a key consideration is financing reclamation efforts that can occur over scale and here markets for carbon mitigation that are developing in response to managing climate change may have a major future role, particularly given the scale of mitigation investment that will be required to contain climate change in coming decades.

References

Abrol, I.P., Yadav, J.S.P., Massoud, F.I., 1988. Salt-affected soils and their management. FAO Soils Bulletins 39. Food and Agricultre Organization of the United Nations, Rome.

Archibald, R.D., Harper, R.J., Fox, J.E.D., Silberstein, R.P., 2006. Tree performance and root-zone salt accumulation in three dryland Australian plantations. Agroforestry Syst. 66, 191–204.

Bari, M.A., Ruprecht, J.K., 2003. Water yield response to land use change in south-west Western Australia. Salinity and Land Use Impacts Series, Report SLUI 31. Department of Environment, Perth.

Barrett-Lennard, E.G., 2002. Restoration of saline land through revegetation. Agric. Water Manage. 53, 213–226.

Barrett-Lennard, E.G., Malcolm, C.V., 2000. Increased concentrations of chloride beneath stands of saltbushes (*Atriplex* species) suggest substantial use of groundwater. Aust. J. Exp. Agric. 39, 949–955.

Burn, S., Hoang, M., Zarzo, D., Olewniak, F., Campos, E., Bolto, B., et al., 2015. Desalination techniques—a review of the opportunities for desalination in agriculture. Desalination 364, 2−16.

Cañedo-Argüelles, M., Kefford, B.J., Piscart, C., Prat, N., Schäfer, R.B., Schulz, C.-J., 2013. Salinisation of rivers: an urgent ecological issue. Environ. Pollut. 173, 157−167.

Cheeseman, J.M., 1988. Mechanisms of salinity tolerance in plants. Plant Physiol. 87, 547−550.

Church, J.A., Clark, P.U., Cazenave, A., Gregory, J.M., Jevrejeva, S., Levermann, A., et al., 2013. Sea level change. In: Stocker, T.F., Qin, D., Plattner, G.-K., Tignor, M., Allen, S.K., Boschung, J., Nauels, A., Xia, Y., Bex, V., Midgley, P.M. (Eds.), Climate Change 2013: The Physical Science Basis. Contribution of Working Group I to the Fifth Assessment Report of the Intergovernmental Panel on Climate Change. Cambridge University Press, Cambridge and New York.

Datta, K.K., Jong, C., 2002. Adverse effect of waterlogging and soil salinity on crop and land productivity in northwest region of Haryana, India. Agric. Water Manage. 57, 223−238.

Degens, B.P., Muirden, P.D., Kelly, B., Allen, M., 2012. Acidification of salinised waterways by saline groundwater discharge in south-western Australia. J. Hydrol. 470-471, 111−123.

Doolittle, J.A., Brevik, E.C., 2014. The use of electromagnetic induction techniques in soils studies. Geoderma 223−225, 33−45.

Eaton, F.M., 1950. Significance of carbonates in irrigation waters. Soil Sci. 69, 123−134.

Egamberdiyeva, D., Garfurova, I., Islam, K.R., 2007. Salinity effects on irrigated soil chemical and biological properties in the Aral Sea basin of Uzbekistan. In: Lal, R., Suleimenov, M., Stewart, B.A., Hansen, D.O., Doraiswamy, P. (Eds.), Climate Change and Terrestrial Carbon Sequestration in Central Asia. Taylor & Francis, London.

FAO, ITPS, 2015. Status of the World's Soil Resources (SWSR) − Main Report. Food and Agriculture Organization of the United Nations and Intergovernmental Technical Panel on Soils, Rome.

Furby, S., Caccetta, P., Wallace, J., 2010. Salinity monitoring in Western Australia using remotely sensed and other spatial data. J. Environ. Qual. 39, 16−25.

García, C., Hernández, T., 1996. Influence of salinity on the biological and biochemical activity of a calciorthird soil. Plant Soil 178, 255−263.

Gaybullaev, B., Chen, S.-C., Gaybullaev, D., 2012. Changes in water volume of the Aral Sea after 1960. Appl. Water Sci. 2, 285−291.

George, R.J., Mcfarlane, D.J., Nulsen, R.A., 1997. Salinity threatens the viability of agriculture and ecosystems in Western Australia. Hydrogeol. J. 5, 6−21.

George, R.J., Nulsen, R.A., Ferdowsian, R., Raper, G.P., 1999. Interactions between trees and groundwaters in recharge and discharge areas − a survey of Western Australian sites. Agric. Water Manage. 39, 91−113.

George, S.J., Harper, R.J., Tibbett, M., Hobbs, R.J., 2012. A sustainable agricultural landscape for Australia: interlacing carbon sequestration, biodiversity and salinity management in agroforestry systems. Agric. Ecosyst. Environ. 163, 28−36.

Gerritsen, H., 2005. What happened in 1953? The Big Flood in the Netherlands in retrospect. Philos. Trans. R. Soc. A: Math. Phys. Eng. Sci. 363.

Ghassemi, F., Jakeman, A.J., Nix, H.A., 1995. Salinization of Land and Water Resources. University of New South Wales Press, Canberra.

Halse, S.A., Ruprecht, J.K., Pinder, A.M., 2003. Salinization and prospects for biodiversity in rivers and wetlands of southwest Western Australia. Aust. J. Bot. 51, 673–688.

Halvorson, A.D., Reule, C.A., 1980. Alfalfa for hydrologic control of saline seeps. Soil Sci. Soc. Am. J. 44, 370–374.

Harper, R.J., Sochacki, S.J., Smettem, K.R.J., Robinson, N., 2014. Managing water in agricultural landscapes with short-rotation biomass plantations. GCB Bioenergy 6, 544–555.

Harper, R.J., Sochacki, S.J., Mcgrath, J.F., 2017. The development of reforestation options for dryland farmland in south-western Australia: a review. South. For. 79, 185–196.

Heuperman, A., 1999. Hydraulic gradient reversal by trees in shallow water table areas and repercussions for the sustainability of tree-growing systems. Agric. Water Manage. 39, 153–167.

Heuperman, A.F., Kapoor, A.S., Denecke, H.W., 2002. Biodrainage: principles, experiences and applications. Knowledge Synthesis Report No 6. International Programme for Technology and Research in Irrigation and Drainage. Secretariat, Food and Agriculture Organization of the United Nations, Rome.

Hillel, D.J., 1991. Out of the Earth: Civilization and the Life of the Soil. The Free Press, New York.

Hunter, M.C., Smith, R.G., Schipanski, M.E., Atwood, L.W., Mortensen, D.A., 2017. Agriculture in 2050: recalibrating targets for sustainable intensification. Bioscience 67, 386–391.

Imadi, S.R., Shah, S.W., Kazi, A.G., Azooz, M.M., Ahmad, P., 2016. Phytoremediation of saline soils for sustainable agricultural productivity. In: Ahmad, P. (Ed.), Plant Metal Interaction: Emerging Remediation Techniques. Elsevier.

Ladeiro, B., 2012. Saline agriculture in the 21st century: using salt contaminated resources to cope food requirements. J. Bot. 2012, 7.

Lambert, M., Turner, J., 2000. Commercial Forest Plantations on Saline Lands. CSIRO Publishing, Melbourne.

Lantz, T.C., Kokelj, S.V., Fraser, R.H., 2015. Ecological recovery in an Arctic delta following widespread saline incursion. Ecol. Appl. 25.

Lesch, S.M., Rhoades, J.D., Lund, L.J., Corwin, D.L., 1992. Mapping soil salinity using calibrated electromagnetic measurements. Soil. Sci. Soc. Am. J. 56, 540–548.

Lesch, S.M., Strauss, D.J., Rhoades, J.D., 1995. Spatial prediction of soil salinity using electromagnetic induction techniques. 1. Statistical prediction models: a comparison of multiple linear regression and cokriging. Water Resour. Res. 31, 373–386.

Levers, L.R., Schwabe, K.A., 2017. Biofuel as an integrated farm drainage management crop: a bioeconomic analysis. Water Resour. Res. 53, 2940–2955.

Marcar, N., Crawford, D., 2004. Trees for Saline Landscapes. Rural Industries Research and Development Corporation, Canberra.

McFarlane, D.J., George, R.J., Barrett-Lennard, E.G., Gilfedder, M., 2016. Salinity in dryland agricultural systems: challenges and opportunities. In: Farooq, M., Siddique, K.H.M. (Eds.), Innovations in Dryland Agriculture. Springer Nature, Cham, Switzerland.

McLeod, M.K., Slavich, P.G., Irhas, Y., Moore, N., Rachman, A., Ali, N.,T.,I., et al., 2010. Soil salinity in Aceh after the December 2004 Indian Ocean tsunami. Agric. Water Manage. 97, 605–613.

Micklin, P., 2007. The Aral Sea disaster. Annu. Rev. Earth Planet Sci. 35, 47–72.

Morris, J.D., Collopy, J.J., 1999. Water use and salt accumulation by *Eucalyptus camaldulensis* and *Casuarina cunninghamiana* on a site with shallow saline groundwater. Agric. Water Manage. 39, 205–227.

Mujeeb-Kazia, A., Munns, R., Rasheed, A., Ogbonnaya, F.C., Ali, N., Hollington, P., et al., 2019. Breeding strategies for structuring salinity tolerance in wheat. Adv. Agron. 155, 121−187.

Myers, N., Mittermeier, R.A., Mittermeier, C.G., Da Fonseca, G.A.B., Kent, J., 2000. Biodiversity hotspots for conservation priorities. Nature 403, 853−858.

National Land and Water Resources Audit, 2001. Dryland salinity in Australia. A Summary of the National Land and Water Resources Audit's Australian Dryland Salinity Assessment 2000. National Land and Water Resources Audit, Canberra.

Nikalje, G.C., Srivastava, A.K., Pandey, G.K., Suprasanna, P., 2018. Halophytes in biosaline agriculture: mechanism, utilization, and value addition. Land. Degrad. Dev. 29, 1081−1095.

Nosetto, M.D., Jobbágy, E.G., Tóth, T., Jackson, R.B., 2008. Regional patterns and controls of ecosystem salinization with grassland afforestation along a rainfall gradient. Global Biogeochem. Cycles 22, GB2015. Available from: https://doi.org/10.1029/2007GB003000.

Peck, A.J., Williamson, D.R., 1987. Effects of forest clearing on groundwater. J. Hydrol. 94, 47−65.

Qadir, M., Quillérou, E., Nangia, V., Murtaza, G., Singh, M., Thomas, R.J., et al., 2014. Economics of salt-induced land degradation and restoration. Nat. Resour. Forum 38, 282−295.

Rahman, M.M., Penny, G., Mondal, M.S., Zaman, M.H., Kryston, A., Salehin, M., et al., 2019. Salinization in large river deltas: drivers, impacts and socio-hydrological feedbacks. Water Secur. 6, 100024.

Rengasamy, P., 2006. World salinization with emphasis on Australia. J. Exp. Bot. 57, 1017−1023.

Rhoades, J.D., 1993. Electrical conductivity methods for measuring and mapping soil salinity. Adv. Agron. 49, 201−251.

Rhoades, J.D., Miyamoto, S., 1990. Testing soils for salinity and sodicity. In: Westerman, R.L. (Ed.), Soil Testing and Plant Analysis, third ed. Soil Science Society of America, Madison, WI.

Rhoades, J.D., Kandiah, A., Mashali, A.M., 1992. The use of saline waters for crop production. FAO Irrigation and Drainage Paper No. 48. Food and Agricultural Organization of the United Nations, Rome.

Ruprecht, J.K., Dogramaci, S., 2005. Salinization. In: Anderson, M.G. (Ed.), Encyclopedia of Hydrological Sciences. John Wiley and Sons.

Ruprecht, J.K., Sparks, T., Liu, N., Dell, B., Harper, R.J., 2019. Using reforestation to reverse salinisation in a large watershed. J. Hydrol. 577, 123976.

Schofield, N.J., 1992. Tree planting for dryland salinity control in Australia. Agroforestry Syst. 20, 1−23.

Shahbaz, M., Ashraf, M., 2013. Improving salinity tolerance in cereals. Crit. Rev. Plant Sci. 32, 237−249.

Sochacki, S.J., Harper, R.J., Smettem, K.R.J., 2012. Bio-mitigation of carbon from reforestation of abandoned farmland. GCB Bioenergy 4, 193−201.

Squires, V.R., Ayoub, A.T., 1994. Halophytes as a Resource for Livestock and for Rehabilitation of Degraded Lands. Springer Netherlands, Dordrecht.

Stanturf, J.A., Botman, E., Kalachev, A., Borissova, Y., Kleine, M., Rajapbaev, M., et al., 2020. Dryland forest restoration under a changing climate in central Asia and Mongolia. Mong. J. Biol. Sci. 18, 3−18.

Stolte, W.J., McFarlane, D.J., George, R.J., 1997. Flow systems, tree plantations, and salinisation in a Western Australian catchment. Aust. J. Soil Res. 35, 1213–1229.

Tillman, D., Blazer, C., Hill, J., Befort, B.L., 2011. Global food demand and the sustainable intensification of agriculture. Proc. Natl. Acad. Sci. U.S.A. 108, 20260–20264.

Townsend, P.V., Harper, R.J., Brennan, P.D., Dean, C., Wu, S., Smettem, K.R.J., et al., 2012. Multiple environmental services as an opportunity for watershed restoration. For. Policy Econ. 17, 45–58.

United Nations, 2017. World Population Prospects: The 2017 Revision, Key Findings and Advance Tables. Department of Economic and Social Affairs, Population Division, United Nations.

Vengosh, A., 2003. Salinization and saline environments. In: Lollar, B.S., Holland, H.D., Turekian, K.K. (Eds.), Treatise on Geochemistry. Elsevier.

Volkov, V., Beilby, M.J., 2017. Editorial: Salinity tolerance in plants: mechanisms and regulation of ion transport. Front. Plant. Sci. 8.

Walden, L.L., Harper, R.J., Sochacki, S.J., Montagu, K.D., Wocheslander, R., Clarke, M., et al., 2017. Mitigation of carbon following *Atriplex nummularia* revegetation in southern Australia. Ecol. Eng. 106, 253–262.

Weersink, A., Wossink, A., 2005. Lessons from agri-environmental policies in other countries for dealing with salinity in Australia. Aust. J. Exp. Agric. 45, 1481–1493.

Wiebe, B.H., Eilers, W.D., Brierley, J.A., 2010. Soil salinity. In: Eilers, W., Mackay, R., Graham, L., Lefebvre, A. (Eds.), Environmental Sustainability of Canadian Agriculture: Agri-Environmental Indicator Report Series – Report No. 3. Agriculture and Agri-Food Canada, Ottawa, ON.

Williams, B.G., Baker, G.C., 1982. An electromagnetic induction technique for reconnaissance surveys of soil salinity hazards. Aust. J. Soil Res. 20, 107–118.

Yao, R., Yang, J.S., 2010. Quantitative evaluation of soil salinity and its spatial distribution using electromagnetic induction method. Agric. Water Manage. 97, 1961–1970.

Biochar amendments show potential for restoration of degraded, contaminated, and infertile soils in agricultural and forested landscapes

8

Rachel L. Brockamp[1] and Sharon L. Weyers[2]

[1]*Department of Soil Science, University of Saskatchewan, Saskatoon, SK, Canada*
[2]*U.S. Department of Agriculture, Agricultural Research Service, North Central Soil Conservation Research Laboratory, Morris, MN, United States*

Biochar use as a soil amendment arose from the influx of new understanding of the origin of the "Terra Preta" soils rediscovered in the Amazon. Current theory assumes that active measures were taken centuries ago (CE 450–950) to enrich soils with burnt biomass (char), bone, and manure (Sohi et al., 2010; Novotny et al., 2007). These active measures over time resulted in extremely fertile soils rich in plant nutrients, microbial activity, and reactive functional groups that influence cation exchange capacity and a host of other physicochemical soil properties (Glaser et al., 2001). A wealth of research resulting from this rediscovery has demonstrated that laboratory and commercial grade biochar may enhance soil health and plant growth and can be used as a soil amendment in forestry, agriculture, degraded landscapes, and contaminated soils. However, not all biochars are alike or have the same uses, as variable feedstocks and thermoconversion conditions influence biochar properties. Similarly, the landscape to which they are applied will express variable reactions due to differences in soil type, climate, vegetation, and management. In this chapter the occurrence, production, and properties of biochar, and how these properties influence specific soil conditions and reactions is described. We then discuss the proposed, prescribed, and applied uses for biochar on the landscape. We finish with mention of potential hazards from biochar use and gaps in our knowledge.

Main points are:

1. biochar terminology and historical aspects regarding natural occurrence;
2. production practices and properties of industrially produced biochar;

Soils and Landscape Restoration. DOI: https://doi.org/10.1016/B978-0-12-813193-0.00008-4

3. biochar effects in the environment and potential modes of action; and

4. applied and prescribed uses for biochar.

8.1 Overview

In a broad sense the term "biochar" has been used to refer to any carbon-rich material falling within a continuum of thermolytically converted organic materials below a 0.6 O:C ratio that can be separated into slightly charred biomass or char, charcoal, soot, and graphite categories (Hedges et al., 2000; Spokas, 2010; Fig. 8.1). The terms "pyrogenic carbon" and "black carbon" have also been used to broadly refer to this continuum. Pyrogenic C is the term most often invoked in studies evaluating the occurrence, characterization, and persistence of these materials in the environment, particularly along geologic timescales or as found in the prehistoric record (e.g., Hedges et al., 2000; Knicker, 2011; Bird et al., 2015). In all cases pyrogenic C retains its broader meaning. However, black C has been used to narrowly define the soot or aerosolized particulates resulting from the burning of fossil fuels and causing environmental pollution (e.g., Booth and Bellouin, 2015). "Biochar" was originally intended to narrowly define the C-rich pyrogenic material specifically applied to land for soil improvement (Novotny et al., 2015). In this vein, "natural biochar" might be used as a synonym to "charcoal" the material produced by wildfire or prescribed burns also known to improve soil (DeLuca and Aplet, 2008; Knicker, 2011).

Wildfire in producing pyrogenic C has contributed to the formation of the most productive and valuable soils in the world, particularly chernozems (FAO) in Germany and across the Russian Steppe and Mollisols (United States) of North American prairies (Schmidt et al., 1999; Schmidt and Noack, 2000; Skjemstad et al., 2002; Glaser and Amelung, 2003). The importance of wildfire in shaping our soils is evident throughout the Phanerozoic geologic record (Schmidt and Noack, 2000; Knicker, 2011), which supports that the most recalcitrant forms of pyrogenic C persist for millennia (Hedges et al., 2000; Schmidt and Noack, 2000). Prehistoric human modification of the landscape by fire (Caldararo, 2002), more likely through low-heat smoldering fires but also possibly slash-and-burn (Glaser et al., 2001), has improved fertility of soils classified as anthrosols (FAO; i.e., Anthreps, United States), such as the "Terra Preta" soils found in Central and South America, and similar soils found in Africa (Fairhead and Leach, 2009; Woods et al., 2009).

Globally, natural and anthropogenic wildfires are estimated to produce $40-240$ Tg year^{-1} of pyrogenic C, as solid residuals including black carbon and charcoal (Preston and Schmidt, 2006). Pyrogenic C made through wildfire contains everything from slightly burned wood to aerosolized particulates and varies spatially with available fuel loads, plant community (i.e., fuel) composition, moisture conditions, and fire intensity (Preston and Schmidt, 2006; McElligott et al.,

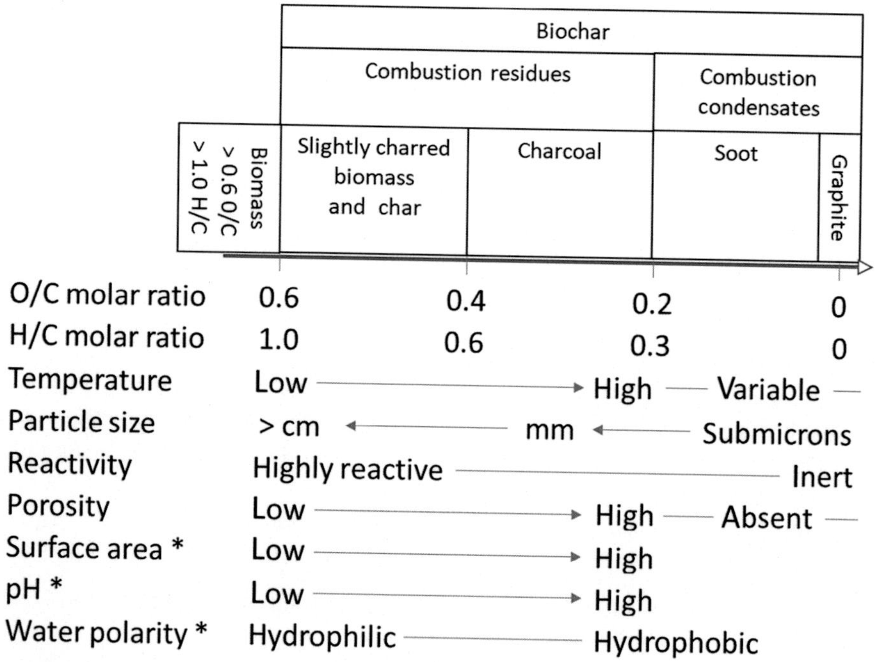

		Biochar			
		Combustion residues		Combustion condensates	
> > Biomass > 1.0 H/C > 0.6 O/C		Slightly charred biomass and char	Charcoal	Soot	Graphite
O/C molar ratio		0.6	0.4	0.2	0
H/C molar ratio		1.0	0.6	0.3	0
Temperature		Low ⟶ High — Variable —			
Particle size		> cm ⟵ mm ⟵ Submicrons			
Reactivity		Highly reactive ⟶ Inert			
Porosity		Low ⟶ High — Absent —			
Surface area *		Low ⟶ High			
pH *		Low ⟶ High			
Water polarity *		Hydrophilic ⟶ Hydrophobic			

FIGURE 8.1

Select properties of materials falling along the combustion continuum defined by O/C and H/C molar ratio. H/C scale is nonlinear. * Applies to combustion residues only.

Modified from Hedges, J.I., Eglinton, G., Hatcher, P.G., Kirchman, D.L., Arnosti, C., Derenne, S., et al., 2000. The molecularly-uncharacterized component of nonliving organic matter in natural environments. Org. Geochem., 31(10), 945–958; Hammes, K., Schmidt, M.W., Smernik, R.J., Currie, L.A., Ball, W.P., Nguyen, T.H., et al., 2007. Comparison of quantification methods to measure fire-derived (black/elemental) carbon in soils and sediments using reference materials from soil, water, sediment and the atmosphere. Glob. Biogeochem. Cycles 21(3), GB3016 https://doi.org/10.1029/2006GB002914; Spokas, K.A., 2010. Review of the stability of biochar in soils: predictability of O:C molar ratios. Carbon Manage., 1(2), 289–303; Bird, M.I., Wynn, J.G., Saiz, G., Wurster, C.M., McBeath, A., 2015. The pyrogenic carbon cycle. Annu. Rev. Earth Planet. Sci. 43, 273–298.

2011). Uncontrolled wildfires produce substantial amounts of ash, but the more stable components contribute from 10% to 50% of the total carbon found in forest soils, with greater content of pyrogenic C occurring where fire return intervals are shorter, or more frequent prescribed burns are conducted (DeLuca and Aplet, 2008; Pingree et al., 2012). Comparatively in grassland systems, only 4%–18% of soil organic C was of pyrogenic origin (Glaser and Amelung, 2003). In forests, charcoal formation is estimated at 1%–10% of the total fuel consumed in a burn but decreases as the amount of fuel consumed increases in relation to fire intensity; prescribed burns consume 60% less fuel than wildfires due to lower fire intensity and thus could result in less charcoal formation (DeLuca and Aplet, 2008; Pingree et al., 2012). Unfortunately, fire suppression reduces recalcitrant pyrogenic C inputs and threatens the C balance of these ecosystems (Deluca and Aplet, 2008; Thomas and Gale, 2015).

8.2 Production and characterization of biochar

The utility of biochar rests in its chemical and physical characteristics. In general, as temperature during thermolytic conversion increases, so does porosity, surface area, and pH; however, particle size, reactivity, and polarity decrease (Fig. 8.1). Reactivity is due to the chemical composition of the three carbon pools, labile, semilabile, and resistant, that comprise all biochars (Bird et al., 2015). This composition affects the half-life, or persistence in the environment, and the nature of interactions with soil chemical, physical, and biological properties (Table 8.1). As a result, biochar can aid in the stabilization of soil organic matter (SOM)

Table 8.1 Characterization and environmental interactions specific to the three pools of carbon likely to comprise biochar.

Pools	Half-life	Primary composition	Interactions
Labile (active) <5% of total C[a]	Weeks to months	Anhydrosugars and methoxylated phenols	Easily mineralized C substrates; potential priming of native organic matter
Semi-labile 10%–62% of total C	Years to decades	Polyaromatic C composed of small PACs with a carbon ring size less than seven	Microbial mineralization processes
Resistant (non-labile) 38%–90% of total C	Centuries to millennia	Polycyclic aromatic C with carbon ring sizes greater than seven	Highly resistant to mineralization

PACs, Polycyclic aromatic compounds.
Compiled from Woolf, D., Amonette, J.E., Street-Perrott, F.A., Lehmann, J., Joseph, S., 2010. Sustainable biochar to mitigate global climate change. Nat. Commun., 1, 56; Mašek, O., Brownsort, P., Cross, A., Sohi, S., 2013. Influence of production conditions on the yield and environmental stability of biochar. Fuel, 103, 151–155; Bird, M.I., Wynn, J.G., Saiz, G., Wurster, C.M., McBeath, A., 2015. The pyrogenic carbon cycle. Annu. Rev. Earth Planet. Sci. 43, 273–298; Fang, Y., Singh, B., Singh, B.P., Krull, E., 2014. Biochar carbon stability in four contrasting soils. Eur. J. Soil Sci., 65(1), 60–71; McBeath, A.V., Wurster, C.M., Bird, M.I., 2015. Influence of feedstock properties and pyrolysis conditions on biochar carbon stability as determined by hydrogen pyrolysis. Biomass Bioenergy 73, 155–173.
[a]Proportions vary with production temperature, whereby increasing temperature increases resistant C content.

(Nair et al., 2017). Any form of organic matter containing C can be converted thermolytically into biochar, including forest wood products and slash; agricultural residues such as crop residue including straw and stover; animal and human waste; industrial waste including paper mill waste/sludge; food production waste including bagasse, pomace, and nutshells; and more (Sohi et al., 2009; Ahmad et al., 2014). In nature, biochar produced by wildfire or prescribed burns is influenced by spatial variability of fuel loads, fuel type, moisture conditions, temperature, and fire intensity (Preston and Schmidt, 2006). In practice, biochar produced through a wide variety of thermochemical conversion is similarly influenced not only by fuel type or feedstock, moisture, and temperature but also by the rate of temperature increase, residence time, oxygen, and pressure conditions. The selective use of feedstocks and operating conditions of the thermolytic conversion process can be used to design biochars for specific purposes (Ahmad et al., 2014; Mohan et al., 2014).

8.2.1 Industrial production

Biochar production occurs via several methods that thermolytically decompose organic materials to obtain bio-oils, syngases, and biochar (Ahmad et al., 2014; Cha et al., 2016; Nanda et al., 2016). Thermal conversion systems range from low temperature processes such as torrefaction (50°C−300°C) and pyrolysis (300°C−700°C) yielding 12%−84% biochar, to higher temperature gasification (700°C−900°C) to combustion (800°C−950°C) systems yielding <10% biochar or mostly ash (Novotny et al., 2015; Cha et al., 2016). The products of conversion can be further refined for other uses, for example, bio-oils refined into biofuels (No, 2014); syngases recycled back into the pyrolysis process for heat production (Sohi et al., 2010); and biochar used in fuel cells (Cha et al., 2016). Biochar and ash have other industrial uses including water and air purification and as catalysts in a myriad of chemical reactions (Cha et al., 2016). Ash, including wood ash, or fly ash, and bottom ash from the combustion of coal, has been used as a fertility amendment, delivering micro-nutrients lacking volatile forms, including K, Na, Ca, and others (Demeyer et al., 2001; Ram and Masto, 2014; Merino et al., 2016).

8.2.1.1 Pyrolysis

The most common method used to produce biochar for land application is pyrolysis as it typically generates the most biochar. Various pyrolysis platforms are available including slow, vacuum, intermediate, flash, and fast that are distinguished mainly by heating rate (0.1−200°C s^{-1}) and residence time (hours down to seconds); all require an oxygen free environment and can be operated at any temperature, but typically between 300°C and 900°C (No, 2014; Nanda et al., 2016). Slow pyrolysis, particularly at lower temperatures, favors the production of biochar over bio-oil and syngas at a 35:30:35 ratio, whereas intermediate or fast pyrolysis favors bio-oil, ranging 50%−75% of total product (Ahmad et al.,

2014; Novotny et al., 2015). More importantly, slow pyrolysis biochars are C dense at 95% C with a feedstock conversion rate of 58%, compared to fast pyrolysis biochars at 74% C and conversion rate of only 20%−26% (Cha et al., 2016). Feedstock, temperature, heating rate, and residence time can be modified to influence chemical and physical properties of the resultant biochar; however, temperature has the greater controlling effect (Nanda et al., 2016). An important aspect of the feedstock is the relative proportions of hemicellulose, cellulose, and lignin, as well as occurrence of major (e.g., Na, Mg, K, Ca, and Si) and minor mineral elements (e.g., Al, Fe, Mn, P, and S). These feedstock components degrade, respectively, at increasing temperatures, influencing chemical structure and reactions during pyrolysis (Nanda et al., 2013). A comparison of the general, chemical, and morphological properties of biochar produced under low and high temperatures (independent of a slow or fast process) is presented in Table 8.2.

Low-temperature pyrolysis favors the formation of biochar over ash, bio-oils or syngas, where biochar output is typically >30%. Low temperatures prevent loss of volatiles, moisture, CO, CO_2, and H_2, which maintains higher O/C and H/C ratios and contributes to surface O−H, C−O, and C−H functional group binding sites. These functional groups create a negatively charged surface that binds well with positively charged cationic compounds, thus increasing cation exchange capacity (CEC) (Ahmad et al., 2014). However, these biochars also have a lower pH, due to the formation of organic acids and phenolics from degradation of hemicellulose and cellulose (Cao and Harris, 2010). These low temperature biochars typically have lower pore volume, possibly due to hydrocarbons condensing as tars that fill pore spaces. These biochars also have less microporosity and lower total surface area. Although their original structure and particle size is better maintained, they are more hydrophilic and possess higher water-holding capacity.

High-temperature pyrolysis favors the formation of bio-oils, syngas and ash, whereas biochar output is typically <30% (Nanda et al., 2016). However, biochar yield can be increased using feedstocks with higher lignin content and moisture. Higher temperatures also drive off more O and H, due in part to dehydration and decarboxylation, which decreases the H/C and O/C ratios, making the biochar more C dense and giving it a higher heating value. The lignin is transformed into stable aromatic C ring structures and surface functional groups become dominated with C−C bonds. This structural aspect lowers the CEC and electrical conductivity, the opposite of pH. Extreme alkalinity (pH > 10) can result in some high-temperature biochars due to loss of protons but also concentration of the alkali and alkaline earth metals (e.g., Na, K, Ca, and Mg) from the starting feedstock (Cao and Harris, 2010). The surface of these biochars are less polar, which is one reason they are hydrophobic, but they have ionic charges that are better for organic contaminant absorption. The C density and greater stability (i.e., recalcitrance) makes high-temperature biochars better for C sequestration in amended soils (Spokas, 2010). At higher production temperatures volatile loss contributes to formation of micropores that increase total surface area (Tan et al., 2015), relative to low temperature biochars. However, extremely high temperatures can cause biomass to

Table 8.2 Effect of low and high temperature production on biochar properties and utility.

Low temperature ($\sim$ <500°C)	High temperature ($\sim$ >500°C)
General composition	
Biochar production favored; volatile matter retained [moisture, hydrocarbons (e.g., tars), CO, CO_2, H_2] Utility: Greater biochar output may better serve land application needs	Ash, syngas, and bio-oils favored (particularly at 500°C and with fast pyrolysis); volatiles lost; biochars with higher heating value (loss of H and O) Utility: Products can be refined into other energy products; adds heating value
Physicochemical composition	
Lower pH (related to organic acids, phenolics); higher H/C and O/C ratios; higher CEC and EC; hydrophilic negatively charged surface due to O–H, C–O, and C–H groups, and higher water-holding capacity Utility: Enhanced nutrient cycling capacity due to high CEC; useful for delivering fertilizers and other agricultural chemicals; better for inorganic contaminant absorption (O–H groups)	Carbon dense, more fixed, and recalcitrant forms (aromatic structures; C–C bonds); higher pH >9, due to condensation of alkali, alkaline earth metals (particularly at low heating rates) related to ash content; hydrophobic, less polar, more ionic surfaces, and lower water-holding capacity Utility: More recalcitrant in the environment, useful for C-sequestration; effective as liming agents; better for organic contaminant absorption
Morphological composition	
Low surface area (surface area also decreases with increasing pressure, slow heating rates, or blockage by tars); lower pore volume, larger particle size Utility: Due to larger particle size, improved soil tilth characteristics such as bulk density and water retention	High surface area particularly above 700°C (however, lowest surface area above 1000°C, pore structure destroyed); higher pore volume due to microporosity, smaller particle size Utility: More surface area for (organic) contaminant absorption

CEC, Cation exchange capacity; *EC*, electrical conductivity.
Summarized from Nanda et al. (2016).

undergo plastic deformation, that is, they melt, but macropores develop with high surface area, particularly under low-pressure conditions (Cetin et al., 2004).

8.2.1.2 Activated biochar

Activated carbon refers to a variety of carbon-based material (e.g., peat, lignite, coal, charcoal, and carbonized plant matter) treated in a physical or chemical way to increase its absorptive properties by increasing surface area and porosity (Kyotani, 2000). Activated carbon has many industrial, pharmaceutical, and environmental applications, the latter including water and air purification (Marsh and Reinoso, 2006). Cha et al. (2016) review many similarities between biochar and activated carbon, particularly for environmental applications. This was echoed by evaluation of biochar in soil contaminant remediation (Lone et al., 2015). Although biochar already has high porosity and similar absorptive properties of activated carbon (Ahmad et al., 2012; Cha et al., 2016), chemical activation might increase porosity 50-fold (Azargohar and Dalai, 2006). Ahmad et al. (2012) went further to suggest biochar was more advantageous than activated carbon because it can be produced as a coproduct of energy production making it more cost-effective to produce.

8.2.2 Pyrogenic C formation by wildfire and prescribed burn

The process of generating pyrogenic C from a wildfire or prescribed burn is much like that occurring in the reaction vessel of a pyrolysis unit, as changes in temperature and progression of thermochemical reactions occur as fire moves across the ground surface. Four stages of combustion are defined by the diffusion flame model along the cross-section of an advancing fireline: preheating (or preignition), flaming (or ignition), glowing, and smoldering (Ward, 2001; Pyne et al., 1996). During preheating water is driven off at 100°C, and pyrolysis reactions generate combustible volatiles from hydrocarbons c. 325°C. The flaming process begins when the heat rises above 425°C, the temperature at which combustible gases can burn and celluloses and lignins also undergo pyrolysis releasing more volatiles. During this phase, carbon and ash can build up at the fuel surface, and oxygen depleted regions develop. Diffusion of oxygen is necessary at this point to maintain the flaming reactions; if not, the fire moves into the glowing phase, which creates more charcoal, whereas the smoldering phase generates larger amounts of particulate matter, that is, smoke.

Presence and characterization of pyrogenic C generated by recent or ancient wildfires has been conducted by both visible and physical separation of charcoal and by elemental analysis, mass spectrometry or gas chromatography for the non-visible finer fractions comprising SOM (Preston and Schmidt, 2006; Hammes et al., 2007). Commonly a bulk analysis is performed on the organic and mineral soil layers before, after, and along chronosequences of wildfire, or to simply confirm the presence of pyrogenic C from wildfire (e.g., Skjemstad et al., 2002; Hart and Luckai, 2014; Santín et al., 2016). Many analyses focus more on the pyogenic C chemistry rather than other independent properties such as pH or porosity. In this sense pyrogenic C from wildfire has been characterized in a similar manner as industrial biochar. For example, a decrease in H/C and O/C ratios (indicating increased aromatization) occurred as burn severity increased (Merino et al., 2015), which indicates heat and residence time affect the degree of organic matter degradation (Vega et al., 2013). Santín et al. (2016) directly confirmed the impact of increasing temperature with increased occurrence of resistant aromatic-C structures of charred material following an experimental burn. As with industrial biochar, characteristics can be influenced by fuel (i.e., feedstock) composition, at the individual level from plant species differing in composition of cellulose, hemicellulose, and lignin, to entire habitat composition, ranging from grasslands to woodlands high in plant diversity and comingled litter residues (McKendry, 2002; Yang et al., 2007). Influence of feedstock was demonstrated by a particle size analysis of charcoal sampled from a pine-dominated forest that indicated greater aromatic C forms of coarse fragments originating from wood residues, compared to greater aliphatic C and N containing components with higher reactivity of finer fragments, resulting from pine needles (Nocentini et al., 2010).

Pyrogenic C from wildfire has pronounced impacts on soil characteristics that are difficult to address as a direct function of the physicochemical nature of the pyrogenic C itself, as only visible charcoal has been distinctly analyzed separately from the soil matrix. However, numerous relationships have been drawn from laboratory-based investigations by comparing pre- to postfire conditions, or pyrogenic C containing versus noncontaining vertically or horizontally adjacent soils. Similar biological, physical, and chemical outcomes of industrial biochar addressed previously (Table 8.2) have been ascribed to pyrogenic C of wildfires (Zackrisson et al., 1996; DeLuca et al., 2006; Hart and Luckai, 2014). Additional aspects of pyrogenic C will be discussed in the following section.

8.3 Prescribed and applied uses of biochar in the environment

8.3.1 Forest soils

Prescribing biochar application in forested systems has come about with the recognition of benefits natural and prescribed fire have in maintaining biodiversity, regulating nutrient cycling, and other fire-induced impacts on soil physicochemical properties (Certini, 2005; Alcañiz et al., 2018). The primary reason for prescribing biochar is to improve the production of forested, logged, or reclaimed forest landscapes. In application, biochar can be obtained through the management of the slash left behind from thinning or logging or from utilizing wood-processing waste. Each have logistical issues to overcome, with contrasts that can be made between on- and off-site production that relate to the economics of redeployment and/or transport of the feedstock as well as the biochar product.

8.3.1.1 Tree growth and forest restoration

Several studies evaluated in a metaanalysis (Thomas and Gale, 2015) have demonstrated positive impact of biochar application for seedling growth in pot or field studies with boreal through tropical deciduous and coniferous species (e.g., Wardle et al., 1998; Sovu et al., 2012; Fagbenro et al., 2013; Scharenbroch et al., 2013). Growth was increased by 41% on average with deciduous species (i.e., angiosperms) demonstrating a higher growth response to biochar application than coniferous species (i.e., gymnosperms). Though growth response declined over time, this metaanalysis only addressed very short-term (<4 years) pot and field trials. Proposed mechanisms for increased growth are wide ranging from general effects on soil fertility, to placement of biochar to feedstock. For example, increased 4-month biomass of alder was correlated with increased pH, CEC, and Ca and Na content after a 10% application of a pine chip biochar (Robertson et al., 2012). Increased growth of larch seedlings after 4 months was attributed to P delivery through mycorrhizal associations stimulated by mid-profile burial compared to mixed or surface placed field-collected charcoal from a prescribed fire in a larch, birch, and bamboo system (Makoto et al., 2010). Findings presented by Pluchon et al. (2014) identified links between P content of biochar (related to feedstock) and P content in the soil, with greater growth response in relation to biochar with high P, particularly when applied to the low P soils.

Other aspects associated with positive seedling growth were the ability of the added charcoal to absorb phenolic compounds that inhibit nutrient cycling and tree growth (Wardle et al., 1998). However, the positive effects gained in the short-term may decline as properties change over time as the added biochar or charcoal ages or becomes redistributed into the soil matrix. For example, the absorptive capacity of wildfire charcoal for a phenolic compound was variable within a 115-years chronosequence of wildfire history (Pingree et al., 2016) but declined when extended to a 350-years chronosequence (Zackrisson et al., 1996). The same study demonstrated that tree basal area plateaued between 100 and 200 years followed by decline up to 350-year, suggesting that reamendment might be necessary.

8.3.1.2 Slash and waste management

Fire suppression, achieved with prescribed burning, selective harvesting, and other methods that limit the build-up of fuels, has altered natural fire regimes (Brown et al., 2004) and might lead to loss of ecosystem resiliency, and other negative effects on forest ecosystem function and C balance (Hart and Luckai, 2014; Thomas and Gale, 2015). As logging and selective harvest residues (i.e., slash) might still pose a fire risk, on- or off-site conversion of slash into biochar, with intent to redeploy, has been proposed as an alternative to burning slash gathered into piles (which are prone to erosion and nutrient loss) or leaving material to decompose (Page-Dumroese et al., 2009, 2016; Scott and Page-Dumroese, 2016). Use of mobile and stationary pyrolysis units showed promise for the generation of heat, electricity, and marketable biochar and activated charcoal products from forest slash and wood processing wastes, but economic viability including transportation costs was questionable (Anderson et al., 2013). Sorenson (2010) suggested that the value of biochar might range from $100-500 \text{ Mg}^{-1}$ depending on emerging carbon markets, compared to a value of 150 Mg^{-1} if used for heating or $120-180 \text{ Mg}^{-1}$ if used for horticulture. Other studies indicated the economic viability as well as competing interests for converting slash and processing waste into energy products by pyrolysis units, as the biochar coproduct (i.e., charcoal, char, or torrefied wood) was valued for energy production not for land application (Brown et al., 2013, 2014; Carrasco et al., 2017). One field study indicated no benefits to tree growth from applying a biochar made from mixed conifer mill waste at a rate of 2.5 Mg ha^{-1}, possibly because the soil was already enriched by volcanic ash (Sherman et al., 2018). In order to avoid environmental risks or competing interests of thinning or slash harvesting for bioenergy (Page-Dumroese et al., 2010; Kimsey et al., 2011), and perhaps avoid the risks of traditional slash pile burning, recent alternatives were proposed whereby mobilized burning units (e.g., mini kilns) and grapplers can be deployed within logged areas most needing restoration followed by spreading of the charcoal in the surrounding area (Page-Dumroese et al., 2017).

8.3.2 Agricultural production systems

Global agroecosystems are under stress from unsustainable management, thus lending to loss of vital ecosystem services that support a rapidly growing human population (Nair et al., 2017). An overwhelming amount of existing biochar research has focused on its use in agricultural production systems to mitigate harmful effects and improve soil quality. Biochar use has been tested worldwide in slash-and-burn and subsistence farming systems, production management systems where removal or burning of crop residues are frequent (e.g., bioenergy systems and grass seed production systems), and in heavily mechanized farming systems where use of mineral fertilizers and other chemical amendments are common. Biochar use results in definitive changes in the physical, chemical, and biological properties of soils, thoroughly reviewed by Atkinson et al. (2010). These changes can be distilled down to three main purposes for using biochar in agricultural systems: (1) improving crop yields, (2) improving soil health, and (3) improving environmental externalities. The mechanisms that lead to these positive outcomes are likely extensive, but in many cases not well understood.

8.3.2.1 Yields

Two recent metaanalyses revealed that the impacts on crop yields and soil properties vary by soil type, climate, cropping practices, and the nature of the biochar used (Crane-Droesch et al., 2013; Jeffery et al., 2011, 2017). Both studies determined a global mean increase in crop yields of 10% from biochar application greater than 3 Mg ha^{-1}, with no specific response to increasing application rates; however, variability was high and the range included studies showing negative yield responses. Crane-Droesch et al. (2013) demonstrated that yield response increased overtime with continued application of biochar and noted a positive yield response with initially low soil CEC and organic C content, but no significant association with soil pH or clay content. Conversely, Jeffery et al. (2011), who evaluated the pH effect in more detail, determined that a positive yield response was associated with neutral (pH > 6) and acidic ($5 \geq pH \leq 6$) soils, but not very acidic (pH < 5) soil, and soils in which biochar increased pH by >1.0 U; additionally, demonstrating that yields responded positively in coarse and medium textured soils, but not fine-textured soils.

The two metaanalyses also differed in associating yield response to biochar feedstock type. Jeffery et al. (2011) found that yields responded positively to hardwood-based and poultry litter-based biochars, but negatively to biosolids. On the other hand, Crane-Droesch et al. (2013) determined that variability was too high to indicate any biochar effect; however, comparatively yields responded more positively to manure-based than wood-based biochars. Jeffery et al. (2017) expanded on their 2011 metaanalyses to perform a more inclusive analysis of biochar type and soil pH across a broader climate and soil gradient. They concluded that yield benefits were generally restricted to tropical soils inherently low in fertility, and pH, and indicated the primary mechanisms were likely the liming and fertilizing effects of high pH and nutrient laden (i.e., manure-based) biochars. These findings echoed the global projection performed by Crane-Droesch et al. (2013) indicating that highly weathered soils, typical of the humid tropics, were more likely to respond to biochar application than the more fertile soils already extensively used for agricultural production in temperate regions.

8.3.2.2 Soil health

The integration of biological, chemical, and physical components in the soil is the basis of measuring soil health. Research has demonstrated that biochar can improve the following aspects within all three components, respectively: microbes and other soil biota; nutrient availability, cation exchange, and pH; and bulk density, aggregation, and water relations. Acting together, these components define the underlying mechanisms supporting soil fertility and sustaining the production and harvest of agricultural commodities. For example, biochar strongly influenced SOM stabilization by improving aggregate stabilization and C incorporation into macroaggregates in fine-textured soils relative to coarse-textured soils where no significant changes in SOM were seen (Wang et al., 2017). However, as addressed previously, measured improvements in soil health do not necessarily translate into increased crop yields. Different outcomes could be due to differences in preexisting soil properties. For example, nutrient release from biochar could be better in a sandy soil than a clayey or organic soil with a higher CEC (Nair et al., 2017).

Bacteria, fungi, and other soil biota are important drivers of nutrient turnover and C management in the soil environment. Increasing biomass and activity of these organisms can increase the rates of nutrient release into the soil, priming plant uptake; however, a certain level of these nutrients can also become tied-up in organism biomass. Warnock et al. (2007) proposed four mechanisms by which biochar might benefit soil mycorrhizal fungi, which might also explain patterns observed with other soil organisms. These were (1) physicochemical changes in the environment (e.g., optimizing pH, improving water content, and enhancing nutrient availability); (2) suppression of competing biota or conversely promotion of "helper" bacteria; (3) inducing plant-fungal chemical signaling that interferes with or detoxifies allelochemicals attached to the biochar; and (4) providing refugia [e.g., protection from predators/grazers, or more simply surfaces on which to attach (Lehmann et al., 2011)]. Additional avenues for biotic benefits are the provision of labile carbon substrates, and additive effects that enhance the ability of biota to resist disease (e.g., in the case of mycorrhizal fungi, resistance to fungal pathogens when root colonization is enhanced) (Warnock et al., 2007; Atkinson et al., 2010). Little is known about biochar effects on nematodes and microarthropods (Lehmann et al., 2011). However, earthworms have been shown to respond both positively and negatively to biochar and or wood-ash additions and are likely responding negatively due to toxic compounds or physical irritation, or positively due to microbial stimulation, increased availably of nutrients, or consuming biochar directly for nutritional or digestive reasons (Liesch et al., 2010; Weyers and Spokas, 2011).

8.3.2.3 Environmental externalities

Water quality and climate change are major environmental concerns, and agricultural processes have been blamed for both. Agriculture is considered a primary source of N and P loads in water ways and water bodies that leads to eutrophication and other negative impacts such as the anoxic zone in the Gulf of Mexico. As of 2012, agriculture was estimated to contribute up to 29% of the world's anthropogenic production of greenhouse gases, including CO_2, CH_4, and N_2O (Vermeulen et al., 2012). Biochar use in agricultural systems might help regulate the environment by improving water quality, sequestering carbon, and reducing greenhouse gases, due to its physicochemical nature as well industrial process of production, given that it is made with an energy capturing platform.

Nonlabile fractions in biochar have a half-life of centuries to millennia due to a recalcitrant nature that prevents conversion C into CO_2, one of the greenhouse gases contributing to climate change (see Section 8.1.1, Table 8.1). The production of CO_2 can also be limited by replacing fossil fuels with bioenergy platforms, for example, bio-oils produced by pyrolysis. Other greenhouse gases, N_2O and CH_4, can also be reduced by selective application of biochar to land, where it may inhibit microbial processes producing these gases. In particular, Brassard et al. (2016) identified that significant reductions in soil N_2O emissions resulted from amendment with biochars that had higher C:N ratios and lower molar O:C ratios than biochars that either increased or caused no change in emissions. Given that carbon mineralization from dead plant matter is generally faster in warm and moist conditions, biochar may represent a more effective means of carbon sequestration in C-poor soils where it has the most agricultural benefit, than in soils where it has less benefit, for example, carbon-saturated soils with no more potential for SOM stabilization (Castellano et al., 2015). This alignment, however, would depend on the degree to which biochar mineralization is dependent on environmental characteristics versus on its intrinsic chemistry, which for biochar remains poorly understood (Singh et al., 2012; Schmidt et al., 2011).

The production of N_2O is formed as a partial final step in the nitrification to denitrification processes that first converts NH_4^+ into NO_3^- substrates then into N_2 gas (Paul and Clark, 1996). A prominent greenhouse gas, N_2O, has a 20-year global warming potential of 268 compared to CO_2 (IPCC, 2014). Numerous studies have measured the reduction in N_2O flux in soils to which biochar has been applied including in acidic soils, paddy soils, soybean crops, and various other cropping systems (Yanai et al., 2007; Steiner, 2010; Zhang et al., 2010; Stavi and Lal, 2013; Aguilar-Chávez et al., 2012; Liu et al., 2014). Reported reductions can range from 50% to 94% compared to nonamended soils (Rondon et al., 2005; Spokas and Reicosky, 2009; Spokas et al., 2009; Case et al., 2015), with the magnitude of differences influenced by the type of biochar used (Fungo et al., 2014). However, N_2O reduction is not ubiquitous, for example, an aerobic incubation of rice paddy soil amended with swine manure biochar showed increased N_2O production (Yoo and Kang, 2012).

The mechanisms controlling N_2O emissions are not well understood, likely because mechanisms are dependent on soil and biochar characteristics, and many observations were made on short-term laboratory incubations or limited field observations. One notable study conducted in the field over five growing seasons demonstrated the consistent reduction in N_2O emissions with an initial wheat-based biochar application at $40\,t\,ha^{-1}$ to a corn—wheat rotation (Liu et al., 2014). This longer term impact is in line with short-term (5-month) indication that aging of a poultry manure + rice hull or wood-based biochar ($>400°C$) increases the absorption capacity for NH_4^+, thus decreasing its availability for denitrification processes (Singh et al., 2010). In contrast an initial 3-month increase in N_2O flux from a $>30\,t\,ha^{-1}$ application of a wood-based biochar ($500°C$) to a wheat cropping system fertilized with ammonium—nitrate then urea was followed 14 months later with no difference in flux from the biochar-free control (Castaldi et al., 2011). Similarly, a short-term analysis indicated decreased N_2O flux in the first 14 days with no further benefit in a 45-d greenhouse incubation of kiln-combusted, wood charcoal added at 1.5%—4.5% with wastewater sludge as a fertilizer for wheat (Aguilar-Chávez et al., 2012).

Across these studies, soil textures varied from loam to silty-loam, and pH from 5.4 to 8.8 before biochar addition, pointing to inhibition or stimulation of biologically based mechanisms producing N_2O. For example, activity of denitrifiers might decline due to loss of anaerobic microsites as biochar can increase soil aeration (Yanai et al., 2007). Limited availability of NH_4^+ or NO_3^- by absorption onto biochar surfaces will inhibit both nitrification and denitrification (Chintala et al., 2013a,b, 2014; Van Zwieten et al., 2010). Conversely, increased NO_3^- availability could result in increased N_2O rates where biochar stimulates the nitrifier community, as demonstrated by increased nitrification potential in forest soils amended with wood-based wildfire charcoal addition as compared to grassland soils already populous with nitrifiers (DeLuca et al., 2006). Alternatively, biochar addition may be inducing changes that allow completion of the denitrification process to the N_2 end product, such as an increase in pH (DeLuca et al., 2009; Van Zwieten et al., 2010), which could be related to provision of electrons via an "electron shuttle" to promote the reduction of N_2O to N_2 by denitrifiers (Cayuela et al., 2013). However, Case et al. (2015) were unable to confirm any of the preceding hypotheses and instead suggested further research into the existence of inhibitory compounds.

8.3.3 Contaminated soils

Organic and inorganic soil contaminants are a global issue and easy remediation solutions are rare. Common inorganic contaminants include lead (Pb), chromium (Cr), arsenic (As), zinc (Zn), cadmium (Cd), copper (Cu), mercury (Hg), and nickel (Ni) (Evanko and Dzombak, 1997) that result from mining, smelting, and other manufacturing processes and are also found in fertilizers, animal manures, pesticides, wastewater, and sewage sludge (Adriano, 2001; Ok et al., 2011; Usman et al., 2012; Lim et al., 2013). Common organic contaminants such as hydrocarbons (alkanes, alkenes), gasoline additives (methyl *tert*-butyl ether), polycyclic aromatic hydrocarbons (PAHs), polychlorinated biphenyls, chemical dyes and atrazine, result from fossil fuel refining and combustion, manufacturing and industrial dumping, waste disposal and incineration, land development, and overuse of agricultural and household pesticides and fertilizers (Valentín et al., 2013). Soils are a sink for both organic and inorganic contaminants, as they sorb or form complexes with organic matter, including humus (Kirpichtchikova et al., 2006; Adriano, 2003). Once in the soil, contaminants interfere with biodegradation, are transported on eroded sediments and in water, are taken up by plants where they can cause phytotoxicity, and have other negative environmental consequences for human and animal safety (Maslin and Maier, 2000; McLaughlin et al., 2000a, b; Ling et al., 2007; Wuana and Okieimen, 2011).

Due to reactive surfaces and a porous nature, biochar can serve as a means for mitigating soil and water contamination (Ahmad et al., 2014; Mohan et al., 2014; Tan et al., 2015) from heavy metals (e.g., Beesley et al., 2010; Mohan et al., 2011; Uchimiya et al., 2011a,b; Park et al., 2011); and organic chemicals (e.g., Cao and Harris, 2010; Cao et al., 2011; Ahmad et al., 2012; Xu et al., 2011; 2012). Biochar surfaces are comprised of reactive carboxyl, hydroxyl, and phenolic functional groups containing $-OH$ bonds, and other remnant components of the feedstock including exchangeable metals (e.g., Ca^{2+}, K^{+}, Mg^{2+}, and Na^{+}) with their mineral components (e.g., CO_3^{2-}, PO_4^{3-}, and SiO_4^{2-}). These surface components furnish the chemical covalent, ionic, hydrogen, and $\pi-\pi$ (via graphitic residues) bonding mechanisms for metal or organic contaminant absorption (Fig. 8.2). Other physical features of biochar, including porosity, surface area, and noncarbonized fractions also influence contaminant absorption through pore-filling, surface precipitation, and physical absorption mechanisms.

Despite seemingly straightforward mechanisms, absorption activity of soil and biochar is interdependent and changes in these conditions when biochar is added might help or hinder these processes. Changes in soil pH are particularly relevant (Ashworth and Alloway, 2008). For example, biochar with O-containing surface functional groups increases the overall negative charge of the soil that facilitates heavy metal cation binding (Peng et al., 2011; Uchimiya et al., 2011a). However,

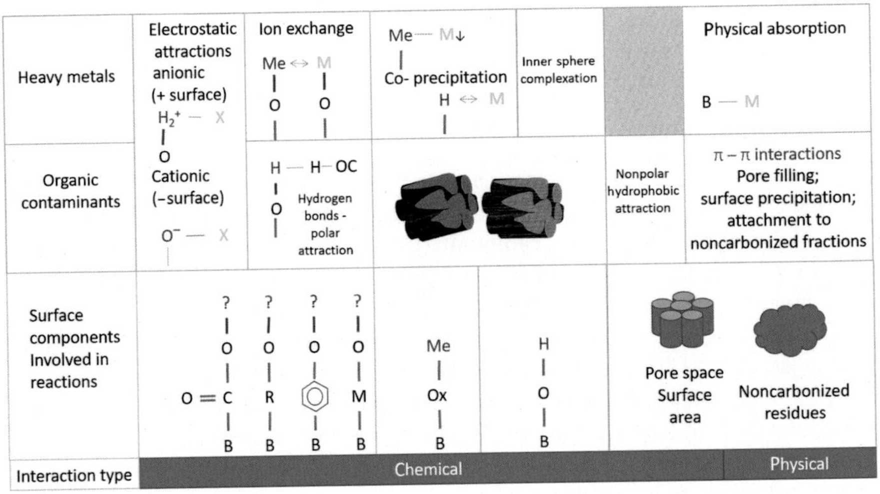

Heavy metals	Electrostatic attractions anionic (+ surface) $H_2^+ - X$ $\vert$ O $\vert$	Ion exchange $Me \leftrightarrow M$ $\vert$ $\vert$ O O $\vert$ $\vert$	$Me - M\downarrow$ $\vert$ Co- precipitation $H \leftrightarrow M$ $\vert$	Inner sphere complexation		Physical absorption $B - M$
Organic contaminants	Cationic (−surface) $O^- - X$ $\vert$	$H - H - OC$ $\vert$ O Hydrogen bonds - polar attraction			Nonpolar hydrophobic attraction	$\pi - \pi$ interactions Pore filling; surface precipitation; attachment to noncarbonized fractions
Surface components Involved in reactions	? ? ? ? $\vert$ $\vert$ $\vert$ $\vert$ O O O O $\vert$ $\vert$ $\vert$ $\vert$ $O = C$ R ⬡ M $\vert$ $\vert$ $\vert$ $\vert$ B B B B		Me $\vert$ Ox $\vert$ B	H $\vert$ O $\vert$ B	Pore space Surface area	Noncarbonized residues
Interaction type	Chemical					Physical

FIGURE 8.2

Chemical and physical interactions on the surface of biochar and proposed modes of action for the absorption of heavy metals and organic contaminants. B—biochar surface; R—hydrocarbon;⬡—phenolic compound; Me—exchangeable metal cations (e.g., Ca, K, Mg, and Na); M—metal; Ox—oxides (e.g., PO_4, CO_3, and SiO_2); ?/H/Me—chemical interaction sites; X—metal (M) or organic chemical (OC) contaminant.

Modified from Ahmad, M., Rajapaksha, A.U., Lim, J.E., Zhang, M., Bolan, N., Mohan, D., et al., 2014. Biochar as a sorbent for contaminant management in soil and water: a review. Chemosphere 99, 19–33 and Tan, X., Liu, Y., Zeng, G., Wang, X., Hua, X., Gu, Y., et al., 2015. Application of biochar for the removal of pollutants from aqueous solutions. Chemosphere 125, 70–85.

if the biochar increases soil pH or supplies dissolved organic carbon, absorption benefits can be hindered by mobilization of nontarget metals such as As and Cu (Park et al., 2011; Hartley et al., 2009; Beesley et al., 2010). Beesley et al. (2010) suggest that preamending with compounds such as iron oxides before adding biochar can immobilize soil As. Hydroxides or anionic mineral components in soils, for example, phosphates, sulfates, or carbonates, can precipitate and immobilize heavy metals (Basta and Tabatabai, 1992; Basta et al., 2005; Sarkar et al., 2017). Conversely, certain metals must undergo additional conformity changes before they bind to the biochar, such as toxic Cr(VI) reducing to essential nutrient Cr(III) (Dong et al., 2011; Choppala et al., 2012; Bolan et al., 2013). As with heavy metals, biochar might reduce bioavailability of toxic organic compounds but also nontargeted pesticides and herbicides, which could result in other negative outcomes such as from increased application rates and accumulation in soil with potential for becoming available as biochar ages (Spokas et al., 2009; Kookana, 2010).

Biochars do not have a universal function for absorption of metal or organic chemical contaminants. But, in general, low temperature biochars provide more ion-exchange, electrostatic attraction through O-containing functional groups, and precipitation mechanisms that favor greater inorganic or polar organic contaminant absorption; and high temperature biochars offer greater surface area, microporosity and smaller particle size, along with a less polar, aromatic surface with greater carbonization into graphitic structures (supporting $\pi-\pi$ bonds) that favor greater organic contaminant absorption (Ahmad et al., 2014; Tan et al., 2015) (Table 8.2). The activation of biochar, as with activated carbon, might also be necessary for absorbing compounds such as PAHs and pesticides (Beesley et al., 2010). The limited knowledge on absolute binding mechanisms for specific contaminants, on interactions between cooccurrence or multiple occurrence of heavy metals and organic chemicals, and of longevity of binding sites or impacts of biochar aging in situ require more careful consideration of production processes, feedstocks, characterization, and field testing of biochar (Ahmad et al., 2014; Lone et al., 2015).

8.4 Potential drawbacks of biochar application

Biochar source material and pyrolysis conditions are likely to influence the occurrence and impact of potentially dangerous and carcinogenic substances (including PAHs), volatile organic carbons (VOCs), and heavy metals, found in biochars or other pyrolysis products (Schimmelpfennig and Glaser, 2012; Hospido et al., 2005). PAHs and VOCs naturally occur but can form under pyrolysis conditions depending on feedstock (Chagger et al., 1998; Conesa et al., 2009). The inhibition of microbial mineralization of pyrolyzed red pine wood was linked to thermal changes typical of cellulose conversion into furan-like materials, possibly benzofuran or dibenzofuran (Baldock and Smernik, 2002), which are related to toxic polychlorinated dibenzofurans (Conesa et al., 2009). Heavy metals found in biochar are primarily due to initial content in the feedstock and can be particularly high in waste materials, including animal manures and sewage sludge (Hospido et al., 2005). Although human and animal wastes are often converted into biochars, biological pathogens such as *Escherichia coli* or *Campylobacter* are unlikely to survive the temperatures of thermolytic processes (Lehmann and Joseph, 2009).

Schimmelpfennig and Glaser (2012) identified several PAHs, including naphthalene, phenanthrene, cresols (components of creosote), and others, in various biochars produced with different source materials and thermal conversion processes. They found higher PAH content in biochar produced in wood gasifiers and traditional charcoal stack production systems, and lower content in biochar from commercialized energy platforms (e.g., Pyreg), hydrothermal carbonization (HTC) processes, and rotary kilns. Their results indicated that PAH content was more related to the whole production process than specific feedstock or temperatures. However, they explained the higher PAH under gasification was due to production and condensation of tars as well as higher production temperatures ($>700°C$) that reform alkyl-C into aromatic structures. In contrast, Keiluweit et al. (2012) associated PAH content directly to pyrolysis temperature conditions with mid-temperatures ranging $400°C-600°C$ producing more PAHs than higher or lower temperatures with the same feedstocks. These latter findings matched those of Hale et al. (2012) for slow pyrolysis chars produced at temperatures between $350°C$ and $550°C$. But again, PAH content may still be dictated by conditions that allow for condensation of volatile or liquid materials conveying PAHs back onto the biochar (Dutta et al., 2017).

Spokas et al. (2012) identified 75 different VOCs with acetone, benzene, methyl ethyl ketone, toluene, and methyl acetate occurring in over half of the 77 different biochars tested. Similar to PAHs, VOC concentrations were more likely to be higher in fast pyrolysis biochars; but conversely VOCs were higher than PAHs under HTC processes and lower with gasification or kiln mound systems (akin to charcoal stacks) (Spokas et al., 2011). Ethylene has been considered by Spokas et al. (2010) as one of the more important VOCs found in biochar, as it can result in various negative responses in the soil environment, including mimicking plant hormones (Insam and Seewald, 2010) or suppressing nitrification and methanotrophic activities (Spokas and Reicosky, 2009). Other unidentified VOCs occurring in biochar, removable with solvents, inhibit seed germination (Rombolà et al., 2015).

Although negative effects on plants and soil microbes, reported after soil biochar amendments, have been attributed to the occurrence of PAHs and VOCs (Ward et al., 1997; Gell et al., 2011; Spokas and Reicosky, 2009; Dempster et al., 2012), little is still understood about persistence of these compounds in the environment (Dutta et al., 2017). Although these compounds can have an initial negative effect, the microbial stimulating aspects of biochar (discussed previously) might eventually counteract this negative effect. Further, PAH- and VOC-laden biochar may not have the same impacts under every soil condition to which these biochars might be applied, whereby desorption from the biochar or reactivity of these compounds in the soil environment might not even occur (Dutta et al., 2017). For instance, a soil high in organic matter might bind and facilitate microbial deactivation of PAHs (Macloed and Semple, 2002) or an already contaminated soil could result in sorbent saturation making the PAHs more readily degradable by existing soil microbes (Cornelissen et al., 2005; Rhodes et al., 2010).

8.5 Review and future directions

Biochar is a term used to define the charcoal-like pyrogenic C generated by controlled (industrial) or uncontrolled (wildfire) conditions of pyrolysis purposely applied to land for improving soil properties. Labile to nonreactive C components comprise all biochars and impart specific functions that define utility and interactions of biochar in the environment. Biochar can alter soil moisture, porosity, pH, CEC, mobilization or immobilization of nutrients, organic compounds and heavy metals, nutrient cycling, biotic community diversity and function, plant growth or inhibition, flux of greenhouse gases, C-sequestration, and so forth. However, biochars do not all function the same way, as the production process and the feedstock used affects the chemical and physical forms of the biochar produced and how it will interact in the environment.

Recent research has come closer to understanding biochar utility and reactivity in the environment than in the preceding decades. In part, this is due to better reporting of production processes, including feedstock and characterization of the biochar before land application. We do know that in nature nonreactive pyrolytic C persists for millennia; we understand that natural fire regimes can restock systems with fresh and revitalized pyrogenic C, and we can characterize and to an extent control the physicochemical properties of industrially produced biochar. The research discussed herein is not an exhaustive analysis of the characterization or use of biochar as a soil amendment or remediation tool. Despite what is known, researchers are still challenged to predict how any specific biochar will react in the environment because soil, climate, vegetation, and management history where the biochar might be applied are so variable. Long-term and thorough field studies are needed to balance the lack of knowledge on how environmental interactions might change as biochar ages over time, and to address aspects of application rate and frequency that may be necessary to achieve specific end goals.

References

Adriano, D.C., 2001. Trace elements in terrestrial environments. Biogeochemistry, Bioavailability, and Risks of Metals. Springer, New York.

Adriano, D.C., 2003. Trace Elements in Terrestrial Environments: Biogeochemistry, Bioavailability and Risks of Metals, second ed. Springer, New York.

Aguilar-Chávez, Á., Díaz-Rojas, M., del Rosario, M., Cárdenas-Aquino, D.L., Luna-Guido, M., 2012. Greenhouse gas emissions from a wastewater sludge-amended soil cultivated with wheat (Triticum spp. L.) as affected by different application rates of charcoal. Soil Biol. Biochem. 52, 90−95.

Ahmad, M., Lee, S.S., Dou, X., Mohan, D., Sung, J.K., Yang, J.E., et al., 2012. Effects of pyrolysis temperature on soybean stover-and peanut shell-derived biochar properties and TCE adsorption in water. Bioresour. Technol. 118, 536−544.

Ahmad, M., Rajapaksha, A.U., Lim, J.E., Zhang, M., Bolan, N., Mohan, D., et al., 2014. Biochar as a sorbent for contaminant management in soil and water: a review. Chemosphere 99, 19−33.

Alcañiz, M., Outeiro, L., Francos, M., Úbeda, X., 2018. Effects of prescribed fires on soil properties: a review. Sci. Total Environ. 613, 944−957.

Anderson, N., Jones, J.G., Page-Dumroese, D., McCollum, D., Baker, S., Loeffler, D., et al., 2013. A comparison of producer gas, biochar, and activated carbon from two distributed scale thermochemical conversion systems used to process forest biomass. Energies 6 (1), 164−183.

Ashworth, D.J., Alloway, B.J., 2008. Influence of dissolved organic matter on the solubility of heavy metals in sewage-sludge-amended soils. Commun. Soil Sci. Plant Anal. 39, 538–550.

Atkinson, C.J., Fitzgerald, J.D., Hipps, N.A., 2010. Potential mechanisms for achieving agricultural benefits from biochar application to temperate soils: a review. Plant Soil. 337, 1–18.

Azargohar, R., Dalai, A.K., 2006. Biochar as a precursor of activated carbon. Appl. Biochem. Biotechnol. 129–132, 762–773.

Baldock, J.A., Smernik, R.J., 2002. Chemical composition and bioavailability of thermally altered Pinus resinosa (Red pine) wood. Org. Geochem. 33 (9), 1093–1109.

Basta, N.T., Tabatabai, M.A., 1992. Effect of cropping systems on adsorption of metals by soils: II. Effect of pH. Soil Sci. 153, 195–204.

Basta, N.T., Ryan, J.A., Chaney, R.L., 2005. Trace element chemistry in residual-treated soil. J. Environ. Qual. 34 (1), 49–63.

Beesley, L., Jiménez, E.M., Eyles, J.L.G., 2010. Effects of biochar and greenwaste compost amendments on mobility, bioavailability and toxicity of inorganic and organic contaminants in a multi-element polluted soil. Environ. Pollut. 158, 2282–2287.

Bird, M.I., Wynn, J.G., Saiz, G., Wurster, C.M., McBeath, A., 2015. The pyrogenic carbon cycle. Annu. Rev. Earth Planet. Sci. 43, 273–298.

Bolan, N.S., Choppala, G., Kunhikrishnan, A., Park, J.H., Naidu, R., 2013. Microbial transformation of trace elements in soils in relation to bioavailability and remediation. Rev. Environ. Contam. Toxicol. 225, 1–56.

Booth, B., Bellouin, N., 2015. Black carbon and atmospheric feedbacks: climate simulations show that interactions between particles of black carbon and convective and cloud processes in the atmosphere must be considered when assessing the full climatic effects of these light-absorbing particulates. Nature 519 (7542), 167–169.

Brassard, P., Godbout, S., Raghavan, V., 2016. Soil biochar amendment as a climate change mitigation tool: key parameters and mechanisms involved. J. Environ. Manage. 181, 484–497. Available from: https://doi.org/10.1016/j.jenvman.2016.06.063.

Brown, R.T., Agee, J.K., Franklin, J.F., 2004. Forest restoration and fire: principles in the context of place. Conserv. Biol. 18 (4), 903–912.

Brown, D., Rowe, A., Wild, P., 2013. A techno-economic analysis of using mobile distributed pyrolysis facilities to deliver a forest residue resource. Bioresour. Technol. 150, 367–376.

Brown, D., Rowe, A., Wild, P., 2014. Techno-economic comparisons of hydrogen and synthetic fuel production using forest residue feedstock. Int. J. Hydrog. Energy 39 (24), 12551–12562.

Caldararo, N., 2002. Human ecological intervention and the role of forest fires in human ecology. Sci. Total Environ. 292 (3), 141–165.

Cao, X., Harris, W., 2010. Properties of dairy-manure-derived biochar pertinent to its potential use in remediation. Bioresour. Technol. 101, 5222–5228.

Cao, X., Ma, L., Liang, Y., Gao, B., Harris, W., 2011. Simultaneous immobilization of lead and atrazine in contaminated soils using dairy-manure biochar. Environ. Sci. Technol. 45, 4884–4889.

Carrasco, J.L., Gunukula, S., Boateng, A.A., Mullen, C.A., DeSisto, W.J., Wheeler, M.C., 2017. Pyrolysis of forest residues: an approach to techno-economics for bio-fuel production. Fuel 193, 477–484.

Case, S.D.C., McNamara, N.P., Reay, D.S., Stott, A.W., Grant, H.K., Whitakera, J., 2015. Biochar suppresses N_2O emissions while maintaining N availability in a sandy loam soil. Soil Biol. Biochem. 81, 178−185.

Castaldi, S., Riondino, M., Baronti, S., Esposito, F.R., Marzaioli, R., Rutigliano, F.A., et al., 2011. Impact of biochar application to a Mediterranean wheat crop on soil microbial activity and greenhouse gas fluxes. Chemosphere 85 (9), 1464−1471.

Castellano, M.J., Mueller, K.E., Olk, D.C., Sawyer, J.E., Six, J., 2015. Integrating plant litter quality, soil organic matter stabilization, and the carbon saturation concept. Glob. Change Biol. 21, 3200−3209.

Cayuela, M.L., Sánchez-Monedero, M.A., Roig, A., Hanley, K., Enders, A., Lehmann, J., 2013. Biochar and denitrification in soils: when, how much and why does biochar reduce N_2O emissions? Sci. Rep. 3, 1732.

Certini, G., 2005. Effects of fire on properties of forest soils: a review. Oecologia 143 (1), 1−10. Available from: https://doi.org/10.1007/s00442-004-1788-8.

Cetin, E., Moghtaderi, B., Gupta, R., Wall, T.F., 2004. Influence of pyrolysis conditions on the structure and gasification reactivity of biomass chars. Fuel 83 (16), 2139−2150.

Cha, J.S., Park, S.H., Jung, S., Ryu, C., Jeon, J., Shin, M., et al., 2016. Production and utilization of biochar: a review. J. Ind. Eng. Chem. 40, 1−15.

Chagger, H.K., Kendall, A., McDonald, A., Pourkashanian, M., Williams, A., 1998. Formation of dioxins and other semi-volatile compounds in biomass combustion. Appl. Energy 60, 101−114.

Chintala, R., Mollinedo, J., Schumacher, T.E., Malo, D.D., Julson, J.L., 2013a. Effect of biochars on chemical properties of acidic soil. Arch. Agron. Soil Sci. 60, 393−404.

Chintala, R., Mollinedo, J., Schumacher, T.E., Malo, D.D., Papiernik, S.K., Clay, D.E., et al., 2013b. Nitrate sorption and desorption by biochars produced from fast pyrolysis. Microporous Mesoporous Mater. 179, 250−257.

Chintala, R., Schumacher, T.E., Kumar, S., Malo, D.D., Rice, J., Bleakley, B., et al., 2014. Molecular characterization of biochar materials and their influence on microbiological properties of soil. J. Hazard. Mater. 279, 244−256.

Choppala, G.K., Bolan, N.S., Mallavarapu, M., Chen, Z., Naidu, R., 2012. The influence of biochar and black carbon on reduction and bioavailability of chromate in soils. J. Environ. Qual. 41, 1−10.

Conesa, J.A., Font, R., Fullana, A., Martín-Gullón, I., Aracil, I., Gálvez, A., et al., 2009. Comparison between emissions from the pyrolysis and combustion of different wastes. J. Anal. Appl. Pyrolysis 84 (1), 95−102.

Cornelissen, G., Haftka, J., Parsons, J., Gustafsson, O., 2005. Sorption to black carbon of organic compounds with varying polarity and planarity. Environ. Sci. Technol. 39, 3688−3694.

Crane-Droesch, A., Abiven, S., Jeffery, S., Torn, M.S., 2013. Heterogeneous global crop yield response to biochar: a meta-regression analysis. Environ. Res. Lett. 8. Available from: https://doi.org/10.1088/1748-9326/8/4/044049.

DeLuca, T.H., Aplet, G.H., 2008. Charcoal and carbon storage in forest soils of the Rocky Mountain West. Front. Ecol. Environ. 6 (1), 18−24.

DeLuca, T.H., MacKenzie, M.D., Gundale, M.J., Holben, W.E., 2006. Wildfire—produced charcoal directly influences nitrogen cycling in ponderosa pine forests. Soil Sci. Soc. Am. J. 70 (2), 448. Available from: https://doi.org/10.2136/sssaj2005.0096.

DeLuca T.H., MacKenzie M.D., Gundale M.J., 2009. Biochar effects on soil nutrient transformations. In: Lehmann, J., Joseph, S. (Eds.), Biochar for Environmental Management: Science and Technology. Earthscan, London, pp. 251–270.

Demeyer, A., Nkana, J.V., Verloo, M.G., 2001. Characteristics of wood ash and influence on soil properties and nutrient uptake: an overview. Bioresour. Technol. 77 (3), 287–295.

Dempster, D.N., Jones, D.L., Murphy, D.M., 2012. Clay and biochar amendments decreased inorganic but not dissolved organic nitrogen leaching in soil. Soil Res. 50, 216–221.

Dong, X.L., Ma, L.Q., Li, Y.C., 2011. Characteristics and mechanisms of hexavalent chromium removal by biochar from sugar beet tailing. J. Hazard. Mater. 190, 909–915.

Dutta, T., Kwon, E., Bhattacharya, S.S., Jeon, B.H., Deep, A., Uchimiya, M., et al., 2017. Polycyclic aromatic hydrocarbons and volatile organic compounds in biochar and biochar-amended soil: a review. GCB Bioenergy 9, 990–1004.

Evanko, C.R., Dzombak, D.A., 1997. Remediation of metals-contaminated soils and groundwater. In: GWRTAC-E Series Tech. Rep. TE-97-01, Ground-Water Remediation Technologies Analysis Center, Pittsburgh, PA.

Fagbenro, J.A., Oshunsanya, S.O., Onawumi, O.A., 2013. Effect of saw dust biochar and NPK 15:15:15 inorganic fertilizer on Moringa oleifera seedlings grown in an oxisol. Agrosearch 13, 57–68.

Fairhead, J., Leach, M., 2009. Amazonian dark earths in Africa? In: Woods, W.I., Teixeira, W.G., Lehmann, J., Steiner, C., WinklerPrins, A., Rebellato, L. (Eds.), Amazonian Dark Earths: Wim Sombroek's Vision. Springer, Dordrecht, pp. 265–278.

Fang, Y., Singh, B., Singh, B.P., Krull, E., 2014. Biochar carbon stability in four contrasting soils. Eur. J. Soil Sci. 65 (1), 60–71.

Fungo, B., Guerena, D., Thiongo, M., Lehmann, J., Neufeldt, H., Kalbitz, K., 2014. N_2O and CH_4 emission from soil amended with steam-activated biochar. J. Plant. Nutr. Soil Sci. 177 (1), 34–38.

Gell, K., van Groenigen, J., Cayuela, M.L., 2011. Residues of bioenergy production chains as soil amendments: immediate and temporal phytotoxicity. J. Hazard. Mater. 186, 2017–2025.

Glaser, B., Amelung, W., 2003. Pyrogenic carbon in native grassland soils along a climosequence in North America. Glob. Biogeochem. Cycles 17 (2).

Glaser, B., Haumaier, L., Guggenberger, G., Zech, W., 2001. The 'Terra Preta' phenomenon: a model for sustainable agriculture in the humid tropics. Naturwissenschaften 88, 37–41.

Hammes, K., Schmidt, M.W., Smernik, R.J., Currie, L.A., Ball, W.P., Nguyen, T.H., et al., 2007. Comparison of quantification methods to measure fire-derived (black/elemental) carbon in soils and sediments using reference materials from soil, water, sediment and the atmosphere. Glob. Biogeochem. Cycles 21 (3), GB3016. Available from: https://doi.org/10.1029/2006GB002914.

Hale, S.E., Lehmann, J., Rutherford, D., Zimmerman, A.R., Bachmann, R.T., Shitumbanuma, V., et al., 2012. Quantifying the total and bioavailable polycyclic aromatic hydrocarbons and dioxins in biochars. Environ. Sci. Technol. 46 (5), 2830–2838.

Hart, S.A., Luckai, N.J., 2014. Charcoal carbon pool in North American boreal forests. Ecosphere 5 (8), 1–14.

Hartley, W., Dickinson, N.M., Riby, P., Lepp, N.W., 2009. Arsenic mobility in brownfield soils amended with green waste compost or biochar and planted with Miscanthus. Environ. Pollut. 157, 2654–2662.

Hedges, J.I., Eglinton, G., Hatcher, P.G., Kirchman, D.L., Arnosti, C., Derenne, S., et al., 2000. The molecularly-uncharacterized component of nonliving organic matter in natural environments. Org. Geochem. 31 (10), 945–958.

Hospido, A., Moreira, T., Martín, M., Rigola, M., Feijoo, G., 2005. Environmental evaluation of different treatment processes for sludge from urban wastewater treatments: anaerobic digestion versus thermal processes. Int. J. Life Cycle Assess. 10, 336–345, 10 pp.

Insam, H., Seewald, M., 2010. Volatile organic compounds (VOCs) in soils. Biol. Fertil. Soils 46, 199–213.

IPCC, 2014. Climate Change 2014: Synthesis, Report Contribution of Working Groups I, II and III to the Fifth Assessment Report of the Intergovernmental Panel on Climate Change. IPCC, Geneva.

Jeffery, S., Verheijen, F.G.A., Van Der Velde, M., Bastos, A.C., 2011. A quantitative review of the effects of biochar application to soils on crop productivity using meta-analysis. Agric. Ecosyst. Environ. 144, 175–187.

Jeffery, S., Abalos, D., Prodana, M., Bastos, A.C., Willem van Groenigen, J., Hungate, B. A., et al., 2017. Biochar boosts tropical but not temperate crop yields. Environ. Res. Lett. 12 (5).

Keiluweit, M., Kleber, M., Sparrow, M.A., Simoneit, B.R.T., Prahl, F.G., 2012. Solvent-extractable polycyclic aromatic hydrocarbons in biochar: influence of pyrolysis temperature and feedstock. Environ. Sci. Technol. 46, 9333–9341.

Kimsey, M., Page-Dumroese, D., Coleman, M., 2011. Assessing bioenergy harvest risks: geospatially explicit tools for maintaining soil productivity in western US forests. Forests 2 (3), 797–813.

Kirpichtchikova, T.A., Manceau, A., Spadini, L., Panfili, F., Marcus, M.A., Jacquet, T., 2006. Speciation and solubility of heavy metals in contaminated soil using X-ray microfluorescence, EXAFS spectroscopy, chemical extraction, and thermodynamic modeling. Geochim. Cosmochim. Acta 70 (9), 2163–2190.

Knicker, H., 2011. Pyrogenic organic matter in soil: its origin and occurrence, its chemistry and survival in soil environments. Quat. Int. 243 (2), 251–263.

Kookana, R.S., 2010. The role of biochar in modifying the environmental fate, bioavailability, and efficacy of pesticides in soils: a review. Soil Res. 48, 627–637.

Kyotani, T., 2000. Control of pore structure in carbon. Carbon 38 (2), 269–286.

Lehmann, J., Joseph, S., 2009. Biochar for environmental management: an introduction. In: Lehmann, J., Joseph, S. (Eds.), Biochar for Environmental Management. Science and Technology, Earthscan, Sterling, pp. 1–12.

Lehmann, J., Rillig, M.C., Thies, J., Masiello, C.A., Hockaday, W.C., Crowley, D., 2011. Biochar effects on soil biota — a review. Soil Biol. Biochem. 43, 1812–1836.

Liesch, A.M., Weyers, S.L., Gaskin, J.W., Das, K.C., 2010. Impact of two different biochars on earthworm growth and survival. Ann. Environ. Sci. 4 (1), 1–9.

Lim, J.E., Ahmad, M., Usman, A.R.A., Lee, S.S., Jeon, W.T., Oh, S.E., et al., 2013. Effects of natural and calcined poultry waste on Cd, Pb and As mobility in contaminated soil. Environ. Earth Sci. 69, 11–20.

Ling, W., Shen, Q., Gao, Y., Gu, X., Yang, Z., 2007. Use of bentonite to control the release of copper from contaminated soils. Aust. J. Soil. Res. 45 (8), 618–623.

Liu, X., Ye, Y., Liu, Y., Zhang, A., Zhang, X., Li, L., et al., 2014. Sustainable biochar effects for low carbon crop production: a 5-crop season field experiment on a low fertility soil from Central China. Agric. Syst. 129, 22–29.

Lone, A.H., Najar, G.R., Ganie, M.A., Sofi, J.A., Ali, T., 2015. Biochar for sustainable soil health: a review of prospects and concerns. Pedosphere 25 (5), 639–653.

Macloed, C.J.A., Semple, K.T., 2002. The adaptation of two similar soils to pyrene catabolism. Environ. Pollut. 119, 357–364.

Makoto, K., Makoto, K., Tamai, Y., Kim, Y.S., Koike, T., 2010. Buried charcoal layer and ectomycorrhizae cooperatively promote the growth of Larix gmelinii seedlings. Plant Soil 327, 143–152.

Marsh, H., Reinoso, F.R., 2006. Activated Carbon. Elsevier.

Mašek, O., Brownsort, P., Cross, A., Sohi, S., 2013. Influence of production conditions on the yield and environmental stability of biochar. Fuel 103, 151–155.

Maslin, P., Maier, R.M., 2000. Rhamnolipid-enhanced mineralization of phenanthrene in organic-metal co-contaminated soils. Biorem. J. 4 (4), 295–308.

McBeath, A.V., Wurster, C.M., Bird, M.I., 2015. Influence of feedstock properties and pyrolysis conditions on biochar carbon stability as determined by hydrogen pyrolysis. Biomass Bioenergy 73, 155–173.

McElligott, K., Page-Dumroese, D., Coleman, M., 2011. Bioenergy Production Systems and Biochar Application in Forests: Potential for Renewable Energy, Soil Enhancement, and Carbon Sequestration. US Forest Service Rocky Mountain Research Station. pp. 1–16.

McKendry, P., 2002. Energy production from biomass (part 1): overview of biomass. Bioresour. Technol. 83, 37–46.

McLaughlin, M.J., Zarcinas, B.A., Stevens, D.P., Cook, N., 2000a. Soil testing for heavy metals. Commun. Soil Sci. Plant. Anal. 31 (11–14), 1661–1700.

McLaughlin, M.J., Hamon, R.E., McLaren, R.G., Speir, T.W., Rogers, S.L., 2000b. Review: a bioavailability-based rationale for controlling metal and metalloid contamination of agricultural land in Australia and New Zealand. Aust. J. Soil. Res. 38 (6), 1037–1086.

Merino, A., Chávez-Vergara, B., Salgado, J., Fonturbel, M.T., García-Oliva, F., Vega, J.A., 2015. Variability in the composition of charred litter generated by wildfire in different ecosystems. Catena 133, 52–63.

Merino, A., Omil, B., Fonturbel, M.T., Vega, J.A., Balboa, M.A., 2016. Reclamation of intensively managed soils in temperate regions by addition of wood bottom ash containing charcoal: SOM composition and microbial functional diversity. Appl. Soil. Ecol. 100, 195–206.

Mohan D., Rajput S., Singh V.K., Steele P.H., Pittman Jr C.U., 2011. Modeling and evaluation of chromium remediation from water using low cost bio-char, a green adsorbent. J. Hazard. Mater., 188, 319–333.

Mohan, D., Sarswat, A., Ok, Y.O., Pittman Jr., C.U., 2014. Organic and inorganic contaminants removal from water with biochar, a renewable, low cost and sustainable adsorbent – a critical review. Bioresour. Technol. 160, 191–202.

Nair, V.D., Ramachandran Nair, P.K., Dari, B., Freitas, A.M., Chatterjee, N., Pinheiro, F.M., 2017. Biochar in the agroecosystem-climate-change-sustainability nexus. Front. Plant. Sci. 8 (2051), 1–9.

Nanda, S., Mohanty, P., Pant, K.K., Naik, S., Kozinski, J.A., Dalai, A.K., 2013. Characterization of North American lignocellulosic biomass and biochars in terms of their candidacy for alternate renewable fuels. Bioenerg. Res. 6, 663–677.

Nanda, S., Dalai, A.K., Berruti, F., Kozinski, J.A., 2016. Biochar as an exceptional bioresource for energy, agronomy, carbon sequestration, activated carbon and specialty materials. Waste Biomass Valoriz. 7, 201–235.

No, S.Y., 2014. Application of bio-oils from lignocellulosic biomass to transportation, heat and power generation—a review. Renew. Sustain. Energy Rev. 40, 1108−1125.

Nocentini, C., Certini, G., Knicker, H., Francioso, O., Rumpel, C., 2010. Nature and reactivity of charcoal produced and added to soil during wildfire are particle-size dependent. Org. Geochem. 41 (7), 682−689.

Novotny, E.H., Deazevedo, E.R., Bonamba, T.J., Cunha, T.J.F., Madari, B.E., Benites, V. M., et al., 2007. Studies of the compositions of humic acids from Amazonian dark earth soils. Environ. Sci. Technol. 41, 400−405.

Novotny, E.H., Maia, C.M.B.D.F., Carvalho, M.T.D.M., Madari, B.E., 2015. Biochar: pyrogenic carbon for agricultural use—a critical review. Rev. BrasileiraCiência do Solo 39 (2), 321−344.

Ok, Y.S., Usman, A.R.A., Lee, S.S., Abd El-Azeem, S.A.M., Choi, B., Hashimoto, Y., et al., 2011. Effects of rapeseed residue on lead and cadmium availability and uptake by rice plants in heavy metal contaminated paddy soil. Chemosphere 85, 677−682.

Page-Dumroese, D., Coleman, M., Jones, G., Venn, T., Dumroese, K.R.; Anderson, N., et al. (2009). Portable in-woods pyrolysis: Using forest biomass to reduce forest fuels, increase soil productivity, and sequester carbon. In: Paper presented at the North American Biochar conference; August 9−12; Boulder, CO,. Center for Energy and Environmental Security. 13 p.

Page-Dumroese, D.S., Jurgensen, M., Terry, T., 2010. Maintaining soil productivity during forest or biomass-to-energy thinning harvests in the western United States. West. J. Appl. Forestry 25 (1), 5−11.

Page-Dumroese, D., Coleman, M., Thomas, S., 2016. Opportunities and uses of biochar on forest sites in North America. In: Bruckman, V., Apaydın Varol, E., Uzun, B., Liu, J. (Eds.), Biochar: A Regional Supply Chain Approach in View of Climate Change Mitigation. Cambridge University Press, Cambridge, pp. 315−335.

Page-Dumroese, D.S., Busse, M.D., Archuleta, J.G., McAvoy, D., Roussel, E., 2017. Methods to reduce forest residue volume after timber harvesting and produce black carbon. Scientifica 2017. Available from: https://doi.org/10.1155/2017/2745764.

Park, J.H., Choppala, G.K., Bolan, N.S., Chung, J.W., Cuasavathi, T., 2011. Biochar reduces the bioavailability and phytotoxicity of heavy metals. Plant Soil 348, 439−451.

Paul, E.A., Clark, F.E., 1996. Soil Microbiology and Biochemistry. Academic Press, California.

Peng, X., Ye, L.L., Wang, C.H., Zhou, H., Sun, B., 2011. Temperature- and duration dependent rice straw-derived biochar: characteristics and its effects on soil properties of an Ultisol in southern China. Soil. Tillage Res. 112, 159−166.

Pingree, M.R.A., Homann, P.S., Morrissette, B., Darbyshire, R., 2012. Long and short—term effects of fire on soil charcoal of a conifer forest in Southwest Oregon. Forests 3 (4), 353−369. Available from: https://doi.org/10.3390/f3020353.

Pingree, M.R., DeLuca, E.E., Schwartz, D.T., DeLuca, T.H., 2016. Adsorption capacity of wildfire-produced charcoal from Pacific Northwest forests. Geoderma 283, 68−77.

Pluchon, N., Gundale, M.J., Nilsson, M.-C., Kardol, P., Wardle, D.A., 2014. Stimulation of boreal tree seedling growth by wood-derived charcoal: effects of charcoal properties, seedling species and soil fertility. Funct. Ecol. 28, 766−775.

Preston, C.M., Schmidt, M.W.I., 2006. Black (pyrogenic) carbon: a synthesis of current knowledge and uncertainties with special consideration of boreal regions. Biogeosciences 3 (4), 397−420.

Pyne, S.J., Andrews, P.L., Laven, R.D., 1996. Introduction to Wildland Fire, second ed. John Wiley and Sons Inc., New York.

Ram, L.C., Masto, R.E., 2014. Fly ash for soil amelioration: a review on the influence of ash blending with inorganic and organic amendments. Earth-Science Rev. 128, 52–74.

Rhodes, A.H., McAllister, L.E., Chen, R., Semple, K.T., 2010. Impact of activated charcoal on the mineralization of 14C-phenanthrene in soils. Chemosphere 79, 463–469.

Rombolà, A.G., Marisi, G., Torri, C., Fabbri, D., Buscaroli, A., Ghidotti, M., et al., 2015. Relationships between chemical characteristics and phytotoxicity of biochar from poultry litter pyrolysis. J. Agric. Food Chem. 63 (30), 6660–6667.

Robertson, S.J., Rutherford, P.M., López-Gutiérrez, J.C., Massicotte, H.B., 2012. Biochar enhances seedling growth and alters root symbioses and properties of subboreal forest soils. Can. J. Soil Sci. 92, 329–340.

Rondon, M., Ramirez, J.A., Lehmann, J., 2005. Charcoal additions reduce net emissions of greenhouse gases to the atmosphere. In: USDA (Ed.), Proceedings of the 3rd USDA Symposium on Greenhouse Gases and Carbon Sequestration in Agriculture and Forestry. USDA, Baltimore, MD, p. 208.

Sarkar, S., Sarkar, B., Basak, B.B., Mandal, S., Biswas, B., Srivastava, P., 2017. Soil mineralogical perspective on immobilization/mobilization of heavy metals. In: Rakshit, A., Abhilash, P., Singh, H., Ghosh, S. (Eds.), Adaptive Soil Management: From Theory to Practices. Springer, Singapore.

Santín, C., Doerr, S.H., Merino, A., Bryant, R., Loader, N.J., 2016. Forest floor chemical transformations in a boreal forest fire and their correlations with temperature and heating duration. Geoderma 264, 71–80.

Scharenbroch, B.C., Meza, E.N., Catania, M., Fite, K., 2013. Biochar and biosolids increase tree growth and improve soil quality for urban landscapes. J. Environ. Qual. 42, 1372–1385.

Schmidt, M.W., Noack, A.G., 2000. Black carbon in soils and sediments: analysis, distribution, implications, and current challenges. Glob. Biogeochem. Cycles 14 (3), 777–793.

Schmidt, M.W.I., Skjemstad, J.O., Gehrt, E., Kögel-Knabner, I., 1999. Charred organic carbon in German chernozemic soils. Eur. J. Soil Sci. 50, 351–365.

Schmidt, M.W.I., Torn, M.S., Abiven, S., Dittmar, T., Guggenberger, G., Janssens, I.A., et al., 2011. Persistence of soil organic matter as an ecosystem property. Nature 478, 49–56.

Schimmelpfennig, S., Glaser, B., 2012. One step forward toward characterization: some important material properties to distinguish biochars. J. Environ. Qual. 41 (4), 1001–1013.

Scott, D.A., Page-Dumroese, D.S., 2016. Wood bioenergy and soil productivity research. BioEnergy Res. 9 (2), 507–517.

Sherman, L.A., Page-Dumroese, D.S., Coleman, M.D., 2018. Idaho forest growth response to post-thinning energy biomass removal and complementary soil amendments. GCB Bioenergy 10 (4), 246–261.

Singh, B.P., Hatton, B.J., Singh, B., Cowie, A.L., Kathuria, A., 2010. Influence of biochars on nitrous oxide emission and nitrogen leaching from two contrasting soils. J. Environ. Qual. 39 (4), 1224–1235. https://doi.org/10.2134/jeq2009.0138.

Singh, N., Abiven, S., Torn, M., Schmidt, M., 2012. Fire-derived organic carbon in soil turns over on a centennial scale. Biogeosciences. 9, 2847–2857.

Skjemstad, J.O., Reicosky, D.C., Wilts, A.R., McGowan, J.A., 2002. Charcoal carbon in U.S. agricultural soils. Soil Sci. Soc. Am. J. 66, 1249–1255.

Sohi, S., Lopez-Capel, E., Krull, E., Bol, R., 2009. Biochar, Climate Change and Soil: A Review to Guide Future Research. CSIRO Land and Water Science Report. CSIRO, Collingwood.

Sohi, S.P., Krull, E., Lopez-Capel, E., Bol, R., 2010. A review of biochar and its use and function in soil. In: Sparks, D.L. (Ed.), Advances in Agronomy. Academic Press, Burlington, pp. 47−82.

Sorenson, C.B., 2010. A Comparative Financial Analysis of Fast Pyrolysis Plants in Southwest Oregon (Doctoral dissertation). The University of Montana, Missoula.

Sovu, M.T., Savadogo, P., Oden, P.C., 2012. Facilitation of forest landscape restoration on abandoned swidden fallows in Laos using mixed-species planting and biochar application. Silva Fennica. 46, 39−51.

Spokas, K.A., 2010. Review of the stability of biochar in soils: predictability of O:C molar ratios. Carbon Manage. 1 (2), 289−303.

Spokas, K., Reicosky, D., 2009. Impacts of sixteen different biochars on soil greenhouse gas production. Ann. Environ. Sci. 3, 179−193.

Spokas K.A., Koskinen W.C., Baker J.M., Reicosky D.C., 2009. Impacts of woodchip biochar additions on greenhouse gas production and sorption/degradation of two herbicides in a Minnesota soil, Chemosphere 77, 574−581.

Spokas, K.A., Baker, J.M., Reicosky, D.C., 2010. Ethylene: potential key for biochar amendment impacts. Plant Soil 333, 443−452.

Spokas, K.A., Novak, J.M., Stewart, C.E., Cantrell, K.B., Uchimiya, M., duSaire, M.G., et al., 2011. Qualitative analysis of volatile organic compounds on biochar. Chemosphere 85, 869−882. Available from: https://doi.org/10.1016/j.chemosphere.

Spokas, K.A., Cantrell, K.B., Novak, J.M., Archer, D.W., Ippolito, J.A., Collins, H.P., et al., 2012. Biochar: a synthesis of its agronomic impact beyond carbon sequestration. J. Environ. Qual. 41, 973−989.

Stavi, I., Lal, R., 2013. Agroforestry and biochar to offset climate change: a review. Agron. Sustain. Dev. 33, 81−96.

Steiner, C., 2010. Biochar in agricultural and forestry applications. In: Biochar From Agricultural and Forestry Residues—A Complimentary Use of Waste Biomass. U.S.-Focused Biochar Report: Assessment of Biochar's Benefits for the United States of America.

Tan, X., Liu, Y., Zeng, G., Wang, X., Hua, X., Gu, Y., et al., 2015. Application of biochar for the removal of pollutants from aqueous solutions. Chemosphere 125, 70−85.

Thomas, S.C., Gale, N., 2015. Biochar and forest restoration: a review and meta-analysis of tree growth responses. N. For. 46, 931−946.

Uchimiya, M., Klasson, K.T., Wartelle, L.H., Lima, I.M., 2011a. Influence of soil properties on heavy metal sequestration by biochar amendment: 1. Copp. Sorpt. Isotherms Rel. Cations. Chemosphere 82, 1431−1437.

Uchimiya, M., Wartelle, L.H., Klasson, T., Fortier, C.A., Lima, I.M., 2011b. Influence of pyrolysis temperature on biochar property and function as a heavy metal sorbent in soil. J. Agric. Food Chem. 59, 2501−2510.

Usman, A.R.A., Lee, S.S., Awad, Y.M., Lim, K.J., Yang, J.E., Ok, Y.S., 2012. Soil pollution assessment and identification of hyperaccumulating plants in chromate copper arsenate (CCA) contaminated sites, Korea. Chemosphere 87, 872−878.

Valentín L., Nousiainen A., Mikkonen A. Introduction to organic contaminants in soil: concepts and risks. In: Vicent T., Caminal G., Eljarrat E., Barceló D. (Eds). Emerging organic contaminants in sludges. The handbook of environmental chemistry, 24, 2013. Springer, Berlin, Heidelberg. Available from: https://doi.org/10.1007/698_2012_208.

Van Zwieten, L., Kimber, S., Morris, S., Downie, A., Berger, E., Rust, J., et al., 2010. Influence of biochars on flux of N_2O and CO_2 from Ferrosol. Soil Res 48 (7), 555—568.

Vega, J.A., Fontúrbel, T., Merino, A., Fernández, C., Ferreiro, A., Jiménez, E., 2013. Testing the ability of visual indicators of soil burn severity to reflect changes in soil chemical and microbial properties in pine forests and shrubland. Plant Soil 369 (1—2), 73—91.

Vermeulen, S.J., Campbell, B.M., Ingram, J.S., 2012. Climate change and food systems. Annu. Rev. Environ. Resour. 37, 195—222.

Wang, D., Xiong, Z., Parikh, S.J., Six, J., Skow, K.M., 2017. Biochar additions can enhance soil structure and the physical stabilization of C in aggregates. Geoderma 303, 110—117.

Ward, D., 2001. Combustion chemistry and smoke. In: Johnson, E.A., Miyanishi, K. (Eds.), Forest Fires: Behavior and Ecological Effects. Academic Press, pp. 55—77.

Ward, B.B., Courtney, K.J., Langenheim, J.H., 1997. Inhibition of Nitrosomonas europaea by monoterpenes from coastal redwood (Sequoia sempervirens) in whole-cell studies. J. Chem. Ecol. 23, 2583—2598.

Wardle, D.A., Zackrisson, O., Nilsson, M.C., 1998. The charcoal effect in Boreal forests: mechanisms and ecological consequences. Oecologia 115, 419—426.

Warnock, D.D., Lehmann, J., Kuyper, T.W., Rillig, M.C., 2007. Mycorrhizal responses to biochar in soil — concepts and mechanisms. Plant Soil 300, 9—20.

Weyers, S.L., Spokas, K.A., 2011. Impact of biochar on earthworm populations: a review. Appl. Environ. Soil Sci. 2011, 1—12.

Woods, W.I., Teixeira, W.G., Lehmann, J., Steiner, C., WinklerPrins, A., Rebellato, L. (Eds.), 2009. Amazonian Dark Earths: Wim Sombroek's Vision. Springer, Berlin.

Woolf, D., Amonette, J.E., Street-Perrott, F.A., Lehmann, J., Joseph, S., 2010. Sustainable biochar to mitigate global climate change. Nat. Commun. 1, 56.

Wuana, R.A., Okieimen, F.E., 2011. Heavy metals in contaminated soils: a review of sources, chemistry, risks and best available strategies for remediation. ISRN Ecol. 2011, 1—20.

Xu, R.K., Xiao, S.C., Yuan, J.H., Zhao, A.Z., 2011. Adsorption of methyl violet from aqueous solutions by the biochars derived from crop residues. Bioresour. Technol. 102, 10293—10298.

Xu, T., Lou, L., Luo, L., Cao, R., Duan, D., Chen, Y., 2012. Effect of bamboo biochar on pentachlorophenol leachability and bioavailability in agricultural soil. Sci. Total Environ. 414, 727—731.

Yanai, Y., Toyota, K., Okazaki, M., 2007. Effects of charcoal addition on N_2O emissions from soil resulting from rewetting air-dried soil in short-term laboratory experiments. Soil Sci. Plant. Nutr. 53, 181—188.

Yang, H., Yan, R., Chen, H., Lee, D.H., Zheng, C., 2007. Characteristics of hemicellulose, cellulose and lignin pyrolysis. Fuel 86, 1781—1788.

Yoo G., Kang H., 2012. Effects of biochar addition on greenhouse gas emissions and microbial responses in a short-term laboratory experiment, J. Environ. Qual 41, 1193—1202.

Zackrisson, O., Nilsson, M.C., Wardle, D.A., 1996. Key ecological function of charcoal from wildfire in the Boreal forest. Oikos 10—19.

Zhang, A.F., Cui, L.Q., Pan, G.X., Li, L.Q., Hussain, Q., Zhang, X.H., et al., 2010. Effect of biochar amendment on yield and methane and nitrous oxide emissions from a rice paddy from Tai Lake plain, China. Agric. Ecosyst. Environ. 139, 469—475.

Bioremediation and soils

9

Ronald S. Zalesny Jr.[1], Michael D. Casler[2], Richard A. Hallett[3], Chung-Ho Lin[4] and Andrej Pilipović[5]

[1]*U.S. Department of Agriculture, Forest Service, Northern Research Station, Rhinelander, WI, United States*
[2]*U.S. Department of Agriculture, Agricultural Research Service, U.S. Dairy Forage Research Center, Madison, WI, United States*
[3]*U.S. Department of Agriculture, Forest Service, Northern Research Station, Durham, NH, United States*
[4]*College of Agriculture, Food, and Natural Resources, University of Missouri, Columbia, MO, United States*
[5]*Institute of Lowland Forestry and Environment, University of Novi Sad, Novi Sad, Serbia*

9.1 Introduction

Land degradation resulting from anthropogenic activities worldwide has multiple and complex impacts on the global environment and public health through direct and indirect processes, which affects a wide array of ecosystem functions and services (Rodríguez-Eugenio et al., 2018). The pollution of soils and water caused by anthropogenic activities is often associated with modern urbanization, industrialization, and agricultural activities such as industrial mining of metals, extraction of petroleum oils and gas, landfill waste, and applications of pesticides and herbicides for food production.

Soil pollution is one of the major effects of human technological advancement. A variety of pollutants affect topsoil and subsoil, including fuel and oil products, heavy metals, hydrocarbon waste, excessive nutrients (e.g., nitrate and phosphate), pesticides, and herbicides. Thousands of chemical pollutants, which are commercially produced on a large scale, are released into terrestrial and aquatic environments on a daily basis, resulting in about 33% of all global soils being at risk of degradation (Rodríguez-Eugenio et al., 2018). For example, agrichemicals, which can help meet the world's growing demand for food, lead to soil pollution and degraded agroecosystems.

According to the Food and Agriculture Organization (FAO) of the United Nations, more than 22 million ha of soil have been affected by soil pollution (Rodríguez-Eugenio et al., 2018). In particular, more than 16% of all Chinese soils and 19% of Chinese agricultural soils are categorized as polluted [China Council for International Cooperation on Environment and Development (CCICED) 2015].

Soils and Landscape Restoration. DOI: https://doi.org/10.1016/B978-0-12-813193-0.00009-6

In Europe, nearly 60% of the top agricultural soils in 11 countries are contaminated with multiple persistent pesticides, and approximately 3 million potentially polluted sites contaminated with industrial pollutants have been identified in the European Economic Area and cooperating countries in the West Balkans (European Environment Agency, 2014). In the United States of America, over 1300 sites are included on the Superfund National Priorities List, with contamination from either heavy metals or hydrocarbon pollutants (US Environmental Protection Agency, 2013). On a larger magnitude the total number of contaminated sites is estimated at 80,000 across Australia (Australia Department of Environment and Conservation, 2010). According to the FAO, approximately 50 million tons of e-waste (i.e., discarded electrical or electronic devices) is generated every year, making it one of the world's fastest growing sources of pollutants that contaminate soil and water (Rodríguez-Eugenio et al., 2018). In addition, land becomes polluted by contaminants not only from industrial waste but also from municipal waste as well. For example, in 2014, Americans produced about 258 million tons of solid waste (US Environmental Protection Agency, 2013). These waste materials release a multitude of hazardous substances (e.g., flame retardants, dioxin-like compounds, polycyclic aromatic hydrocarbons, and heavy metals) that jeopardize environmental quality and human health (Perkins et al., 2014). A little over half of the waste (i.e., 136 million tons) was gathered in landfills, resulting in the soils and leachate at these sites and surrounding areas often being saturated with chemicals and hazardous substances. In addition, according to the National Oceanic and Atmospheric Administration, 80% of pollution in marine environment comes from land through sources such as soil sediments in runoff (Rodríguez-Eugenio et al., 2018).

Furthermore, the aforementioned ecological degradation caused by anthropogenic activities worldwide has resulted in the need to mitigate damage to essential ecosystem services in both rural and urban areas. Phytoremediation is a promising and environmental friendly approach for reclamation of contaminated sites that removes contaminants from systems through enhanced degradation, transformation, extraction, and immobilization (Mirck et al., 2005). There are several advantages of phytoremediation over traditional chemical and physical remediation approaches. In particular, phytoremediation is (1) cost-effective and affordable, (2) easy to implement and maintain, (3) solar-driven, (4) esthetically appealing and socially accepted, (5) minimally invasive, and (6) sustainable in closed-loop systems (Tsao, 2003).

This technical approach has been successfully used to treat soils and surface runoff or leachate that are contaminated with inorganic and organic pollutants (Epps, 2006; Jones et al., 2006; Lin, 2002; Lin et al., 2008, 2011a; Placek et al., 2016; Russell, 2005; Tsao, 2003). Phytoremediation and associated phytotechnologies provide essential ecosystem services during times of accelerated ecological degradation (Epps, 2006). This remediation approach helps restore ecosystem functions, preserve landscapes, and repair degraded lands, while protecting the quality of human life by reducing the risk of exposure to pollutants. Phytoremediation also offers additional ecosystem benefits, such as improved nutrient cycling, carbon sequestration, water flow regulation, and erosion control [Epps, 2006]. As a result, the restoration of contaminated sites offers the highest possible net ecosystem benefit in terms of ecosystem services that are socially, environmentally, and economically beneficial to society.

Phytoremediation is a plant-based technology for restoring contaminated land and water resources (Zalesny et al., 2016b). The success of remediation relies on several fundamental physical, chemical, and biological mechanisms (Fig. 9.1). Transport of the pollutants can be significantly reduced through enhanced infiltration/evapotranspiration, therefore reducing the volume of soil infiltration of the pollutants (Jones et al., 2006; Lin et al., 2011b; Placek et al., 2016; Zupančič-Justin et al., 2010). Organic pollutants can be immobilized and stabilized via enhanced physical adsorption, filtration, or enzymatic conjugation (Chu et al., 2010). A wide range of rhizobacteria have been known to quickly metabolize or transform the contaminants (e.g., explosives, metals, nutrients, herbicides, and pesticides) through biochemical mechanisms, including enzymatic detoxification, nitrification, and denitrification (Lin et al., 2004, 2005, 2008, 2009, 2011a). Direct plant uptake may also help to eliminate herbicides, heavy metals, and nutrients from subsurface flow (Burken and Schnoor, 1997; Lin et al., 2008). Plants have been known to extract and absorb metal contaminants such as Pb, Cd, Cr, Ar, and various radionuclides from soils. One mechanism of phytoremediation, phytoextraction, has been successfully used to remove inorganics from soil through the uptake of heavy metals that are essential for plant growth (e.g., Fe, Mn, Zn, Cu, Mg, Mo, and Ni) (Borghi et al., 2008; Pulford and Watson, 2003). Furthermore, the improvement of soil characteristics by vegetation (e.g., increases in organic matter content) helps enhance the rhizosphere's capacity for adsorption and chemical hydrolysis of pollutants (Chu et al., 2010; Mandelbaum et al., 1993). Recently, many biodefense secondary metabolites released by root exudates, such as benzoxazinones, have been identified as bioactive agents that can rapidly degrade organic pollutants (Willett et al., 2013, 2014, 2016).

The remainder of the chapter addresses the remediation of contaminated soils using plants, focusing on selection of appropriate plant materials and soil factors important for designing remediation systems. The last section of the chapter contains five real-world examples of such systems, including: (1) grasslands used for phytoremediation of soil phosphorus, (2) urban afforestation used to create forests in cities, (3) riparian buffer systems used to reduce agrichemical transport from agroecosystems, (4) short rotation woody crops (SRWCs) used to enhance ecosystem services at landfills, and (5) woody species used for surface mine reclamation.

9.2 Selection of appropriate plant materials

9.2.1 Functional groups

Grasses are probably the most common functional group of herbaceous plants used for phytoremediation, partly because they are highly diverse with a wide range of stress tolerances, they are often capable of forming sod or dense cover that may have multiple

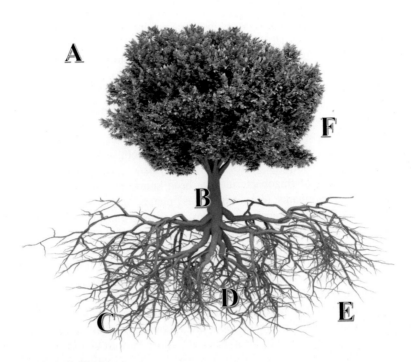

A. **Phytovolatilization**
The uptake and transpiration of a contaminant by a plant. The contaminant, or a modified form of it, is released into the atmosphere.

B. **Phytoextraction**
The uptake and translocation of contaminants by plant roots from the soil into plant parts.

C. **Phytostabilization**
Plants immobilize contaminants in the soil and groundwater through adsorption onto roots or precipitation within the root zone of the plant.

D. **Rhizofiltration**
The use of plant roots to remove pollutants from water or soil solution. Rhizofiltration is similar to phytoextration, but the plant species are used primarily to remediate contaminated groundwater rather than soil.

E. **Rhizodegradation**
The breakdown of contaminants in the soil through microbial activity that is enhanced by the presence of the rhizosphere (i.e., root zone).

F. **Phytodegradation**
The breakdown of contaminants taken up by plants through metabolic processes within the plant or the breakdown of contaminants external to the plant through the effect of compounds produced by the plant, such as enzymes.

FIGURE 9.1

Six processes of phytoremediation that involve contaminant degradation, sequestration, or volatilization in the root zone and in tree roots, wood, and leaves.

Definitions from Mirck, J., Isebrands, J.G., Verwijst, T., Ledin, S., 2005. Development of short-rotation willow coppice systems for environmental purposes in Sweden. Biomass Bioenergy 28, 219–228.

uses, and there is an extensive worldwide seed industry to support commercial distribution of grasses for many purposes. Numerous other monocots include rushes, sedges, and flowering ornamentals for less extensive projects. Dicots may be represented by numerous functional groups of plants that include, but are not limited to agricultural crops, ground covers, or ornamentals that originate from a wide range of habitats.

SRWCs such as poplars (*Populus* spp.), willows (*Salix* spp.), and eucalypts (*Eucalyptus* spp.) are among the most productive temperate forest trees (Zalesny et al., 2011) and, therefore, are the most commonly used trees for phytoremediation (Zalesny et al., 2016b). Selected species and their intra- and interspecific hybrids are phreatophytes, exhibiting extensive root systems, and high biomass production potential relative to other temperate trees, as well as the capability to utilize high volumes of water on moisture-rich sites or exhibit high-water use efficiency on water-limited areas (depending on the genotype selected) (Zalesny et al., 2019a). Given the broad genetic diversity of these genera, there is a high probability of selection gains within and among genomic groups for tolerance and/or uptake of both inorganic and organic pollutants during phytoremediation.

Despite the focus on SRWCs for phytoremediation, slower growing, later successional tree species have gained visibility for phytotechnologies in recent years. In particular, urban greening has become an important policy focus for many cities around the world and, in addition to planting more street trees, cities are also focusing on creating forests in cities (Oldfield et al., 2013). Available land for creating urban forests can include brownfields (Gallagher et al., 2008) and vacant lots (Anderson and Minor, 2017), some of which could benefit from phytoremediation. However, remediation of a site is not the only goal for most urban afforestation projects. This means that even if fast growing "workhorse" trees like poplars and willows are used initially, there is the expectation that longer lived, more esthetically desirable species will be established in the end. However, this does not mean later successional species lack value in situations where soils need remediation. These species can play a role in long-term phytostabilization of a contaminated site (Pulford and Watson, 2003) and can take up metal contaminants. For instance, oak (*Quercus* spp.), maple (*Acer* spp.), and birch (*Betula* spp.) species accumulate Cd, Zn, Cr, and Pb to varying degrees (Evangelou et al., 2015), albeit at much lower concentrations than the SRWCs. A criticism of using later successional species is that the wood is a valuable commodity that could release accumulated metals during processing or expose consumers when they use wood products. This is not a concern for trees growing in urban forests because the primary goal of growing trees in cities is not lumber but rather the ecosystem services they provide (Nowak and Dwyer, 2000). Immobilizing and sequestering contaminants can be added to the long list of ecosystem services provided by trees to cities.

9.2.2 Selection criteria and testing

Multiple criteria are used to select plant materials for phytoremediation projects. Usually, the first decision to be made is with regard to the environment and what functional groups of plants are reasonable choices for that environment. For example, terrestrial versus aquatic environments would lead to completely different functional groups. Within those two broad environmental categories there are many possible subdivisions that will lead to different functional and/or phenotypic differences, for example, savanna, grassland, agricultural, forest, urban (e.g., brownfields and landfills), or ornamental (terrestrial sites) and riverine, estuarial, seashore, or marine (aquatic sites). The question of native versus introduced plants may be the second decision criterion, depending on the needs or desires of the land managers or customer groups. In some cases, natives may not have the desired or necessary traits, so introduced species may be required (Paquin et al., 2004). These decision processes will significantly narrow the range of species under consideration, down to a point where a combination of literature reviews and physical testing and evaluation may be the only additional means of making informed choices.

Testing and evaluation can be conducted at one or more of three levels: bench, pilot, and field—scales. There is no hard-and-fast rule that all three scales be employed for any particular project, but rather the nature of testing will depend on both the results of the literature review and the level and duration of funding available for testing. Bench-scale studies can be conducted at several scales and under a range of conditions that involve the use of pot containers in glasshouses or growth chambers, or small field plots in common-garden experiments that are randomized and replicated to allow statistical comparisons among several species and/or genotypes within species (Zalesny and Bauer, 2007). The fundamental requirement for these studies is to create the appropriate environmental screen that mimics the impacted area or remediation application. For the former, this could be levels of soil contamination with heavy metals, petroleum products, radioactive isotopes, explosives, effluents, or other specific pollutants. For the latter, applications may entail mimicking field phytoremediation applications (e.g., landfill leachate irrigation and wastewater recycling). If the measurements to be made in bench-scale studies are quantitative in nature, statistical rules and principles generally applicable to agronomic field studies should be followed (Casler, 2015). Conversely, if assessments and decisions can be made visually based on vigor and/or survivorship, this may reduce the requirements to follow all of the normal statistical rules of replication and randomization. Bench-scale studies may also be conducted in multiple stages, with visual prescreening of a large number of species, followed by quantitative assessment of a smaller number of candidate species. Bench-scale studies are designed to eliminate unadapted species and focus on a small number of species for pilot-scale (e.g., nursery or small field sizes) or field scale (e.g., large field or production sizes) studies, depending on amount and duration of funding, as well as the level of confidence in the results from the bench-scale studies.

Phyto-recurrent selection is an example of a methodology used for such testing and evaluation (Zalesny et al., 2007b). Phyto-recurrent selection builds upon decades of plant breeding experience in agronomy, horticulture, and other plant sciences to match SRWC genotypes and their tissues of uptake (i.e., root, wood, and leaf) to specific contaminants from soil- and water-based phytoremediation systems. In particular, these genotypes are tested in 2–5 selection cycles in order to make informed decisions on what varieties should be out planted and tested at the field scale. Early selection cycles are short in duration, conducted in controlled conditions, and include hundreds of genotypes that vary greatly in their genetic backgrounds. As phyto-recurrent selection progresses, cycles get longer, trees are grown in nurseries and/or field conditions, and the number of genotypes decreases. In addition, the complexity of data increases with each subsequent selection cycle, with early cycles focusing on survival and biomass traits and later cycles incorporating additional allometric and physiological parameters. The ultimate goal of phyto-recurrent selection is to choose a combination of genotypes with high phytoremediation potential and adequate genetic variation (i.e., selecting a suite of clones rather than just the best clone). In doing so, two different categories of genotypes are identified: (1) generalist clones that perform well across varying site conditions and pollutants and (2) specialist clones that grow well at specific sites and with particular contaminants (Orlović et al., 1998; Zalesny et al., 2016b). Fig. 9.2 illustrates an example of phyto-recurrent selection used for choosing poplar genotypes for landfill phytoremediation.

With similar objectives and needs as for SRWCs and other trees, the selection of grasses, forbs, and other herbaceous species for bioremediation projects often is evaluated by (1) the remediation mechanisms, (2) the plants' tolerance to the pollutants, and (3) other biotic and abiotic stress gradients in the soil microenvironments. Phytoremediation of herbicides with grasses, forbs, and other herbaceous species provides a meaningful example of such selection and testing.

Specifically, the remediation mechanisms are determined by plant-rhizosphere interactions, plant detoxification mechanisms, physiological and morphological characteristics, and chemistry of the herbicides (i.e., their phytotoxicity, solubility, and hydrophobicity) (Lin et al., 2003, 2008, 2011b). More specifically, for mobile- and degradation-resistant organic pollutants, such as herbicides like isoxaflutole and glyphosate, surface- and ground-water mitigation has been achieved through enhanced infiltration/evapotranspiration, therefore reducing the volume of percolating and surface water transport (Lerch et al., 2017; Lin et al., 2008). On the other hand, for organic herbicides or pollutants that are more sensitive to degradation, mitigation has occurred via enhanced biological, chemical, and enzymatic transformation of organic pollutants into less-toxic and less-mobile metabolites in the rhizosphere (Lin et al., 2008, 2009, 2010, 2011a). Many biodefense secondary metabolites, such as benzoxazinones that are released by several native warm season grasses, also play an important role in enhancing the degradation of organic pollutants in the bioremediation systems (Willett et al., 2013, 2014, 2016).

Identification of biotic and abiotic stress gradients in the microenvironment of plants is crucial in the design of tree–shrub–grass multispecies systems for pollutant remediation. This includes knowledge of temporal and spatial characteristics of each stress as well as mechanisms by which specific stresses are reduced

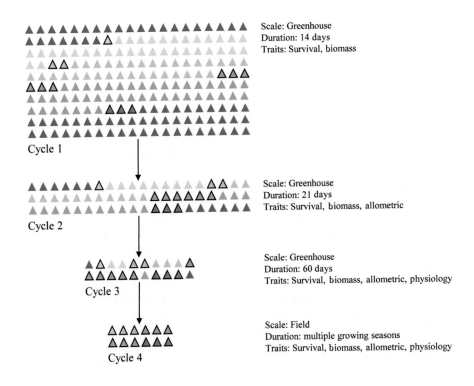

Number of clones tested in four phyto-recurrent selection cycles				
Genomic group / cycle	1	2	3	4
▲ *P. deltoides* Bartr. ex Marsh	27	6	1	0
▲ *P. deltoides* × *P. deltoides*	35	10	3	1
▲ *P. deltoides* × *P. maximowiczii* A. Henry	35	15	5	2
▲ *P. deltoides* × *P. nigra* L.	50	20	7	6
▲ *P. nigra* × *P. maximowiczii*	3	3	3	3
▲ (*P. trichocarpa* Torr. & Gray × *P. deltoides*) × *P. deltoides*	50	6	1	0
Total	200	60	20	12

FIGURE 9.2

Example of phyto-recurrent selection used to choose poplar (*Populus* spp.) genotypes for phytoremediation. Four selection cycles are illustrated showing hypothetical genotypes (represented by *triangles*) belonging to six genomic groups. Clones advancing to subsequent cycles are indicated with bold outlines. Note that testing moves from the greenhouse to the field, duration of testing increases, and data become more complex with later cycles.

(Lin et al., 2004). With regard to bioremediation of organic herbicides by tree–shrub–grass riparian systems, understanding the tolerance of the selected understory grass species to both stresses of shade and high concentrations of pollutants, as well as their detoxification mechanisms, is critical for success.

For example, the C_4 warm season switchgrass (*Panicum virgatum* L.) and eastern gammagrass (*Tripsacum dactyloides* L.) provide an ideal first line of defense along fields of corn (*Zea mays* L.) where atrazine (ATR) concentrations are expected to be highest, thereby providing a sound tree–shrub–grass vegetative buffer system for mitigation of this commonly used herbicide. These grasses not only tolerate high levels of ATR but also have strong capacity to quickly neutralize the ATR through rapid chemical and biological degradation processes (Lin et al., 2003, 2008, 2011a). Switchgrasses and eastern gamagrasses not only help to prevent channelized flow that is generated but they encourage a more uniform sheet flow due to their stem morphology (Lee et al., 1997) and help to decrease the surface transport through encouraged flow infiltration (Lin et al., 2011b). Shade tolerant C_3 species, such as smooth bromegrass (*Bromus inermis* Leyss.) or tall fescue (*Festuca arundinacea* Schreb.), can be ideal choices to be incorporated below the C_4 warm season species near the stream bank where ATR is diluted and tree shading is a concern. These cool-season C_3 species have shown higher annual evapotranspiration rates relative to C_4 grasses (Lin et al., 2003, 2004, 2008, 2011b). As a result, these C_3 species will rapidly remove soil moisture and facilitate the physical trapping of herbicides in the soil. Also, these C_3 species are expected to be tolerant of the moderate shade projected by tree crowns closer to the stream bank (Lin et al., 1998, 2004).

Finally, microbial symbionts may also play a significant role in the success of phytoremediation efforts and in the choice of specific species or genotypes. Arbuscular mycorrhizal (AM) fungi can enhance the adaptation and survivorship of some host plants by enhancing nutrient uptake, absorbing heavy metals, and protecting the host from metal toxicity. Numerous examples of herbaceous monocots and dicots with AM fungal associations acting as hyperaccumulators of heavy metals were provided in the review by Leung et al. (2013); and there are examples from SRWCs, as well (Gunderson et al., 2007; Jordahl et al., 1997). Endophytic fungi (EF) are widely distributed within the grass family and present in many other monocots and dicots. However, the ability of EF to enhance hyperaccumulation characteristics of host plants has received relatively little study to date (Deng and Cao, 2017). There is evidence that EF within the fescues (*Festuca* L. spp.) can lead to increased hyperaccumulation of petroleum pollutants (Soleimani et al., 2010).

9.2.3 Traditional breeding and selection approaches

Traditional breeding and selection approaches are generally conducted under field conditions that are intended to mimic real-world production environments. Generally, this means using appropriate environments, managements, and selection screens. Environments are generally defined by eco-geographic factors such as temperature, precipitation, soil type, and photoperiod. However, when breeding new cultivars for the purpose of phytoremediation, the anthropogenic soil or aquatic factors that have created this demand must be considered when defining both the environment and the selection screen (Zalesny and Bauer, 2007). If the initial germplasm to be used possesses genetic variability for resistance, tolerance, and/or uptake of the targeted anthropogenic element, breeding schemes can be simplified to focus largely on screening plants and progeny for tolerance, uptake, vigor, survivorship, and any other relevant measures necessary to generate superior genotypes (e.g., stable hyperaccumulators, varieties with favorable water use efficiency) (Ernst, 2006).

The rate at which plant breeders can develop new cultivars to solve potential problems is dependent on four general factors: (1) the heritability of the trait, (2) the selection pressure applied, (3) the breeding procedure, and (4) the time required to complete a generation of selection (Falconer and Mackay, 1996). Heritability can generally be increased by the use of replicated families or clonally replicated individuals, obviously with some added expense. Heritability can also be increased by any mechanism that allows the breeder to generate more accurate or precise data used to make selection decisions. Selection pressure is increased by the use of larger population sizes, which tends to lead to space, time, and funding limitations. Different breeding procedures utilize additive and dominance genetic variation to different degrees, which can influence the rate of genetic progress. Lastly, generation time is a major factor determining rate of gain, with woody species, especially forest trees, requiring the longest time, and annual plants or algae species the shortest times. Plant breeders are generally well versed in the myriad of trade-offs that are required to design the most efficient and effective breeding and selection schemes, so project managers who have resources to use these approaches to develop new genotypes or cultivars should consult with a plant breeder who can provide advice and counsel in making many of these decisions. While perennial plants require longer generation times than annual plants, they might have the advantage of clonal propagation for commercial cultivars, allowing the single "best" individual of each generation to be chosen for commercialization and dissemination. Another advantage of clonally propagated species is that they allow the breeder to utilize all forms and amounts of genetic variability within the population; they are very efficient in this regard.

Creation of an effective, efficient, and relevant screen is perhaps the most critical aspect of a breeding and selection component of a phytoremediation project. If the screen is too severe, it might kill all the subjects, while a too-mild screen would not provide sufficient discrimination to be effective in identifying the best genotypes (Ernst, 2006). If improved vigor or survivorship is the goal, screening can be simplified to development of a soil or aqueous medium that provides a concentration of the toxin or pollutant that results in some loss of vigor or some mortality, sufficient to allow discrimination of a small number of individuals with a high level of confidence, for example, often targeting a selection intensity of

0.01%−1% of the population. Conversely, if the goal is hyperaccumulation of a specific element, selection is more complicated, requiring the breeder to measure biomass accumulation, and collect, process, and measure tissue samples using some high-throughput mechanism. Tolerance and hyperaccumulation are not the same trait, hence requiring different measurements and selection approaches (Ernst, 2006). This would add time and expense to the selection process but may be critical or necessary for some project goals.

For woody species, results from tree development programs are often slow or limited given multiyear timeframes between breeding activities and sexual maturity of the trees. Although vegetative propagation can accelerate selection processes of favorable genotypes, information about their full performance (e.g., biomass productivity, disease resistance, and phytoremediation potential) is yet to be obtained until after the end of each production cycle, which can last greater than 10 years for SRWCs. Nevertheless, throughout plantation development, tree performance is typically assessed via growth parameters, which are a reflection of numerous allometric, anatomical, physiological, and biochemical traits (Orlović et al., 1998; Zalesny and Bauer, 2007). Similarly, the effect of different factors (e.g., water and nutrient availability and presence of xenobiotics) on plant growth can be obtained through yield assessment, which is a composite trait that can be tested directly or indirectly via individual traits that affect plant performance (Marron and Ceulemans, 2006). The most common allometric traits include biomass, diameter, height, number of leaves, leaf area, root area, number of roots, and root length (Zalesny and Bauer, 2007), while those related to internal structure and function are parameters related to nitrogen assimilation (Matraszek, 2008; Pilipović et al., 2012a), photosynthesis, transpiration, and water use efficiency (Borghi et al., 2008; Pajević et al., 2009, 2012a, 2019), and biochemical processes (e.g., proline, glutathione, and antioxidant activity) (Di Baccio et al., 2005; Kebert et al., 2017; Nikolić et al., 2008). In addition, morphological changes resulting from variability in these physiological parameters are also useful traits when selecting SRWCs for phytoremediation (Di Baccio et al., 2003; Nikolić et al., 2008; Rogers et al., 2019; Zalesny et al., 2009a). While many of these parameters also are relevant for nonwoody genera, none are more important and cross-cutting than contaminant concentrations in roots, stems, and leaves, which is the greatest measure of remediation success.

The starting plant materials used to develop cultivars for phytoremediation purposes are critical. There must be sufficient genetic variation for the key traits to allow the breeder to reliably choose the "best" individuals and, in so doing, to accumulate the necessary genes in the selected genotypes or their progeny to ensure adequate performance of the new cultivar. There are numerous examples in which heavy metal tolerance and hyperaccumulation abilities have evolved naturally in perennial grasses that are subjected to mine spoils, tailings, Zn-coated electricity pylons, or contaminated soils (Macnair, 1987). Even though the frequency of "tolerant" plants might start out as low as <0.01%, this frequency can increase through long-term on-site exposure or by using artificial selection approaches with large population sizes in the laboratory, for example, screening seedlings on contaminated soil or using a hydroponic system. The low frequency of heavy metal tolerances in these populations, as an example, drives home the critical point that plant breeding is a "numbers game"—the larger the population size that is screened, the greater the likelihood of finding desirable genotypes.

Finally, genotypes or cultivars that are candidates for phytoremediation should have their field performance validated or verified before too many resources are put toward multiplication of seed or vegetative cuttings and before large-scale remediation plantings are initiated. For example, selecting poplar clones that are resistant to diseases such as leaf rust (e.g., *Melampsora* spp.) is important. This phase would involve pilot-scale or field-scale evaluations as with phyto-recurrent selection mentioned earlier in this chapter. Ideally, this would include multiple plantings or environments if there are multiple target sites for phytoremediation. In these cases, pilot-scale trials could be conducted on a small area of each site, for the purposes of confirming that the candidate cultivar has the required levels of tolerance and/or hyperaccumulation ability. If researchers have a high degree of confidence in the future performance of a candidate cultivar, seed or clonal stock multiplication can proceed at the same time as the pilot-scale trials to save time.

9.3 Soil factors important for designing remediation systems

Soil health or quality is a topic of much discussion and research (Bünemann et al., 2018), and the importance of healthy, good quality soils to agricultural and natural systems is not disputed. In fact, definitions of soil quality stress the importance of a soil's ability to buffer "potential pollutants such as agricultural chemicals, organic wastes, and industrial chemicals" (National Research Council, 1993). "Storing, filtering, and transformation of compounds" is listed as one of the seven soil functions by a study group convened by the Royal Academy of Sciences of the Netherlands (Bouma, 2010). Bouma (2010) goes on to discuss a knowledge gap in soil science and highlights the fact that technical "end-of-pipe" solutions to environmental pollution are missing out on the potential for soils to act upstream as a living filter.

Soil degradation from urbanization and industrialization often results in the loss or destruction of plant life from a given site and frequently involves chemical dumping. In addition, severe and long-term agricultural practices often result in nonpoint source pollution of surface waters due to excessive nutrient loads (Sharpley et al., 1994; Sims et al., 1998). Restoration or remediation is needed when human impacts cause the soil system to become compromised or overloaded thereby limiting a soil's ability to filter or transform naturally occurring chemicals (e.g., N and P) or organic wastes and industrial chemicals, all of which can have negative impacts on ecosystem and human health (Galloway et al., 2017; Nieder et al., 2018; Sarwar et al., 2017).

There is plenty of literature on optimal soil physicochemical properties to maximize agricultural yield, promote optimal tree health and growth, or for maintaining healthy ecosystems of many types (Verheye, 2010a–c). In a critical review of soil quality indicators, Bünemann et al. (2018) found that total organic matter and pH are the most frequently used soil quality indicators. In the case of degraded sites the emphasis shifts from optimal conditions to finding genera and/or species that can survive on the site and subsequently help remediate the site so the soil can become functional once again. The key factors to consider when designing phytoremediation systems for degraded sites are highlighted in Table 9.1.

Table 9.1 Soil factors to consider when designing phytoremediation systems for degraded sites.

Scale[a,b]	
Small (<0.25 ha)	Abandoned homes/lots in urban areas, concentrated dumping of chemicals or spills, small agricultural operations.
Medium (>0.25, <10 ha)	Urban construction projects, landfills, brownfields, agricultural operations, chemical disposal.
Large (>10 ha)	Landfills, agricultural operations, mining operations.

Level of physical disturbance[b,c]	
No disturbance	Chemical disposal or leakage. Soil column left intact.
Agricultural	Some agricultural disturbance. Primarily dealing with nonpoint source pollution issues.
Construction	Common in urban settings where construction debris (e.g., concrete, rebar, and asphalt) is deposited and covered with clean fill.
Historical land-use	Abandoned homes and railroad beds.
Landfills	Engineered (e.g., lined and capped) or not engineered will be handled differently.
Mining	Strip mines and tailings will be handled differently.

Chemical contamination[b,c,d,e]	
No contamination	Common in urban setting where clean fill has been used to cover construction debris creating anthrosols. Typically characterized by low levels of organic matter, available nutrients and microbial biomass.
Heavy metals	Factors that impact bioavailability should be considered. Heavy metal bioavailability is influenced by pH, competitive ion concentrations, soil organic matter, cation exchange capacity, and texture.
Mercury	Levels of soil organic matter are important.
Organic contaminants	Includes aromatic hydrocarbons, PAH, PCB, pesticides and polychlorinated dibenzo-*p*-dioxins, among others.
Excessive nutrients	Phosphorus and/or nitrates in soil are often susceptible to erosion or leaching, eventually entering surface waters and causing algae blooms or eutrophication.

(Continued)

Table 9.1 Soil factors to consider when designing phytoremediation systems for degraded sites. *Continued*

Hydrologic flow paths[c,f]	
Contained	No movement of contaminant.
Overland flow	Agricultural systems.
Effluent or leachate	Landfills, agricultural operations, mining operations.
Connected to groundwater	Landfills, agricultural operations, mining operations.

PAH, *Polycylic aromatic hydrocarbons;* PCB, *polychlorinated biphenyls.*
[a]*Anderson and Minor (2017).*
[b]*Oldfield et al. (2014).*
[c]*Gallagher et al. (2008).*
[d]*Gramatica et al. (2002).*
[e]*Lin et al. (2011a).*
[f]*Zalesny et al. (2008).*

9.4 Applications and experiences

9.4.1 Grasslands as a mechanism for phytoremediation of excessive soil phosphorus to reverse eutrophication and improve water quality

Decades of phosphorus fertilization and manure applications from livestock agriculture in excess of crop requirements have resulted in millions of hectares of soils with excessive phosphorus (P) concentrations (Sims et al., 1998). High-P conditions have resulted in increased P loss to surface waters, leading to rapid eutrophication and degradation of water quality, significantly impacting natural resources important for conservation, recreation, drinking water, and fresh and marine water food sources (Steinman et al., 2017). Eutrophication of surface waters leads to inversion of the food pyramid in fresh and marine waters, decreasing the abundance of consumers and predators that are important sources of food and recreation (Binzer et al., 2016).

Phosphorus inputs from point and nonpoint sources can accumulate at many points along the transport pathways from farm fields to surface waters, resulting in numerous sources of legacy P (Sharpley et al., 2013). Accumulated legacy P can be remobilized back into the transport pathway by severe storms or other disturbances or by relaxation of soil conservation practices that may have been put in place decades ago. As a result, many large-scale conservation programs that have been in place for several decades have not delivered the promised increases in water quality within the expected timescales (Sharpley et al., 2013, 2015; Vadas et al., 2018). The only practical way to reduce legacy P in agricultural soils is crop uptake and export. Because annual P removal in crops is generally less than $100 \, kg \, P \, ha^{-1}$, it may require several decades to reduce legacy P to reasonable levels on high-P soils that may have up to $4,000 \, kg \, P \, ha^{-1}$ in the upper 30 cm of soil.

Grasslands are a highly effective mechanism to stabilize soil and reduce soil erosion (Jackson, 2017; Panagos et al., 2015a,b). Grasslands are frequently composed of highly diverse communities of many species with a range of characteristics (Fig. 9.3). Both among species and within species diversity can be used to identify species and genotypes that have the required adaptive characteristics to be used for phytoremediation projects.

Vegetative buffer strips (VBS) or grass margins (GM) are being used with increasing frequency throughout Europe and much of North America to reduce the entry of eroded soil and nutrients from cropland into surface waters (Jackson, 2017; Habibiandehkordi et al., 2019; Panagos et al., 2015a). Traditional use of VBS or GM is based on the use of unharvested perennials, in some cases to the extent of grassland restoration or recreation of seminatural habitats for wildlife and recreation (Holland et al., 2016; Jackson, 2017). Despite the promise of this approach, results have been disappointing, partly due to (1) compromises that can negatively impact the continuity and contiguity of VBS, (2) uncertainty of the best management practices for VBS, and (3) uncertainty of optimal placement to maximize effectiveness and efficiency of

FIGURE 9.3

Thousands of tallgrass prairie and savanna remnants remain throughout the eastern and central United States, providing excellent sources of germplasm for direct release as cultivars or as source materials for selection nurseries to develop cultivars suitable for phytorestoration projects.

Photo by M.D. Casler (USDA Agricultural Research Service).

VBS (Jackson, 2017). While unharvested VBS may have considerable natural appeal as wildlife habitat and for human recreation purposes, this management practice results in no impact on legacy P in high-P soils (Sharpley et al., 2013, 2015; Vadas et al., 2018).

Recent efforts have focused on the use of multifunctional grasslands that are capable of mining and exporting P from high-P soils, providing a crop of economic value to farmers, and (in some cases) providing wildlife habitat and human recreational benefits associated with seminatural habitats. This multifunctionality requires partnerships between agricultural operations and soil conservation organizations, as well as additional partners that are involved in wildlife conservation for some applications. Perennial grasslands can meet these needs in one of the two ways: (1) as a source of forage or feed to support livestock agriculture or (2) as a source of biomass to support renewable energy production systems. Grasslands for use as forage or feed crops can be established not only on VBS or GM scales but also on whole-field scales, but the key element of these systems is to grow productive crops that stabilize the soil and extract P from high-P soils (Fiorellino et al., 2017; Habibiandehkordi et al., 2019; Pant et al., 2004). These feeds must be exported from the farm and their greatest benefit would be derived from sale and feeding to support livestock operations on low-P soils (Vadas et al., 2018). Similarly, a number of perennial grasses are undergoing development as perennial biomass crops for conversion to bioenergy and many of these are also suitable for VBS, GM, or whole-field biomass production systems (Jackson, 2017; Silveira et al., 2013). Biomass crops are generally exported from the farm and used to support conversions systems with energy as the primary product, but always with a coproduct or by-product. In the case of pyrolysis to produce bio-oils, phosphorus from high-P soils ends up in the form of biochar, a soil amendment that is being packaged and used in both farming and gardening applications. Ironically, biochar enhances soil-P uptake and extraction, providing a potential opportunity to improve soil-P management (Biedermann and Harpole, 2013; Gao et al., 2019).

9.4.2 Forest creation in the city: Testing an anthropogenic forest succession strategy

Urban areas around the world are embarking on efforts to increase green space within city limits. Tree planting is one of the areas of focus for these efforts. "Million Trees" programs have been instituted in cities like New York, Los Angeles, London, and Shanghai. Other cities such as Chicago have adopted a tree canopy cover goal. This focus, by cities, on increasing urban tree canopy and maintaining healthy urban trees has been brought about by increasing recognition of the socio economic value trees provide cities (Berman et al., 2008; Nowak et al., 2018). One of the challenges faced by urban land management agencies when trying to implement these programs is finding places to put the trees. One approach is to "restore" city property that has been taken over by exotic invasive plant species. Another approach is to "reclaim" areas that have been severely impacted by construction activities where the soil is primarily made up of fill material of varying quality. These planting goals are particularly important in the light of recent findings showing declining urban and community tree cover in the United States (Nowak and Greenfield, 2018).

The sites that can be reclaimed in the city often bear little resemblance to the places where trees evolved to grow. In fact, the soils on many of these sites are described as human altered and human transported (HAHT) soils (Galbraith, 2018). The combination of these soils in an urban environment and the competition from exotic invasive plant species (Vidra et al., 2006) makes the creation of a forest in the city challenging. These challenges mean that in urban systems human intervention is likely to be a necessary component of a sustainable urban forest starting with establishment and continuing throughout the development of mature canopy dominant native trees (Sasaki et al., 2018; Simmons et al., 2016). Current research on urban afforestation is focused on species palettes, diversity, and soil treatments (Oldfield et al., 2013, 2015). Another element of sustainable forestry is the recruitment of desirable species into the understory. Here again, reclaimed lands in urban areas prove to be challenging. Robinson and Handel (2000) found that on an abandoned municipal landfill that natural recruitment would not support the development of a more diverse woodland and that human intervention would be necessary. Doroski et al. (2018) found that 6 years postplanting, site treatments and conditions strongly influence natural regeneration but that urban sites will need continued human intervention to be sustainable in the long term.

Tree species that are being used to increase urban canopy cover are usually native to the local geographical area (Oldfield et al., 2013). In addition, consideration is given to species that are tolerant of the harsh chemical and climatic conditions of the city where they are being planted. One of the goals of these afforestation efforts is to get the young trees established and to achieve canopy closure as quickly as possible in order to survive amongst exotic invasive plant species that can outcompete native vegetation.

Classical succession theory (Clements, 1916) has been applied to rural forest vegetation dynamics describing the changes in plant communities that occur across time (hundreds to thousands of years) after catastrophic disturbances (e.g., fire and landslides). Despite the extended time trajectory, there are elements of succession theory that can be applied to urban afforestation projects. Anthropogenic succession theory combines elements of classical succession theory with phytoremediation techniques like phyto-recurrent selection of early successional genera (*Populus* and *Salix*) to find fast growing genotypes that are more likely to succeed on HAHT soils in urban areas and perhaps be able to compete with exotic invasive plant species (Fig. 9.4) (Zalesny et al., 2014, 2016b). Once established, these early successional species can improve soil conditions and create an environment where later successional species can thrive. In SRWC

FIGURE 9.4

Freshkills Anthropogenic Succession study, Staten Island, New York, United States. (A) *Salix* spp. planted on legacy dump October 2017. (B) *Salix* spp. early flowering spring 2018 with mugwort (*Artemisia vulgaris* L.) just starting to green up. (C) Mugwort is 2 m tall by August. Can fast growing *Salix* spp. compete?

Photo panel from R.A. Hallett (USDA Forest Service).

and phytoremediation applications, rapid growth and establishment is a priority (Zalesny et al., 2016b). In urban afforestation, this rapid growth and establishment can be leveraged by thinning established early successional trees within a few years and underplanting with slower growing, more desirable, later successional planting palettes including shade tolerant tree and shrub species. This process compresses the time trajectory of natural succession and could result in reduced costs for establishing a forest in the city.

9.4.3 Riparian buffer systems to reduce agrichemical transport from agroecosystems

Well-engineered multispecies riparian buffer strips can be utilized as a cost-effective bioremediation measure to reduce agrichemical transport from agroecosystems and provide a broad range of long-term ecological and environmental benefits.

Over the past decade the University of Missouri's Center for Agroforestry has been dedicated to developing riparian buffer technologies for remediating point and nonpoint sources of pollution (Fig. 9.5). Recent work investigating bioremediation of herbicides and veterinary antibiotics in VBS systems has shown promising results in terms of contaminant load reduction and enhanced degradation (Lin et al., 2003, 2008, 2010, 2011a,b,c; Lin and Thompson, 2013; Lerch et al., 2017). For example, 8 m of riparian buffer strips consisting of native warm season grasses removed 75%–80% of ATR, metolachlor, glyphosate, sulfamethazine,

FIGURE 9.5

The implementation of a vegetative riparian buffer system to remove agrichemicals in soils, surface runoff, and subsurface flow from agricultural fields in Iowa, United States.

Photo from the Center for Agroforestry at University of Missouri, and Department of Natural Resource Ecology and Management at Iowa State University.

tylosin, and enrofloxin in surface runoff. Perennial riparian buffer strips systems tend to harbor soil microbial communities that express greater enzymatic activity that facilitate the degradation of the agrichemicals. Subsequently, ATR degradation was found to be significantly greater in soils previously planted to warm season riparian buffer species relative to bare-soil controls (90% and 24%, respectively), and the half-life of sulfamethazine was 4.25 days shorter in soils collected from the rooting zone of a hybrid poplar (*Populus deltoides* × *Populus nigra*) tree than in control samples (Lin et al., 2010). Enhanced oxytetracycline and sulfadimethoxine sorption to riparian buffer soils has been documented as well (Chu et al., 2010). These results suggested VBS can significantly alter the fate and transport of agrichemicals in agroecosystems, and findings from these studies can be used to design riparian buffer systems that more effectively provide ecosystem services and minimize acreage removed from crop production.

In addition, there are several factors that impact the efficacy of VBS for mitigating surface transport of organic contaminants, including the selection of species, soil type, buffer width, soil erodibility, source to buffer area ratio, buffer placement, runoff flow type (i.e., sheet vs concentrated flow), slope, rainfall intensity, antecedent soil moisture, and the chemical properties of the contaminants (Liu et al., 2008;

Reichenberger et al., 2007; Sabbagh et al., 2009). A broad range of trapping efficiencies resulting from the variation in these factors has been reported in the literature. For example, in a watershed study conducted in central Texas, a 44%—50% reduction in herbicide levels was observed when a filter strip was implemented, while other studies reported 17%—80% removal efficiencies of herbicides in surface runoff (Lin et al., 2011b; Lerch et al., 2017; Hoffman et al., 1995; Mersie et al., 1999; Seybold et al., 2001).

To maximize the removal efficiency, vegetation type, buffer width, and buffer placement are the factors that can be managed. The selection of the plant species and community strongly influences physical, chemical, and biological soil properties that are involved in the bioremediation processes of the pollutants. Many species in various riparian buffer designs have shown the capacity to enhance degradation of herbicides trapped in the rhizospheres because of their ability to stimulate microbial growth and enzyme activities (Lin et al., 2008, 2011a,c; Staddon et al., 2001). Buffer width has been shown to be another important factor to influence the contaminant transport and sediment trapping, with greater buffer widths required to trap fine-grained particles and moderately sorbed pesticides (Liu et al., 2008; Reichenberger et al., 2007). With regard to the placement of the buffer, to prevent the occurrence of concentrated flow through the buffer, the buffer system should be located in close proximity to the source of contamination (Reichenberger et al., 2007). In general, VBS effectiveness will increase with decreasing source to buffer area ratio (Liu et al., 2008). When prioritizing the placement of the system, factors such as the contributing source area, soil wetness, and soil erodibility should be taken into consideration during the design phase (Tomer et al., 2009).

Lastly, there are several physical, chemical, and biological mechanisms involved in the process of bioremediation within the riparian buffer zone. The organic pollutants and nutrients can be intercepted by the roots and residue of the vegetation via the enhanced physical adsorption and filtration (Chu et al., 2010). Rhizobacteria growing in the root zone may have the capacity to metabolize herbicides and nutrients through various biochemical mechanisms, including enzymatic detoxification, nitrification, and denitrification (Lin et al., 2004, 2005, 2008, 2009, 2011a). Direct plant uptake may also help to eliminate the herbicides and nutrients from the subsurface flow (Burken and Schnoor, 1997; Lin et al., 2008). Furthermore, the improvement of soil characteristics by vegetation (e.g., increases in organic matter content and improved porosity) may enhance the rhizosphere's capacity for adsorption and chemical hydrolysis of pollutants (Chu et al., 2010; Mandelbaum et al., 1993).

9.4.4 Using phytoremediation to enhance ecosystem services of landfills

Increasing human population growth and associated industrial development in the last 50 years have contributed to degradation of essential ecosystem services throughout landscapes along the rural to urban continuum (Donohoe, 2003; McDonnell and Pickett, 1990; Wu et al., 2016). For example, human activities ranging from rural farming to industrial production in large cities have greatly impacted soil and water quality in the Great Lakes basin of the United States and Canada (Quinn et al., 2001; Stites and Kraft, 2001). Similar to other areas throughout the world, municipal and industrial landfills in Great Lakes watersheds have contributed to nonpoint source pollution of soils and water, especially given potential impacts of their runoff and leakage (Ferro et al., 2001; Minogue et al., 2012; Zalesny et al., 2016b). Finding methods to reduce these impacts, remediate these sites, and restore these ecosystems is of paramount importance given that the Great Lakes are the largest collection of fresh water in the world and that they provide a tremendous quantity and magnitude of additional ecosystem services (Steinman et al., 2017).

As described previously, selecting appropriate plant materials and identifying key soil factors is essential for designing remediation systems. This is especially true for tree-based phytotechnologies given their broad variation in site conditions, contaminant chemistries, and management objectives (Smesrud et al., 2012; Zupanc and Zupančič-Justin, 2010; Zalesny et al., 2016b). Similar to bioenergy and bioproducts applications, understanding genotype $\times$ environment interactions is crucial for making decisions about what plants to utilize in phytotechnology portfolios and how to maximize biomass productivity from those trees (Headlee et al., 2013; Zalesny et al., 2007a, 2009b). This is especially important at landfills where contaminants range in complexity from inorganics (e.g., heavy metals) to organics (e.g., polycyclic aromatic hydrocarbons), and management priorities are site specific (e.g., landfill leachate recycling, runoff reduction, and contaminant removal) (Christensen and Kjeldsen, 1989; Duggan, 2005; Kjeldsen et al., 2002). While many phytotechnologies have been used at landfills, none have been designed and implemented more than phytoremediation (Zalesny et al., 2016b, 2019b). This frequent use of phytoremediation is due to its broad applicability of having multiple processes that may simultaneously take place in the rhizosphere, roots, wood, and leaves, thus collectively increasing the potential success of the system relative to those that are limited to individual contaminants or plant tissues (Fig. 9.1) (Mirck et al., 2005).

SRWCs such as poplars (*Populus* spp.) and willows (*Salix* spp.) are ideal for phytoremediation of landfills because they grow quickly and have extensive root systems and hydraulic control potential, all of which serve as biological systems that capture and remediate soil and water pollution (Nichols et al., 2014; Nixon et al., 2001; Rockwood et al., 2004). Production gains from poplar and willow breeding programs have been successful in traditional applications given the broad genetic variability of both genera and the subsequent potential to select superior pure species and intra-/interspecies hybrids from within parental and progeny populations (Aravanopoulos et al., 1999; Mahama et al., 2011; Nelson et al., 2018; Rajora and Zsuffa, 1990). Knowledge gained from these traditional tree development activities has translated well into testing and selecting poplar and willow genotypes for phytoremediation (Licht and Isebrands, 2005; Mirck et al., 2005; Zalesny et al., 2016a,b). As described previously, USDA Forest Service researchers have developed phyto-recurrent selection, a tool for choosing generalist tree varieties that remediate a broad range of contaminants, or specialists that are matched to specific pollutants (Zalesny et al., 2007b, 2014). The ability to select varieties across contaminants allows for broad applicability of these phytoremediation systems (Zupančič-Justin et al., 2010).

Recently, phyto-recurrent selection has been used to choose poplar and willow genotypes for phytoremediation buffer systems that are being developed to reduce untreated runoff, recycle wastewater, and groundwater and manage stormwater from landfills within the Lake Superior and Lake Michigan watersheds and, ultimately, to mitigate nonpoint source pollution impacts on nearshore health (Gardiner et al., 2018; GLRI, 2019). In particular, since June 2017, over 20,000 trees have been planted across 16 buffer systems in Wisconsin and Michigan (Fig. 9.6). Key management implications include (1) projecting and measuring the volume of untreated runoff captured or treated, (2) delineating potential

FIGURE 9.6

Poplar (*Populus* spp.) trees 14 months after planting at a landfill in southeastern Wisconsin for runoff reduction and phytoremediation.

Photo by R.S. Zalesny Jr. (USDA Forest Service).

landfill leachate leakage plumes through the use of phytoforensic technologies (Burken and Schnoor, 1998; Limmer et al., 2011), (3) developing a "green tool" to provide site managers with biological treatment options, (4) developing tree health assessment protocols that can be used in phyto-recurrent selection indices, and (5) assessing phytoremediation potential (via phytostabilization and phytovo-latilization). Overall, these phytoremediation activities are reducing uncertainty about the efficacy of using trees to remediate landfills, dumps, and similar sites while improving water quality and soil health, stabilizing stream banks, increasing forest cover, and enhancing ecosystem services.

9.4.5 Surface mine reclamation

Surface mining is one of the most extreme forms of land and soil degradation, being a technology that requires physical removal of overlying soil deposits to access materials such as coal, metals, and minerals (Lima et al., 2016). An example of the significance of such environmental impact is that opencast coal mining damages 2−11 times more land than underground mining (Bai et al., 1999). Such activities result in broad scale disturbance of the landscape, integrity of the habitat, environmental flows, and ecosystem functions (Miller and Zégre, 2014), therefore becoming a continuous source of air and water pollution (Mukhopadhyay et al., 2013). Contemporary mining sites are designed in such a manner to mitigate their impact on the environment. As one of the activities to reduce the risks of mining waste for the environment, Bradshaw and Johnson (1992) recommended revegetation as the most promising approach rather than the application of physicochemical treatments (Ortega-Larrocea et al., 2010).

Unfavorable conditions for plant growth at mining sites present the most crucial limitation in using revegetation for soil remediation (Mulligan et al., 2001). In particular, contamination levels, low-soil fertility, lack of organic matter, disturbance of soil chemical and physical properties, and disappearance of soil microbiota comprise the most common obstacles in revegetation (Borišev et al., 2018). Considering these limitations, selection of proper plant species (and genotypes, where applicable) is a prerequisite for future success of surface mine reclamation. While there are three afforestation strategies for restoration (i.e., pioneer species, climax species, or the biodynamic method combining pioneer and climax species) (Pietrzykowski et al., 2015a), in most cases the use of phytoremediation and erosion control with pioneer species is applied to promote natural succession with more demanding species (Pietrzykowski et al., 2015a,b). Sometimes well-adapted pioneer species like white poplar (*Populus alba* L.) and false indigo-bush (*Amorpha fruticosa* L.) naturally colonize mine sites (Pavlović et al., 2004), which can serve as a meaningful indicator for selection of species for revegetation. Low fertility soils can be enriched by the use of nitrogen fixating species like black locust (*Robinia pseudoacacia* L.) or elder (*Sambucus* L. spp.) (Haynes, 2009). Fast growing species can be used to establish SRWC plantations (Caterino et al., 2017; Quinkenstein and Jochheim, 2016) and improve soil and other ecosystem services at mining sites. However, for better plant survival and performance, soil amendments play a crucial role and are often necessary. For example, amendments such as fertilizers (Hao et al., 2004), seeds of grasses with biosolids (Pietrzykowski et al., 2015a,b), or water holding polymers and microbial fertilizers (Pilipović et al., 2012b) substantially contribute to establishment, survival, and growth of planted seedlings. In addition, esthetic benefits may be achieved through the visual effect of greening the environment with fast growing tree species in relatively short time. However, in the long term, stability of the ecosystem is most often obtained through the use of climax species, which is similar to the urban forests showcased previously. The establishment of plantations with climax or biodynamic species ensures sustainability of the ecosystem. Benefits of such vegetation types can be observed through higher CO_2 sequestration (Brunori et al., 2017; Yuan et al., 2016), enhanced soil health properties (Maiti, 2007), and creation of favorable conditions for soil microbiota (that could last for more than 20 years) (Anderson et al., 2008).

Often, decreased pollution caused by mining activities is considered during the remediation of degraded mine sites, in addition to the restoration of soil and vegetation cover. Copper, Pb, Zn, and other metal mining activities leave land-area footprints that are much larger than the actual size of the mining/disposal sites, which creates the need for application of phytoremediation at these operations. The presence of contamination limits the spectrum of woody species that can be used for this purpose. Most of the research for phytoremediation of heavy metal contaminated sites includes poplars, willows, black locust, and other fast-growing species (Fig. 9.7) (Borghi et al., 2008; Borišev et al., 2016; Di Baccio et al., 2003; Nikolić et al., 2008; Župunski et al., 2016), which can be used for phytoextraction and phytomining of metals. The fast growth of these genera can be combined with their phytoextraction capability in order to obtain both economic and environmental benefits of surface mine reclamation.

The outcomes from research results and practical experience worldwide indicate a high level of complexity associated with remediation of mining sites. In particular, each site has its own soil and contaminant peculiarities that should be analyzed in order to choose the proper remediation technology. First, selection of suitable woody species for mine site recultivation is limited both by environmental factors and objectives of the applied activities. When considering environmental factors, the presence of contamination is most important and, as such, is used to inform what further activities are needed on site. Based on this contamination, selection of species must be matched to their efficiency in phytoremediation, followed by the selection of plantation type. On the other hand, the lack of contamination (e.g., where reduced runoff may be the primary objective) slightly increases the spectrum of potential species, which is then subjected to habitat limitations caused by soil and climate properties (i.e., genotype $\times$ environment interactions). Habitat limitations can be mitigated to some extent by the use of different amendments and measures to promote plant survival during establishment. But in the long run, to establish sustainable ecosystems, there is a need to promote measures aimed at climax phytocenosis. Therefore it is necessary to acquire knowledge about interactions among existing plant species,

FIGURE 9.7

Mine tailings at copper mine RTB "Bor" in Serbia with (A) naturally occurring pioneer tree species of white poplar (*Populus alba* L.), silver birch (*Betula pendula* L.), and black locust (*Robinia pseudoacacia* L.) in the bottom of the photo, and (B) black locust and false indigo-bush (*Amorpha fruticosa* L.) revegetated on the reclaimed plateau in the upper right of the photo.

Photo by A. Pilipović (University of Novi Sad).

microbiota, and soil prior to defining site-specific silvicultural measures. Such measures should be included in all phases of surface mine development to secure benefits of ecosystem services provided by mining activities.

9.5 Summary

Anthropogenic activities worldwide have caused ecological degradation that has resulted in the need to mitigate damage to essential ecosystem services in rural and urban areas. Phytoremediation and associated phytotechnologies are ideal for such applications and require extensive knowledge of soil—plant interactions for restoration to be successful. The information presented previously detailed remediation of contaminated soils using plants, focusing on selection of appropriate plant materials and soil factors important for designing remediation systems. Critical principles and key points include:

- Functional plant groups used for phytoremediation include grasses, SRWCs, and later successional forests species (e.g., oak, maple, and birch). Grasses are used more than other herbaceous species given their broad genetic diversity, wide range of stress tolerances, capability of forming sod or dense cover, and existing worldwide seed industry for commercial distribution. SRWCs are desirable for remediation systems given their extensive root systems, fast growth, and elevated hydraulic control potential, while slower growing species are used given their longevity and esthetics.
- The two primary criteria used to select plant materials for remediation include environment (i.e., terrestrial vs aquatic) and functional group.
- Testing is conducted at bench, pilot, and field scales using methodologies such as phyto-recurrent selection (i.e., using multiple testing cycles to identify superior genotypes with elevated phytoremediation potential).
- Microbial symbionts such as AM fungi can enhance adaptation and survivorship of host plants during phytoremediation.
- While traditional plant breeding approaches are generally conducted under field conditions to mimic real-world production environments, methodologies for phytoremediation may need to include screens of plant growth and development in controlled conditions utilizing soils or other conditions (e.g., wastewater irrigation) from the remediation site to test whether genotypes will survive the contaminants before investing in field trials.
- When advanced to field-scale testing, genotypes should have their field performance validated prior to large-scale deployment to verify efficacy of the system before too many resources are put toward multiplication of seed or vegetative cuttings.
- Similar to traditional plant breeding, development of genotypes for phytoremediation depends on (1) heritability of the traits of interest, (2) selection pressure applied, (3) breeding procedure, and (4) time required to complete a selection generation.
- The primary parameters used to test plant material during phytoremediation include allometric, anatomical, physiological, and biochemical traits, although the most important parameter for measuring remediation success is contaminant concentrations in roots, stems, and leaves.
- There must be sufficient genetic variation in base populations to allow for reliable selection of the "best" individuals, based on specific soil/contaminant conditions and primary traits of interest.
- The key functions of soils related to environmental pollution are to store, filter, and transform compounds—which is complementary to those of plants and should not be overlooked during phytoremediation.

- The most frequently used soil quality indicators are total organic matter and pH, while key factors to consider when designing remediation systems include (1) scale, (2) level of physical disturbance, (3) concentration of contaminants, and (4) hydrologic flow paths.
- Successful remediation technologies have been used across the rural to urban continuum, with examples, including (1) grasslands used for phytoremediation of soil phosphorus, (2) urban afforestation used to create forests in cities, (3) riparian buffer systems used to reduce agrichemical transport from agroecosystems, (4) SRWCs used to enhance ecosystem services at landfills, and (5) woody species used for surface mine reclamation.

Acknowledgments

Collaborations for this chapter were possible through an international scientific exchange as part of the Great Lakes Restoration Initiative (GLRI) Landfill Runoff Reduction Project. We are grateful to Edmund Bauer, Amanda Foust, and Elizabeth Rogers for reviewing earlier versions of this chapter.

References

Anderson, E.C., Minor, E.S., 2017. Vacant lots: an underexplored resource for ecological and social benefits in cities. Urban. For. Urban Green. 21, 146–152.

Anderson, J.D., Ingram, L.J., Stahl, P.D., 2008. Influence of reclamation management practices on microbial biomass carbon and soil organic carbon accumulation in semiarid mined lands of Wyoming. Appl. Soil Ecol. 40, 387–397.

Aravanopoulos, F.A., Kim, K.H., Zsuffa, L., 1999. Genetic diversity of superior *Salix* clones selected for intensive forestry plantations. Biomass Bioenergy 16, 249–255.

Australia Department of Environment and Conservation, 2010. Assessment Levels for Soil, Sediment and Water. Australia Department of Environment and Conservation, Perth, Western Australia. 56 p.

Bai, Z.K., Zhao, J.K., Li, J.C., Wang, W.Y., Lu, C.E., Ding, X.Q., et al., 1999. Ecosystem damage in a large opencast coal mine: a case study on Pingshuo surface coal mine, China. Acta Ecol. Sin. 19, 870–875.

Berman, M.G., Jonides, J., Kaplan, S., 2008. The cognitive benefits of interacting with nature. Psychol. Sci. 19, 1207–1212.

Biedermann, L.A., Harpole, W.S., 2013. Biochar and its effects on plant productivity and nutrient cycling: a meta-analysis. Glob. Change Biol. Bioenergy 5, 202–214.

Binzer, A., Guill, C., Rall, B.C., Brose, U., 2016. Interactive effects of warming, eutrophication and size structure: impacts on biodiversity and food-web structure. Glob. Change Biol. 22, 220–227.

Borghi, M., Tognetti, R., Monteforti, G., Sebastiani, L., 2008. Responses of two poplar species (*Populus alba* and *Populus × canadensis*) to high copper concentrations. Environ. Exp. Bot. 62, 290–299.

Borišev, M., Pajević, S., Nikolić, N., Orlović, S., Župunski, M., Pilipović, A., et al., 2016. Magnesium and iron deficiencies alter Cd accumulation in *Salix viminalis* L. Int. J. Phytoremediat. 18, 164–170.

Borišev, M., Pajević, S., Nikolić, N., Pilipović, A., Arsenov, D., Župunski, M., 2018. Mine site restoration using silvicultural approach. Ch. 7. In: Prasad, M.N.V., de Campos Favas, P.J., Maiti, S.K. (Eds.), Bio-Geotechnologies for Mine Site Rehabilitation. Elsevier, pp. 115–130.

Bouma, J., 2010. Implications of the knowledge paradox for soil science. Ch. 4. In: Sparks, D.L. (Ed.), Advances in Agronomy. Academic Press, London.

Bradshaw, A.D., Johnson, M., 1992. Revegetation of Metalliferous Mine Wastes: The Range of Practical Techniques Used in Western Europe. Elsevier, Manchester.

Brunori, A.M.E., Sdringola, P., Dini, F., Ilarioni, L., Nasini, L., Regni, L., et al., 2017. Carbon balance and life cycle assessment in an oak plantation mined area reclamation. J. Clean. Prod. 144, 69–78.

Bünemann, E.K., Bongiorno, G., Bai, Z., Creamer, R.E., De Deyn, G., de Goede, R., et al., 2018. Soil quality: a critical review. Soil Biol. Biochem. 120, 105–125.

Burken, J.G., Schnoor, J.L., 1997. Uptake and metabolism of atrazine by poplar trees. Environ. Sci. Technol. 31, 1399–1406.

Burken, J.G., Schnoor, J.L., 1998. Predictive relationships for uptake of organic contaminants by hybrid poplar trees. Environ. Sci. Technol. 32, 3379–3385.

Casler, M.D., 2015. Fundamentals of experimental design: guidelines for designing successful experiments. Agron. J. 107, 692–705.

Caterino, B., Schuler, J., Grushecky, S., Skousen, J., 2017. Surface mine to biomass farm: growing shrub willow (*Salix* spp.) in northeast West Virginia: first year results. J. Am. Soc. Min. Reclam. 6, 1–14.

China Council for International Cooperation on Environment and Development (CCICED), 2015. Special policy study on soil pollution management. China Council for International Cooperation on Environment and Development. 47 p.

Christensen, T.H., Kjeldsen, P., 1989. Basic biochemical processes in landfills. Ch. 2.1. In: Christensen, T.H., Cossu, R., Stegmann, R. (Eds.), Sanitary Landfilling: Process, Technology and Environmental Impact. Academic Press, London.

Chu, B., Goyne, K.W., Anderson, S.H., Lin, C.-H., Udawatta, R.P., 2010. Veterinary antibiotic sorption to agroforestry buffer, grass buffer and cropland soils. Agrofor. Syst. 79, 67–80.

Clements, F., 1916. Plant Succession: An Analysis of the Development of Vegetation. Carnegie Institution of Washington, Washington, DC, 652 p.

Deng, Z., Cao, L., 2017. Fungal endophytes and their interactions with plants in phytoremediation: a review. Chemosphere 168, 1100–1106.

Di Baccio, D., Tognetti, R., Sebastiani, L., Vitagliano, C., 2003. Responses of *Populus deltoides* × *Populus nigra* (*Populus* × *euramericana*) clone I-214 to high zinc concentrations. N. Phytol. 159, 443–452.

Di Baccio, D., Kopriva, S., Sebastiani, L., Rennenberg, H., 2005. Does glutathione metabolism have a role in the defence of poplar against zinc excess? N. Phytol. 167, 73–80.

Donohoe, M., 2003. Causes and health consequences of environmental degradation and social injustice. Soc. Sci. Med. 56, 573–587.

Doroski, D.A., Felson, A.J., Bradford, M.A., Ashton, M.P., Oldfield, E.E., Hallett, R.A., et al., 2018. Factors driving natural regeneration beneath a planted urban forest. Urban For. Urban Green. 29, 238–247.

Duggan, J., 2005. The potential for landfill leachate treatment using willows in the UK — a critical review. Resour. Conserv. Recycl. 45, 97–113.

Epps, A.V., 2006. Phytoremediation of Petroleum Hydrocarbons. Washington, DC. 171 p.

Ernst, W.H.O., 2006. Evolution of heavy metal tolerance in higher plants. For. Snow Landsc. Res. 80, 251–274.

European Environment Agency, 2014. Progress in Management of Contaminated Sites. LSI 003. European Environment Agency, Copenhagen.

Evangelou, M.W.H., Papazoglou, E.G., Robinson, B.H., Schulin, R., 2015. Phytomanagement: phytoremediation and the production of biomass for economic revenue on contaminated land. In: Ansari, A.A., Gill, S.S., Gill, R., Lanza, G.R., Newman, L. (Eds.), Phytoremediation: Management of Environmental Contaminants, Volume I. Springer, Switzerland, pp. 115–132.

Falconer, D.S., Mackay, T.F.C., 1996. Introduction to Quantitative Genetics, fourth ed. Pearson, London, 475 p.

Ferro, A., Chard, J., Kjelgren, R., Chard, B., Turner, D., Montague, T., 2001. Groundwater capture using hybrid poplar trees: evaluation of a system in Ogden, Utah. Int. J. Phytoremediat. 3, 87–104.

Fiorellino, N., Kratochvil, R., Coale, F., 2017. Long-term agronomic drawdown of soil phosphorus in mid-Atlantic coastal plain soils. Agron. J. 109, 455–461.

Galbraith, J.M., 2018. Human-altered and human-transported (HAHT) soils in the U.S. soil classification system. Soil Sci. Plant Nutr. 64, 190–199.

Gallagher, F.J., Pechmann, I., Bogden, J.D., Grabosky, J., Weis, P., 2008. Soil metal concentrations and vegetative assemblage structure in an urban brownfield. Environ. Pollut. 153, 351–361.

Galloway, J.N., Leach, A.M., Erisman, J.W., Bleeker, A., 2017. Nitrogen: the historical progression from ignorance to knowledge, with a view to future solutions. Soil Res. 55, 417–424.

Gao, S., DeLuca, T.H., Cleveland, C.C., 2019. Biochar additions alter phosphorus and nitrogen availability in agricultural ecosystems: a meta-analysis. Sci. Total Environ. 654, 463–472.

Gardiner, E.S., Ghezehei, S.B., Headlee, W.L., Richardson, J., Soolanayakanahally, R.Y., Stanton, B.J., et al., 2018. The 2018 Woody Crops International Conference, Rhinelander, Wisconsin, USA, 22–27 July 2018. Forests 9, 693–727.

Gramatica, P., Pozzi, S., Consonni, V., Di Guardo, A., 2002. Classification of environmental pollutants for global mobility potential. SAR QSAR Environ. Res. 13, 205–217.

Great Lakes Restoration Initiative (GLRI), 2019. <https://www.glri.us/> (last accessed 03.03.19.).

Gunderson, J.J., Knight, J.D., Van Rees, K.C.J., 2007. Impact of ectomycorrhizal colonization of hybrid poplar on the remediation of diesel-contaminated soil. J. Environ. Qual. 36, 927–934.

Habibiandehkordi, R., Lobb, D.A., Owens, P.N., Flaten, D.N., 2019. Effectiveness of vegetated buffer strips in controlling legacy phosphorus exports from agricultural land. J. Environ. Qual. 48, 314–321.

Hao, X.-Z., Zhou, D.-M., Si, Y.-B., 2004. Revegetation of copper mine tailings with ryegrass and willow. Pedosphere 14, 283–288.

Haynes, R.J., 2009. Reclamation and revegetation of fly ash disposal sites: challenges and research needs. J. Environ. Manage. 90, 43–53.

Headlee, W.L., Zalesny Jr., R.S., Donner, D.M., Hall, R.B., 2013. Using a process-based model (3-PG) to predict and map hybrid poplar biomass productivity in Minnesota and Wisconsin, USA. Bioenergy Res. 6, 196–210.

Hoffman, D.W., Gerik, T.J., Richardson, C.W., 1995. Use of contour strip cropping as a best management practice to reduce atrazine contamination of surface water. In: Proceedings of the Second International Conference IAWQ Specialized Conference on Diffuse Pollution, Brno and Prague, Czech Republic, pp. 595–596.

Holland, J.M., Bianchi, F.J.J.A., Entling, M.H., Moonen, A., Smith, B.M., Jeanneret, P., 2016. Structure, function and management of semi-natural habitats for conservation biological control: a review of European studies. Pest. Manage. Sci. 72, 1638–1651.

Jackson, R.D., 2017. Targeted use of perennial grass biomass crops in and around annual crop production fields to improve soil health. In: Al-Kaisi, M.M., Lowery, B. (Eds.), Soil Health and Intensification of Agroecosystems. Academic Press, London, pp. 335–352.

Jones, D.L., Williamson, K.L., Owen, A.G., 2006. Phytoremediation of landfill leachate. Waste Manage. 26, 825–837.

Jordahl, J.L., Foster, L., Schnoor, J.L., Alvarez, P.J.J., 1997. Effect of hybrid poplar trees on microbial populations important to hazardous waste bioremediation. Environ. Toxicol. Chem. 16, 1318–1321.

Kebert, M., Rapparini, F., Neri, L., Bertazza, G., Orlović, S., Biondi, S., 2017. Copper-induced responses in poplar clones are associated with genotype- and organ-specific changes in peroxidase activity and proline, polyamine, ABA, and IAA levels. J. Plant. Growth Regul. 36, 131–147.

Kjeldsen, P., Barlaz, M.A., Rooker, A.P., Baun, A., Ledin, A., Christensen, T.H., 2002. Present and long-term composition of MSW landfill leachate: a review. Crit. Rev. Environ. Sci. Technol. 32, 297–336.

Lee, K.H., Isenhart, T.M., Schultz, R.C., Mickelson, S.K., 1997. Nutrient and sediment removal by switchgrass and cool-season grass filter strips. In: Buck, L.E., Lassoie, J.P. (Eds.), Exploring the Opportunity for Agroforestry in Changing Rural Landscapes: Proceedings of the Fifth Conference on Agroforestry in North America. Department of Natural Resources, College of Agriculture and Life Sciences, Cornell University, Ithaca, NY, pp. 68–73.

Lerch, R.N., Lin, C.-H., Goyne, K.W., Kremer, R.J., Anderson, S.H., 2017. Vegetative buffer strips for reducing herbicide transport in runoff: effects of season, vegetation, and buffer width. J. Am. Water Resour. Assoc. 53, 667–683.

Leung, H., Wang, Z., Ye, Z., Yung, K., Peng, X., Cheung, K., 2013. Interactions between arbuscular mycorrhizae and plants in phytoremediation of metal-contaminated soils: a review. Pedosphere 23, 549–563.

Licht, L.A., Isebrands, J.G., 2005. Linking phytoremediated pollutant removal to biomass economic opportunities. Biomass Bioenergy 28, 203–218.

Lima, A.T., Mitchell, K., O'Connell, D.W., Verhoeven, J., Cappellen, P.V., 2016. The legacy of surface mining: remediation, restoration, reclamation and rehabilitation. Environ. Sci. Policy 66, 227–233.

Limmer, M.A., Balouet, J.C., Karg, F., Vroblesky, D.A., Burken, J.G., 2011. Phytoscreening for chlorinated solvents using rapid *in vitro* SPME sampling: application to urban plume in Verl, Germany. Environ. Sci. Technol. 45, 8276–8282.

Lin, C.-H., 2002. Bioremediation Capacity of Five Forage Grasses for Atrazine, Isoxaflutole (Balance) and Nitrate Removal (Doctoral dissertation). University of Missouri Columbia, Columbia, MO.

Lin, C.-H., Thompson, B., 2013. Development of Nanocarbon-Based Biocatalyst for Remediation of Environmental Pollutants. USA Patent 13/448,065. USA Patent.

Lin, C.-H., McGraw, R.L., George, M.F., Garrett, H.E., 1998. Shade effects on forage crops with potential in temperate agroforestry practices. Agrofor. Syst. 44, 109–119.

Lin, C.-H., Lerch, R.N., Johnson, W.G., Jordan, D., Garrett, H.E., George, M.F., 2003. The effect of five forage species on transport and transformation of atrazine and isoxaflutole (balance) in lysimeter leachate. J. Environ. Qual. 32, 1992–2000.

Lin, C.-H., Lerch, R.N., Garrett, H.E., George, M.F., 2004. Incorporating forage grasses in riparian buffers for bioremediation of atrazine, isoxaflutole and nitrate in Missouri. Agrofor. Syst. 63, 91–99.

Lin, C.-H., Lerch, R.N., Kremer, R.J., Garrett, H.E., Udawatta, R.P., George, M.F., 2005. Soil microbiological activities in vegetative buffer strips and their association with herbicides degradation. In: Brooks, K.N., Folliott, P.F. (Eds.), Moving Agroforestry Into The Mainstream: Proceedings of the Ninth Conference on Agroforestry in North America. Department of Forest Resources, University of Minnesota, St. Paul, MN, pp. 1–10.

Lin, C.-H., Lerch, R.N., Garrett, H.E., George, M.F., 2008. Bioremediation of atrazine-contaminated soil by forage grasses: transformation, uptake, and detoxification. J. Environ. Qual. 37, 196–206.

Lin, C.-H., Thompson, B.M., Hsieh, H.Y., Lerch, R.N., Kremer, R.J., Garrett, H.E., 2009. Introduction of atrazine-degrading *Pseudomonas* sp. strain ADP to enhance rhizodegradation of atrazine. In: Gold, M.A., Hall, M.M. (Eds.), Agroforestry Comes of Age: Putting Science into Practice. Proceedings of the 11th North American Agroforestry Conference, Columbia, MO, pp. 183–190.

Lin, C.-H., Goyne, K.W., Kremer, R.J., Lerch, R.N., Garrett, H.E., 2010. Dissipation of sulfamethazine and tetracycline in the root zone of grass and tree species. J. Environ. Qual. 39, 1269–1278.

Lin, C.-H., Lerch, R.N., Kremer, R.J., Garrett, H.E., 2011a. Stimulated rhizodegradation of atrazine by selected plant species. J. Environ. Qual. 40, 1113–1121.

Lin, C.-H., Lerch, R.N., Goyne, K.W., Garrett, H.E., 2011b. Reducing herbicides and veterinary antibiotics losses from agroecosystems using vegetative buffers. J. Environ. Qual. 40, 791–799.

Lin, C.-H., Thompson, B.M., Hsieh, H.-Y., Lerch, R.N., 2011c. Chapter 10: Introduction of atrazine degrader to enhance rhizodegradation of atrazine. In: Goh, K.S., Bret, B.L., Potter, T.L., Gan, J. (Eds.), Pesticide Mitigation Strategies for Surface Water Quality. American Chemical Society, Washington, DC, pp. 139–154. , ACS Symposium Series.

Liu, X., Zhang, X., Zhang, M., 2008. Major factors influencing the efficacy of vegetated buffers on sediment trapping: a review and analysis. J. Environ. Qual. 37, 1667–1674.

Macnair, M.R., 1987. Heavy metal tolerance in plants: a model evolutionary system. Trends Ecol. Evol. 2, 354–359.

Mahama, A.A., Hall, R.B., Zalesny Jr., R.S., 2011. Differential interspecific incompatibility among *Populus* hybrids in sections *Aigeiros* Duby and *Tacamahaca* Spach. For. Chron. 87, 790–796.

Maiti, S.K., 2007. Bioreclamation of coalmine overburden dumps: with special emphasis on micronutrients and heavy metals accumulation in tree species. Environ. Monit. Assess. 125, 111−122.

Mandelbaum, R.T., Wackett, L.P., Allan, D.L., 1993. Rapid hydrolysis of atrazine to hydroxyatrazine by soil bacteria. Environ. Sci. Technol. 27, 1943−1944.

Marron, N., Ceulemans, R., 2006. Genetic variation of leaf traits related to productivity in a *Populus deltoides* × *Populus nigra* family. Can. J. For. Res. 36, 390−400.

Matraszek, R., 2008. Nitrate reductase activity of two leafy vegetables as affected by nickel and different nitrogen forms. Acta Physiologiae Plant. 30, 361−370.

McDonnell, M.J., Pickett, S.T.A., 1990. Ecosystem structure and function along urban-rural gradients: an unexploited opportunity for ecology. Ecology 71, 1232−1237.

Mersie, W., Seybold, C.A., McNamee, C., Huang, J., 1999. Effectiveness of switchgrass filter strips in removing dissolved atrazine and metolachlor from runoff. J. Environ. Qual. 28, 816−821.

Miller, A., Zégre, N., 2014. Mountaintop removal mining and catchment hydrology. Water 6, 472−499.

Minogue, P.J., Miwa, M., Rockwood, D.L., Mackowiak, C.L., 2012. Removal of nitrogen and phosphorus by *Eucalyptus* and *Populus* at a tertiary treated municipal wastewater sprayfield. Int. J. Phytoremediat. 14, 1010−1023.

Mirck, J., Isebrands, J.G., Verwijst, T., Ledin, S., 2005. Development of short-rotation willow coppice systems for environmental purposes in Sweden. Biomass Bioenergy 28, 219−228.

Mukhopadhyay, S., Maiti, S.K., Masto, R.E., 2013. Use of reclaimed mine soil index (RMSI) for screening of tree species for reclamation of coal mine degraded land. Ecol. Eng. 57, 33−142.

Mulligan, C.N., Yong, R.N., Gibbs, B.F., 2001. Remediation technologies for metal-contaminated soils and groundwater: an evaluation. Eng. Geol. 60, 193−207.

National Research Council, Board on Agriculture, Committee on Long-Range Soil and Water Conservation Policy, 1993. Soil and Water Quality: An Agenda for Agriculture. National Academies Press, Washington, DC.

Nelson, N.D., Berguson, W.E., McMahon, B.G., Cai, M., Buchman, D.J., 2018. Growth performance and stability of hybrid poplar clones in simultaneous tests on six sites. Biomass Bioenergy 118, 115−125.

Nichols, E.G., Cook, R., Landmeyer, J., Atkinson, B., Shaw, G., Malone, D., et al., 2014. Phytoremediation of a petroleum-hydrocarbon contaminated shallow aquifer in Elizabeth City, North Carolina, USA. Remediat. J. 24, 29−46.

Nieder, R., Benbi, K.D., Reichl, F.X., 2018. Reactive water-soluble forms of nitrogen and phosphorus and their impacts on environment and human health. In: Nieder, R., Benbi, K.D., Reichl, F.X. (Eds.), Soil Components and Human Health. Springer, Dordrecht, pp. 223−255. , 886 p.

Nikolić, N., Kojić, D., Pilipović, A., Pajević, S., Krstić, B., Borišev, M., et al., 2008. Responses of hybrid poplar to cadmium stress: photosynthetic characteristics, cadmium and proline accumulation, and antioxidant enzyme activity. Acta Biol. Cracov. Bot. 50, 95−103.

Nixon, D.J., Stephens, W., Tyrrel, S.F., Brierly, E.D.R., 2001. The potential for short rotation energy forestry on restored landfill caps. Bioresour. Technol. 77, 237−245.

Nowak, D.J., Dwyer, J.F., 2000. Understanding the benefits and costs of urban forest ecosystems. In: Kuser (Ed.), Handbook of Urban and Community Forestry in the Northeast. Kluwer Academic/Plenum, New York, pp. 11–25.

Nowak, D.J., Greenfield, E.J., 2018. Declining urban and community tree cover in the United States. Urban For. Urban Green. 32, 32–55.

Nowak, D.J., Bodine, A.R., Hoehn, R.E., Ellis, A., Hirabayashi, S., Coville, R., et al., 2018. The Urban Forest of New York City. Resource Bulletin NRS-117, U.S. Department of Agriculture, Forest Service, Newtown Square, PA, 82 p.

Oldfield, E., Warren, R., Felson, A.J., Bradford, M.A., 2013. FORUM: challenges and future directions in urban afforestation. J. Appl. Ecol. 50, 1169–1177.

Oldfield, E.E., Felson, A.J., Wood, S., Hallett, R., Strickland, M.S., Bradford, M., 2014. Positive effects of afforestation efforts on the health of urban soils. For. Ecol. Manage. 313, 266–273.

Oldfield, E.E., Felson, A.J., Auyeung, D.S.N., Crowther, T.W., Sonti, N.F., Harada, Y., et al., 2015. Growing the urban forest: tree performance in response to biotic and abiotic land management. Restor. Ecol. 23, 707–718.

Orlović, S., Guzina, V., Krstić, B., Merkulov, L., 1998. Genetic variability in anatomical, physiological and growth characteristics of hybrid poplar *(Populus* × *euramericana* DODE (GUINIER)) and eastern cottonwood (*Populus deltoides* BARTR.) clones. Silvae Genetica 47, 183–190.

Ortega-Larrocea, Md.P., Xoconostle-Cázares, B., Maldonado-Mendoza, I.E., Carrillo-González, R., Hernández-Hernández, J., Díaz Garduño, M., et al., 2010. Plant and fungal biodiversity from metal mine wastes under remediation at Zimapan, Hidalgo, Mexico. Environ. Pollut. 158, 1922–1931.

Pajević, S., Borišev, M., Nikolić, N., Krstić, B., Pilipović, A., Orlović, S., 2009. Phytoremediation capacity of poplar (*Populus* spp.) and willow (*Salix* spp.) clones in relation to photosynthesis. Arch. Biol. Sci. 61, 239–247.

Panagos, P., Borrelli, P., Meusburger, K., Alewell, C., Lugato, E., Montanarella, L., 2015a. Estimating the soil erosion cover-management factor at the European scale. Land Use Policy 48, 38–50.

Panagos, P., Borrelli, P., Meusburger, K., van der Zanden, E.H., Poesen, J., Alewell, C., 2015b. Modelling the effect of support practices (P-factor) on the reduction of soil erosion by water at European scale. Environ. Sci. Policy 51, 23–34.

Pant, H.K., Adjei, M.B., Scholberg, J.M.S., Chambliss, C.G., Rechcigl, J.E., 2004. Forage production and phosphorus phytoremediation in manure-impacted soils. Agron. J. 96, 1780–1786.

Paquin, D.G., Campbell, S., Li, Q.X., 2004. Phytoremediation in subtropical Hawaii — a review of over 100 plant species. Remediation 13, 127–139.

Pavlović, P., Mitrović, M., Djurdjević, L., 2004. An ecophysiological study of plants growing on fly ash deposits from the "Nikola Tesla—A" thermal power station in Serbia. J. Environ. Manage. 33, 654–663.

Perkins, D.N., Brune Drisse, M.N., Nxele, T., Sly, P.D., 2014. E-waste: a global hazard. Ann. Glob. Health 80, 286–295.

Pietrzykowski, M., Krzaklewski, W., Likus, J., Wos, B., 2015a. Assessment of English oak (*Quercus robur* L.) growth in varied soil-substrate conditions of reclaimed Piaseczno sulfur mine dump. Folia Forestalia Polonica, Ser. A 57, 28–32.

Pietrzykowski, M., Krzaklewski, W., Wos, B., 2015b. Preliminary assessment of growth and survival of green alder (*Alnus viridis*), a potential biological stabilizer on fly ash disposal sites. J. For. Res. 26, 131–136.

Pilipović, A., Orlović, S., Nikolić, N., Borišev, M., Krstić, B., Rončević, S., 2012a. Growth and plant physiological parameters as markers for selection of poplar clones for crude oil phytoremediation. Sumarski List. 136, 273–281.

Pilipović, A., Orlović, S., Stojnić, S., Vasić, V., 2012b. Effects of supplements on vitality of black locust used for recultivation of copper mine tailings. International Scientific Conference: Forests in Future-Sustainable Use, Risks and Challenges. Institute of Forestry Belgrade, Belgrade, October 4–5, 2012.

Pilipović, A., Zalesny Jr., R.S., Rončević, S., Nikolić, N., Orlović, S., Beljin, J., et al., 2019. Growth, physiology, and phytoextraction potential of poplar and willow established in soils amended with heavy-metal contaminated, dredged river sediments. J. Environ. Manage. 239, 352–365.

Placek, A., Grobelak, A., Kacprzak, M., 2016. Improving the phytoremediation of heavy metals contaminated soil by use of sewage sludge. Int. J. Phytoremediat. 18, 605–618.

Pulford, I.D., Watson, C., 2003. Phytoremediation of heavy metal-contaminated land by trees: a review. Environ. Int. 29, 529–540.

Quinkenstein, A., Jochheim, H., 2016. Assessing the carbon sequestration potential of poplar and black locust short rotation coppices on mine reclamation sites in Eastern Germany: model development and application. J. Environ. Manage. 168, 53–66.

Quinn, J.J., Negri, M.C., Hinchman, R.R., Moos, L.P., Wozniak, J.B., Gatliff, E.G., 2001. Predicting the effect of deep-rooted hybrid poplars on the groundwater flow system at a large-scale phytoremediation site. Int. J. Phytoremediat. 3, 41–60.

Rajora, O.P., Zsuffa, L., 1990. Allozyme divergence and evolutionary relationships among *Populus deltoides*, *P. nigra*, and *P. maximowiczii*. Genome 33, 44–49.

Reichenberger, S., Bach, M., Skitschak, A., Frede, H.-G., 2007. Mitigation strategies to reduce pesticide inputs into ground- and surface-water and their effectiveness: a review. Sci. Total Environ. 384, 1–35.

Robinson, G.R., Handel, S.N., 2000. Directing spatial patterns of recruitment during an experimental urban woodland reclamation. Ecol. Appl. 10, 174–188.

Rockwood, D.L., Naidu, C.V., Carter, D.R., Rahmani, M., Spriggs, T.A., Lin, C., et al., 2004. Short-rotation woody crops and phytoremediation: opportunities for agroforestry? Agrofor. Syst. 61, 51–63.

Rodríguez-Eugenio, N., McLaughlin, M., Pennock, D., 2018. Soil Pollution: A Hidden Reality. Food and Agriculture Organization of the United Nations, Rome, 156 p.

Rogers, E.R., Zalesny Jr., R.S., Hallett, R.A., Headlee, W.L., Wiese, A.H., 2019. Relationships among root-shoot ratio, early growth, and health of hybrid poplar and willow genotypes grown in different landfill soils. Forests 10, 49–67.

Russell, K., 2005. The Use and Effectiveness of Phytoremediation to Treat Persistent Organic Pollutants. Washington, DC. 44 p.

Sabbagh, G.J., Fox, G.A., Kamanzi, A., Roepke, B., Tang, J.-Z., 2009. Effectiveness of vegetative filter strips in reducing pesticide loading: quantifying pesticide trapping efficiency. J. Environ. Qual. 38, 762–771.

Sarwar, N., Imran, M., Shaheen, M.R., Ishaque, W., Kamran, M.A., Matloob, A., et al., 2017. Phytoremediation strategies for soils contaminated with heavy metals: modifications and future perspectives. Chemosphere 171, 710–721.

Sasaki, T., Ishii, H., Morimoto, Y., 2018. Evaluating restoration success of a 40-year-old urban forest in reference to mature natural forest. Urban For. Urban Green 32, 123–132.

Seybold, C., Mersie, W., Delorem, D., 2001. Removal and degradation of atrazine and metolachlor by vegetative filter strips on clay loam soil. Commun. Soil Sci. Plant Anal. 32, 723–737.

Sharpley, A., Jarvie, H.P., Buda, A., May, L., Spears, B., Kleinman, P., 2013. Phosphorus legacy: overcoming the effects of past management practices to mitigate future water quality impairment. J. Environ. Qual. 42, 1308–1326.

Sharpley, A.N., Chapra, S.C., Wedepohl, R., Sims, I.T., Daniel, T.C., Reddy, K.R., 1994. Managing agricultural phosphorus for protection of surface waters: issues and options. J. Environ. Qual. 23, 437–451.

Sharpley, A.N., Bergstrom, L., Aronsson, H., Bechmann, M., Bolster, C.H., Borling, K., et al., 2015. Future agriculture with minimized phosphorus losses to waters: research needs and direction. Ambio 44, S163–S179.

Silveira, M.L., Vendramini, J.M.B., Sui, X., Sollenberger, L., O'Connor, G.A., 2013. Screening perennial warm-season bioenergy crops as an alternative for phytoremediation of excess soil P. Bioenergy Res. 6, 469–475.

Simmons, B.L., Hallett, R.A., Sonti, N.F., Auyeung, D.S.N., Lu, J.W.T., 2016. Long-term outcomes of forest restoration in an urban park. Restor. Ecol. 24, 109–118.

Sims, J.T., Simard, R.R., Joern, B.C., 1998. Phosphorus loss in agricultural drainage: historical perspective and current research. J. Environ. Qual. 27, 277–293.

Smesrud, J.K., Duvendack, G.D., Obereiner, J.M., Jordahl, J.L., Madison, M.F., 2012. Practical salinity management for leachate irrigation to poplar trees. Int. J. Phytoremediat. 14 (S1), 26–46.

Soleimani, M., Afyuni, M., Hajabbasi, M.A., Nourbakhsh, F., Sabzalian, M.R., Christensen, J.H., 2010. Phytoremediation of an aged petroleum contaminated soil using endophyte infected and non-infected grasses. Chemosphere 81, 1084–1090.

Staddon, W.J., Locke, M.A., Zablotowicz, R.M., 2001. Microbiological characteristics of a vegetative buffer strip soil and degradation and sorption of metolachlor. Soil Sci. Soc. Am. J. 65, 1136–1142.

Steinman, A.D., Cardinale, B.J., Munns Jr., W.R., Ogdahl, M.E., Allan, J.D., Angadi, T., et al., 2017. Ecosystem services in the Great Lakes. J. Gt. Lakes Res. 43, 161–168.

Stites, W., Kraft, G.J., 2001. Nitrate and chloride loading to groundwater from an irrigated north-central U.S. sand-plain vegetable field. J. Environ. Qual. 30, 1176–1184.

Tomer, M.D., Dosskey, M.G., Burkart, M.R., James, D.E., Helmers, M.J., Eisenhauer, D. E., 2009. Methods to prioritize placement of riparian buffers for improved water quality. Agrofor. Syst. 75, 17–25.

Tsao, D.T., 2003. Overview of phytotechnologies. In: Tsao, D.T. (Ed.), Phytoremediation. Springer, Berlin, Heidelberg, pp. 1–50. , Advances in Biochemical Engineering/ Biotechnology.

United States Environmental Protection Agency (USEPA), 2013. Protecting and Restoring Land: Making a Visible Difference in Communities. OSWER FY13 End of Year Accomplishments Report. 47 p.

Vadas, P.A., Fiorellino, N.M., Coale, F.J., Kratochvil, R., Mulkey, A.S., McGrath, J.M., 2018. Estimating legacy soil phosphorus impacts on phosphorus loss in the Chesapeake Bay watershed. J. Environ. Qual. 47, 480–486.

Verheye, W.H., 2010a. Soils, Plant Growth and Crop Production — Volume I. Encyclopedia of Life Support Systems (EOLSS) Publishers/UNESCO, United Kingdom, 438 p.

Verheye, W.H., 2010b. Soils, Plant Growth and Crop Production — Volume II. Encyclopedia of Life Support Systems (EOLSS) Publishers/UNESCO, United Kingdom, 454 p.

Verheye, W.H., 2010c. Soils, Plant Growth and Crop Production — Volume III. Encyclopedia of Life Support Systems (EOLSS) Publishers/UNESCO, United Kingdom, 492 p.

Vidra, R.L., Shear, T.H., Wentworth, T.R., 2006. Testing the paradigms of exotic species invasion in urban riparian forests. Nat. Areas J. 26, 339—350.

Willett, C.D., Lerch, R.N., Lin, C.-H., Goyne, K.W., Leigh, N.D., Roberts, C.A., 2013. Identification of an atrazine-degrading benzoxazinoid in Eastern gamagrass (*Tripsacum dactyloides*). J. Agric. Food Chem. 34, 8026—8033.

Willett, C.D., Lerch, R.N., Goyne, K.W., Leigh, N.D., Lin, C.-H., Roberts, C.A., 2014. A simple method for isolation and purification of DIBOA-Glc from *Tripsacum dactyloides*. Nat. Product. Commun. 9, 1283—1286.

Willett, C.D., Lerch, R.N., Lin, C.-H., Goyne, K.W., Leigh, N.D., Roberts, C.A., 2016. Benzoxazinone-mediated triazine degradation: a proposed reaction mechanism. J. Agric. Food Chem. 24, 4858—4865.

Wu, Q., Zhou, H., Tam, N.F.Y., Tian, Y., Tan, Y., Zhou, S., et al., 2016. Contamination, toxicity and speciation of heavy metals in an industrialized urban river: implications for the dispersal of heavy metals. Mar. Pollut. Bull. 104, 153—161.

Yuan, Y., Zhao, Z., Bai, Z., Wang, H., Wang, Y., Niu, S., 2016. Reclamation patterns vary carbon sequestration by trees and shrubs in an opencast coal mine, China. Catena 147, 404—410.

Zalesny, J.A., Zalesny Jr., R.S., Coyle, D.R., Hall, R.B., 2007a. Growth and biomass of *Populus* irrigated with landfill leachate. For. Ecol. Manage. 248, 143—152.

Zalesny, J.A., Zalesny Jr., R.S., Wiese, A.H., Hall, R.B., 2007b. Choosing tree genotypes for phytoremediation of landfill leachate using phyto-recurrent selection. Int. J. Phytoremediat. 9, 513—530.

Zalesny, J.A., Zalesny, R.S., Wiese, A.H., Sexton, B.T., Hall, R.B., 2008. Uptake of macro- and micro-nutrients into leaf, woody, and root tissue of *Populus* after irrigation with landfill leachate. J. Sustain. For. 27, 303—327.

Zalesny, J.A., Zalesny Jr., R.S., Coyle, D.R., Hall, R.B., Bauer, E.O., 2009a. Clonal variation in morphology of *Populus* root systems following irrigation with landfill leachate or water during 2 years of establishment. Bioenergy Res. 2, 134—143.

Zalesny Jr., R.S., Bauer, E.O., 2007. Selecting and utilizing *Populus* and *Salix* for landfill covers: implications for leachate irrigation. Int. J. Phytoremediat. 9, 497—511.

Zalesny Jr., R.S., Hall, R.B., Zalesny, J.A., Berguson, W.E., McMahon, B.G., Stanosz, G.R., 2009b. Biomass and genotype × environment interactions of *Populus* energy crops in the Midwestern United States. Bioenergy Res. 2, 106—122.

Zalesny Jr., R.S., Cunningham, M.W., Hall, R.B., Mirck, J., Rockwood, D.L., Stanturf, J.A., et al., 2011. Chapter 2: Woody biomass from short rotation energy crops. In: Zhu, J.Y., Zhang, X., Pan, X. (Eds.), Sustainable Production of Fuels, Chemicals, and Fibers from Forest Biomass. ACS Symposium Series; American Chemical Society, pp. 27—63. , 526 p.

Zalesny Jr., R.S., Hallett, R.A., Falxa-Raymond, N., Wiese, A.H., Birr, B.A., 2014. Propagating native Salicaceae for afforestation and restoration in New York City's five boroughs. Native Plants J. 15, 29–41.

Zalesny Jr., R.S., Stanturf, J.A., Gardiner, E.S., Perdue, J.H., Young, T.M., Coyle, D.R., et al., 2016a. Ecosystem services of woody crop production systems. Bioenergy Res. 9, 465–491.

Zalesny Jr., R.S., Stanturf, J.A., Gardiner, E.S., Bañuelos, G.S., Hallett, R.A., Hass, A., et al., 2016b. Environmental technologies of woody crop production systems. Bioenergy Res. 9, 492–506.

Zalesny, R.S. Jr., Berndes, G., Dimitriou, I., Fritsche, U., Miller, C., Eisenbies, M., et al., 2019a. Positive water linkages of producing short rotation poplars and willows for bioenergy and phytotechnologies. In: WIREs Energy and Environment, e345.

Zalesny, R.S. Jr., Headlee, W.L., Gopalakrishnan, G., Bauer, E.O., Hall, R.B., Hazel, D.W., et al., 2019b. Ecosystem services of poplar at long-term phytoremediation sites in the Midwest and Southeast, United States. In: WIREs Energy and Environment, e349.

Zupanc, V., Zupančič-Justin, M., 2010. Changes in soil characteristics during landfill leachate irrigation of *Populus deltoides*. Waste Manage. 30, 2130–2136.

Zupančič-Justin, M., Pajk, N., Zupanc, V., Zupančič, M., 2010. Phytoremediation of landfill leachate and compost wastewater by irrigation of *Populus* and *Salix*: biomass and growth response. Waste Manage. 30, 1032–1042.

Župunski, M., Borišev, M., Orlović, S., Arsenov, D., Nikolić, N., Pilipović, A., et al., 2016. Hydroponic screening of black locust families for heavy metal tolerance and accumulation. Int. J. Phytoremediat. 18, 583–591.

Adaptive management of landscapes for climate change: how soils influence the assisted migration of plants

10

R.S. Winder[1], J.M. Kranabetter[2] and J.H. Pedlar[3]

[1]*Natural Resources Canada, Canadian Forest Service, Pacific Forestry Centre, Victoria, BC, Canada*

[2]*B.C. Ministry of Forests, Lands, and Natural Resource Operations, Victoria, BC, Canada*

[3]*Natural Resources Canada, Canadian Forest Service, Great Lakes Forestry Centre, Sault Ste. Marie, ON, Canada*

10.1 Introduction

As climate change proceeds apace, the rate of change may exceed historical rates of natural migration for forest species, particularly for tree species and plants at northern latitudes (Winder et al., 2011). This raises questions for the survival and productivity of plants experiencing rapid shifts in climate envelopes. Human-assisted migration of plants is one option to address this climate adaptation challenge in forests around the world. Humans have long aided the transportation of plants (Rossetto et al., 2017; Winder et al., 2011); even today, foresters consider the movement of tree provenances to be one of the tools that they can use to ensure forest resilience to various environmental factors (Pedlar et al., 2012). Current assisted migration efforts focus on changes in the atmospheric climate, such as mean annual temperature, minimum winter temperature, and seasonal precipitation levels. Plants, however, must also adapt to soil environments, which may differ markedly between locations—even those that are relatively proximate (Loescher et al., 2014). This chapter will cover what we know about this other important consideration in assisted migration (AM).

10.1.1 Types of assisted migration

Here, we briefly outline several types of assisted migration that can be discerned along a gradient of translocation distance (Fig. 10.1). As such, each type involves different risks (Aubin et al., 2011; Winder et al, 2011) and may be undertaken for a variety of

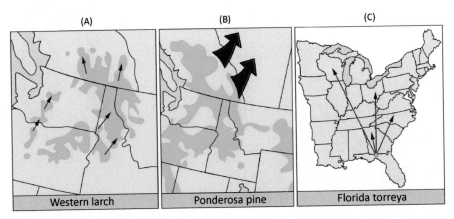

FIGURE 10.1

Three modes of assisted migration that could potentially be employed to conserve tree species: (A) APM of western larch provenances within the range of the species in North America, (B) Hypothetical ARE of Ponderosa pine just beyond its current range in North America, and (C) LAM of *Torreya taxifolia* well beyond its limited range in North America. Risks for successful migration and ecological outcomes increase as distance increases. *APM*, Assisted population migration; *ARE*, assisted range expansion; *LAM*, long-range assisted migration.

Modified from Dumroese, R.K., Williams, M.I., Stanturf, J.A., 2015. Considerations for restoring temperate forests of tomorrow: forest restoration, assisted migration, and bioengineering. N. For. 46, 947—964.

reasons, including both conservation and forestry-related objectives (Pedlar et al., 2012). The types are operative at different scales, with different environmental factors being relevant at each scale (Fig. 10.2). Although the debate around AM remains primarily conceptual, there are a growing number of real-world applications. Examples of some of these applications are provided in the following sections. Note that, while there are many instances of historical plant and animal translocations—some of which have ended poorly (Ricciardi and Simberloff, 2009)—we restrict our discussion here to movements undertaken as an intentional response to climate change.

Assisted population migration (APM) refers to translocation within the historic range of a plant species to improve resilience and adaptability of forests [e.g., moving populations to higher (poleward) latitudes or upwards in elevation; Figs. 10.1A and 10.3A]. A number of APM projects have been undertaken in the context of commercial forestry, where efforts are facilitated by existing provenance trial data, seed zones and transfer rules, and seed procurement, propagation, and planting systems. In British Columbia (BC), Canada, a significant effort was recently made to transition the province to a climate-based seed transfer system (O'Neill et al., 2017). This new approach, known as a focal zone seed transfer system, employs biogeoclimatic ecosystem classification (BEC) units as spatial elements for seed procurement and deployment actions.

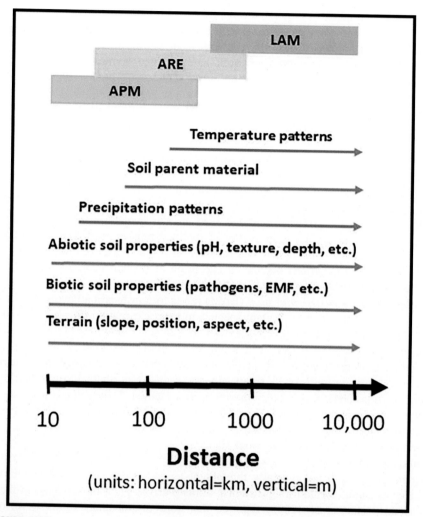

FIGURE 10.2

Approximate spatial scale at which various soil and climate attributes may affect assisted migration outcomes. *APM*, Assisted population migration; *ARE*, assisted range expansion; *LAM*, long-range assisted migration.

Specifically, provenance data from several major timber species were employed to calculate critical, climate-based seed transfer distances, which were then used to identify all pairs of BEC units between which seed could be safely transferred under climate change (O'Neill et al., 2017; O'Neill et al., 2014). In support of this effort and subsequent seed transfer system modifications, the BC Assisted Migration Adaptation Trial (AMAT) has also been established. It is designed to test 48 seed sources from populations of trees in breeding programs throughout BC and adjacent states, at 48 long-term test locations between southern Yukon Territory and northern California (O'Neill et al., 2013; Pickles et al., 2015b). Finally, landscape restoration research has provided some results applicable to assisted migration of grasses and other herbaceous plants. For example, *Pseudoroegneria spicata* is an important grass used in environmental restoration efforts. Populations of *P. spicata* from different climates have been planted in common gardens, and the resulting data used to delineate seed transfer zones in restoration efforts (St. Clair et al., 2013).

Assisted range expansion (ARE) refers to translocation just beyond the current range (but possibly within the historic range) of a plant species in anticipation of future habitat suitability (Figs. 10.1B and 10.3B). Here again, an example is found in the context of BC forestry operations. Based on the findings of Rehfeldt and Jaquish (2010), the province modified seed transfer guidelines for western larch (*Larix occidentalis*) to allow up to 10% of seedlings to be planted north of the current species' range limits (Klenk and Larson, 2015). The objective of this policy is to allow limited range and population expansion of western larch into areas projected to be climatically suitable in the year 2030.

Long-range assisted migration (LAM) refers to translocation beyond the current and historic ranges of a plant species, to avert extirpation (Ste-Marie et al., 2011; Figs. 10.1C and 10.3C). Perhaps the longest running AM program in North America is an effort by a group of naturalists (known as the Torreya Guardians) to move seeds and seedlings of the endangered conifer *Torreya taxifolia* to locations north of its small endemic range in northern Florida (Barlow and Martin, 2004). Since 2005, the group has planted *T. taxifolia* at dozens of locations across the eastern United States (Torreya Guardians, 2018; Fig. 10.4). In recent years, they have started to explore the northern limits of the species, reporting successful plantings from states as far north as Wisconsin, Michigan, Ohio, and Maine—movements of more than 1000 km north of current *T. taxifolia* range. Though not typically thought of as assisted migration, horticultural activities can also entail long-distance, climate-driven plant movements. In fact, Van der Veken et al. (2008) reported that horticultural range limits were, on average, about 1000 km further north than natural range limits for 357 native European plant species.

In forestry research, there are many experimental sites addressing a wide variety of questions that might also provide serendipitous information concerning assisted migration and site conditions, including the influence of edaphic conditions and soil biota. Common gardens, multisite and multisource provenance trials, plantations of "exotic" species, long-term trials, etc. may all either directly or indirectly provide valuable information concerning the suitability of local sites for translocated trees. Winder (2017) has cataloged the metadata for research sites throughout Canada and Alaska that have some relevance to climate change research. Compilation of similar information in other areas of the world is needed.

FIGURE 10.3

(Continued)

FIGURE 10.4

A seedling of *Torreya taxifolia* growing in a forest in northern Alabama. This was planted as a seed, well north of its current range.

Photo courtesy TorreyaGuardians.org.

◄ Examples of the three modes of assisted migration. Panel (A) shows a "common garden" established through APM of coastal Douglas-fir (*Pseudotsuga menziesii* var. *menziesii*) provenances collected throughout coastal BC, Canada. Plantings such as this one at Dove Creek, BC improve our understanding of the site and genetic interactions that can determine the success of APM. Panel (B) illustrates a successful example of ARE; Western Larch (*Larix occidentalis*) from NW and W Montana, United States was planted between Burns Lake and Terrace, BC and produced better than predicted growth rates. Panel (C) shows an example of LAM. In this case, a small (c. 100 tree) Douglas-fir plantation was established in the Hakalau Forest National Forest Refuge on Mauna Kea in Hawaii, United States, as a memorial to the botanist Sir David Douglas. The trees were planted approximately 70 years prior to this photograph. Although the site supported very good growth, seeds of the Douglas-fir planted here appeared to be sterile, possibly due to exposure to intense tropical sunlight. *APM*, Assisted population migration; *ARE*, assisted range expansion; *LAM*, long-range assisted migration.

Photo B from LePage, P., McCulloch, L., 2011. Assessment of Off-Site Tree Plantations in the Northwest Interior of British Columbia — Project Summary. Report Prepared for Forest Genetics Council of B.C. by B.C. Ministry of Natural Resource Operations and LM Forest Resource Solutions, Ltd., 28pp.

10.2 **The interplay between soils and other factors in assisted migration**

10.2.1 **Abiotic factors**

A range of physical edaphic properties may constrain attempts to translocate tree species, particularly when the translocated plants face competition with established plant species. In forests of northeastern North America, soil pH, low nutrient availability, waterlogged conditions, variable soil moisture, and dry soils are factors that may constrain the translocation of a number of conifer and hardwood species (Lafleur et al., 2010; Pinto et al., 2016). For example, a study of white ash (*Fraxinus americana*) and gray birch (*Betula populifolia*) forests in Quebec underplanted with white pine (*Pinus strobus*), red oak (*Quercus rubra*), bitternut hickory (*Carya cordiformis*), and black walnut (*Juglans nigra*) migrated from other areas revealed a significant correlation between establishment success and soil fertility (Truax et al., 2018).

As climate change proceeds, conditions at potential translocation sites may develop differently. For example, tundra in the northern hemisphere has been predicted to provide new forest habitat with warming, but the soil in some areas is expected to become wetter rather than drier due to permafrost melt, suggesting that there would be mixed opportunities and challenges for northward expansion of the boreal forest, assisted or otherwise (Winder et al., 2011).

There are also genetic adaptations to abiotic gradients within the native habitat of tree species that may affect the success of assisted migration. For example, areas in Spain with severe soil phosphorus deficiency have imposed selective pressures on local *Pinus* genotypes (Martins et al., 2009), which suggests provenances from similarly constrained edaphic conditions would be best suited for assisted migration. Xie et al. (1998) showed that some spruce provenances are adapted to the higher pH of calcareous soils; these provenances would not be expected to be as productive on more acidic soils. Other examples would include genetic adaptations among conifer populations to inherent soil fertility, particularly the forms of organic and inorganic nitrogen (Hawkins, 2007; Kranabetter, 2014). Differences in soil moisture may also have implications for resistance to soil pathogens. Cruickshank et al. (2010) reported that Douglas-fir accessions from warmer, drier areas were more resistant to *Armillaria* root rot. In general, genotype × environment interactions are common in forest genetic trials, underlining the possible local-adaptations by trees to a number of abiotic conditions (e.g., nitrogen, phosphorus, drought, pH) that are not fully understood but ideally would be incorporated into assisted migration strategies.

In some areas, climate may interact with the edaphic properties of soils to provide specialized niches for plant species that would be difficult to replicate in assisted migration scenarios. For example, the Coastal Redwood of California and Oregon (*Sequoia sempervirens*) is a tree species that is expected to be threatened by climate change (Fernández et al., 2015; Roberts and Hamann, 2016). It produces seedlings that have radicles lacking root hairs. For successful establishment, these seedlings require continuously moist and deep organic soils typical of areas near coastal fog belts. However, high levels of magnesium and sodium in soils are detrimental to the growth of the species; optimal habitat is often on slopes near, but facing away from the Pacific Ocean (Florin, 1963; Roy, 1966; Olsen et al., 1990; Snyder, 1992). Soil characteristics thus have implications for assisted migration of species associated with foggy coastal zones; the ability of a site to support the growth of transplanted or cultivated plants may not equate to an ability to support early seedling growth and natural recruitment.

10.2.2 Biotic factors

10.2.2.1 Establishment

At a biogeographic level, soil biotas are a key determinant of plant community structure and biodiversity (Reynolds et al., 2003; Winder and Shamoun, 2006). For example, soil pathogens that cause damping-off of plant seedlings may constrain the species that can grow in some areas (Packer and Clay, 2000; Augspurger and Wilkinson, 2007). Conversely, some microbiota may be required for seedling growth in herbaceous species. In orchids, for example, successful seedling growth can be related to particular lineages of fungal symbionts (Bidartondo and Read, 2008).

Other plants competing for soil resources also affect the success of establishment efforts. Where competing vegetation is significant, the size, quality, and condition of stock types are important in initial establishment efforts (Pinto et al., 2018). Of course, stock properties do not come into consideration during natural establishment and recruitment of subsequent generations.

All of the preceding has implications for the assisted migration of many plant species. Whether the concern is initial establishment of seedlings or continuing natural recruitment of established populations, the impact of soil biota and competing species on migrated species may need to be considered (Winder et al., 2011).

10.2.2.2 Tree growth and the importance of mycorrhizae

Once a translocated plant is established, successful growth can be regarded as an interaction between many physical and biological factors: soil properties, climate, pests, and other agents of disturbance, local biodiversity, and genetic variability. One of the most important interactions occurs between climate, soil biota, and genetics, exemplified by mycorrhizal fungi. An essential component of forest ecosystems, mycorrhizae are the symbiosis between fungi and tree roots (Fig. 10.5), supporting nutrient capture, water uptake and other essential services (Van Der Heijden et al., 2015). In the absence of preferred mycorrhizal partners, tree growth can be stunted and unproductive (Björkman, 1970). Indeed, the association of symbiotic fungi with roots is so fundamental to forest biology that the perception of individual trees has been supplanted by a broader concept of a meta-organism or "holobiont" that explicitly includes symbionts when describing tree physiology

(A)

(B)

FIGURE 10.5

(A) A mycorrhizal root tip colonized by a *Lactarius* species. The branches of the root tip are shortened and covered in a fungal mantle. Other fungal species, such as (B) *Hebeloma* can have copious emanating hyphae that extend into the environment to capture and redistribute nutrients.

(Vandenkoornhuyse et al., 2015). For many tree species of commercial importance, such as *Abies, Larix, Pinus, Picea, Tsuga*, and *Pseudotsuga*, the symbiotic fungi form a hyphal mantle on the outside surface of the feeder roots (ectomycorrhizae), along with a Hartig net penetrating between cortical cells, to enable nutrient exchange collected by the fungi for photosynthate carbon supplied by the tree. A striking attribute of ectomycorrhizal fungi (EMF) is the tremendous diversity typically displayed in forest ecosystems, both within stands and across landscapes (Tedersoo et al., 2012). This diverse and essential community of fungi may, however, be a source of ecosystem vulnerability under climate change and assisted migration; to what degree will future forest productivity reflect the underlying influence of mycorrhizae if these fungal communities are diminished or maladapted?

Numerous studies have explored the potential for soil and root inoculations of mycorrhizal fungi to improve plant establishment and productivity at restoration sites (Castellano, 1996). Generally, these studies have reported a positive impact on survival and growth, particularly for non-N-fixing forbs and woody plants at P-limited sites (Hoeksema et al., 2010). EMF can also ameliorate soil environments by protecting plant roots from contaminants (Schützendübel and Polle, 2002). It is possible to deliberately inoculate nursery-produced stock with some species of EMF (Trappe, 1977; Molina, 1979; Kottke et al., 1987). Nursery seedlings often already contain "weedy" EMF such as *Thelephora* spp. or *Amphinema* spp. (Menkis et al., 2005), and these are often eventually displaced by endemic EMF (Dahlberg and Stenström, 1991). Nevertheless, inoculation of nursery stock and monitoring the incidence of EMF on root tips might help to establish a more diverse set of EM species in restoration areas low in diversity of soil fungi (Kernaghan et al., 2002; Henry et al., 2015) or in other conservation situations requiring the migration of mycorrhizal partners.

10.2.2.3 Vulnerability of mycorrhizae versus site and soil characteristics

The vulnerability of mycorrhizae can be inferred by their site specificity and adaptive traits of these fungal communities across contemporary landscapes (e.g. Winder, 2013). Gradient analysis of climatic or edaphic factors such as nitrogen (N) availability has commonly demonstrated strong patterns in community assembly (e.g., Jarvis et al., 2015; Suz et al., 2014). The close alignment of mycorrhizal communities with abiotic conditions would, at least conceptually, reflect an "optimal forage strategy" that enables maximum fitness for a single host species over highly variable landscapes, either through stress tolerance (e.g., drought) or effective exploitation of scarce resources (e.g., phosphorus) (Johnson et al., 2013). The best evidence for a spatially distinct adaptive strategy is perhaps found in studies of N dynamics, where certain mycorrhizal species demonstrate superior abilities in exploiting recalcitrant, organic N on low fertility sites (Bödeker et al., 2014; Heinonsalo et al., 2015). Conversely, other species are much more effective in the uptake of inorganic N (particularly NH_4^+) on productive soils (Kranabetter et al.,

2015a; Leberecht et al., 2016). Rapid changes in environmental conditions and species composition could therefore decouple native EMF communities from current site conditions, thereby undermining locally based adaptive strategies. At the same time, there may be considerable resilience and adaptability of mycorrhizal communities under warming due to functional redundancy and trait plasticity among EMF species (Jones et al., 2010; Rineau and Courty, 2011). Experimental evidence to date has been largely mixed and of limited geographic distribution, with some evidence of increased mycorrhizal abundance but decreased activity under warmer temperatures (Pickles et al., 2012; Mohan et al., 2014).

Another source of vulnerability regarding mycorrhizae may be found in the genetic linkages between tree host populations and EMF communities (Korkama et al., 2006; Velmala et al., 2013; Lamit et al., 2016). Native forests have been associated with site-adapted EMF communities for multiple generations under reasonably stable climates, leading to close, possibly coevolved affiliations with local tree populations and fungal species (Hoeksema et al., 2012; Kranabetter et al., 2012, 2015b; Gehring et al., 2017). The onset of rapid climate change has affected, and will continue to affect, the suitability of current range distributions in host tree species, while increases in disturbance severity and frequency, such as wildfire, could promote faster migration of host populations or abrupt changes in tree species composition (Aitken et al., 2008; Park et al., 2014). Under these circumstances the degree of heritability that has developed among local EMF communities with native host populations would be increasingly displaced, possibly undermining the health and productivity of these more novel forest ecosystems.

10.2.2.4 Other vulnerabilities of soil biota

The introduction of different plant genotypes (Schweitzer et al., 2008) and species (Chapman et al., 2013) can lead to changes in the structure and diversity of microbial communities in forest soils. While these communities include mycorrhizal fungi as an important component, soil biotas also provide a wide variety of other key functions with their own vulnerabilities in the context of assisted migration. For example, dynamic communities of free-living and symbiotic bacteria fix N in forest soils (Levy-Booth and Winder, 2010). At biogeographic scales, work with latitudinal and altitudinal gradients has shown that these N-fixing communities may also be affected by climate change (Shay et al., 2015). After translocation, assisted migration efforts might therefore consider the ongoing impact of climate change for both plants and for the local soil biota.

There are also other microbial functions to consider. Genetics and local adaptation of tree populations can affect the community structure of mycorrhizal and other fungi providing drought protection (Gehring et al., 2017; Tíscar et al., 2018). Mycorrhizal fungi also ameliorate soil conditions by scavenging heavy metals and transporting them away from plant roots (Wilkinson and Dickinson, 1995). How well these functions continue for translocated plants in ex situ soils is a question likely to be answered on a case-by-case basis.

10.3 Assisted population migration and soils

10.3.1 Risks

In the case of APM, many of the risks typically associated with assisted migration are reduced (Pedlar et al., 2012; Winder et al., 2011). Chief among these is the concern that an invasive species may arise at the recipient location due to a lack of ecological controls, such as predation and competition, on the migrated species. When movements are within existing range limits, as is the case with APM, this scenario appears highly unlikely as ecological checks and bounds on the migrated population would already be in place. Similarly, the potential for APM to cause the loss of certain genotypes/phenotypes at the recipient location through hybridization (i.e., genetic swamping) is negligible as within-range movements would not be expected to create new hybridization opportunities. APM does have some potential for inadvertently introducing harmful organisms, such as insects and fungal pathogens, to the recipient location. However, the risks involved are similar to those associated with current forestry and horticultural practices, wherein hundreds of millions of seedlings are moved annually from nurseries (where pests might be acquired) to planting sites. Phytosanitary practices are in place at nurseries, but there is always a small risk of disease transmission. The potential disease risks have led to calls for policy development that would improve risk assessments and regulations to address this (Simler et al., 2018).

Another factor to consider for APM is that response to site factors (different soil types, soil biota, and other site characteristics) may have a larger influence than climate. There is growing evidence that spatial variability of soil biota and associated plant—soil feedback effects are drivers of differences in plant development in restoration situations (Wubbs and Bezemer, 2016; Koziol and Bever, 2017; Michaelis and Diekmann, 2018). Locally adapted mycorrhizal fungi have even been shown to affect levels of herbivory (Middleton et al., 2015). While adaptation to climate in local ecotypes may play a role in their growth and survival, this factor could possibly be overridden by other drivers of local adaptation (soil and associated biota). Concern about this possibility has led to a call for large-scale multispecies experiments to guide management decisions concerning assisted migration (Bucharova, 2016).

10.3.2 Mycorrhizae

As touched upon earlier, spatially based adaptations, where native trees and EMF populations are genetically and functionally well suited to the edaphic and climatic attributes of a particular area, may be disrupted to some degree through assisted migration of the host. The potential risk of such displacement would be a mismatch in host genetics with native mycorrhizae that results in maladaptation for the holobiont (i.e., relative decline in host fitness survival, growth, or nutrition). Evidence for maladaptation has been found with arbuscular mycorrhizae and the superior fitness of grassland plants grown with locally derived soil and fungal inocula (Rúa et al., 2016), or in the drought tolerance of *Pinus edulis* with native EMF communities (Gehring et al., 2014; Gehring et al., 2017). Direct tests of assisted migration effects on trees and mycorrhizae are rare; Kranabetter et al. (2015b) examined a 40-year-old genetics trial of *Pseudotsuga menziesii* and found transplanted southern populations declined in height by up to 15% with the extent of their dissimilarity in EMF communities relative to local hosts. Pickles et al. (2015a) tested transfer effects for *P. menziesii* in a glasshouse environment and also found local adaptation patterns, specifically in the reduced survival of seedlings from increasingly distant seed and soil origins. In both studies, the mycorrhizal maladaptation effect was most pronounced with increasing transfer distances (from seed source to test site), suggesting a prudent strategy in any range expansion would be to move multiple host populations incrementally northward or higher in elevation.

10.4 Assisted range expansion and long-range assisted migration and soils

10.4.1 Risks

Risks associated with ARE are expected to be somewhat greater than those described above for APM, though still modest compared to long-range and/or intercontinental (LAM) movements (Pedlar et al., 2012; Winder et al., 2011). Mueller and Hellmann (2008) examined the continental origin of 468 invasive species in the United States and reported that there were far fewer invasions of intracontinental (14%) than intercontinental (86%) origin, despite a (presumably) much higher frequency of intracontinental movements. Furthermore, of the five taxonomic groups under study, plants had the lowest proportion (7.5%) of intracontinental invaders. These findings generally support the notion that short-distance, intracontinental movement of trees carries a relatively low level of invasion risk.

Assisted migration beyond the range or current habitat of a species is still a consideration where there is a threat to biodiversity; there can also be perceived risks in failing to respond in order to conserve a species. One example would be the *Abies religiosa* forests of central Mexico, which form the overwinter refuge for the monarch butterfly but occupy an increasingly smaller area. Studies of the community of EMF found in the soils of these and adjacent forests have suggested that nearby *Pinus montezumae* forests have the potential to share these fungal communities. This provides a potential area where *A. religiosa* could be migrated, beyond its current habitat, to maintain critical habitat for the butterfly (Argüelles-Moyao and Baribay-Orijel, 2018).

10.4.2 Abiotic and biotic factors

Beyond the native range of a species, one can expect a much larger discrepancy in site characteristics, whether the shift was primarily latitudinal or altitudinal in nature. Differences already important within range, such as changes in soil fertility and soil pH, different soil moisture profiles, etc. become even more important because they may, in fact, help to constrain the current range of a plant. Communities of soil microbes are influenced by soil age and parent materials (Turner et al., 2017; Alfaro et al., 2017; Ulrich and Becker, 2006); these can be expected to change significantly over larger geographic distances, potentially becoming a constraining factor.

These factors may be particularly important at ecotone transitions where soil conditions are contributing to existing biome boundaries. For example, at the boreal-temperate forest boundary in eastern North America, soil transition from predominantly acidic, lower nutrient soils underlain by granitic parent material in the north to lower pH, higher nutrient soils derived from calcareous bedrock in the south. Such a transition may make assisted migration of temperate forest species particularly challenging in a region that has been projected to exhibit significant declines under climate change (Pedlar and McKenney, 2017).

For LAM of trees at larger scales, another consideration is the degree to which abiotic soil properties may be altered or impacted by the biotic component. In Portugal, for example, the commercial establishment of large (c. 300 km^2) plantations of *Eucalyptus* spp. has resulted in the accumulation of leaf litter causing deleterious impacts on soils, including increased hydrophobicity (Doerr et al., 1998) and nutrient losses following fire (Thomas et al., 2000).

10.4.3 Mycorrhizae

In the Northern hemisphere, where ectomycorrhizal forests are widely distributed, there appear to be no strong impediments in obtaining mycorrhizal colonization for EMF trees planted beyond their current range due to the existing diverse communities of nonhost-specific fungal species (O'Hanlon and Harrington, 2012; Trocha et al., 2012; Bahram et al., 2013; Rudawska et al., 2016; Garcia et al., 2016). For example, Pickles et al. (2015b) found transplanting interior Douglas-fir into soils from *Pinus contorta* stands was not limited by EMF colonization, although there were shifts in taxonomic and/or functional traits of fungi in these novel environments. In contrast, for parts of the globe where ectomycorrhizal forests are less common, there are examples of a necessary coinvasion by EMF for the successful establishment of nonnative conifers (Dickie et al., 2010; Hayward et al., 2015). It is possible the likewise expansion of tree ranges into currently unforested ecosystems (e.g., alpine meadows, tundra) could lag because of the paucity of EMF inocula in these soils, but the presence of ectomycorrhizal shrubs, such as *Salix*, appears to mitigate these limitations (Kernaghan and Harper, 2001; Reithmeier and Kernaghan, 2013).

While the ectomycorrhizal symbiosis is certainly robust enough to ensure the survival and growth of nonnative trees, it is unclear whether the fitness of the "holobiont" would be as optimal in these novel habitats as might be found under native host conditions. Interestingly, there is some evidence for performance to increase when host trees are exposed to fungal partners without a recent shared evolutionary history (Gundale et al., 2014; Karst et al., 2018). An enhancement in growth response would certainly facilitate range expansion by trees, both naturally and by assisted migration. It should be recognized these types of extreme transfer studies (between continents or with allopatric symbionts) are likely beyond the scope of most assisted migration strategies, and lack the complexity of natural ecosystems where the long-term interaction of abiotic stresses, pests, and competition can influence mycorrhizal effectiveness (Gehring et al., 2014). Further experimental evidence, especially under long-term field conditions, would provide insights as to the full implications of range expansion by trees (Rúa et al., 2018).

10.5 **Conclusions and future research needs**

As a strategy for adapting to climate change, assisted migration is still developing, with clear needs for further research in many different areas. One of those areas involves the need for landscape-level trials with appropriate controls for a full evaluation of fitness and productivity of the introduced tree species. Most experimental trials to-date are either at the site level, or involve common-garden trials on particular sites. In terms of forestry practice, examples of significant assisted migration have involved ad-hoc APM and LAM of commercial forest species established in plantations or replanted after timber harvests. But that activity is dwarfed by the size and scale of efforts that would likely be needed to fully address climate change at landscape and regional levels. Experiments dispersed at the landscape level could contribute valuable information on the impact of different soil types on establishment, growth, and eventual recruitment of migrated provenances. These experiments can be used to inform multivariate analyses of species and provenances that incorporate tolerances to climate, disease, soil conditions, and other factors (Park et al., 2018). Such analysis would improve our ability to predict the success of assisted migration efforts.

An important need going forward will be to understand the best strategy for dealing with changes in soil biota responding to climate change (and impacting plant growth). Given the emphasis here on assisted migration, it is perhaps counterintuitive to put forward strategies regarding the retention of undisturbed forest ecosystems. However, late-seral ecosystems host a great diversity of soil bacteria, fungi, and fauna (mites, collembolans, etc.) that likely promote resilience in the face of climate change, particularly where better-adapted soil organisms thrive under future soil conditions as less-adapted organisms retreat. We posit that an important strategy going forward is a robust network of retention systems, leave-areas and old-growth/mature forest management areas to maintain diverse late-seral communities of soil biota (Lindenmayer et al., 2012). Even modest amounts of refugia can be very effective at capturing landscape-level species diversity, and could greatly facilitate the "unassisted" migration of soil biota in response to new climatic and vegetative environments (Outrebridge and Trofymow, 2004; Cline et al., 2005; Luoma et al., 2006; Jones et al., 2008; Kranabetter et al., 2018; Kitsberger et al., 2000).

Internationally, there is a need to catalog existing provenance trials and experiments relevant to assisted migration and climate-change adaptation. Especially where they are long-established, these trials can provide information on plant growth and recruitment rates versus different soil environments. In cases where whole plants have been translocated rather than seeds, it could be possible to determine the occurrence of mycorrhizal fungal partners colocating and establishing along with the plant host. Knowing something about the movement of mycorrhizal partners has implications for both forest and conservation of biodiversity.

In the context of soils and assisted migration, restoration sites provide a particular challenge and opportunity. Restoring areas that have been mined or disturbed by human activities may not be successful if climate change ultimately causes the restored vegetation to be maladapted for the site; using plant stock adapted for the future climate of a restoration area is therefore an important consideration. Restoration sites are in many ways opportune areas to focus assisted migration efforts. They may not face the same planting restrictions that might be in place for forestry operations, and disturbance may have "reset" successional trajectories to earlier seral stages more amenable to establishment. On the other hand, the soil environment in restoration environments (if present) is sometimes highly disturbed or contaminated in various ways, and soil biota in such areas may be depauperate (Henry et al., 2015) . So, there is an additional challenge beyond those already noted. We need more information and provenance/common garden trials concerning the performance of migrated species on a variety of disturbed soils. We also need more information on the best ways, if any, to migrate associated EMF in such situations. For example, Flores-Rentería et al. (2018) have shown that pre-inoculation of tree seedlings with EMF can have either positive or negative effects on water relations and growth in restoration areas, with the introduced fungi being replaced by native fungi after a year. Clearly, AM of EM fungal symbionts will require field-testing with specific plant partners (provenance as well as species) to determine the scope of any potential benefit.

There is more to consider than the atmosphere in assisted migration of plants. Suitable sites for relocation will also require suitable soils and their associated biota. The soil environment has a profound impact on the establishment and growth of plants, but its influence on translocation success needs to be better understood if assisted migration is to be implemented at larger scales.

References

Aitken, S.N., Yeaman, S., Holliday, J.A., Wang, T., Curtis-McLane, S., 2008. Adaptation, migration or extirpation: climate change outcomes for tree populations. Evol. Appl. 1, 95–111.

Alfaro, F.D., Manzano, M., Marquet, P.A., Gaxiola, A., 2017. Microbial communities in soil chronosequences with distinct parent material: the effect of soil pH and litter quality. J. Ecol. 105, 1709–1722.

Argüelles-Moyao, A., Baribay-Orijel, R., 2018. Ectomycorrhizal fungal communities in high mountain conifer forests in central Mexico and their potential use in the assisted migration of *Abies religiosa*. Mycorrhiza 28, 509–521.

Aubin, I., Garbe, C.M., Colombo, S., Derver, C.R., McKenney, D.W., Messier, C., et al., 2011. Why we disagree about assisted migration: ethical implications of a key debate regarding the future of Canada's forests. For. Chron. 87, 755–765.

Augspurger, C., Wilkinson, H., 2007. Host specificity of pathogenic *Pythium* species: implications for tree species diversity. Biotropica 39, 702–708.

Bahram, M., Kõljalg, U., Kohout, P., Mirshahvaladi, S., Tedersoo, L., 2013. Ectomycorrhizal fungi of exotic pine plantations in relation to native host trees in Iran: evidence of host range expansion by local symbionts to distantly related host taxa. Mycorrhiza 23, 11–19.

Barlow, C., Martin, P.S., 2004. Bring *Torreya taxifolia* North Now. Wild Earth Winter/Spring, pp. 52–56.

Bidartondo, M.I., Read, D.J., 2008. Fungal specificity bottlenecks during orchid germination and development. Mol. Ecol. 17, 3707–3716

Björkman, E., 1970. Forest tree mycorrhiza: the conditions for its formation and the significance for tree growth and afforestation. Plant Soil 32, 589–610.

Bödeker, I.T.M., Clemmensen, K.E., de Boer, W., Martin, F., Olson, Å., Lindahl, B.D., 2014. Ectomycorrhizal *Cortinarius* species participate in enzymatic oxidation of humus in northern forest ecosystems. N. Phytol. 203, 245–256.

Bucharova, A., 2016. Assisted migration within species range ignores biotic interaction and lacks evidence. Rest. Ecol. 25, 14–18.

Castellano, M.A., 1996. Outplanting performance of mycorrhizal inoculated seedlings. In: Mukerji, K.G. (Ed.), Concepts in Mycorrhizal Research. Handbook of Vegetation Science, vol. 19/2. Springer Science + Business Media, Dordrecht, pp. 223–301.

Chapman, S.K., Newman, G.S., Schweitzer, J.A., Hart, S.C., Koch, G.W., 2013. Leaf litter mixtures alter microbial community development: mechanisms for non-additive effects in litter decomposition. PLoS One 8 (4), e62671.

Cline, E.T., Ammirati, J.F., Edmonds, R.L., 2005. Does proximity to mature trees influence ectomycorrhizal fungus communities of Douglas-fir seedlings? N. Phytol. 166, 993–1009.

Cruickshank, M., Jaquish, B., Nemec, F., 2010. Resistance of half-sib interior Douglas-fir families to *Armillaria ostoyae* in British Columbia following artificial inoculation. Can. J. For. Res. 40, 155–166.

Dahlberg, A., Stenström, E., 1991. Dynamic changes in nursery and indigenous mycorrhiza of *Pinus sylvestris* seedlings planted out in forest and clearcuts. Plant Soil 136, 73.

Dickie, I.A., Bolstridge, N., Cooper, J.A., Peltzer, D.A., 2010. Co-invasion by Pinus and its mycorrhizal fungi. N. Phytol. 187, 475–484.

Doerr, S.H., Shakesby, R.A., Walsh, R.P.D., 1998. Spatial variability of hydrophobicity in fire-prone Eucalyptus and pine forests, Portugal. Soil Sci. 163, 313–324.

Dumroese, R.K., Williams, M.I., Stanturf, J.A., 2015. Considerations for restoring temperate forests of tomorrow: forest restoration, assisted migration, and bioengineering. N. For. 46, 947–964.

Fernández, M., Hamilton, H.H., Kueppers, L.M., 2015. Back to the future: using historical climate variation to project near-term shifts in habitat suitable for coast redwood. Global Change Biol. 21, 4141−4152.

Flores-Rentería, D., Barradas, V.L., Álvarez-Sánchez, J., 2018. Ectomycorrhizal pre-inoculation of *Pinus hartwegii* and *Abies religiosa* is replaced by native fungi in a temperate forest of central Mexico. Symbiosis 74, 133−144.

Florin, R., 1963. The distribution of conifer and taxad genera in time and space. Acta Horti Bergiani 20 (4), 121−312.

Garcia, M.O., Smith, J.E., Luoma, D.L., Jones, M.D., 2016. Ectomycorrhizal communities of ponderosa pine and lodgepole pine in the south-central Oregon pumice zone. Mycorrhiza 26, 275−286.

Gehring, C., Flores-Renter, D., Sthultz, C.M., Leonard, T.M., Fores-Renter, L., Whipple, A.V., et al., 2014. Plant genetics and interspecific competitive interactions determine ectomycorrhizal fungal community responses to climate change. Mol. Ecol. 23, 1379−1391.

Gehring, C.A., Sthultz, C.M., Fores-Rentería, L., Whipple, A.V., Thitham, T.G., 2017. Tree genetics defines fungal partner communities that may confer drought tolerance. PNAS 114 (42), 11169−11174.

Gundale, M.J., Kardol, P., Nilsson, M.-C., Nilsson, U., Lucas, R.W., Wardle, D.A., 2014. Interactions with soil biota shift from negative to positive when a tree species is moved outside its native range. N. Phytol. 202, 415−421.

Hawkins, B.J., 2007. Family variation in nutritional and growth traits in Douglas fir seedlings. Tree Physiol. 27, 911−919.

Hayward, J., Horton, T.R., Nunez, M.A., 2015. Ectomycorrhizal fungal communities coinvading with Pinaceae host plants in Argentina: Gringos bajo el bosque. N. Phytol. 208, 497−506.

Heinonsalo, J., Sun, H., Santalahti, M., Bäcklund, K., Hari, P., Pumpanen, J., 2015. Evidences on the ability of mycorrhizal genus *Piloderma* to use organic nitrogen and deliver it to Scots pine. PLoS One 10 (7), e0131561.

Henry, C., Raivoarisoa, J.-F., Razafimamonjy, A., Ramanankiernana, H., Andrianaivomahefa, P., Selosse, M.-A., et al., 2015. *Asteropeia mcphersonii*, a potential mycorrhizal facilitator for ecological restoration in Madagascar wet tropical rainforests. For. Ecol. Manage. 358, 202−211.

Hoeksema, J.D., Chaudhary, V.B., Gehring, C.A., Johnson, N.C., Karst, J., Koide, R.T., et al., 2010. A meta-analysis of context-dependency in plant response to inoculation with mycorrhizal fungi. Ecol. Lett. 13 (3), 394−407.

Hoeksema, J.D., Hernandez, J.V., Rogers, D.L., Mendoza, L.L., Thompson, J.N., 2012. Geographic divergence in a species-rich symbiosis: interactions between Monterey pines and ectomycorrhizal fungi. Ecology 93, 2274−2285.

Jarvis, S.G., Woodward, S., Taylor, A.F.S., 2015. Strong altitudinal partitioning in the distributions of ectomycorrhizal fungi along a short (300 m) elevation gradient. N. Phytol. 206, 1145−1155.

Johnson, N.C., Angelard, C., Sanders, I.R., Kiers, T., 2013. Predicting community and ecosystem outcomes of mycorrhizal responses to global change. Ecol. Lett. 16, 140−153.

Jones, M.D., Twieg, B.D., Durall, D., Berch, S.M., 2008. Location relative to a retention patch affects the ECM fungal community more than patch size in the first season after timber harvesting on Vancouver Island, British Columbia. For. Ecol. Manage. 255, 1342−1352.

Jones, M.D., Twieg, B.D., Ward, V., Barker, J., Durall, D.M., Simard, S., 2010. Functional complementarity in Douglas-fir ectomycorrhizas for extracellular enzyme activity after wildfire or clearcut logging. Funct. Ecol. 24, 1139–1151.

Karst, J., Burns, C., Cale, J.A., Antunes, P.M., Woods, M., Lamit, L.J., et al., 2018. Tree species with limited geographical ranges show extreme responses to ectomycorrhizas. Global Ecol. Biogeogr. 27, 839–848.

Kitsberger, T., Steinaker, D.F., Veblen, T.T., 2000. Effects of climatic variability on facilitation of tree establishment in northern Patagonia. Ecology 81, 1914–1924.

Kernaghan, G., Harper, K.A., 2001. Community structure of ectomycorrhizal fungi across an alpine/subalpine ecotone. Ecograph 24, 181–188.

Kernaghan, G., Hambling, B., Fung, M., Khasa, D., 2002. In vitro selection of boreal ectomycorrhizal fungi for usein reclamation of saline-alkaline habitats. Restor. Ecol. 10, 43–51.

Klenk, N.L., Larson, B.M., 2015. The assisted migration of western larch in British Columbia: a signal of institutional change in forestry in Canada? Global Environ. Change 31, 20–27.

Korkama, T., Pakkanen, A., Pennanen, T., 2006. Ectomycorrhizal community structure varies among Norway spruce (*Picea abies*) clones. N. Phytol. 171, 815–824.

Kottke, I., Guttenberger, M., Hampp, R., Nilsson, C., 1987. An in vitro method for establishing mycorrhizae on coniferous tree seedlings. Trees 1, 191–194.

Koziol, L., Bever, J.D., 2017. The missing link in grassland restoration: arbuscular mycorrhizal fungi inoculation increases plant diversity and accelerates succession. J. Appl. Ecol. 54, 1301–1309.

Kranabetter, J.M., 2014. Ectomycorrhizal fungi and the nitrogen economy of conifers—implications for genecology and climate change mitigation. Botany 92, 417–423.

Kranabetter, J.M., Stoehr, M.U., O'Neill, G.A., 2012. Divergence in ectomycorrhizal communities with foreign Douglas-fir populations and implications for assisted migration. Ecol. Appl. 22, 550–560.

Kranabetter, J.M., Hawkins, B.J., Jones, M.D., Robbins, S., Dyer, T., Li, T., 2015a. Species turnover (β diversity) in ectomycorrhizal fungi linked to NH4 + uptake capacity. Mol. Ecol. 24, 5992–6005.

Kranabetter, J.M., Stoehr, M., O'Neill, G.A., 2015b. Ectomycorrhizal fungal maladaptation and growth reductions associated with assisted migration of Douglas-fir. N. Phytol. 206 (3), 1135–1144.

Kranabetter, J.M., Berch, S.M., MacKinnon, J.A., Ceska, O., Dunn, D.E., Ott, P.K., 2018. Species-area curve and distance-decay relationship indicate habitat thresholds of ectomycorrhizal fungi in an old-growth *Pseudotsuga menziesii* landscape. Divers. Dist. 24, 755–764.

Lafleur, B., Paré, D., Munson, A.D., Bergeron, Y., 2010. Response of northeastern North American forests to climate change: will soil conditions constrain tree species migration? Environ. Rev. 18, 279–289.

Lamit, L.J., Holeski, L.M., Flores-Rentería, L., Whitham, T.G., Gehring, C.A., 2016. Tree genotype influences ectomycorrhizal fungal community structure: ecological and evolutionary implications. Fungal Ecol. 24, 124–134.

Leberecht, M., Dannenmann, M., Tejedor, J., Simon, J., Rennenberg, H., Polle, A., 2016. Segregation of nitrogen use between ammonium and nitrate of ectomycorrhizas and beech trees. Plant, Cell Environ. 39, 2691–2700.

LePage, P., McCulloch, L., 2011. Assessment of Off-Site Tree Plantations in the Northwest Interior of British Columbia — Project Summary. Report prepared for Forest

Genetics Council of B.C. by B.C. Ministry of Natural Resource Operations and LM Forest Resource Solutions, Ltd., 28pp.

Levy-Booth, D.J., Winder, R.S., 2010. Quantification of nitrogen reductase and nitrite reductase genes in soil of thinned and clear-cut Douglas-fir stands by using real-time PCR. Appl. Environ. Microbiol. 76 (21), 7116–7125.

Lindenmayer, D.B., Franklin, J.F., Lohmus, A., Baker, S.C., Bauhus, J., Beese, W., et al., 2012. A major shift to the retention approach for forestry can help resolve some global forest sustainability issues. Conserv. Lett. 5, 421–431.

Loescher, H., Ayres, E., Duffy, P., Luo, H., Brunke, M., 2014. Spatial variation in soil properties among North American ecosystems and guidelines for sampling designs. PLoS One 9 (1), e83216.

Luoma, D.L., Stockdale, C.A., Molina, R., Eberhart, J.L., 2006. The spatial influence of *Pseudotsuga menziesii* retention trees on ectomycorrhizal diversity. Can. J. For. Res. 36, 2561–2573.

Martins, P., Sampedro, L., Moreira, X., Zas, R., 2009. Nutritional status and genetic variation in the response to nutrient availability in *Pinus pinaster*. A multisite field study Northwest. Spain. For. Ecol. Manage. 258, 1429–1436.

Menkis, A., Vasiliauskas, R., Taylor, A.F.S., Stenlid, J., Finlay, R., 2005. Fungal communities in mycorrhizal roots of conifer seedlings in forest nurseries under different cultivation systems, assessed by morphotyping, direct sequencing and mycelial isolation. Mycorrhiza 16, 33.

Michaelis, J., Diekmann, N., 2018. Effects of soil types and bacteria inoculum on the cultivation and reintroduction success of rare plant species. Plant Ecol. 219 (4), 441–453.

Middleton, E.L., Richardson, S., Koziol, L., Palmer, C.E., Yermakov, Z., Henning, J.A., et al., 2015. Locally adapted arbuscular mycorrhizal fungi improve vigor and resistance to herbivory of native prairie plant species. Ecosphere 6 (12), 1–16.

Mohan, J.E., Cowden, C.C., Baas, P., Dawadi, A., Frankson, P.T., Helmick, K., et al., 2014. Mycorrhizal fungi mediation of terrestrial ecosystem responses to global change: mini-review. Fungal Ecol. 10, 3–19.

Molina, R., 1979. Ectomycorrhizal inoculation of containerized Douglas-fir and lodgepole pine seedlings with six isolates of *Pisolithus tinctorius*. For. Sci. 25, 585–590.

Mueller, J.M., Hellmann, J.J., 2008. An assessment of invasion risk from assisted migration. Conserv. Biol. 22 (3), 562–567.

Olsen Jr., D., Douglass, F., Walters, G., 1990. *Sequoia sempervirens* (D. Don.) Endl., Redwood, Taxiodiaceae, Redwood Family. In: Burns, R., Honkala, B. (Eds.), Silvics of North America: Volume 1, Conifers. Agriculture Handbook 654. USDA Forest Service, Washington, DC, pp. 541–551. , 675pp.

O'Hanlon, R., Harrington, T.J., 2012. Similar taxonomic richness but different communities of ectomycorrhizas in native forests and nonnative plantation forests. Mycorrhiza 22, 371–382.

O'Neill, G., Carlson, M., Berger, V., Ukrainetz, N., 2013. Assisted migration adaptation trial: workplan. <https://www2.gov.bc.ca/assets/gov/farming-natural-resources-and-industry/forestry/forest-genetics/amat_workplan_22.pdf> (accessed 02.03.18).

O'Neill, G., Stoehr, M., Jaquish, B., 2014. Quantifying safe seed transfer distance and impacts of tree breeding on adaptation. For. Ecol. Manage. 328, 122–130.

O'Neill, G., Wang, T., Ukrainetz, N., Charleson, L., McAuley, L., Yanchuk, A., et al., 2017. A proposed climate-based seed transfer system for British Columbia. Prov. B.C.,

Victoria, B.C. Tech. Rep. 099. www.for.gov.bc.ca/hfd/pubs/Docs/Tr/Tr099.htm and Roy, Douglass F. 1966. Silvical characteristics of redwood (Sequoia sempervirens [D. Don] Endl.). Res. Paper PSW-RP-28. Berkeley, CA: Pacific Southwest Forest & Range Experiment Station Forest Service, U. S. Department of Agriculture; 20 p

Outrebridge, R.A., Trofymow, J.A., 2004. Diversity of ectomycorrhizae on experimentally planted Douglas-fir seedlings in variable retention forestry sites on southern Vancouver Island. Can. J. Bot. 82, 1671–1681.

Packer, A., Clay, K., 2000. Soil pathogens and spatial patterns of seedling mortality in a temperate tree. Nature 404, 278–281.

Park, A., Puettmann, K., Wilson, E., Messier, C., Kames, S., Dhar, A., 2014. Can boreal and temperate forest management be adapted to the uncertainties of 21st century climate change? Crit. Rev. Plant Sci. 33 (4), 251–285.

Park, A., Talbot, C., Smith, R., 2018. Trees for tomorrow: an evaluation framework to assess potential candidates for assisted migration to Manitoba's forests. Clim. Change 148, 591–606.

Pedlar, J.H., McKenney, D.W., 2017. Assessing the anticipated growth response of northern conifer populations to a warming climate. Sci. Rep. 7, 43881.

Pedlar, J.H., McKenney, D.W., Aubin, I., Beardmore, T., Bealieu, J., Iverson, L., et al., 2012. Placing forestry in the assisted migration debate. BioScience 62 (9), 835–842.

Pickles, B.J., Egger, K.N., Massicotte, H.B., Green, D.S., 2012. Ectomycorrhizas and climate change. Fungal Ecol. 5, 73–84.

Pickles, B.J., Gorzelak, M.A., Green, D.S., Egger, K.N., Massicotte, H.B., 2015a. Host and habitat filtering in seedling root-associated fungal communities: taxonomic and functional diversity are altered in 'novel' soils. Mycorrhiza 25, 517–531.

Pickles, B.J., Twieg, B.D., O'Neill, G.A., Mohn, W.W., Simard, S.W., 2015b. Local adaptation in migrated interior Douglas-fir seedlings is mediated by ectomycorrhizas and other soil factors. N. Phytol. 207 (3), 858–871.

Pinto, J.R., McNassar, B.A., Kildisheva, O.A., Davis, A.S., 2018. Stocktype and vegetative competition influences on *Pseudotsuga menziesii* and *Larix occidentalis* seedling establishment. Forests 9, 228.

Pinto, J.R., Marshall, J.D., Dumroese, R.K., Davis, A.S., Cobos, D.R., 2016. Seedling establishment and physiological responses to temporal and spatial soil moisture changes. N. For. 47, 223–241.

Rehfeldt, G.E., Jaquish, B.C., 2010. Ecological impacts and management strategies for western larch in the face of climate-change. Mitigat. Adapt. Strat. Global Change 15 (3), 283–306.

Reynolds, H.L., Packer, A., Bever, J.D., Clay, K., 2003. Grassroots ecology: plant-microbe-soil interactions as drivers of plant community structure and dynamics. Ecology 84, 2281–2291.

Reithmeier, L., Kernaghan, G., 2013. Availability of ectomycorrhizal fungi to Black spruce above the present treeline in eastern Labrador. PLoS One 8 (10), e77527.

Ricciardi, A., Simberloff, D., 2009. Assisted colonization is not a viable conservation strategy. Trends Ecol. Evol. 24 (5), 248–253.

Rineau, F., Courty, P.-E., 2011. Secreted enzymatic activities of ectomycorrhizal fungi as a case study of functional diversity and functional redundancy. Ann. For. Sci. 68, 69–80.

Roberts, D.R., Hamann, A., 2016. Climate refugia and migration requirements in complex landscapes. Ecography 39 (12), 1238–1246.

Rossetto, M., Ens, E.J., Honings, T., Wilson, P.D., Yap, J.Y.S., Costello, O., et al., 2017. From songlines to genomes: prehistoric assisted migration of a rain forest tree by Australian Aboriginal people. PLoS One 12 (11), e0186663.

Roy, D.F., 1966. Silvical characteristics of redwood (*Sequoia sempervirens* [D. Don] Endl.). In: U.S. Forest Serv. Res. Paper PSW-28. Pacific SW Forest & Range Exp. Sta., Berkeley, CA.

Rúa, M.A., Antoninka, A., Antunes, P.M., Chaudhary, V.B., Gehring, C., Lamit, L.J., et al., 2016. Home-field advantage? Evidence of local adaptation among plants, soil, and arbuscular mycorrhizal fungi through meta-analysis. BMC Evol. Biol. 16, 122.

Rúa, M.A., Lamit, L.J., Gehring, C., Antunes, P.M., Hoeksema, J.D., Zabinski, C., et al., 2018. Accounting for local adaptation in ectomycorrhizas: a call to track geographical origin of plants, fungi, and soils in experiments. Mycorrhiza 28, 187−195.

Rudawska, M., Pietras, M., Smutek, I., Strzeliński, P., Leski, T., 2016. Ectomycorrhizal fungal assemblages of *Abies alba* Mill. outside its native range in Poland. Mycorrhiza 26, 57−65.

Schützendübel, A., Polle, A., 2002. Plant responses to abiotic stresses: heavy metal-induced oxidative stress and protection by mycorrhization. J. Exp. Bot. 53, 1351−1365.

Schweitzer, J.A., Bailey, J.K., Fischer, D.G., LeRoy, C.J., Lonsdorf, E.V., Witham, T.G., et al., 2008. Plant-soil microorganism interactions: heritable relationship between plant genotype and associated soil microorganisms. Ecology 89, 773−781.

Shay, P.-E., Winder, R.S., Trofymow, J.A., 2015. Nutrient-cycling microbes in coastal Douglas-fir forests: regional-scale correlation between communities, in situ climate, and other factors. Front. Microbiol. 6, 1097.

Simler, A.B., Williamson, M.A., Schwartz, M.W., Rizzo, D.M., 2018. Amplifying plant disease risk through assisted migration. Conserv. Lett. 2018, e12605.

Snyder, J.A., 1992. The Ecology of *Sequoia sempervirens*: an Addendum to "On the Edge: Nature's Last Stand for Coast Redwoods" (M.A. thesis). Department of Biological Sciences, San Jose State University, San Jose, CA.

Ste-Marie, C., Nelson, E.A., Dabros, A., Bonneau, M.-E., 2011. Assisted migration: introduction to a multifaceted concept. For. Chron. 87, 724−730.

St. Clair, J.B., Kilkenny, F.F., Johnson, R.C., Shaw, N.L., Weaver, G., 2013. Genetic variation in adaptive traits and seed transfer zones for *Pseudoroegneria spicata* (Bluebunch wheatgrass) in the northwestern United States. Evol. Appl. 6, 933−948.

Suz, L.M., Barsoum, N., Benham, S., Dietrich, H.-P., Fetzer, K.D., Fischer, R., et al., 2014. Environmental drivers of ectomycorrhizal communities in Europe's temperate oak forests. Mol. Ecol. 23, 5628−5644.

Tedersoo, L., Bhram, M., Toots, M., Diédhiou, A.G., Henkel, T.W., Kjøller, R., et al., 2012. Towards global patterns in the diversity and community structure of ectomycorrhizal fungi. Mol. Ecol. 21, 4160−4170.

Thomas, A.D., Walsh, R.P.D., Shakesby, R.A., 2000. Solutes in overland flow following fire in Eucalyptus and pine forests, northern Portugal. Hydrol. Processes 14, 971−985.

Tíscar, P.A., Lucas-Borja, M.E., Candel-Pérez, D., 2018. Lack of local adaptation to the establishment conditions limits assisted migration to adapt drought-prone *Pinus nigra* populations to climate change. For. Ecol. Manage. 409, 719−728.

Torreya Guardians. 2018. Torreya Guardians.<http://www.torreyaguardians.org> (accessed 24.08.18.).

Trappe, J.M., 1977. Selection of fungi for ectomycorrhizal inoculation in nurseries. Ann. Rev. Phytopathol. 15, 203−222.

Trocha, L.K., Kałucka, I., Stasińska, M., Nowak, W., Dabert, M., Leski, T., et al., 2012. Ectomycorrhizal fungal communities of native and non-native *Pinus* and *Quercus* species in a common garden of 35-year-old trees. Mycorrhiza 22, 121−134.

Truax, B., Gagnon, D., Fortier, J., Lambert, F., Pétrin, M.-A., 2018. Ecological factors affecting white pine, red oak, bitternut hickory and black walnut underplanting success in a northern temperate post-agricultural forest. Forests 9, 499.

Turner, S., Mikutta, R., Meyer-Stüve, S., Guggenberger, G., Schaarschmidt, F., Lazar, C. S., et al., 2017. Microbial community dynamics in soil depth profiles over 120,000 years of ecosystem development. Front. Microbiol. 8, 874.

Ulrich, A., Becker, R., 2006. Soil parent material is a key determinant of the bacterial community structure in arable soils. FEMS Microbiol. Ecol. 56, 430−443.

Van der Veken, S., Hermy, M., Vellend, M., Knapen, A., Verheyen, K., 2008. Garden plants get a head start on climate change. Front. Ecol. Environ. 6 (4), 212−216.

Van Der Heijden, M.G.A., Martin, F., Selosse, M.A., Sanders, I.R., 2015. Mycorrhizal ecology and evolution: the past, the present, and the future. N. Phytol. 205, 1406−1423.

Vandenkoornhuyse, P., Quaiser, A., Duhamel, M., Le Van, A., Dufresne, A., 2015. The importance of the microbiome of the plant holobiont. N. Phytol. 206, 1196−1206.

Velmala, S.M., Rajala, T., Haapanen, M., Taylor, A.F.S., Pennanen, T., 2013. Genetic host-tree effects on the ectomycorrhizal community and root characteristics of Norway spruce. Mycorrhiza 23, 21−33.

Wilkinson, D.M., Dickinson, N.M., 1995. Metal resistance in trees: the role of mycorrhizae. Oikos 72 (2), 298−300.

Winder, R., 2017. Catalog of provenance trials applicable to climate change adaptation research. In: Information Report BC-X-441. Natural Resources Canada, Canadian Forest Service, Pacific Forestry Centre, Victoria.

Winder, R., Shamoun, S., 2006. Forest pathogens: friend or foe to biodiversity? Can. J. Plant Pathol. 28, S221−S227.

Winder, R., Nelson, E.A., Beardmore, T., 2011. Ecological implications for assisted migration in Canadian forests. For. Chron. 87 (6), 731−744.

Winder, R., 2013. The impact of climate change on the distribution of mycorrhizal mushroom species. In: Proceeding of the International Symposium on Forest Mushrooms, Korea Forest Research Institute, August 6, 2013, Seoul. Korea Forest Research Inst., Seoul, pp. 127−137.

Wubbs, E.R.J., Bezemer, T.M., 2016. Effects of spatial plant-soil feedback heterogeneity on plant performance in monocultures. J. Ecol. 104 (2), 364−376.

Xie, C.-Y., Yanchuk, A.D., Kiss, G.K., 1998. Genetics of interior spruce in British Columbia: performance and variability of open-pollinated families in the East Kootenays. In: Research Report 07. B.C. Ministry of Forests Research Branch, Victoria.

Soils and restoration of forested landscapes

11

C.E. Prescott[1], K. Katzensteiner[2] and C. Weston[3]

[1]*Faculty of Forestry, University of British Columbia, Vancouver, BC, Canada*
[2]*Institute of Forest Ecology, University of Natural Resources and Life Sciences (BOKU), Vienna, Austria*
[3]*University of Melbourne, School of Ecosystem and Forest Sciences, Melbourne, Australia*

11.1 Introduction

Forest landscape restoration (FLR) has the potential to return degraded sites to functional ecosystems, which are able to provide the ecosystem services on which human societies depend, including production of food, forage and timber, flood control and water supply, C sequestration, and climate modulation. The success of FLR can also rehabilitate degraded soils; indeed the success of FLR efforts will largely depend on the degree to which they improve soil fertility. In many areas undergoing FLR, soils have been degraded through unsustainable agricultural practices, and areas with the most favorable soil conditions will usually be reserved for agricultural purposes. As a result, the sites on which forests are being reestablished will often have characteristics that are challenging for young trees. Rehabilitation of the soil may be a prerequisite for restoring forests on these sites. In this chapter, we consider the soil conditions that could constrain tree growth and survival and suggest restoration practices that could be employed to overcome these soil constraints.

11.2 Unstable, erosive soils

Although the root systems of trees stabilize the soil once the site is afforested, unstable soils can make it difficult for trees to become established. Soil instability is a common feature of degraded soils that have lost much of their structure, leaving soil particles vulnerable to loss by erosion by wind and water (Lal, 2001). The loss of fine soil particles (which can reach $3-80$ tons $(ha\ year)^{-1}$ in highly vulnerable regions) by water or wind erosion is a major component of desertification (Chang et al., 2015).

Sloping ground is particularly vulnerable to loss of soil through water erosion and slope stabilization is a necessary starting point for restoration. Terraces have been employed throughout the world for millennia to allow for agriculture on sloping ground (Tarolli et al., 2014; Arnáez et al., 2015; Fig. 11.1). Much effort is required to construct

FIGURE 11.1

Afforestation with terraces in a semiarid region in Turkey.

Colak, A.H., Kirca, S., Rotherham, I.D., Ince, A., 2010. Restoration and Rehabilitation of Deforested and Degraded Forest Landscapes in Turkey. Springer, New York (Colak et al., 2010). Photo credit: R. Çetiner.

terraces and to maintain them (Liu et al., 2011); increased erosion has followed land abandonment of terraced land in many parts of Europe (Koulouri and Giourga, 2007). In a metaanalysis of terraces and water erosion in China, Chen et al. (2017) found that bench terraces were superior to other types with respect to runoff and sediment reductions and that terraces associated with tree crops and forests conserved the greatest amount of soil and water.

Construction of bunds can reduce soil erosion by water. Contour bunds are embankments of stones or soil built along the contour of sloping land in order to intercept surface runoff and reduce the velocity of overland flow (Gebremichael et al., 2005). In the Ethiopian Highlands, stone bunds provided a 68% reduction in annual soil loss due to water erosion (Gebremichael et al., 2005). Nardi/Vallerani trenches can also be dug perpendicular to the slope, with a ridge on the downhill side to trap and store the runoff water (Abdo, 2014). Soil or stone bunds combined with trenches and integrated with trees have proven to be an effective soil conservation measure in Ethiopia (Haregeweyn et al., 2015). Chirwa and Mahamane (2017) provide an overview of management practices for reducing soil erosion in the degraded landscapes of the Sahelian and dryland forests and woodlands of East and southern Africa (Fig. 11.2).

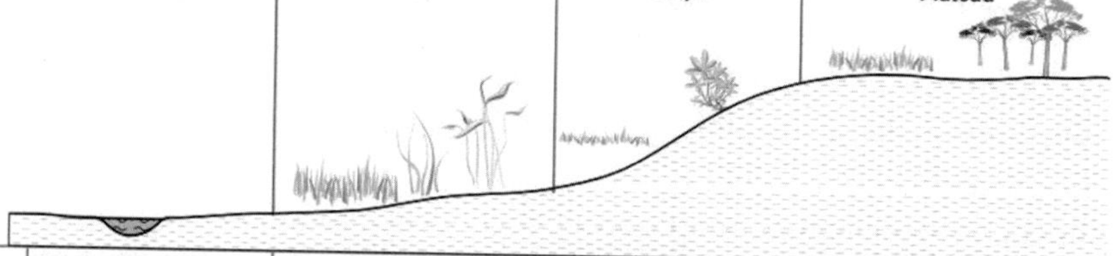

	Valley bottom	Pediment	Slope	Plateau
f soil	Fertile alluvial soil	Deep, fairly fertile colluvial soil	Shallow stony soil (or sandy soil in the case of dune stabilisation)	Shallow, infertile soil, duricrust outcrops, barren areas with hardened soil crust
Use	Individual plots with: • irrigated crops • market gardens Communal grazing and watering areas	Individual plots with rain-fed crops	Communal land with some grazing areas	Communal land for: • grazing • collection of wood and other products (fruits, medicinal plants)
Risks	Gully erosion Siltation Flooding	Gully erosion Sheet erosion	Gully erosion Landslides	Sheet erosion Gully erosion Wind erosion
SLM practices	Water-spreading weirs Small-scale dams Village irrigation schemes Assisted natural regéneration Permeable rock dams Contour stone bunds	Contour stone bunds Permeable rock dikes Zaï planting pits Manure/compost Mulching Grass strips Permeable rock dams	Hand-dug trenches Permeable rock dams Contour stone bunds Dune stabilisation	Semicircular bunds Nardi/Vallerani trenches Contour bunds (firebreaks)

FIGURE 11.2

Overview of management practices for reducing soil erosion in the degraded landscapes of the Sahelian and dryland forests and woodlands of East and southern Africa.

Chirwa, P.W., Mahamane, L., 2017. Overview of restoration and management practices in the degraded landscapes of the Sahelian and dryland forests and woodlands of East and southern Africa. South. For. J. For. Sci. 79, 87–94. https://doi.org/10.2989/20702620.2016.1255419.

Where water erosion has led to gully formation, check dams, permeable rock dams, or similar impediments may be necessary to reduce water flow and soil losses (Zhao et al., 2013; Gabou and Maisharou, 2015; Quiñonero-Rubio et al., 2016). Once these measures are in place, rapid planting of live stakes or seedlings of fast growing trees and/or other plants assists in holding the soil on-site. As Quiñonero-Rubio et al. (2016) point out, check dams provide an immediate reduction in sediment yield over a restricted time period, while reforestation has a sustained benefit. Wattle fences are a particularly effective means of slope stabilization as they provide both mechanical stabilization and rapid root system development (Fig. 11.3).

Mulch is a layer of material (organic or inorganic) that covers the soil surface; common mulch materials are cereal straw, weeds and manure, processing residues such as sawdust or husks, or plastic films or polymers (Kader et al., 2017). Surface mulches of various compositions can also reduce losses of soil in water, as they protect the soil from the impact of raindrops, provide surface roughness,

FIGURE 11.3

Wattle fences used for slope stabilization in British Columbia, Canada.

Photo credit: D. Polster.

which reduces overland flow generation rates and velocity, and improve infiltration capacity (Prosdocimi et al., 2016). A mulch cover of 60% is usually considered as the minimum necessary to significantly reduce soil loss (Prosdocimi et al., 2016). Straw mulch was efficient in reducing soil erosion rates in a persimmon plantation in eastern Spain (Cerdà et al., 2016).

Wind erosion can be alleviated by planting species that develop extensive roots systems that stabilize the soil. Examples include sand dunes stabilized by lyme grass [*Leymus arenarius* (L.) Hochst] in Iceland (Runólfsson, 1987), mountain pine (*Pinus mugo* Turra) in Denmark (Madsen et al., 2005), and xerophytic shrubs in China (Yang et al., 2014). Once established, vegetation decreases soil erodibility and reduces the wind speed at the soil surface, which allows other plant species to become established (Ma et al., 2017). Windbreaks in the forms of shelterbelts and hedgerows have been used for centuries to reduce wind speeds and soil loss from agricultural land (Brandle et al., 2000). In semiarid and dry temperate areas of Australia, Bird et al. (1992) estimated that planting of 5% of the land to shelter could reduce wind speed by 30%−50% and soil loss by up to 80%. Windbreaks also lead to increase in SOC stocks nearby (Cardinael et al., 2017; Wiesmeier et al., 2018). Windbreaks and hedges may also provide the nucleus for colonization of native trees to establish, the seeds being dropped by wind or birds/mammals attracted to the hedge (Harvey, 2000).

In extreme situations, such as shifting sand dunes, artificial barriers or covers are necessary for stabilization either before or concurrent with establishment of vegetation (Tibke, 1988). In the Horqin Sandy Land in northern China, wind erosion of soil and sand dune mobility are treated by exclosing the area to grazing, followed by burying wheat straw or other residues in a checkerboard pattern and/or planting sand-fixing shrubs such as *Artemisia halodendron* Turcz. Ex Besser (Zhang et al., 2004; Shirato et al., 2005; Kang et al., 2017; Zhang et al., 2018). Shirato et al. (2005) found that sand dunes were stabilized within 1 year of installing straw checkerboard or planting *A. halodendron*, and a biological soil crust formed within 3 years, which reduces wind erodibility of soil (Belnap and Gillette, 1997). In Niger, sand dunes are stabilized by first erecting wind barriers made of millet stalks, followed by planting trees within the squares (Gabou and Maisharou, 2015). Tree species with symbiotic N-fixation such as *Casuarina equisetifolia* L. have been successfully used for dune stabilization in the tropics (National Research Council, 1984a,b). The possibility of vegetative propagation and fast growth rates under harsh conditions make them suitable candidates for soil stabilization and soil improvement. *Casuarinas* can, however, be invasive (e.g., Wheeler et al., 2011) and negatively affect biological diversity.

Application of microbial biopolymers (polysaccharides secreted by microorganisms) can reduce soil erosion by binding together soil particles. Chang et al. (2015) found that application of β-glucan secreted by *Aureobasidium pullulans* and xanthan gum secreted by *Xanthomonas campestris* reduced erodibility and increased soil water retention efficiency, seed germination, and growth of oat plants (Fig. 11.4).

Erosive situations can be prevented by maintaining forest or other vegetation cover on steep slopes, by not cultivating sloping land, and by managing grazing intensity to maintain cover and particularly rooting systems of plants.

11.3 Inadequate water

Water availability can constrain plant growth and survival; if too low there is insufficient water for plant metabolic processes, if too high plant metabolism is constrained by insufficient oxygen or redox conditions. Insufficient water-holding capacity can result from soils being too shallow (i.e., depth to bedrock or root-restricting layer) such there is insufficient storage capacity. This situation may arise in young soils, alpine soils, on steep slopes, and in areas where erosion or other disturbances have resulted in loss of topsoil. Water-holding capacity can also be low in soils that are coarse-textured and/or with high coarse-fragment content, or in degraded soils in which organic matter and structure have been lost. Extreme examples occur in saprolites, in which the soil has been lost through erosion and only weathered bedrock material remains.

Soil properties and/or other site conditions that result in conditions of inadequate water availability for restoring forests can be offset to variable degrees by

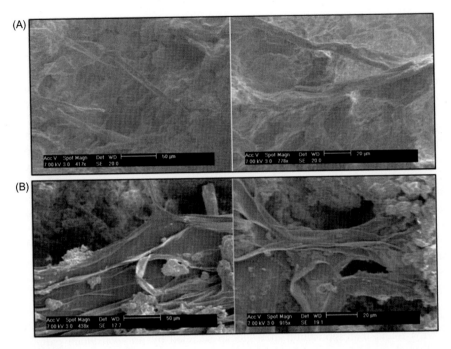

FIGURE 11.4

SEM images of soil with oat roots: (A) natural soil and (B) biopolymer (i.e., beta-glucan) treated soil.

Chang, I., Prasidhi, A.K., Im, J., Shin, H.-D., Cho, G.-C., 2015. Soil treatment using microbial biopolymers for anti-desertification purposes. Geoderma 253–254, 39–47. https://doi.org/10.1016/j. geoderma.2015.04.006; https://ars.els-cdn.com/content/image/1-s2.0-S0016706115001123-gr4.jpg.

(1) constructing bunds, swales, or terraces that retain water, (2) planting trees in pits where water accumulates, and (3) planting species that are efficient in using water or have deep or extensive root systems or other adaptations that enable them to survive under xeric conditions. Amending soils with organic matter or biochar can also improve their capacity to hold water, and surface mulches can reduce evaporative losses of water. In extremely coarse-textured soils, some compaction can lead to an improvement in water-holding capacity.

Terraces have been used in hilly and mountainous regions throughout the world (Chen et al., 2017; Wei et al., 2016) to increase infiltration of water into the soil (Arnáez et al., 2015). Terraces associated with tree crops and forests are particularly effective at conserving soil and water (Chen et al., 2017). Contour bunds of stone or soil, sometimes associated with ditches or plants, are common rainwater-harvesting techniques in sub-Saharan Africa; dams of various sizes are used to collect and store runoff from ephemeral streams (Biazin et al., 2012).

Planting pits are a common means of fostering tree regeneration in arid regions (Fig. 11.5). The pits retain water, allowing plants to survive long dry

FIGURE 11.5

Planting pits.

spells Roose et al., 1999). For example, farmers in Burkina Faso construct Zaï pits, often adding organic matter such as manure to the base of the pit (Reij et al., 2009). Similar microcatchment rainwater-harvesting systems are employed on sloping land—semicircular holes are dug into a slope, and the excavated earth used to form a bund downslope of the hole, which prevents water running off (Biazin et al., 2012). Trees are planted in the pits and the pits may be enriched with organic matter such as manure. Examples of this practice include "fish-scale" pits common on sloping land in China and "half-moon" pits in Niger (Gabou and Maisharou, 2015). Mechanical preparation of planting holes (spot tillage) has been shown to improve the performance of planted seedlings in arid environments (Querejeta et al., 2000; Löf et al., 2012). Subsoiling may also be effective in improving infiltration and enabling seedlings to root deeply and access additional water reserves (Löf et al., 2012).

Soil organic matter (SOM) can be a limiting factor for plant establishment in arid and semiarid lands on degraded soils, and vegetation survival, growth and cover can be enhanced by amending soils with organic matter such as crop residues, sewage sludge, or animal manure (Hueso-Gonzalez et al., 2016). Mulches are commonly applied to foster plant survival and growth in arid regions (Coello et al., 2018). Mulches reduce water loss rates, topsoil temperature, and evaporation and increase infiltration, water intake, and water storage (Prosdocimi et al., 2016).

Soil conditioners are synthetic compounds such as superabsorbent polymers (SAP) (also referred to as hydrogels or superabsorbers), which retain water thereby reducing evaporation and increasing soil water availability (Coello et al.,

2018). The use of SAP—alone or combined with fertilizers and other components—has proven effective in agriculture and forestry, increasing soil water availability, reducing evaporation, and enhancing early survival and growth in a wide range of species (Huttermann et al., 2009; Coello et al., 2018). Hueso-Gonzalez et al. (2016) found the addition of mulch or a hydroabsorbent polymer increased soil water content and growth of Aleppo pine (*Pinus halepensis* Mill.) saplings in southern Spain.

If low water availability is a consequence of a particularly arid climate, planting trees, even those that may survive due to superior water acquisition abilities, is not recommended, as transpiration water loss can exacerbate water shortage and cause salinization of soils (Cao et al., 2011). Dryland vegetation such as grasses is better suited to arid conditions.

11.4 Excess moisture

Excessive levels of water can constrain tree growth through insufficient oxygen availability to roots. Excess moisture is usually a consequence of climate and/or slope position or inundation (flooding), but soil factors can also contribute. For example, drainage may be impeded by the presence of a root-restricting layer, by very fine soil texture (high clay content and certain clays) or as a result of compaction. Sobanski and Marques (2014) found that restoration of forests on previous pasture sites in the Brazilian Atlantic Forest region was more challenging on sites with gleysols than on cambisols, due to the high water table and frequent flooding. On northern Vancouver Island, Canada, subtle differences in soil moisture conditions explained pronounced differences in forest regeneration success on adjacent sites, with poor regeneration on some sites attributable to excess moisture (Sajedi et al., 2012). Mechanical site preparation practices such as ditching, subsoiling, and/or mounding prior to planting can increase tree survival rates and the productivity of such sites (Löf et al., 2012). However, if excess moisture is a consequence of climate and/or landscape position, ecosystems other than forests may be more appropriate for the site.

11.5 Low organic matter content

Organic matter is critical to soil function—it contributes to water-holding capacity and cation-exchange capacity (CEC), but it is particularly beneficial as the fuel that supports the belowground ecosystem that is responsible for the recycling of nutrients and the development of stable SOM. The amount of organic matter present in soil is a function of the amount of organic material that enters the soil as plant, faunal, or microbial residues and the degree to which it is either decomposed or transformed into more stable compounds (Prescott, 2010). Historically most attention has been paid to rates of input and decomposition of aboveground plant litter, but there is increasing recognition of the importance of roots as a source of SOM (Rasse et al., 2005). Mycorrhizal fungi also appear to be an important source of SOM. In a poplar plantation, Godbold et al. (2006) found that mycorrhizal external mycelium was the dominant pathway through which carbon entered the SOM pool, exceeding the input via leaf litter and fine root turnover. Clemmenson et al. (2015) found that 50%—70% of stored carbon in the humus layer of boreal forests was derived from roots and root-associated microorganisms, including fungal residues. There is also evidence that most stable SOM has been transformed by soil microorganisms into microbial residues and by-products (Kallenbach et al., 2016; Liang et al., 2017). Likewise, transformations of litter during gut passage in soil macrofauna such as earthworms and millipedes may render the material more stable than plant litter (Frouz, 2018; Morrien and Prescott, 2018).

Degraded soils are usually depleted in organic matter, usually as a consequence of prolonged cultivation or overgrazing. Unless compensated for by organic amendments, cultivation reduces SOM through removal of biomass, tilling which promotes decomposition of SOM, and use of annual crops and fertilizer, which reduce root biomass, mycorrhization, and root exudation. Likewise, if grazed in manner that does not allow plants to regrow and replenish both shoot and root systems between grazing events, pastures may become depleted in SOM.

Once forests are established they ameliorate the soil by contributing organic matter both on the surface via litterfall and throughfall, and in the soil through litter and exudates from roots and mycorrhizal fungi. However, inadequate SOM can limit tree growth and survival during the establishment phase, due to inadequate moisture retention, nutrient supply or lack of beneficial soil organisms such as mycorrhizal fungi. Applications of organic matter can kick-start the process of building SOM and immediately assist with moisture retention and nutrient supply as well as providing necessary resources for soil organisms. A wide variety of organic amendments have been used for rehabilitating soils, including agricultural (livestock manure and crop residues), forestry (wood chips and deinking sludges), urban wastes (biosolids from wastewater treatment plants and the biodegradable component of municipal solid waste), and food processing waste (Larney and Angers, 2012).

The choice of tree species to plant can also influence the rate of SOM development. Planting tree species that associate with arbuscular mycorrhizal fungi (AMF) may assist in building SOM as trees with AMF tend to have greater mineral soil C stocks (Vesterdal et al., 2013; Craig et al., 2018; Wiesmeier et al., 2018; Augusto et al., 2014). SOM development can be promoted by planting tree species with N-fixing root associates (Chaer et al., 2011; Prescott, 2010). Several studies on postmining sites have reported SOM levels increasing faster under N-fixing trees than under other tree species (Frouz et al., 2009; Schiavo et al., 2009; Kuznetsova et al., 2011). A metaanalysis of north-temperate forests reported a significant increase in mineral soil C storage in response to N-fixing vegetation (Nave et al., 2009). Of course, the tree species selected must be appropriate for the site conditions—they must survive and grow well in order to generate SOM.

11.6 Loss of soil structure

Soil structure refers to the proportions of solids and voids in a soil, the voids providing the space for both water and air. A key aspect of soil structure is the aggregation of individual mineral and organic particles into larger units (Binkley and Fisher, 2013). Soil structure is influenced by characteristics of the soil (especially texture and clay mineralogy), climatic factors (especially wet—dry and freeze—thaw cycles), characteristics of the plants and organic matter, and activities of soil biota (Daynes et al., 2013; Rashid et al., 2016; Bronick and Lal, 2005; Regelink et al., 2015). The soil biota promotes soil structure by contributing organic matter in the form of plant litter (especially root litter; Rasse et al., 2005), faunal fecal material (Frouz, 2018), and microbial residues (Abiven et al., 2009; Kallenbach et al., 2016). Roots and fungi also physically bind soil particles together (i.e., enmeshment) and release substances such as the glycoprotein glomalin that act as binding agents, which promote soil aggregate stability (Rillig and Mummey, 2006; Rashid et al., 2016; Fig. 11.6). Several studies have reported strong positive relationships among abundance of AMF, glomalin, soil aggregation, and aggregate water stability (Jastrow et al., 1998; Rillig et al., 2002; Wright and Upadhyaya, 1998; Kohler et al., 2017). The movements of burrowing earthworms create biopores that increase porosity and water infiltration (Marashi and Scullion, 2003) and promote the formation of organo-mineral and water-stable aggregates that increase water-holding capacity (Frouz and Kuráž, 2014; Lavelle, 1988; Scullion and Malik, 2000).

Soil structure influences many of the processes critical to soil functioning, including density, porosity, infiltration, drainage, aeration, water-holding capacity, and resistance to erosion (Barto et al., 2010). A decline in soil structure is a key facet of soil degradation and leads to issues such as crusting, compaction, anaerobiosis, erosion, reduced CEC, loss of fertility, and desertification (Lal, 1997). Soil structure can be lost through land use practices that reduce organic matter inputs, lower the abundance of roots and soil organisms, or compact the soil. Such practices include deforestation, overgrazing, and agriculture practices such as tillage, bare fallows, applications of fungicides and insecticides, and substitution of organic fertilizers with chemical ones. Poor soil structure is a common characteristic of soils in arid and semiarid regions, which contributes to their high susceptibility to erosion (Caravaca et al., 2002a; Cerdà et al., 2010).

Afforestation of sites with degraded soils assists in the development of soil structure within the bounds set by soil texture and climate; indeed rapid rates of organic matter accumulation and aggregate formation have been reported in postmining sites in the first two to three decades following rehabilitation and revegetation (Wick et al., 2009a,b; Shrestha and Lal, 2010; Vinduskova and Frouz, 2013). However, especially in arid and semiarid environments, improvement of soil structure may be a precondition for successful afforestation (Caravaca et al., 2002a). If soils are low in OM, soil structure can be encouraged by application of exogenous organic matter

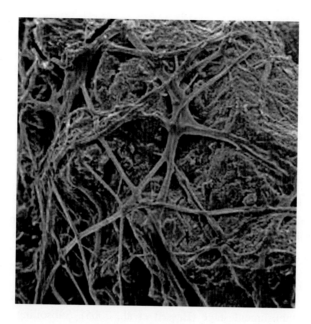

FIGURE 11.6

Net-like fungal mycelia can stabilize microaggregates.

University of Bremen micropedology SEM website, https://www.google.ca/search?biw = 1745&bih = 853&
tbm = isch&sa = 1&ei = IXb3W56_K_TLOPEPopKDwAQ&q = fungal + mycelia + soil + aggregate + photo&
oq = fungal + mycelia + soil + aggregate + photo&gs_l = img.12...115824.121981..123440...0.0..0.48.
452.12......1....1..gws-wiz-img.coLhW4kMVtY#imgrc = 7H_eYzs4wyPLBM.

such as compost, manure, or municipal waste (sewage sludge, garden, and municipal solid waste). Applications of organic wastes to degraded soils have been demonstrated to reduce bulk density and increase plant-available water, infiltration rates, nutrient supply, microbial biomass, extracellular enzyme activity, aggregation, and aggregate stability (Chambers et al., 2002; Tejada et al., 2006; Larney and Angers, 2012; Bastida et al., 2015), in addition to fostering plant development and tree growth. Biochar, especially when applied with a green manure, can also increase soil aggregation (Fungo et al., 2017). Several studies in Spain have demonstrated increased soil aggregate stability following addition of municipal solid wastes (Roldán et al., 1996; Annabi et al., 2007; Caravaca et al., 2002b; Bastida et al., 2007). Positive effects on soil porosity, available water content, infiltration, soil aggregation, and bulk density have been reported following application of biochar, domestic organic waste, sewage sludge, and composts (Garcia-Orenes et al., 2005; Hernandez et al., 2015; Luna et al., 2018).

Once there is sufficient organic matter, soil structure development can be further promoted by enhancing the abundance and activity of soil organisms, particularly mycorrhizal fungi and soil macrofauna. The effectiveness of AMF in the improvement of the structure of degraded semiarid soils has been demonstrated, as has variation among AMF species in promoting aggregation (Caravaca et al., 2002a,b; Alguacil et al., 2011; Rillig et al., 2010). Colonization by soil macrofauna, particularly burrowing earthworms, also promotes the development of structure in degraded soil. Nine years after reclamation of an opencast coal mine, plots with earthworms had higher amounts of carbon associated with the clay-sized fraction, greater microbial biomass, and higher carbohydrates, which promote aggregate stability (Scullion and Malik, 2000).

11.7 Compaction

Many degraded soils have compacted layers resulting from long-term cropping or grazing. The increased bulk density and decreased number of larger pores lead to reduced infiltration and groundwater recharge, and increased surface runoff (Batey, 2009). Compaction also hinders gas exchange and so is problematic for roots and soil organisms. Once established, trees increase soil permeability through their root systems, but loosening of soil may be necessary to get seedlings through the establishment stage (Bassett et al., 2005). Subsoiling loosens compacted soil beneath the surface, which promotes water infiltration, aeration, and root growth (Löf et al., 2012). Decompaction through mechanical site preparation has also been shown to increase mycorrhizal colonization (Bauman et al., 2013). Tree species that tolerate prolonged anoxic soil conditions can improve soil structure and aeration (Fountain, 2011). For example, Meyer et al. (2014) reported rapid regeneration of soil structure and air permeability in heavily compacted skid lanes replanted with black alder [*Alnus glutinosa* (L.) Gaertn.]. Subsoiling in Vertisols requires least 15 cm of rainfall after slitting to fill trenches with silt before planting can begin; otherwise, root systems are in danger of losing contact with shrinking and swelling of the soil (Stanturf et al., 2001).

11.8 Poor nutrient supply

Aboveground productivity of most forests is limited to some extent by the availabilities of nutrients (usually N and/or P), and nutrient deficiencies can be accentuated in degraded soils. Severe nutrient deficiencies or imbalances of nutrients can also hinder tree establishment in highly degraded soils. Severe P deficiencies that occur naturally can severely limit seedling establishment (Pritchett and Comerford, 1982). Nutrient losses can be avoided through forest management practices such as bole-only harvesting, on-site processing and residue retention, and reduced slash burning (Berthrong et al., 2009). Afforestation has the potential to rebuild stocks of nutrients in soils and restore the nutrient cycling processes responsible for maintaining the supply of nutrients in forms that are available for plant uptake. In a global metaanalysis, soil C and N stocks were elevated 30 and 50 years after afforestation, respectively, except under pines (Li et al., 2012). In severely degraded soils, nutrient amendment may be required for successful establishment and growth of trees.

In soils in which low nutrient stocks are associated with low stocks of SOM, amendment with various organic wastes would be most effective in improving fertility (as discussed earlier). Organic materials can be applied as mulch or incorporated into the soil, and either broadcast or placed into planting pits. SOM and N contents are also enhanced by planting N-fixing species, and these have been widely used for restoring degraded land (Fisher, 1995; Franco and De Faria, 1997; Macedo et al., 2008; Wang et al., 2010). N-fixing plants can be used either as a nurse crop; for example, in the tropics, dipterocarps are grown in gaps of *Acacia* sp. (Norisada et al., 2005), or in mixed plantations. Mixtures of N-fixing *Acacia* sp. and *Eucalyptus* sp. were more effective in increasing soil C stocks relative to monoculture *Eucalyptus* plantations (Forrester et al., 2013; Koutika et al., 2014). In addition to C and N, N-fixing tree species can increase the availability of P; for example, Bini et al. (2018) reported an increase in AMF root colonization of *Eucalyptus grandis* W. Hill and an increase in phosphatase activity when intercropped with *Acacia mangium* Willd.

Inoculation with mycorrhizal fungal symbionts increases the availability of nutrients, especially N and P to roots, by exploiting a greater volume of soil and by secreting enzymes that aid in liberating nutrients in plant-available forms. Ectomycorrhizal fungi produce enzymes that render the organic N compounds common in surface humus layers available for plant uptake, while AMF release phosphatase enzymes with liberate P in plant-available forms. Mycorrhizal inoculation is discussed in more detail later in this chapter.

In established stands, deficiencies of nutrients may arise as consequence of increasing immobilization of nutrients in plant biomass. Deficiencies or imbalances of nutrients can be diagnosed through foliar analysis (Binkley and Fisher, 2013) and nutrients found to be inadequately available can be augmented through fertilization. Spot fertilization in close proximity of trees reduces the risk of nutrient loss or stimulation of unwanted vegetation but can negatively affect root distribution and shoot—root ratio, which may reduce seedling performance at a later development stage (Glatzel et al., 1990/91).

11.9 Excessive nutrient supply

Restoration of certain ecosystems on former cropland can be challenging where the soil contains high concentrations of residual P from previous fertilizer additions (N may also be high but tends to not be as persistent as P; Smits et al., 2008; Ceulemans et al., 2014). Topsoil removal is an effective but expensive measure (Schelfhout et al., 2015). Topsoil removal followed by soil transfer from a target ecosystem site has been shown to be effective in restoration of species-rich grasslands on former arable land (Carbajo et al., 2011). Growing, mowing, and removal of hay eventually reduce nutrient levels in soil but may become less effective as concentrations of nutrients other than P become limiting to plant growth (Oelmann et al., 2009). Supplementing the availability of these other nutrients may be needed to maintain productivity of the hay and thus the effectiveness of the "P-mining" treatment (Schelfhout et al., 2015).

11.10 Soil acidification

Though soil acidification is a natural process in humid environments, high rates of biomass extraction or the input of acids or acidifying compounds such as reduced N in fertilizers or atmospheric deposition may accelerate acidification, cause base cation depletion, and alter the bioavailability of trace elements and heavy metals (Bolan et al., 2003). Under acidic soil conditions, aluminum, in particular in the form of Al^{3+}, will be present in the soil solution. Free Al^{3+} may be highly toxic to plant roots. Though forest tree species may be adapted to high Al^{3+} concentrations (Brunner and Sperisen, 2013), high acidification rates due to deposition of acidifying compounds, in particular at sites depleted in base cations, may have detrimental effects on species performance. The molar Ca/Al ratio in soil solution or fine roots is a useful indicator for estimating the risk of Al stress to tree seedlings (Cronan and Grigal, 1995; Vanguelova et al., 2007).

Liming is one of the oldest management practices to counteract soil acidification and related problems in agriculture. Bolan et al. (2003) give a broad review on liming effects on soil physical, chemical, and biological characteristics. Positive effects on soil structure, improvement of the pH value, increased Ca/Al ratios in the soil solution, increased CEC, and positive effects on soil fauna and flora are reported. In postmining site reclamation and reforestation liming is a common practice to counteract acidification caused by sulfur oxidation (e.g., Hüttl, 1998). Liming (usually with dolomitic lime) is a common amelioration practice for acidified forest soils in Europe, with positive effects on tree growth, vitality, and resistance to pathogens (Court et al., 2018; Katzensteiner et al., 1995; Halmschlager and Katzensteiner, 2017), but liming may enhance nitrification and alter root distribution (Huettl and Zoettl, 1993). The decision on liming for restoration measures requires careful analyses of soil conditions and an evaluation of species requirements regarding soil pH and Ca/Al ratios.

11.11 Salinity

Salinity and sodicity are two categories of salt-effected soils caused by the accumulation of salts in the soil profile, or on the soil exchange complex, to the point where they may alter soil properties and the suitability of the site for plant growth. Saline–sodic soils exhibit both saline and sodic properties. Definitions of salinity and sodicity are given in Table 11.1, along with key associated impacts on soil properties that influence plant growth (after Munns, 2005). Saline and sodic soils result from natural processes (referred to as primary salinity) and impact around 0.9–1.128 M ha globally, mostly in arid and semiarid areas (FAO, 2008; Wicke et al., 2011). While most sodic soils are natural, human activity has created around 100 M ha of saline soils by altering soil hydrology, either through clearing for agriculture, or due to irrigation (referred to as secondary salinity) (Levy and Shainberg, 2005). Secondary salinity is often the result of vegetation clearing or thinning to the point where more water (either precipitation or irrigation water) enters the soil, causing groundwater tables to rise and mobilizing salts that can be drawn to the soil surface through capillary action (Salama et al., 1999). Secondary salinity impacts around 11% of global agricultural land, more so in productive irrigated areas (about 45 M ha) than in dryland areas (32 M ha) (FAO, 2005).

Our focus here is on salinity in landscapes for restoration—typically agricultural, deforested, or degraded sites, rather than sodicity in landscapes of naturally low productivity. Saline soils have high concentrations of salts in the soil profile—the major soluble cations are Na^+, Ca^{2+}, Mg^{2+}, and K^+ and the anions Cl^-, SO_4^{2-}, HCO_3^-, CO_3^-, and NO_3^- (Shi and Wang, 2005). This salinity reduces microbial activity in soil by causing osmotic stress (increased osmotic potential draws water out of cells causing dying and lysis) and specific ion effects (ions in high concentration become toxic) resulting in reduced rates of microbial transformations that maintain key ecosystem processes (Yan et al., 2015). Differences in salinity tolerance between fungi (less tolerant) and bacteria (more tolerant), and among bacteria (tolerant to nontolerant strains), result in changes to community structure relative to nonsaline soils (Pankhurst et al., 2001).

High salts also negatively impact soil chemical and physical properties (Wong et al., 2010). With increasing salinity vegetation growth is reduced due to the detrimental osmotic potential impacts and the toxic effect of salts. The severity of salinity in soils is usually determined by the electrical conductivity (EC_e in $dS\ m^{-1}$) of a saturated soil extract—ranging from EC_e <4 (low) to EC_e >16 (extreme). Assessing the severity and nature of salinity in an area for restoration is a crucial step in developing a plan. Salt-adapted trees can grow in saline soils up to about EC_e 8–10, beyond which only highly salt-tolerant plants or halophytes can grow. Because of the adverse soil–air and soil–water relations of saline soils, the trajectory of restoration is lower than in nonsaline soils. At a landscape scale the amelioration of detrimental soil properties through the introduction of Ca^{2+} to the exchange complex to replace Na^+ is usually restricted to saline–sodic (and sodic soils) rather than saline soils; techniques include chemical amendments (lime) and phytoremediation (Qadir et al., 2007).

The restoration or introduction of trees either directly in saline soils or regionally in recharge zones to lower the water table in saline areas of the landscape is a key measure to reduce advancing salinity risk (Bell, 1999; Cramer and Hobbs,

Table 11.1 A classification of salt-affected soils with comments on soil properties and diagnostic features.

Salt-affected soil type	Definition and effect on plant growth	EC_e (dS m^{-1})	pH	Sodium adsorption ratio	Soil physical condition
Saline	Saline soils have a high concentration of soluble salts. Most crop yields are reduced in saline soils. Plant growth is reduced by osmotic stress and salt-specific effect.	>4	<8.5	<13	Normal
Saline–sodic	Saline–sodic soils have a high concentration of soluble salts and high exchangeable sodium (Na +) percentage (ESP). They have properties similar to both saline and sodic soils—but are usually grouped with sodic soils. Unlike sodic soils they have more normal soil physical properties due to high salt concentration.	>4	<8.5	>13	Normal
Sodic	Sodic soils have a low concentration of soluble salts and a high ESP that causes degradation of the structure of clays. The clay particles separate (disperse) when the soil is wet resulting in poor soil drainage and waterlogging; when the soil is dry it becomes hard (massive).	<4	>8.5	>13	Poor—soil structure inhibits root growth

EC_e is the electrical conductivity of a saturated soil paste extract—it reflects the concentration of salts in saturated soil. Osmotic stress reduces the plants ability to take up water, leading to slower growth. Salt-specific effect results from salt entering the transpiration stream and injuring cells in transpiring leaves, leading to slower growth. ESP, Exchangeable sodium percentage.

2005). By lowering the water table the leaching of salts from surface soil with good quality water can proceed, resulting in the removal of salts from reclaimed soil (Yan and Marschner, 2013). Apart from reestablishing hydrological function to the landscape and ameliorating the causes of salinity, the restoration of trees in saline landscapes can provide ecosystem services from land otherwise not utilized. There are good examples of tree restoration in saline landscapes in India (e.g., with *Prosopis* sp.); Australia (e.g., with *Eucalyptus occidentalis* Endl., *E. grandis*, and *Melaleuca halmaturorum* F. Muel. ex. Miq.); and China (*Populus euphratica* Oliv.) (Abassi et al., 2014; Dagar and Minhas, 2016). In all these examples land that is not suited to traditional agriculture provides ecosystem services such as biomass energy, carbon sequestration, and biodiversity benefits (Polle and Chen, 2015). On extremely saline sites, where ECe exceeds about 16 dS m^{-1} only highly salt-tolerant halophytes, which are of much lower productivity than trees, are suitable for restoration; typically grasses [e.g., *Thinopyrum ponticum*—*Elymus elongatus* (Host) Runemark] and succulents (e.g., *Atriplex* sp.) are planted.

11.12 Vertisols

A particular challenge for restoration is degraded Vertisols (black cotton soils). Vertisols develop in tropical and subtropical, semiarid to subhumid, and even humid climates with pronounced wet and dry seasons, mainly in lower landscape positions (FAO, 2015). The soils swell and shrink according to moisture levels due to a high content of smectite clay minerals. When dry, the soils form deep, wide (>1 cm) cracks from the surface downward, and crumb topsoil may slide downward. When wet, aggregates swell and subsurface soil is pushed upward. A vertic subsoil with shear planes and wedge-shaped structures develops. Over time, pedoturbation may lead to the concentration of coarser fragments at the surface—a self-mulching effect, and a pronounced microrelief with mounds and depressions (Deckers et al., 2001; Moeyersons et al., 2006). Drainage of Vertisols is usually poor. In that way, Vertisols show both features described previously: inadequate water during the dry season and excess water during the wet season. Natural vegetation would in many instances be savanna, but frequently the systems are used for intensive cattle grazing or converted to croplands. Although Vertisols are mostly located in depressions, along rivers or on only slightly sloping terrain, erosion, in particular gully formation, can be a severe problem. In such situations check dams can be constructed in order to keep the groundwater table high and control erosion (Deckers et al., 2001). The use of coarse fragments from the topsoil for the construction of stone bunds may, however, reduce the self-mulching effect and increase erosion and soil evaporation (Nyssen et al., 2001).

Trees face particular problems in Vertisols: during crack formation roots may be mechanically damaged or be exposed to desiccation, whereas during the wet season waterlogging leads to anaerobic conditions (Schmidt et al., 2008). As Vertisols are usually nutrient-rich, competition from ground vegetation may also hamper tree establishment (Stanturf et al., 2004). Trees adapted to these soil conditions show specific functional traits (Schmidt et al., 2008; Werden et al., 2018), including strong roots with high tolerance to mechanical stress, pronounced phenological seasonality (deciduousness), high photosynthetic rates and water use efficiency, and eventually aerial tissue in the wood (an adaptation to waterlogging that can increase seedling survival). Vertisols may be depleted in nutrients after prolonged continuous cultivation; in this situation N-fixing species may improve their fertility (Mekonnen et al., 2006).

Soil management methods developed for agriculture are also useful for afforestation measures. Mechanical site preparation, for example, the so-called broadbed and furrow technique—a tillage system where c. 80-cm-wide ridges alternate with 20-cm-wide and 15- to 20-cm-deep furrows have been successfully applied in afforestation trials in Ethiopia (Mekonnen et al., 2006). In reforesting bottomland hardwoods in Louisiana, bedding (molding soil in parallel ridges) effects on the performance of planted or direct seeded trees were species specific. Where labor is not a limiting factor, raised beds can be prepared manually and circular basins around planting spots may be used for water harvesting (Mekonnen et al., 2008). Planting spots should be prepared at the end of the rainy season, when soils are neither too wet nor too hard. Direct sowing or, in particular in sites with prolific competing vegetation, planting of seedlings, must be done immediately after the first rain (Schmidt et al., 2008). Soil amendments such as cow dung or manure (Mekonnen et al., 2008), sand or hydrogel (Werden et al., 2018), or biochar may increase seedling survival. Mulching (eventually also with stones) will reduce soil and root desiccation and may help to control competing vegetation.

11.13 Impoverished soil biota

In highly degraded or disturbed soils the abundance and diversity of soil microorganisms and fauna may be inadequate to support soil processes. Plant establishment can also benefit from the presence of appropriate soil organisms, particularly symbiotic mycorrhizal fungi.

11.13.1 Mycorrhizal fungi

Most tree species form associations with mycorrhizal fungi, which enhance their survival and growth on difficult sites by assisting with uptake of nutrients and water, by providing protection against drought, high temperatures, pathogens, heavy metals, alkalinity, and salts; and by increasing soil aggregation (Maherali and Klironomos, 2007; Treseder, 2013; Hawkins et al., 2015). Inoculation with mycorrhizal fungi is effective in restoring forests on degraded soils, as plant responses to mycorrhizal inoculation are often greatest under extreme conditions such as infertile soils and drought (Chen et al., 2014). In a metaanalysis across several degraded ecosystem types (Maltz and Treseder, 2015), inoculation consistently increased the abundance of mycorrhizal fungi and improved the establishment of plants. In a metaanalysis of experiments on degraded dryland restoration, inoculation with mycorrhizal fungi in the nursery was the most effective treatment for enhancing both the survival and growth of the planted seedlings (Piñeiro et al., 2013).

Arbuscular mycorrhizal fungal associates are particularly important on P-deficient soils, as they release phosphatase enzymes, which solubilize inorganic phosphate in the root zone. On severely degraded lands, the abundance and diversity of AMF is low, and regenerating trees and shrubs are likely to benefit from inoculation with AMF (Asmelash et al., 2016). Plants that form ectomycorrhizae (ECM) generally dominate forests with surface organic (humus) layers in which N occurs primarily in organic forms and tree growth is primarily limited by low availability of inorganic forms of N (ammonium and nitrate) or colimited by N and P (Lambers et al., 2008; Hawkins et al., 2015). Ectomycorrhizae are particularly important in N uptake but also contribute to P nutrition (Jones et al., 1998; Hawkins et al., 2015).

During reforestation, abundance and diversity of ECM can be promoted by avoiding severe practices such as clear-cutting followed by burning or heavy site preparation, which remove the forest floor and thereby reduce ECM inoculum potential (Hawkins et al., 2015). During afforestation and on severely degraded soils, inoculation with AMF or ECM may be necessary. Sources, techniques, and effectiveness of both AMF and ECM inoculants have been thoroughly reviewed by Chen et al. (2014). Inoculations with multiple species usually result in greater plant response (Caravaca et al., 2002a; Hoeksema et al., 2010). Inoculation with native forest soil (if available) is effective at increasing mycorrhizal infection and plant establishment and growth (Moynahan et al., 2002; Saxerud and Funke, 1991). Comparative studies have shown native soil inocula to be more beneficial than commercial inocula for increasing plant biomass and nutrient uptake and mycorrhizal colonization on degraded sites (Asmelash et al., 2016; Maltz and Treseder, 2015; Rowe et al., 2007; Urgiles et al., 2014; Wubs et al., 2016). However, if the afforestation program includes plantations of nonnative species, it will be necessary to inoculate with fungal species known to associate with that tree species (Chen et al., 2014), which may not be found locally (and may present an invasion risk).

Although it is presumed that a diverse mycorrhizal fungal community is most beneficial for plant growth, there is not a consistent positive relationship between the diversity of the AMF community and the aboveground growth of the plant (Holste et al., 2016). The identity of the fungal partner seems to be more important than the diversity of fungi, as particular fungus—plant associations appear to have a greater impact on the growth of specific plant species than others

(Klironomos, 2003). Ectomycorrhizal fungal species have significant host-plant specificity, which must be taken into account in the reforestation of deforested areas (Chen et al., 2014; Ding et al., 2011).

An important point when considering the effectiveness of mycorrhizal fungi in forest restoration is that assessments of effectiveness of a given tree—fungus association should be based on measurements of survival and whole-plant growth. Abundance and diversity of mycorrhizal fungi may be negatively related to aboveground plant biomass, because root and fungal growth rely on carbohydrates imported from the shoot. While aboveground growth is obviously necessary, the roots are of fundamental importance for stabilizing severely degraded and erosive soils. Roots and associated mycorrhizal fungi are also critical to building SOM and soil structure, which increase water-holding capacity and infiltration. Therefore any assessment of the benefit of mycorrhizae should include measurements of above- and belowground plant biomass, as well as SOM content and soil structure.

Inoculation with other soil microorganisms such as nonmycorrhizal fungi and bacteria is comparatively rare. It is generally presumed that these organisms generally make their way to a site and will colonize once the soil is sufficiently developed. This of course depends on the availability of propagules in the surrounding area, which may be insufficient in landscapes in which forests are rare. Inoculation with soil from an established or target forest could assist in this regard, and organic amendments would make the soil more hospitable for introduced soil organisms. Inoculation of soil with microorganisms from different habitats has been found to influence the plant community that develops on the site (Wubs et al., 2016), stressing the importance of inoculating with soil from the desired forest type. It must be kept in mind that soil and organic amendments may also introduce potential pathogens (Chen et al., 2014).

11.13.2 Soil fauna

Populations of soil fauna can also be low in degraded soils and their colonization can be impeded if the degraded soil is not adjacent to less degraded soils. On degraded postmining sites the abundance of soil fauna is closely linked to the development of the habitat, including vegetation and litter layer (Dunger et al., 2001; Moradi et al., 2018). For example, in the Alberta's oil sands region, oribatid mite abundance on reclaimed sites reached levels found in nearby natural forests once the forest floor was 2 cm thick (McAdams et al., 2018). Colonization by soil fauna also depends on the distance from suitable refugia for each species (Frouz et al., 2014) and their mode of transit (winged or surface movers). Like their aboveground counterparts, landscape connectivity may be important for migration of soil fauna to restored sites. Some fauna, especially earthworms, have been deliberately introduced into degraded soils (Baker et al., 2006; Butt, 2008), and litter-degrading fauna can be promoted by planting tree species that promote their abundance (Frouz et al., 2013; Schelfhout et al., 2017).

References

Abassi, M., Mguis, K., Bejaoui, Z., Albouchi, A., 2014. Morphogenetic responses of *Populus alba* L. under salt stress. J. For. Res. 25, 155–161. Available from: https://doi.org/10.1007/s11676-014-0441-6.

Abdo, M., 2014. Practices, techniques and technologies for restoring degraded landscapes in the Sahel. African Forest Forum, Working Paper Series vol. (2) (3), 42 pp. African Forest Forum, Nairobi, Kenya.

Abiven, S., Menassari-Aubry, S., Chenu, C., 2009. The effects of organic inputs over time on soil aggregate stability – a literature analysis. Soil Biol. Biochem. 41, 1–12. Available from: https://doi.org/10.1016/j.soilbio.2008.09.015.

Alguacil, M.M., Torres, M.P., Torrecillas, E., Diaz, G., Roldan, A., 2011. Plant type differently promote the arbuscular mycorrhizal fungi biodiversity in the rhizosphere after revegetation of a degraded, semiarid land. Soil Biol. Biochem. 43, 167–173. Available from: https://doi.org/10.1016/j.soilbio.2010.09.029.

Annabi, M., Houot, S., Francou, F., Poitrenaud, M., Le Bissonnais, Y., 2007. Soil aggregate stability improvement with urban composts of different maturities. Soil Sci. Soc. Am. J. 71, 413–423. Available from: https://doi.org/10.2136/sssaj2006.0161.

Arnáez, J., Lana-Renault, N., Lasanta, T., Ruiz-Flaño, P., Castroviejo, J., 2015. Effects of farming terraces on hydrological and geomorphological processes. A review. Catena 128, 122–134. Available from: https://doi.org/10.1016/j.catena.2015.01.021.

Asmelash, F., Bekele, T., Birhane, E., 2016. The potential role of arbuscular mycorrhizal fungi in the restoration of degraded lands. Front. Microbiol. 7, 1095. Available from: https://doi.org/10.3389/fmicb.2016.01095.

Augusto, L., De Schrijver, A., Vesterdal, L., Smolander, A., Prescott, C.E., Ranger, J., 2014. Influences of evergreen gymnosperm and deciduous angiosperm tree species on the functioning of temperate and boreal forests. Biol. Rev. 90, 444–466. Available from: https://doi.org/10.1111/brv.12119.

Baker, G.H., Brown, G., Butt, K., Curry, J.P., Scullion, J., 2006. Introduced earthworms in agricultural and reclaimed land: their ecology and influences on soil properties, plant production and other soil biota. Biol. Invasions 8, 1301–1316. Available from: https://doi.org/10.1007/s10530-006-9024-6.

Barto, E.K., Alt, F., Oelmann, Y., Wilcke, W., Rillig, M.C., 2010. Contributions of biotic and abiotic factors to soil aggregation across a land use gradient. Soil Biol. Biochem. 42, 2316–2324. Available from: https://doi.org/10.1016/j.soilbio.2010.09.008.

Bassett, I.E., Simcock, R.C., Mitchell, N.D., 2005. Consequences of soil compaction for seedling establishment: implications for natural regeneration and restoration. Austral Ecol. 30, 827–833. Available from: https://doi.org/10.1111/j.1442-9993.2005.01525.x.

Bastida, F., Moreno, J.L., Garcia, C., Hernandez, T., 2007. Addition of urban waste to semiarid degraded soil: long-term effect. Pedosphere 17, 557–567. Available from: https://doi.org/10.1016/S1002-0160(07)60066-6.

Bastida, F., Selevsek, N., Torres, I.F., Hernandez, T., Garcia, C., 2015. Soil restoration with organic amendments: linking cellular functionality and ecosystem processes. Sci. Rep. 5, 15550. Available from: https://doi.org/10.1038/srep15550.

Batey, T., 2009. Soil compaction and soil management – a review. Soil Use Manage. 25, 335–345. Available from: https://doi.org/10.1111/j.1475-2743.2009.00236.x.

Bauman, J.M., Keiffer, C.H., Hiremath, S., McCarthy, B.C., Kardol, P., 2013. Soil preparation methods promoting ectomycorrhizal colonization and American chestnut *Castanea dentata* establishment in coal mine restoration. J. Appl. Ecol. 50, 721–729. Available from: https://doi.org/10.1111/1365-2664.12070.

Bell, D.T., 1999. Australian trees for the rehabilitation of waterlogged and salinity-damaged landscapes. Aust. J. Bot. 47, 697–716. Available from: https://doi.org/10.1071/BT96110.

Belnap, J., Gillette, D.A., 1997. Disturbance of biological soil crusts: impacts on potential wind erodibility of sandy desert soils in southeastern Utah. Land Degrad. Dev. 8, 355–362. Available from: https://doi.org/10.1002/(SICI)1099-145X(199712)8:4 < 355::AID-LDR266 > 3.0.CO;2-H.

Berthrong, S.T., Jobbagy, E.G., Jackson, R.B., 2009. A global meta-analysis of soil exchangeable cations, pH, carbon, and nitrogen with afforestation. Ecol. Appl. 19, 2228–2241. Available from: https://doi.org/10.1890/08-1730.1.

Biazin, B., Sterk, G., Temesgen, M., Abdulkedir, A., Stroosnijder, L., 2012. Rainwater harvesting and management in rainfed agricultural systems in sub-Saharan Africa – a review. Phys. Chem. Earth 139–151. Available from: https://doi.org/10.1016/j.pce.2011.08.015.

Bini, D., dos Santos, C.A., da Silva, M., Bonfim, J.A., Cardoso, E., 2018. Intercropping *Acacia mangium* stimulates AMF colonization and soil phosphatase activity in *Eucalyptus grandis*. Sci. Agricola 75, 102–110. Available from: https://doi.org/10.1590/1678-992X-2016-0337.

Binkley, D., Fisher, R.F., 2013. Ecology and Management of Forest Soils. Wiley-Blackwell, Hoboken, NJ.

Bird, P.R., Bicknell, D., Bulman, P.A., Burke, S.J.A., Leys, J.F., Parker, J.N., et al., 1992. The role of shelter in Australia for protecting soils, plants and livestock. Agroforestry Syst. 20, 59–86. Available from: https://doi.org/10.1007/bf00055305.

Bolan, N.S., Adriano, D.C., Curtin, D., 2003. Soil acidification and liming interactions with nutrient and heavy metal transformation and bioavailability. Advances in Agronomy. Academic Press, San Diego, CA, pp. 215–272.

Brandle, J.R., Hodges, L., Wight, B., 2000. Windbreak practices. In: Garrett, H.E., Rietveld, W.J., Fisher, R.F. (Eds.), North American Agroforestry: An Integrated Science and Practice. American Society of Agronomy, pp. 79–118.

Bronick, C.J., Lal, R., 2005. Soil structure and management: a review. Geoderma 124, 3–22. Available from: https://doi.org/10.1016/j.geoderma.2004.03.005.

Brunner, I., Sperisen, C., 2013. Aluminum exclusion and aluminum tolerance in woody plants. Front. Plant Sci. 4, 172. Available from: https://doi.org/10.3389/fpls.2013.00172.

Butt, K.R., 2008. Earthworms in soil restoration: lessons learned from United Kingdom case studies of land reclamation. Restor. Ecol. 16, 637–641. Available from: https://doi.org/10.1111/j.1526-100X.2008.00483.x.

Cao, S., Chen, L., Shankman, D., Wang, C., Wang, X., Zhang, H., 2011. Excessive reliance on afforestation in China's arid and semi-arid regions: lessons in ecological restoration. Earth-Sci. Rev. Available from: https://doi.org/10.1016/j.earscirev.2010.11.002.

Caravaca, F., Barea, J.M., Figueroa, D., Roldan, A., 2002a. Assessing the effectiveness of mycorrhizal inoculation and soil compost addition for enhancing reafforestation with *Olea europaea* subsp. sylvestris through changes in soil biological and physical parameters. Appl. Soil Ecol. 20, 107−118. Available from: https://doi.org/10.1016/S0929-1393(02)00015-X.

Caravaca, F., Hernandez, T., Garcia, C., Roldan, A., 2002b. Improvement of rhizosphere aggregate stability of afforested semiarid plant species subjected to mycorrhizal inoculation and compost addition. Geoderma 108, 133−144. Available from: https://doi.org/10.1016/S0016-7061(02)00130-1.

Carbajo, V., den Braber, B., van der Putten, W.H., De Deyn, G.B., 2011. Enhancement of late successional plants on ex-arable land by soil inoculations. PLoS One 6. Available from: https://doi.org/10.1371/journal.pone.0021943.

Cardinael, R., Chevallier, T., Cambou, A., Beral, C., Barthes, B.G., Dupraz, C., et al., 2017. Increased soil organic carbon stocks under agroforestry: a survey of six different sites in France. Agric. Ecosyst. Environ. 236, 243−255. Available from: https://doi.org/10.1016/j.agee.2016.12.011.

Cerdà, A., Gonzalez-Pelayo, O., Gimelnez-Morera, A., Jordan, A., Pereira, P., Novara, A., et al., 2016. The use of barley straw residues to avoid high erosion and runoff rates on persimmon plantations in Eastern Spain under low frequency-high magnitude simulated rainfall events. Soil Res. 54, 154−165. Available from: https://doi.org/10.1071/SR15092.

Cerdà, A., Lavee, H., Romero-Díaz, A., Hooke, J., Montanarella, L., 2010. Soil erosion and degradation on Mediterranean type ecosystems. Preface to special issue. Land Degrad. Dev. 21, 71−74. Available from: https://doi.org/10.1002/ldr.968.

Ceulemans, T., Stevens, C.J., Duchateau, L., Jacquemyn, H., Gowing, D.J., Merckx, R., et al., 2014. Soil phosphorus constrains biodiversity across European grasslands. Global Change Biol. 20, 3814−3822. Available from: https://doi.org/10.1111/gcb.12650.

Chaer, G.M., Resende, A.S., Campello, E., de Faria, S.M., Boddey, R.M., 2011. Nitrogen-fixing legume tree species for the reclamation of severely degraded lands in Brazil. Tree Physiol. 31, 139−149. Available from: https://doi.org/10.1093/treephys/tpq116.

Chambers, B., Royle, S., Hadden, S., Maslen, S., 2002. The use of biosolids and other organic substances in the creation of soil-forming materials. Water Environ. J. 16, 34−39. Available from: https://doi.org/10.1111/j.1747-6593.2002.tb00365.x.

Chang, I., Prasidhi, A.K., Im, J., Shin, H.-D., Cho, G.-C., 2015. Soil treatment using microbial biopolymers for anti-desertification purposes. Geoderma 253−254, 39−47. Available from: https://doi.org/10.1016/j.geoderma.2015.04.006.

Chen, Y.L., Liu, R.J., Bi, Y.L., Feng, G., 2014. Use of mycorrhizal fungi for forest plantations and minesite rehabilitation. In: Solaiman, Z.M., Abbott, L.K., Varma, A. (Eds.), Mycorrhizal Fungi: Use in Sustainable Agriculture and Land Restoration. Soil Biology, vol. 41. . Available from: http://doi.org/10.1007/978-3-662-45370-4_21.

Chen, D., Wei, W., Chen, L., 2017. Effects of terracing practices on water erosion control in China: a meta-analysis. Earth-Sci. Rev. 173, 109−121. Available from: https://doi.org/10.1016/j.earscirev.2017.08.007.

Chirwa, P.W., Mahamane, L., 2017. Overview of restoration and management practices in the degraded landscapes of the Sahelian and dryland forests and woodlands of East and southern Africa. South. For. J. For. Sci. 79, 87−94. Available from: https://doi.org/10.2989/20702620.2016.1255419.

Clemmensen, K.E., Finlay, R.D., Dahlberg, A., Stenlid, J., Wardle, D.A., Lindahl, B.D., 2015. Carbon sequestration is related to mycorrhizal fungal community shifts during long-term succession in boreal forests. New Phytol 205, 1525−1536. Available from: https://doi:10.1111/nph.13208.

Coello, J., Ameztegui, A., Rovira, P., Fuentes, C., Pique, M., 2018. Innovative soil conditioners and mulches for forest restoration in semiarid conditions in northeast Spain. Ecol. Eng. 118, 52–65. Available from: https://doi.org/10.1016/j.ecoleng.2018.04.015.

Colak, A.H., Kirca, S., Rotherham, I.D., Ince, A., 2010. Restoration and Rehabilitation of Deforested and Degraded Forest Landscapes in Turkey. Springer, New York.

Court, M., van der Heijden, G., Didier, S., Nys, C., Richter, C., Pousse, N., et al., 2018. Long-term effects of forest liming on mineral soil, organic layer and foliage chemistry: insights from multiple beech experimental sites in northern France. For. Ecol. Manage. 409, 872–889. Available from: https://doi.org/10.1016/j.foreco.2017.12.007.

Craig, M.E., Turner, B.L., Liang, C., Clay, K., Johnson, D.J., Phillips, R.P., 2018. Tree mycorrhizal type predicts within-site variability in the storage and distribution of soil organic matter. Global Change Biol. 24, 3317–3330. Available from: https://doi.org/10.1111/gcb.14132.

Cramer, V.A., Hobbs, R.J., 2005. Assessing the ecological risk from secondary salinity: a framework addressing questions of scale and threshold responses. Austral Ecol. 30, 537–545. Available from: https://doi.org/10.1111/j.1442-9993.2005.01468.x.

Cronan, C.S., Grigal, D.F., 1995. Use of calcium aluminum ratios as indicators of stress in forest ecosystems. J. Environ. Qual. 24, 209–226. Available from: https://doi.org/10.2134/jeq1995.00472425002400020002x.

Dagar, J.C., Minhas, P., 2016. Agroforestry for the Management of Waterlogged Saline Soils and Poor-Quality Waters. Springer India, New Delhi. Available from: https://doi.org/10.1007/978-81-322-2659-8.

Daynes, C.N., Field, D.J., Saleeba, J.A., Cole, M.A., McGee, P.A., 2013. Development and stabilisation of soil structure via interactions between organic matter, arbuscular mycorrhizal fungi and plant roots. Soil Biol. Biochem. 57, 683–694. Available from: https://doi.org/10.1016/j.soilbio.2012.09.020.

Deckers, J., Spaargaren, O., Nachtergaele, F., 2001. Vertisols: genesis, properties and soilscape. The Sustainable Management of Vertisols. IWMI/CABI Publishing, Wallingford, pp. 3–20.

Ding, Q., Liang, Y., Legendre, P., He, X., Pei, K., Du, X., et al., 2011. Diversity and composition of ectomycorrhizal community on seedling roots: the role of host preference and soil origin. Mycorrhiza 21, 669–680. Available from: https://doi.org/10.1007/s00572-011-0374-2.

Dunger, W., Wanner, M., Hauser, H., Hohberg, K., Schulz, H.J., Schwalbe, T., et al., 2001. Development of soil fauna at mine sites during 46 years after afforestation. Pedobiologia 45, 243–271. Available from: https://doi.org/10.1078/0031-4056-00083.

FAO, 2005. Global Network on Integrated Soil Management for Sustainable Use of Salt-Affected Soils. FAO Land and Plant Nutrition Management Service, Rome.

FAO, 2008. Global Network on Integrated Soil Management for Sustainable Use of Salt Affected Soils. Food and Agriculture Organization of the United Nations, Rome.

FAO, 2015. World Reference Base for Soil Resources 2014, update 2015 International Soil Classification System for Naming Soils and Creating Legends for Soil Maps. FAO, Rome. Available from: http://www.fao.org/3/a-i3794en.pdf.

Fisher, R.F., 1995. Amelioration of degraded rain-forest soils by plantations of native trees. Soil Sci. Soc. Am. J. 59, 544–549.

Forrester, D.I., Pares, A., O'Hara, C., Khanna, P.K., Bauhus, J., 2013. Soil organic carbon is increased in mixed-species plantations of *Eucalyptus* and nitrogen-fixing *Acacia*. Ecosystems 16, 123–132. Available from: https://doi.org/10.1007/s10021-012-9600-9.

Fountain, W.M., 2011. Trees and Compacted Soils. Cooperative Extension Service. University of Kentucky College of Agriculture. Available from: http://www2.ca.uky.edu/agcomm/pubs/ho/ho93/ho93.pdf.

Franco, A.A., De Faria, S.M., 1997. The contribution of N_2-fixing tree legumes to land reclamation and sustainability in the tropics. Soil Biol. Biochem. 29, 897−903. Available from: https://doi.org/10.1016/S0038-0717(96)00229-5.

Frouz, J., 2018. Effects of soil macro- and mesofauna on litter decomposition and soil organic matter stabilization. Geoderma 332, 161−172. Available from: https://doi.org/10.1016/j.geoderma.2017.08.039.

Frouz, J., Kuráž, V., 2014. Soil fauna and soil physical properties. In: Frouz, J. (Ed.), Soil Biota and Ecosystem Development in Post Mining Sites. CRC Press, Boca Raton, FL, London, pp. 265−278.

Frouz, J., Pizl, V., Cienciala, E., Kalcik, J., 2009. Carbon storage in post-mining forest soil, the role of tree biomass and soil bioturbation. Biogeochemistry 94, 111−121. Available from: https://doi.org/10.1007/s10533-009-9313-0.

Frouz, J., Livečková, M., Albrechtova, J., Chronakova, A., Cajthaml, T., Pižl, V., et al., 2013. Is the effect of trees on soil properties mediated by soil fauna? A case study from post-mining sites. For. Ecol. Manage. 309, 87−95. Available from: https://doi.org/10.1016/j.foreco.2013.02.013.

Frouz, J., Pižl, V., Tajovský, K., Starý, J., Holec, M., Materna, J., 2014. Soil macro- and mesofauna succession in post-mining sites and other disturbed areas. In: Frouz, J. (Ed.), Soil Biota and Ecosystem Development in Post Mining Sites. CRC Press, Boca Raton, FL, London, pp. 216−235.

Fungo, B., Lehmann, J., Kalbitz, K., Tenywa, M., Thiongo, M., Neufeldt, H., 2017. Emissions intensity and carbon stocks of a tropical Ultisol after amendment with Tithonia green manure, urea and biochar. Field Crops Res. 209, 179−188. Available from: https://doi.org/10.1016/j.fcr.2017.05.013.

Gabou, M.H., Maisharou, A., 2015. Management practices/techniques commonly used in Niger Republic, West Africa. In: Lal, R., Singh, B.R., Mwaseba, D.L. (Eds.), Sustainable Intensification to Advance Food Security and Enhance Climate Resilience in Africa. Springer International Publishing, Switzerland, pp. 305−314.

Garcia-Orenes, F., Guerrero, C., Mataix-Solera, J., Navarro-Pedreno, J., Gomez, I., Mataix-Beneyto, J., 2005. Factors controlling the aggregate stability and bulk density in two different degraded soils amended with biosolids. Soil Tillage Res. 82, 65−76. Available from: https://doi.org/10.1016/j.still.2004.06.004.

Gebremichael, D., Nyssen, J., Poesen, J., Deckers, J., Haile, M., Govers, G., et al., 2005. Effectiveness of stone bunds in controlling soil erosion on cropland in the Tigray Highlands, northern Ethiopia. Soil Use Manage. 21, 287−297. Available from: https://doi.org/10.1079/SUM2005321.

Glatzel, G., Haselwandter, K., Katzensteiner, K., Sterba, H., Weißbacher, J., 1990. The use of organic and mineral fertilizers in reforestation and in revitalization of declining protective forests in the Alps. Water Air Soil Pollut. 54, 567−576.

Godbold, D.L., Hoosbeek, M.R., Lukac, M., Cotrufo, M.F., Janssens, I.A., Ceulemans, R., et al., 2006. Mycorrhizal hyphal turnover as a dominant process for carbon input into soil organic matter. Plant Soil 281, 15−24. Available from: https://doi.org/10.1007/s11104-005-3701-6.

Halmschlager, E., Katzensteiner, K., 2017. Vitality fertilization balanced tree nutrition and mitigated severity of *Sirococcus* shoot blight on mature Norway spruce. Forest Ecol. Manage. 389, 96–104. Available from: https://doi.org/10.1016/j.foreco.2016.12.019.

Haregeweyn, N., Tsunekawa, A., Nyssen, J., Poesen, J., Tsubo, M., Meshesha, D.T., et al., 2015. Soil erosion and conservation in Ethiopia. Prog. Phys. Geogr. 39, 750–774. Available from: https://doi.org/10.1177/0309133315598725.

Harvey, C.A., 2000. Colonization of agricultural windbreaks by forest trees: effects of connectivity and remnant trees. Ecol. Appl. 10, 1762–1773. Available from: https://doi.org/10.1890/1051-0761(2000)010[1762:COAWBF]2.0.CO;2.

Hawkins, B.J., Jones, M.D., Kranabetter, J.M., 2015. Ectomycorrhizae and tree seedling nitrogen nutrition in forest restoration. New For. 46, 747–771. Available from: https://doi.org/10.1007/s11056-015-9488-2.

Hernandez, T., Garcia, E., Garcia, C., 2015. A strategy for marginal semiarid degraded soil restoration: a sole addition of compost at a high rate. A five-year field experiment. Soil Biol. Biochem. 89, 61–71. Available from: https://doi.org/10.1016/j.soilbio.2015.06.023.

Hoeksema, J.D., Chaudhary, V.B., Gehring, C.A., Johnson, N.C., Karst, J., Koide, R.T., et al., 2010. A meta-analysis of context-dependency in plant response to inoculation with mycorrhizal fungi. Ecol. Lett. 13, 394–407. Available from: https://doi.org/10.1111/j.1461-0248.2009.01430.x.

Holste, E.K., Holl, K.D., Zahawi, R.A., Kobe, R.K., 2016. Reduced aboveground tree growth associated with higher arbuscular mycorrhizal fungal diversity in tropical forest restoration. Ecol. Evol. 6, 7253–7262. Available from: https://doi.org/10.1002/ece3.2487.

Hueso-Gonzalez, P., Martinez-Murillo, J.F., Ruiz-Sinoga, J.D., 2016. Effects of topsoil treatments on afforestation in a dry Mediterranean climate (southern Spain). Solid Earth 7, 1479–1489. Available from: https://doi.org/10.5194/se-7-1479-2016.

Huettl, R.F., Zoettl, H.W., 1993. Liming as a mitigation tool in Germany declining forests – reviewing results from former and recent trials. For. Ecol. Manage. 61, 325–338.

Huttermann, A., Orikiriza, L., Agaba, H., 2009. Application of superabsorbent polymers for improving the ecological chemistry of degraded or polluted lands. Clean—Soil Air Water 37, 517–526. Available from: https://doi.org/10.1002/clen.200900048.

Hüttl, R.F., 1998. Ecology of post strip-mining landscapes in Lusatia, Germany. Environ. Sci. Policy 1, 129–135. Available from: https://doi.org/10.1016/S1462-9011(98)00014-8.

Jastrow, J.D., Miller, R.M., Lussenhop, J., 1998. Contributions of interacting biological mechanisms to soil aggregate stabilization in restored prairie. Soil Biol. Biochem. 30, 905–916. Available from: https://doi.org/10.1016/S0038-0717(97)00207-1.

Jones, M.D., Durall, D.M., Tinker, P.B., 1998. A comparison of arbuscular and ectomycorrhizal *Eucalyptus coccifera*: growth response, phosphorus uptake efficiency and external hyphal production. New Phytol. 140, 125–134. Available from: https://doi.org/10.1046/j.1469-8137.1998.00253.x.

Kader, M.A., Senge, M., Mojid, M.A., Ito, K., 2017. Recent advances in mulching materials and methods for modifying soil environment. Soil Tillage Res. 168, 155–166. Available from: https://doi.org/10.1016/j.still.2017.01.001.

Kallenbach, C.M., Frey, S.D., Grandy, A.S., 2016. Direct evidence for microbial-derived soil organic matter formation and its ecophysiological controls. Nat. Commun. 7, 13630. Available from: https://doi.org/10.1038/ncomms13630.

Kang, J., Zhao, M., Tan, Y., Zhu, L., Bing, D., Zhang, Y., et al., 2017. Sand-fixing characteristics of *Carex brunnescens* and its application with straw checkerboard technique in restoration of degraded alpine meadows. J. Arid Land 9, 651–665. Available from: https://doi.org/10.1007/s40333-017-0066-7.

Katzensteiner, K., Eckmuellner, O., Jandl, R., Glatzel, G., Sterba, H., Wessely, A., et al., 1995. Revitalization experiments in magnesium deficient Norway spruce stands in Austria. Plant Soil 489–500. Available from: https://doi.org/10.1007/BF00029361.

Klironomos, J.N., 2003. Variation in plant response to native and exotic arbuscular mycorrhizal fungi. Ecology 84, 2292–2301. Available from: https://doi.org/10.1890/02-0413.

Kohler, J., Roldan, A., Campoy, M., Caravaca, F., 2017. Unraveling the role of hyphal networks from arbuscular mycorrhizal fungi in aggregate stabilization of semiarid soils with different textures and carbonate contents. Plant Soil 410, 273–281. Available from: https://doi.org/10.1007/s11104-016-3001-3.

Koulouri, M., Giourga, C., 2007. Land abandonment and slope gradient as key factors of soil erosion in Mediterranean terraced lands. Catena 69, 274–281. Available from: https://doi.org/10.1016/j.catena.2006.07.001.

Koutika, L.-S., Epron, D., Bouillet, J.-P., Mareschal, L., 2014. Changes in N and C concentrations, soil acidity and P availability in tropical mixed *Acacia* and Eucalypt plantations on a nutrient-poor sandy soil. Plant Soil 379, 205–216. Available from: https://doi.org/10.1007/s11104-014-2047-3.

Kuznetsova, T., Lukjanova, A., Mandre, M., Lohmus, K., 2011. Aboveground biomass and nutrient accumulation dynamics in young black alder, silver birch and Scots pine plantations on reclaimed oil shale mining areas in Estonia. For. Ecol. Manage. 262, 56–64. Available from: https://doi.org/10.1016/j.foreco.2010.09.030.

Lal, R., 1997. Degradation and resilience of soils. Philos. Trans. R. Soc. London Ser. B Biol. Sci. 352, 997–1010. Available from: https://doi.org/10.1098/rstb.1997.0078.

Lal, R., 2001. Soil degradation by erosion. Land Degrad. Dev. 12, 519–539. Available from: https://doi.org/10.1002/ldr.472.

Lambers, H., Raven, J.A., Shaver, G.R., Smith, S.E., 2008. Plant nutrient-acquisition strategies change with soil age. Trends Ecol. Evol. 23, 95–103. Available from: https://doi.org/10.1016/j.tree.2007.10.008.

Larney, F.J., Angers, D.A., 2012. The role of organic amendments in soil reclamation: a review. Can. J. Soil Sci. 92, 19–38. Available from: https://doi.org/10.4141/CJSS2010-064.

Lavelle, P., 1988. Earthworm activities and the soil system. Biol. Fertil. Soils 6. Available from: https://doi.org/10.1007/BF00260820.

Levy, G.J., Shainberg, I., 2005. Sodic Soils. Encycl. Soils Environ. . Available from: https://doi.org/10.1016/B0-12-348530-4/00218-6.

Li, D., Niu, S., Luo, Y., 2012. Global patterns of the dynamics of soil carbon and nitrogen stocks following afforestation: a meta-analysis. New Phytol. 195, 172–181. Available from: https://doi.org/10.1111/j.1469-8137.2012.04150.x.

Liang, C., Schimel, J.P., Jastrow, J.D., 2017. The importance of anabolism in microbial control over soil carbon storage. Nat. Microbiol. 2, 17105. Available from: https://doi.org/10.1038/nmicrobiol.2017.105.

Liu, X., Zhang, S., Zhang, X., Ding, G., Cruse, R.M., 2011. Soil erosion control practices in Northeast China: a mini-review. Soil Tillage Res. 117, 44–48. Available from: https://doi.org/10.1016/j.still.2011.08.005.

Löf, M., Dey, D.C., Navarro, R.M., Jacobs, D.F., 2012. Mechanical site preparation for forest restoration. New For. 43, 825–848. Available from: https://doi.org/10.1007/s11056-012-9332-x.

Luna, L., Vignozzi, N., Miralles, I., Sole-Benet, A., 2018. Organic amendments and mulches modify soil porosity and infiltration in semiarid mine soils. Land Degrad. Dev. 29, 1019–1030. Available from: https://doi.org/10.1002/ldr.2830.

Ma, Q., Fehmi, J.S., Zhang, D., Fan, B., Chen, F., 2017. Changes in wind erosion over a 25-year restoration chronosequence on the south edge of the Tengger Desert, China: implications for preventing desertification. Environ. Monit. Assess. 189, 1–14. Available from: https://doi.org/10.1007/s10661-017-6183-0.

Macedo, M.O., Resende, A.S., Garcia, P.C., Boddey, R.M., Jantalia, C.P., Urquiaga, S., et al., 2008. Changes in soil C and N stocks and nutrient dynamics 13 years after recovery of degraded land using leguminous nitrogen-fixing trees. For. Ecol. Manage. 255, 1516–1524. Available from: https://doi.org/10.1016/j.foreco.2007.11.007.

Madsen, P., Jensen, F.A., Fodgaard, S., 2005. Afforestation in Denmark. In: Stanturf, J., Madsen, P. (Eds.), Restoration of Boreal and Temperate Forests. CRC Press, Boca Raton, FL.

Maherali, H., Klironomos, J.N., 2007. Influence of phylogeny on fungal community assembly and ecosystem functioning. Science 316, 1746–1748. Available from: https://doi.org/10.1126/science.1143082.

Maltz, M.R., Treseder, K.K., 2015. Sources of inocula influence mycorrhizal colonization of plants in restoration projects: a meta-analysis. Restor. Ecol. 23, 625–634. Available from: https://doi.org/10.1111/rec.12231.

Marashi, A.R.A., Scullion, J., 2003. Earthworm casts form stable aggregates in physically degraded soils. Biol. Fertil. Soils 37, 375–380. Available from: https://doi.org/10.1007/s00374-003-0617-2.

McAdams, B.N., Quideau, S.A., Swallow, M.J.B., Lumley, L.M., 2018. Oribatid mite recovery along a chronosequence of afforested boreal sites following oil sands mining. For. Ecol. Manage. 422, 281–293. Available from: https://doi.org/10.1016/j.foreco.2018.04.034.

Mekonnen, K., Yohannes, T., Glatzel, G., Amha, Y., 2006. Performance of eight tree species in the highland Vertisols of central Ethiopia: growth, foliage nutrient concentration and effect on soil chemical properties. New For. 32, 285–298. Available from: https://doi.org/10.1007/s11056-006-9003-x.

Mekonnen, K., Glatzel, G., Kidane, B., Alebachew, M., Bekele, K., Tsegaye, M., 2008. Processes, lessons and challenges from participatory tree species selection, planting and management research in the highland Vertisol areas of central Ethiopia. For. Trees Livelihoods 18, 151–164. Available from: https://doi.org/10.1080/14728028.2008.9752626.

Meyer, C., Luscher, P., Schulin, R., 2014. Recovery of forest soil from compaction in skid tracks planted with black alder *Alnus glutinosa* (L.) Gaertn. Soil Tillage Res. 143, 7–16. Available from: https://doi.org/10.1016/j.still.2014.05.006.

Moeyersons, J., Nyssen, J., Poesen, J., Deckers, J., Haile, M., 2006. On the origin of rock fragment mulches on Vertisols: a case study from the Ethiopian highlands. Geomorphology 76, 411–429. Available from: https://doi.org/10.1016/j.geomorph.2005.12.005.

Moradi, J., Vicentini, F., Simackova, H., Pizl, V., Tajovsky, K., Stary, J., et al., 2018. An investigation into the long-term effect of soil transplant in bare spoil heaps on survival and migration of soil meso and macrofauna. Ecol. Eng. 110, 158–164. Available from: https://doi.org/10.1016/j.ecoleng.2017.11.012.

Morrien, E., Prescott, C.E., 2018. Pellets or particles? How can we predict the effect of soil macro-arthropods on litter decomposition? Funct. Ecol. 32, 2480−2482. Available from: https://doi.org/10.1111/1365-2435.13217.

Moynahan, O.S., Zabinski, C.A., Gannon, J.E., 2002. Microbial community structure and carbon-utilization diversity in a mine tailings revegetation study. Restor. Ecol. 10, 77−87. Available from: https://doi.org/10.1046/j.1526-100X.2002.10108.x.

Munns, R., 2005. Genes and salt tolerance: bringing them together. New Phytol. 167, 645−663. Available from: https://doi.org/10.1111/j.1469-8137.2005.01487.x.

National Research Council (U.S.A), 1984a. *Casuarinas*, Nitrogen-Fixing Trees for Adverse Sites: Report of an Ad Hoc Panel of the Advisory Committee on Technology Innovation, Board on Science and Technology for International Development, Office of International Affairs, National Research Council. National Academy Press, Washington, DC.

National Research Council, U.S.A., 1984b. *Casuarinas*: Nitrogen-Fixing Trees for Adverse Sites. The National Academies Press, Washington, DC.

Nave, L.E., Vance, E.D., Swanston, C.W., Curtis, P.S., 2009. Impacts of elevated N inputs on north temperate forest soil C storage, C/N, and net N-mineralization. Geoderma 153, 231−240. Available from: https://doi.org/10.1016/j.geoderma.2009.08.012.

Norisada, M., Hitsuma, G., Kuroda, K., Yamanoshita, T., Masumori, M., Tange, T., et al., 2005. *Acacia mangium*, a nurse tree candidate for reforestation on degraded sandy soils in the Malay Peninsula. For. Sci. 51, 498−510.

Nyssen, J., Haile, M., Poesen, J., Deckers, J., Moeyersons, J., 2001. Removal of rock fragments and its effect on soil loss and crop yield, Tigray, Ethiopia. Soil Use Manage. 17, 179−187. Available from: https://doi.org/10.1079/SUM200173.

Oelmann, Y., Broll, G., Holzel, N., Kleinebecker, T., Vogel, A., Schwartze, P., 2009. Nutrient impoverishment and limitation of productivity after 20 years of conservation management in wet grasslands of north-western Germany. Biol. Conserv. 142, 2941−2948. Available from: https://doi.org/10.1016/j.biocon.2009.07.021.

Pankhurst, C.E., Yu, S., Hawke, B.G., Harch, B.D., 2001. Capacity of fatty acid profiles and substrate utilization patterns to describe differences in soil microbial communities associated with increased salinity or alkalinity at three locations in South Australia. Biol. Fertil. Soils 33, 204−217. Available from: https://doi.org/10.1007/s003740000309.

Piñeiro, J., Maestre, F.T., Bartolomé, L., Valdecantos, A., 2013. Ecotechnology as a tool for restoring degraded drylands: a meta-analysis of field experiments. Ecol. Eng. 61, 133−144. Available from: https://doi.org/10.1016/j.ecoleng.2013.09.066.

Polle, A., Chen, S., 2015. On the salty side of life: molecular, physiological and anatomical adaptation and acclimation of trees to extreme habitats. Plant Cell Environ. 38, 1794−1816. Available from: https://doi.org/10.1111/pce.12440.

Prescott, C.E., 2010. Litter decomposition: what controls it and how can we alter it to sequester more carbon in forest soils? Biogeochemistry 101, 133−149. Available from: https://doi.org/10.1007/s10533-010-9439-0.

Pritchett, W., Comerford, N., 1982. Long-term response to phosphorus fertilization on selected southeastern coastal plain soils. Soil Sci. Soc. Am. J. 46, 640−644.

Prosdocimi, M., Tarolli, P., Cerda, A., 2016. Mulching practices for reducing soil water erosion: a review. Earth-Sci. Rev. . Available from: https://doi.org/10.1016/j.earscirev.2016.08.006.

Qadir, M., Oster, J.D., Schubert, S., Noble, A.D., Sahrawat, K.L., 2007. Phytoremediation of Sodic and Saline-Sodic Soils. Elsevier Science & Technology, San Diego, CA. Available from: http://dx.doi.org/10.1016/S0065-2113(07)96006-X.

Querejeta, J.I., Roldan, A., Albaladejo, J., Castillo, V., 2000. Soil physical properties and moisture content affected by site preparation in the afforestation of a semiarid rangeland. Soil Sci. Soc. Am. J. 64, 2087−2096. Available from: https://doi.org/10.2136/sssaj2000.6462087x.

Quiñonero-Rubio, J.M., Nadeu, E., Boix-Fayos, C., Vente, J., 2016. Evaluation of the effectiveness of forest restoration and check-dams to reduce catchment sediment yield. Land Degrad. Dev. 27, 1018−1031. Available from: https://doi.org/10.1002/ldr.2331.

Rashid, M.I., Mujawar, L.H., Shahzad, T., Almeelbi, T., Ismail, I.M.I., Oves, M., 2016. Bacteria and fungi can contribute to nutrients bioavailability and aggregate formation in degraded soils. Microbiol. Res. 183, 26−41. Available from: https://doi.org/10.1016/j.micres.2015.11.007.

Rasse, D.P., Rumpel, C., Dignac, M.-F., 2005. Is soil carbon mostly root carbon? Mechanisms for a specific stabilisation. Plant Soil 269, 341−356. Available from: https://doi.org/10.1007/s11104-004-0907-y.

Regelink, I.C., Stoof, C.R., Rousseva, S., Weng, L., Lair, G.J., Kram, P., et al., 2015. Linkages between aggregate formation, porosity and soil chemical properties. Geoderma 24−37. Available from: https://doi.org/10.1016/j.geoderma.2015.01.022.

Reij, C., Tappan, G., Smale, M., 2009. Re-greening the Sahel: farmer-led innovation in Burkina Faso and Niger. In: Spielman, D.J., Pandya-Lorch, R. (Eds.), Millions Fed: Proven Successes in Agricultural Development. International Food Policy Research Institute (IFPRI), Washington, DC, pp. 53−58. Available from: http://ebrary.ifpri.org/cdm/ref/collection/p15738coll2/id/130817.

Rillig, M.C., Mummey, D.L., 2006. Mycorrhizas and soil structure. New Phytol. 171, 41−53. Available from: https://doi.org/10.1111/j.1469-8137.2006.01750.x.

Rillig, M.C., Wright, S.F., Eviner, V.T., 2002. The role of arbuscular mycorrhizal fungi and glomalin in soil aggregation: comparing effects of five plant species. Plant Soil 238, 325−333. Available from: https://doi.org/10.1023/A:1014483303813.

Rillig, M.C., Mardatin, N.F., Leifheit, E.F., Antunes, P.M., 2010. Mycelium of arbuscular mycorrhizal fungi increases soil water repellency and is sufficient to maintain water-stable soil aggregates. Soil Biol. Biochem. 42, 1189−1191. Available from: https://doi.org/10.1016/j.soilbio.2010.03.027.

Roldán, A., Albaladejo, J., Thornes, J.B., 1996. Aggregate stability changes in a semiarid soil after treatment with different organic amendments. Arid Soil Res. Rehab. 10, 139−148. Available from: https://doi.org/10.1080/15324989609381428.

Roose, E., Kabore, V., Guenat, C., 1999. Zaï practice: a West African traditional rehabilitation system for semiarid degraded lands, a case study in Burkina Faso. Arid Soil Res. Rehabil. 13, 343−355. Available from: https://doi.org/10.1080/089030699263230.

Rowe, H.I., Brown, C.S., Claassen, V.P., 2007. Comparisons of mycorrhizal responsiveness with field soil and commercial inoculum for six native montane species and Bromus tectorum. Restor. Ecol. 15, 44−52. Available from: https://doi.org/10.1111/j.1526-100X.2006.00188.x.

Runólfsson, S., 1987. Land reclamation in Iceland. Arct. Alp. Res. 19, 514−517. Available from: https://doi.org/10.2307/1551418.

Sajedi, T., Prescott, C.E., Seely, B., Lavkulich, L.M., 2012. Relationships among soil moisture, aeration and plant communities in natural and harvested coniferous forests in coastal British Columbia, Canada. J. Ecol. 100, 605–618. Available from: https://doi.org/10.1111/j.1365-2745.2011.01942.x.

Salama, R., Hatton, T., Dawes, W., 1999. Predicting land use impacts on regional scale groundwater recharge and discharge. J. Environ. Qual. 28, 446–460. Available from: https://doi.org/10.2134/jeq1999.00472425002800020010x.

Saxerud, M.H., Funke, B.R., 1991. Effects on plant growth of inoculation of stored stripmining topsoil in North Dakota with mycorrhizal fungi contained in native soils. Plant Soil 131, 135–141. Available from: https://doi.org/10.1007/BF00010428.

Schelfhout, S., De Schrijver, A., De Bolle, S., De Gelder, L., Demey, A., Du Pre, T., et al., 2015. Phosphorus mining for ecological restoration on former agricultural land: phosphorus-mining for ecological restoration. Restor. Ecol. 23, 842–851. Available from: https://doi.org/10.1111/rec.12264.

Schelfhout, S., Mertens, J., Verheyen, K., Vesterdal, L., Baeten, L., Muys, B., et al., 2017. Tree species identity shapes earthworm communities. Forests 8, 85. Available from: https://doi.org/10.3390/f8030085.

Schiavo, J.A., Busato, J.G., Martins, M.A., Canellas, L.P., 2009. Recovery of degraded areas revegetated with *Acacia mangium* and *Eucalyptus* with special reference to organic matter humification. Sci. Agric. 66, 353–360.

Schmidt, L., Musokony, C., Meke, G., 2008. Tree species for cracking clay soils. In: Development Briefs. Technical, vol. 1. http://tinyurl.com/y3fsdd3m.

Scullion, J., Malik, A., 2000. Earthworm activity affecting organic matter, aggregation and microbial activity in soils restored after opencast mining for coal. Soil Biol. Biochem. 32, 119–126. Available from: https://doi.org/10.1016/S0038-0717(99)00142-X.

Shi, D., Wang, D., 2005. Effects of various salt-alkaline mixed stresses on *Aneurolepidium chinense* (Trin.) Kitag. Plant Soil 271, 15–26. Available from: https://doi.org/10.1007/s11104-004-1307-z.

Shirato, Y., Zhang, T.-H., Ohkuro, T., Fujiwara, H., Taniyama, I., 2005. Changes in topographical features and soil properties after exclosure combined with sand-fixing measures in Horqin Sandy Land, northern China. Soil Sci. Plant Nutr. 51, 61–68. Available from: https://doi.org/10.1111/j.1747-0765.2005.tb00007.x.

Shrestha, R.K., Lal, R., 2010. Carbon and nitrogen pools in reclaimed land under forest and pasture ecosystems in Ohio, USA. Geoderma 157, 196–205. Available from: https://doi.org/10.1016/j.geoderma.2010.04.013.

Smits, N.A.C., Willems, J.H., Bobbink, R., 2008. Long-term after-effects of fertilisation on the restoration of calcareous grasslands. Appl. Veg. Sci. 11, 279–286. Available from: https://doi.org/10.3170/2008-7-18417.

Sobanski, N., Marques, M.C.M., 2014. Effects of soil characteristics and exotic grass cover on the forest restoration of the Atlantic Forest region. J. Nat. Conserv. 22, 217–222. Available from: https://doi.org/10.1016/j.jnc.2014.01.001.

Stanturf, J.A., Van Oosten, C., Netzer, D.A., Coleman, M.D., Portwood, C.J., 2001. Ecology and silviculture of poplar plantations. Poplar Culture in North America. NRC Research Press, Ottawa, Canada, pp. 153–206.

Stanturf, J.A., Conner, W.H., Gardiner, E.S., Schweitzer, C.J., Ezell, A.W., 2004. Recognizing and overcoming difficult site conditions for afforestation of bottomland hardwoods. Ecol. Restor. 22, 183–193.

Tarolli, P., Preti, F., Romano, N., 2014. Terraced landscapes: from an old best practice to a potential hazard for soil degradation due to land abandonment. Anthropocene 6, 10–25. Available from: https://doi.org/10.1016/j.ancene.2014.03.002.

Tejada, M., Hernandez, M.T., Garcia, C., 2006. Application of two organic amendments on soil restoration: effects on the soil biological properties. J. Environ. Qual. 35, 1010–1017. Available from: https://doi.org/10.2134/jeq2005.0460.

Tibke, G., 1988. Basic principles of wind erosion control. Agric. Ecosyst. Environ. 22, 103–122. Available from: https://doi.org/10.1016/0167-8809(88)90011-4.

Treseder, K.K., 2013. The extent of mycorrhizal colonization of roots and its influence on plant growth and phosphorus content. Plant Soil 371, 1–13. Available from: https://doi.org/10.1007/s11104-013-1681-5.

Urgiles, N., Strauss, A., Loján, P., Schüssler, A., 2014. Cultured arbuscular mycorrhizal fungi and native soil inocula improve seedling development of two pioneer trees in the Andean region. New For. 45, 859–874. Available from: https://doi.org/10.1007/s11056-014-9442-8.

Vanguelova, E.I., Hirano, Y., Eldhuset, T.D., Sas-Paszt, L., Bakker, M.R., Püttsepp, et al., 2007. Tree fine root Ca/Al molar ratio — indicator of Al and acidity stress. Plant Biosyst. 141, 460–480. Available from: https://doi.org/10.1080/11263500701626192.

Vesterdal, L., Clarke, N., Sigurdsson, B.D., Gundersen, P., 2013. Do tree species influence soil carbon stocks in temperate and boreal forests? For. Ecol. Manage. 309, 4–18. Available from: https://doi.org/10.1016/j.foreco.2013.01.017.

Vinduskova, O., Frouz, J., 2013. Soil carbon accumulation after open-cast coal and oil shale mining in Northern Hemisphere: a quantitative review. Environ. Earth Sci. 69, 1685–1698. Available from: https://doi.org/10.1007/s12665-012-2004-5.

Wang, F., Li, Z., Xia, H., Zou, B., Li, N., Liu, J., et al., 2010. Effects of nitrogen-fixing and non-nitrogen-fixing tree species on soil properties and nitrogen transformation during forest restoration in southern China. Soil Sci. Plant Nutr. 56, 297–306. Available from: https://doi.org/10.1111/j.1747-0765.2010.00454.x.

Wei, W., Chen, D., Wang, L., Daryanto, S., Chen, L., Yu, Y., et al., 2016. Global synthesis of the classifications, distributions, benefits and issues of terracing. Earth-Sci. Rev. . Available from: https://doi.org/10.1016/j.earscirev.2016.06.010.

Werden, L.K., Alvarado, P., Zarges, S., Calderon, M.E., Schilling, E.M., Gutierrez, M., et al., 2018. Using soil amendments and plant functional traits to select native tropical dry forest species for the restoration of degraded Vertisols. J. Appl. Ecol. 55, 1019–1028. Available from: https://doi.org/10.1111/1365-2664.12998.

Wheeler, G.S., Taylor, G.S., Gaskin, J.F., Purcell, M.F., 2011. Ecology and management of Sheoak (Casuarina spp.), an invader of coastal Florida, U.S.A. J. Coastal Res. 27, 485–492. Available from: https://doi.org/10.2112/JCOASTRES-D-09-00110.1.

Wicke, B., Smeets, E.M.W., Dornburg, V., Vashev, B., Gaiser, T., Turkenburg, W.C., et al., 2011. The global technical and economic potential of bioenergy from salt-affected soils. Energy Environ. Sci. 4, 2669–2681. Available from: https://doi.org/10.1039/c1ee01029h.

Wick, A.F., Ingram, L.J., Stahl, P.D., 2009a. Aggregate and organic matter dynamics in reclaimed soils as indicated by stable carbon isotopes. Soil Biol. Biochem. 41, 201–209. Available from: https://doi.org/10.1016/j.soilbio.2008.09.012.

Wick, A.F., Stahl, P.D., Ingram, L.J., 2009b. Aggregate-associated carbon and nitrogen in reclaimed sandy loam soils. Soil Sci. Soc. Am. J. 73, 1852–1860. Available from: https://doi.org/10.2136/sssaj2008.0011.

Wiesmeier, M., Lungu, M., Cerbari, V., Boincean, B., Habner, R., Kogel-Knabner, I., 2018. Rebuilding soil carbon in degraded steppe soils of Eastern Europe: the importance of windbreaks and improved cropland management. Land Degrad. Dev. 29, 875–883. Available from: https://doi.org/10.1002/ldr.2902.

Wong, V.N.L., Greene, R.S.B., Dalal, R.C., Murphy, B.W., 2010. Soil carbon dynamics in saline and sodic soils: a review: soil carbon dynamics in saline and sodic soils. Soil Use Manage. 26, 2–11. Available from: https://doi.org/10.1111/j.1475-2743.2009.00251.x.

Wright, S.F., Upadhyaya, A., 1998. A survey of soils for aggregate stability and glomalin, a glycoprotein produced by hyphae of arbuscular mycorrhizal fungi. Plant Soil 198, 97–107. Available from: https://doi.org/10.1023/A:1004347701584.

Wubs, E., van der Putten, W.H., Bosch, M., Bezemer, T.M., 2016. Soil inoculation steers restoration of terrestrial ecosystems. Nat. Plants 2 (8). Available from: https://doi.org/10.1038/NPLANTS.2016.107.

Yang, H., Li, X., Li, G., Wang, Z., Jia, R., Liu, L., et al., 2014. Carbon sequestration capacity of shifting sand dune after establishing new vegetation in the Tengger Desert, northern China. Sci. Total Environ. 478, 1–11. Available from: https://doi.org/10.1016/j.scitotenv.2014.01.063.

Yan, N., Marschner, P., 2013. Microbial activity and biomass recover rapidly after leaching of saline soils. Biol. Fertil. Soils 49, 367–371. Available from: https://doi.org/10.1007/s00374-012-0733-y.

Yan, N., Marschner, P., Cao, W., Zuo, C., Qin, W., 2015. Influence of salinity and water content on soil microorganisms. Int. Soil Water Conserv. Res. 3, 316–323. Available from: https://doi.org/10.1016/j.iswcr.2015.11.003.

Zhang, T.-H., Zhao, H.-L., Li, F.-R., Li, S.-G., Shirato, Y., Ohkuro, T., et al., 2004. A comparison of different measures for stabilizing moving sand dunes in the Horqin Sandy Land of Inner Mongolia, China. J. Arid Environ. 58, 203–214. Available from: https://doi.org/10.1016/j.jaridenv.2003.08.003.

Zhang, S., Ding, G.-D., Yu, M.-H., Gao, G.-L., Zhao, Y.-Y., Wu, G.-H., et al., 2018. Effect of straw checkerboards on wind proofing, sand fixation, and ecological restoration in shifting sandy land. Int. J. Environ. Res. Public Health 15, 2184. Available from: https://doi.org/10.3390/ijerph15102184.

Zhao, G., Mu, X., Wen, Z., Wang, F., Gao, P., 2013. Soil erosion, conservation, and eco-environment changes in the Loess Plateau of China. Land Degrad. Dev. 24, 499–510. Available from: https://doi.org/10.1002/ldr.2246.

Restoring fire to forests: Contrasting the effects on soils of prescribed fire and wildfire

Daniel G. Neary and Jackson M. Leonard

U.S. Department of Agriculture, Forest Service, Rocky Mountain Research Station, Flagstaff, AZ, United States

12.1 Introduction

12.1.1 Fire-evolved ecosystems

Fire is a dynamic process, predictable but uncertain, that varies over time and landscape space. It is an integral component of most temperate wildland ecosystems and has shaped plant communities for as long as vegetation and lightning have existed on earth (Pyne, 1982; Scott, 2000). Wildland fire occurs across a spectrum ranging from low severity, localized prescribed fires to landscape-level, high-severity wildfires that affect vegetation, soils, water, fauna, air, and cultural resources (Brown and Smith, 2000; Smith, 2000; Sandberg et al., 2002; Neary et al., 2005; Zouhar et al., 2008; Bento-Gonçalves et al., 2012). Knowledge of fire effects has risen in importance to land managers because fire, as a disturbance process, is an integral part of the concept of ecosystem management and restoration ecology. Fire is an intrusive disturbance in both managed and wildland forests. It initiates changes in ecosystems that affect the composition, structure, and patterns of vegetation on the landscape. It also affects the soil and water resources of ecosystems that are critical to overall functions and processes (DeBano et al., 1998).

Most forest, shrubland, and grassland ecosystems have evolved with wildfire to some degree or another. Wildfire is a frequent occurrence in natural forest, grassland, and shrubland ecosystems (Pyne, 1982). It occurs less frequently in forest ecosystems such as jack pine (*Pinus banksiana*), lodgepole pine (*Pinus contorta*), but is more common in longleaf pine (*Pinus palustris*), and ponderosa pine (*Pinus ponderosa*) (White and Harley, 2016). Fire is an important ecological factor in those latter two ecosystems. Concerns exist among wildland managers that introduction of prescribed fire as a management

Soils and Landscape Restoration. DOI: https://doi.org/10.1016/B978-0-12-813193-0.00012-6

tool will cause unintended and deleterious disturbances to ecosystem functions and processes (Neary and Leonard, 2015). Risks certainly exist, but the reintroduction of prescribed fire to forest ecosystems is a critical component of management-directed restoration activities needed to prevent catastrophic, high-severity wildfire that poses an even greater risk to forests. The objective of this paper is to evaluate and contrast the risks and impacts to forest soils created by the increased use of prescribed fire in forest management, especially where it is a component of restoration programs.

12.1.2 Fire classification

The general character of fire that occurs within a particular vegetation type or ecosystem across long successional time frames, typically centuries, is commonly defined as the characteristic fire regime. The fire regime describes the typical or modal fire severity that occurs. But it is recognized that, on occasion, fires of greater or lesser severity also occur within a vegetation type. For example, a stand-replacing crown fire is common in long fire-return-interval forests (Fig. 12.1).

The fire regime concept is useful for comparing the relative role of fire between ecosystems and for describing the degree of departure from historical conditions (Hardy et al., 2001; Schmidt et al., 2002). Discussion of the development of fire regime classifications based on fire characteristics and effects has been presented by several authors (Agee, 1993; Brown, 2000). Combinations of factors, including fire frequency, periodicity, intensity, size, pattern, season, and

FIGURE 12.1 High severity, stand-replacing wildfire.

Photo courtesy USDA Forest Service.

Table 12.1 Comparison of fire regime classifications (Hardy et al., 1998, 2001; Brown, 2000).

Hardy et al. (1998, 2001)			Brown (2000)	
Fire regime group	Frequency (years)	Severity	Severity and effects	Fire regime
I	0–35	Low	Understory fire	1
II	0–35	Stand replacement	Stand replacement	2
III	35–100+	Mixed	Mixed	3
IV	35–100+	Stand replacement	Stand replacement	2
V	>200	Stand replacement	Stand replacement	2
			Nonfire regime	4

depth of burn severity, and fire periodicity, season, frequency, and effects, have been used to map fire regimes in the Western United States (Table 12.1) (Heinselman, 1978; Hardy et al., 1998).

The fire regimes described in Table 12.1 are defined as follows:

- *Understory fire regime:* These fires are generally nonlethal to the dominant vegetation and do not substantially change dominant vegetation structure. About 80% of the aboveground dominant vegetation survives fires. This fire regime applies to fire-resistant forest and woodland vegetation (Fig. 12.2).
- *Stand replacement fire regime:* Fires of this type are lethal to most of the dominant aboveground vegetation. About >80% of the aboveground dominant vegetation is either consumed or dies, substantially changing the aboveground vegetative structure. This regime applies to fire-susceptible forests and woodlands, shrublands, and grasslands (Fig. 12.3).
- *Mixed fire regime:* In this fire regime class, severity of fires varies between nonlethal understory and lethal stand replacement fires. Variations could occur in space or time and within the same fire. Spatial variability occurs when fire severity varies, producing a spectrum from understory burning to stand replacement within an individual fire. Small-scale changes in the fire environment (fuels, terrain, or weather) and random changes in weather and plume dynamics produce this type of variability. While this type of fire regime has not been explicitly described in previous classifications, it commonly occurs in some ecosystems because of fluctuations in the fire environment (DeBano et al., 1998; Ryan, 2002). Complex terrain favors mixed severity fires because fuel moisture and wind vary on small spatial scales. Second, temporal variation in fire severity occurs when individual fires alternate over time between infrequent surface fires and long-interval stand replacement fires (Brown and Smith, 2000; Ryan, 2002). Temporal variability also occurs when periodic cool-moist climate cycles are followed by warm dry periods leading to cyclic (multiple decade level) changes in the role of fire

FIGURE 12.2

Understory fire regime: grass and emory oak savanna fire in a southern Arizona woodland, Coronado National Forest.

Photo by Daniel G. Neary, USDA Forest Service, Rocky Mountain Research Station.

in ecosystem dynamics. In moist upland forests, reduced fire occurrence during a wet cycle leads to increased stand density and fuel build-up. Fires that occur during the transition between cool-moist and warm-dry periods can be expected to be more severe and have long-lasting effects on patch and stand dynamics (Kauffman et al., 2003).

- *Nonfire regime:* Fire is not likely to occur in this vegetation type because of wet conditions (Fig. 12.4). However, fires have been documented in swamps and riparian areas previously thought to be fire proof.

Fire-related changes associated with fire severity contribute to diverse responses in the water, soil, floral, and faunal components of the affected ecosystems because of the interdependency between fire severity and ecosystem response. Both immediate and long-term responses to fire occur (Fig. 12.5). Immediate effects occur as a result of direct heating (e.g., damage to living cells) and from chemicals mobilized in ash created by biomass combustion. The response of biological components can be both dramatic and rapid. Another immediate effect of fire is the release of gases and other

FIGURE 12.3

Stand replacement fire regime: *Eucalyptus* stand replacement fire, Victoria, Australia.

Photo courtesy Chris Dicus, California Polytechnic State University.

FIGURE 12.4

Nonfire Regime: East Fork of the Black River, Apache-Sitgreaves National Forest, Arizona.

Photo by Alvin L. Medina, USDA Forest Service, Rocky Mountain Research Station.

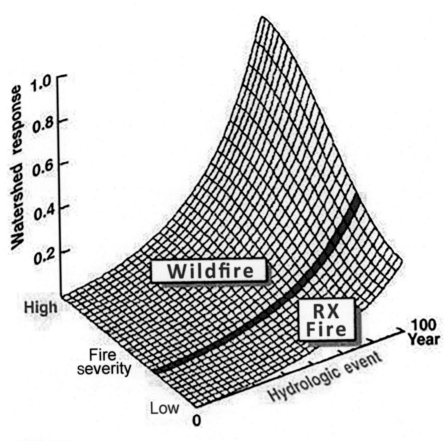

FIGURE 12.5

A conceptual model of watershed responses to wildfire severity.

From Neary, D.G., Ryan, K.C., DeBano, L.F. (Eds.), 2005. Wildland Fire in Ecosystems: Effects of Fire on Soil and Water. General Technical Report RMRS-GTR-42 Volume 4. USDA Forest Service, Rocky Mountain Research Station, Ogden, UT, p. 250.

air pollutants produced by the burning of vegetation and soil organic matter (Dicosty et al., 2006). The long-term fire effects on soils and water are usually subtle, can persist for years following the fire, or may even be permanent as occurs when soils are disturbed by post–fire erosion (DeBano et al., 1998; Ryan et al., 2012). Other long-term fire effects arise from the relationships between fire, vegetation, soils, hydrology, nutrient cycling, and site productivity (Wright and Bailey, 1982).

12.1.3 Prescribed fire

An example of prescribed fires that are low in severity is shown in Fig. 12.2. They occur on the lower zones of the fire severity and resource response curve (Fig. 12.5,

Neary et al., 2005). The weather conditions for prescribed fires in forests are typically lower air temperature, higher relative humidity, lower wind speeds, and higher soil moisture. Also, prescribed fires are usually conducted when fuel loading is low and fuel moisture is moderate to high. These conditions produce lower fire intensities and, as a consequence, lower fire severity, leading to a reduced potential for damage to soil and water resources. Prescribed fire, by its design, usually has minor impacts on vegetation, soil, water, and other resources. However, if fire is applied to maintain a desired type of vegetation, the impact on other vegetation could be major. For example, prescribed fire has a major impact on woody vegetation in mesic prairies, where it prevents a state change to shrubland or forest. Likewise, in longleaf pine, prescribed fires have a major impact on vegetation, including higher plant community diversity and altered community structure, since they are designed to maintain a grass—longleaf pine savanna (Brown and Smith, 2000).

The difference between wildfire and prescribed fires in shrublands is mainly the ignition source, the amount of fuel, weather, winds, resultant fire severity, and the rate at which the plant canopy is consumed. During wildfires the entire plant canopy can be consumed within a matter of seconds, and large amounts of heat are transmitted to the soil surface and into the underlying soil. The contrast in soil surface temperatures between wildfires and prescribed fires in shrubland ecosystems is remarkable, with peak soil temperatures of about 450°C in prescribed fires, but nearly 1500°C in high severity wildfires (DeBano et al., 1998). Not surprisingly, the difference in soil resource damage between the two conditions is significant. In contrast to wildfires, brush can be prescribe-burned under cooler and wetter conditions (e.g., higher fuel moisture contents, lower wind speeds, higher humidity, and lower ambient temperatures) such that fire behavior is less explosive and heat output is less. Under these cooler burning conditions the shrub canopy may be not be entirely consumed, and in some cases, a mosaic burn pattern may be created (particularly on north-facing slopes).

The general effects of fire in grassland on soil physical properties depend on a number of factors and range from very minor to serious (Pyne, 1982; DeBano et al., 1998). Since grassland fires are often rapidly moving with the wind and have much less fuel than that in brush and forest ecosystems, soil heating is significantly lower, and therefore physical damage is much less than what occurs during crown, surface, or smoldering fires. In general, physical impacts of fire on soil systems include alterations in soil physical properties, development of water repellency, and erosion (DeBano, 2000; Doerr et al., 2000). Comparisons of wildfire and prescribed fire in grassland ecosystems show few differences in soil impacts, particularly when wildfire severities are low (DeBano et al., 1998; Neary and Leonard, 2015). On the other hand, high severity wildfires that occur under adverse climate conditions and fuel loads can produce strong negative effects on soil properties.

12.1.4 **Wildfire**

All fires produce a range of severities. Wildfire that is at the far end of the severity spectrum (left side of Fig. 12.5) more nearly represents conditions that are present during a wildfire. Severity is frequently high where temperatures, wind speeds, and fuel loadings are high, and humidity and fuel moisture are low.

In contrast to prescribed fires, wildfires often have a major effect on vegetation, soil, and watershed processes, leading to increased sensitivity of burned sites to significant vegetative loss, increased runoff, erosion, reduced land stability, and adverse aquatic ecosystem impacts (Agee, 1993; DeBano et al., 1998; Scott, 2000). Fuels such as slash piles, decomposing logs, and thick layers of litter burn for long periods in wildfires and accentuate soil resource damage (Rhoades et al., 2004, 2015).

Another difference between prescribed fires and wildfires is the scale of burning. Prescribed fires are usually less than 400 ha in size although prescribed natural fires, managed like normal prescribed fires but initiated naturally, sometimes exceed 10 times that size. Small wildfires range in size from 400 to 40,000 ha. Megafires are those in excess of 40,000 ha and can exceed 1×10^6 ha (Stephens et al., 2014). Several individual forest wildfires in North America in the 19th Century and early 20th Century burned over 1.2×10^6 ha. The size of fires prior to the historical ones of the 19th Century is unknown, but wildfires and deliberate ignitions by Native Americans were definitely part of the presettlement North American environment (Frost, 1998). More recently, a series of fires in 2011 in Texas consumed over 1.6×10^6 ha of grassland and savanna, and in 2019, wildfires consumed in excess of 1.0×10^7 ha of land area in Australia. Fires of this scale clearly have the potential to produce serious effects on natural resources and human infrastructure over large landscapes (Wright and Bailey, 1982).

12.2 Fire effects

12.2.1 Combustion

Energy generated as heat during the combustion of aboveground and surface fuels provides the driving force that causes a wide range of changes in soil properties during a fire (DeBano et al., 1998). *Combustion* is the rapid physical–chemical decomposition of organic matter that releases the large amounts of energy stored in fuels. These fuels consist of dead and live standing biomass, fallen logs, surface litter (including bark, leaves, stems, and twigs), humus, soil organic matter, and roots. Importantly, the material that soil scientists consider the topmost genetic horizon in a soil profile (the O horizon) is composed entirely of these combustible fuels. During the combustion process, heat and a mixture of gaseous and particulate byproducts are vented into the air. When ignited, flames are the visual evidence of the combustion process.

Three components are needed in order for a fire to ignite and initiate the combustion process (DeBano et al., 1998; Neary et al., 2005). First, burnable fuel must be available. Second, sufficient heat must be applied to the fuel to raise its temperature to point where flammable organics volatilize. Lastly, sufficient oxygen (O_2) must be present to keep the combustion process going and to maintain the heat supply and oxidative source necessary for fuel ignition.

A common sequence of physical processes occurs in fuels before the energy contained in them is released and transferred upward, laterally, or downward. There are five physical phases during the course of a fire, namely, preignition, ignition, flaming, smoldering, and glowing (DeBano et al., 1998). Preignition occurs first when fuels are heated sufficiently to dry out and start the initial thermal decomposition (pyrolysis). After ignition the three final phases are flaming, smoldering, and glowing combustion in the presence of O_2. When active flaming begins to diminish, smoldering increases, and combustion diminishes to the glowing phase, which finally leads to extinction of the fire.

12.2.2 Soil heating

Heat produced during the combustion of aboveground fuels is transferred to the soil surface and downward through the soil by several heat transfer processes (radiation, convection, conduction, vaporization, and condensation) (DeBano et al., 1998; Neary and Leonard, 2015). *Radiation* is the transfer of heat from bodies not in contact with one another. *Conduction* is the transfer of heat by molecular activity from one part of a substance to another, or between substances in contact. *Convection* is a process whereby heat is transferred from one point to another by the mixing of one portion of a fluid (in this case combustion gases) with another fluid. *Vaporization* of water occurs when it is heated to a temperature at which it changes from a liquid to a gas. *Condensation* occurs when water changes from a gas to a liquid with the simultaneous release of heat. The combined mechanisms of vaporization and condensation provide a conduit for the transfer of both water and organic materials through the soil during fires (DeBano et al., 1998). Vaporization and condensation are important in fire behavior since they function in providing a more rapid transfer of heat through soils.

The mechanisms for the movement of heat into different ecosystems components vary considerably. Although heat is transferred in all directions, large amounts are preferentially lost into the atmosphere by radiation, convection, and mass transfer (DeBano et al., 1998). It has been estimated that only about 10%−15% of the heat energy released during the combustion of aboveground fuels is absorbed and transmitted by radiation directly downward to the litter, or mineral soil if surface organic layers are absent (DeBano, 1974; Raison et al., 1986). Within the fuels, most of the heat transfer is by radiation, convection, and mass transfer. In dry soils, convection and vaporization and condensation are the most important mechanisms for heat transfer. In a wet soil, conduction can contribute significantly to heat transfer. Movement of heat through mineral soil is an important thermodynamic process because it produces heating and changes in soil physical, chemical, and biological properties

12.2.3 Severity

In any discussion of the effects of fire on soils, it is important to differentiate between fire intensity and fire severity. The terms are not the same and are frequently confused (Hartford and Frandsen, 1992; DeBano et al., 1998; Neary and Leonard, 2015). *Fire intensity* is used to describe the rate at which a fire produces thermal energy. It is most quantified in terms of heat release per unit length of fire line. *Fire severity*, on the other hand, is a more qualitative term that is used to describe ecosystem impacts to fire and is especially useful for describing the effects of fire on soil and water components (Simard, 1991). Severity reflects the amount of heat energy that is released by a fire and the degree that it affects the soil and water resources. Fire severity has been classified based on visual postfire criteria at burned sites and has been categorized simply as low, moderate, and high-fire severity. Detailed descriptions of the appearances of these three severity levels for timber, shrublands, and grasslands can be found elsewhere (DeBano et al., 1998; Neary et al., 2005; Neary and Leonard, 2015). The level of fire severity depends upon:

- Length of time fuel accumulates between fires and the amount combusted (fuel load and fuel consumption).
- Properties of the fuels (size, flammability, moisture content, mineral content, etc.).
- The effect of fuels on fire behavior during ignition and combustion phases.
- Heat transfer to the soil during combustion of aboveground and surface fuels.

High-intensity fires can result in high severity changes in the soil, but this is not always the case. Low-intensity smoldering fires in roots or litter can cause extensive soil heating and produce large changes in the adjacent mineral soil. In contrast, high severity crown fires may not cause substantial heating at the soil surface because they sweep so rapidly over a landscape. Under these conditions, little heat generated during combustion is transferred downward to the soil surface.

12.2.4 Water repellency

The creation of *water repellency* in soils involves both physical and chemical processes (Fig. 12.6). It modifies physical processes such as infiltration and water movement in soils. Fire-induced water repellency was first identified on severely burned chaparral watersheds in southern California in the early 1960s (DeBano, 2000). Both the production of fire-induced water repellency and the loss of protective vegetative cover have played a major role in postfire runoff and erosion in some areas where large human population centers are situated immediately below steep, unstable shrubland watersheds. High severity wildfires are much more likely to produce extensive areas of water repellent soil layers than prescribed fires. Very localized zones of water repellency occur on prescribed fires, but these "spots" are unlikely to have any effect of watershed-level response and erosion.

Normally dry soils have an affinity for adsorbing liquid and vapor water. In water repellent soils, however, the water forms beads, ponds, and rills on soil surfaces where it can remain for long periods and in some cases will evaporate before being absorbed by the soil (DeBano et al., 1998). Water will not infiltrate

FIGURE 12.6

Rain ponding on a water repellent soil in the Rodeo—Chediski Wildfire of 2002, Apache-Sitgreaves National Forest, Arizona.

Photo by Daniel G. Neary, Rocky Mountain Research Station, USDA Forest Service.

into soils where the mineral particles are coated with hydrophobic organic compounds that repel water. Water repellency is produced by soil organic matter and can be found in both fire and nonfire environments (DeBano, 2000). Water repellency can result from the following processes involving organic matter:

- irreversible drying of soil organic matter;
- coating of mineral soil particles with leachates from organic materials;
- coating of soil particles with hydrophobic microbial byproducts;
- intermixing of dry mineral soil particles and dry organic matter; and
- vaporization of organic matter and condensation of these substances on mineral particles.

The magnitude of fire-induced water repellency depends upon a number of fire, edaphic, and climate parameters (Simard, 1991; Doerr et al., 2000). Although these occur with most fires, severity is mostly problematic with wildfires, not prescribed fires.

- the severity of the fire
- type and amount of organic matter present
- temperature gradients in the upper mineral soil
- texture of the soil
- water content of the soil

The more severe the fire, the deeper the formation of the water repellent layer, unless the fire temperatures are so hot that they destroy the surface organic matter. Most vegetation and fungal mycelium contain hydrophobic compounds that induce water repellency. Steep temperature gradients in dry soil enhance the downward movement of volatilized hydrophobic substances. Finally, soil water affects the movement of hydrophobic substances during a fire because it affects heat transfer and the formation of steep temperature gradients. Some studies in chaparral vegetation indicated that sandy and coarse-textured soil was the most susceptible to fire-induced water repellency (DeBano, 1981). However, other research has indicated that water repellency frequently occurs in soils other than coarse-textured ones (Doerr et al., 2000).

12.2.5 Effect of water repellency on post–fire erosion

Fire affects the processing of water and interactions with the soil in two ways. First, if the burned soil surface is unprotected due to complete combustion of the organic horizons, rain drop impact can loosen and disperse fine soil and ash particles. This situation can result in sealing of soil surface pores, reducing infiltration. The material that soil scientists consider the topmost genetic horizon in a soil profile (the O horizon) is composed entirely of combustible fuels. Second, soil heating during a fire can produce a water-repellent layer at or near the soil surface, lowering the saturated hydraulic conductivity, and further impeding water infiltration into the soil (Neary, 2011).

The severity of water repellency in surface soil horizons decreases over time due to repeated raindrop impacts, soil heating and cooling, freeze/thaw cycles, rodent activity, soil arthropod actions, and root growth (DeBano et al., 1998; Neary and Leonard, 2015). Water repellency deeper in the soil (10−15 cm) is more problematic since some of the mechanisms that produce surface repellency breakdown do not function at depth. It is only high-severity wildfires that have the energy release needed to form deep water repellency. Prescribed fires do not have the heat output that wildfires generate. Localized spots of very limited size such as under downed logs may occur. On a landscape scale, prescribed fires do not pose the risks that wildfires do with respect to the formation of water repellent layers and the associated erosion problems.

Sediment yields usually are the highest the first year after a fire and then decline in subsequent years (Neary et al., 2005). However, if precipitation is below normal, the peak sediment delivery year might be delayed until years 2 or 3. In semiarid areas, postfire sediment transport is episodic in nature, and the delay may be longer. All fires increase sediment yield, but it is wildfire that produces the largest amounts. Slope is a major factor in determining the amount of sediment yielded during periods of rainfall following fire (Table 12.2). There is growing evidence that short-duration, high-intensity rainfall (greater than $50 \, \text{mm} \, \text{h}^{-1}$ in 10- to-15 minute bursts) over areas of about $1 \, \text{km}^2$ often produce the flood flows that result in large amounts of sediment transport (Robichaud et al., 2009).

Table 12.2 Sediment losses the first year after prescribed fires.

Location	Management activity	First year sediment loss
		Mg ha^{-1}
South Carolina	Unburned (*loblolly pine*)	0.027
	Understory prescribed fire	0.042
North Carolina	Unburned (*southern hardwoods*)	0.002
	Semiannual prescribed fire	6.899
Mississippi	Unburned (*scrub Oak*)	0.470
	3-year prescribed fire	1.142
Texas	Unburned level terrain (*juniper*)	0.025
	Prescribed fire level terrain	0.029
	Unburned 15%–20% slope	0.076
	Prescribed Fire 15%–20% slope	1.874
	Unburned 43%–54% slope	0.013
	Prescribed fire 43%–54% slope	8.443
Oklahoma	Unburned (*mixed hardwoods*)	0.022
	Annual prescribed fire	0.246
Arkansas	Unburned (*shortleaf pine*)	0.036
	Slash prescribed fire	0.237
California	Unburned (*ponderosa pine*)	0.001
	Understory prescribed fire	0.001
Montana	Unburned (*larch, douglas fir*)	0.001
	Slash prescribed fire	0.150
Arizona	Unburned (*chaparral*)	0.001
	Prescribed fire	3.778

Adapted from Neary, D.G., Ryan, K.C., DeBano, L.F. (Eds.), 2005. Wildland Fire in Ecosystems: Effects of Fire on Soil and Water. General Technical Report RMRS-GTR-42 Volume 4. USDA Forest Service, Rocky Mountain Research Station, Ogden, UT, p. 250.

Soil erosion following fires can vary from under 0.1–15 Mg ha^{-1} year^{-1} in prescribed fires (Table 12.2), and from <0.1 Mg ha^{-1} year^{-1} in low severity wildfire, to more than 369 Mg ha^{-1} year^{-1} in high-severity wildfires on steep slopes (Table 12.3) (DeBano et al., 1998; Neary et al., 2005; Robichaud et al., 2009). Best Management Practices (BMPs) certainly have value in reducing sediment losses from prescribed fires. Wildfires, on the other hand, due to their very nature of being unplanned and often unmanageable, they do not lend themselves to being mitigated by BMPs.

The use of prescribed fire presents managers with alternatives for minimizing the damage done to the soil. The least amount of damage occurs during cool-burning, low-severity fires. These fires do not heat the soil substantially, and the changes in most soil properties are only minor and are of short duration. However, the burning of concentrated fuels (e.g., slash, large woody debris) can

Table 12.3 Sediment losses the first year after wildfires.

Location	Management activity	1st year sediment loss
		Mg ha^{-1}
Washington	Unburned (*mixed conifer*)	0.238
	Wildfire	2.353
California	Unburned (*chaparral*)	0.043
	Wildfire	28.605
California	Unburned (*chaparral*)	5.530
	Wildfire	55.300
Arizona	Unburned (*chaparral*)	0.096
	Wildfire	28.694
Arizona	Unburned (*chaparral*)	0.175
	Wildfire	204.000
Arizona	Unburned (*mixed conifer*)	0.001
	Wildfire 43% slope	71.680
	Wildfire 66% slope	201.600
	Wildfire 78% slope	369.600
Arizona	Unburned (*Pinus ponderosa*)	0.003
	Wildfire low severity	0.080
	Wildfire moderate severity	0.300
	Wildfire high severity	1.254
Idaho	Unburned (*grass and brush*)	0.030
	Wildfire moderate severity	0.125
	Wildfire high severity	1.538

Adapted from Neary, D.G., Ryan, K.C., DeBano, L.F. (Eds.), 2005. Wildland Fire in Ecosystems: Effects of Fire on Soil and Water. General Technical Report RMRS-GTR-42 Volume 4. USDA Forest Service, Rocky Mountain Research Station, Ogden, UT, p. 250.

cause substantial damage to the soil resource, although these long-term effects are limited to only a small proportion of the landscape where the fuels are piled. These types of fire use should be avoided whenever possible. The burning of organic soils is also a special case where extensive damage can occur unless burning prescriptions are carefully planned. The effects of prescribed fires on soils in the eastern part of North America (where the majority of prescribed fires take place on the continent) have been reviewed by Callaham et al. (2012).

12.2.6 Climate

Recent climate change analyses suggest that much of the Western United States and other regions world-wide have a high probability of continuing or developing and future drought into the next 20–30 years, with associated increase in risk for wildfire. The cumulative impacts of warmer and drier climate, with ancillary disturbances such as insect outbreaks, urbanization, and the spread of exotic grasses will potentially increase the risk of wildfire. Prescribed fire will continue to be an important tool in reducing the risks of wildfire in this drought environment (Neary and Leonard, 2015). Fire regimes are representative of long-term patterns in the severity and occurrence of fires (Brown, 2000). They reflect the interdependence of several factors in the fire nexus, including climate, fuel accumulation, variability of soil and topographic properties, and ignition sources. Climate will remain a neutral discriminating factor between prescribed fires and wildfires.

12.2.7 The fire nexus

A "nexus" is defined as a connected group or series of factors which can be linked to cause an action or disturbance of some consequence. For prescribed fires and wildfires the fire nexus leading to soil disturbance consists of severity, scale, slope, rainfall, and infiltration conditions. These factors, in and of themselves, do not necessarily lead to soil disturbance. But when they combine into the fire nexus, they might produce profound disturbances in forest grassland, and shrubland ecosystems. Knowledge of disturbance thresholds is important for understanding and management of postfire vegetation and soils.

The severity threshold is moderate severity (DeBano et al., 1998). Prescribed fires rarely have large areas of moderate fire severity. High fire severity is common in wildfires and definitely a warning sign that major impacts to soil integrity and sustainability may occur.

Fires occur at all sizes and scales. Prescribed fires are generally <400 ha so size is not a major contributor to the fire nexus. Size and accompanying soil disturbance becomes problematic in the midrange of small wildfires (>4000 ha), and becomes critical factor for landscapes in the zone of megafires (>40,000 ha) (Adams, 2013; Stephens et al., 2014; Jones et al., 2016).

The disturbance threshold for slope is 30%. Above that, Burned Area Emergency Response measures become less effective and soil disturbance becomes more severe (Robichaud et al., 2009). Low slopes (<10%) do not generate high velocity peak flows that cause significant soil erosion. Even with high severity wildfires, low slopes limit the damage to soil systems. However, mobilization of nutrients can be a factor in soil desertification on low gradient (DeBano et al., 1998). Slope is a neutral discriminating factor between prescribed fires and wildfires. Prescribed fires on steep slopes might respond like wildfires if severities are high and the forest floor is combusted.

The peak rainfall intensities that mark the threshold for setting off surface runoff, erosion, and flooding are variable depending on other contributing factors and the soils. Peak rainfall intensities are reported as the I_{30p} where I is the intensity, 30 is the time period, and p is peak rainfall. An I_{30p} of 10 mm h^{-1} has been reported by some investigators as a critical threshold while others reported an I_{30p} threshold value of 15 mm h^{-1} (Moody and Martin, 2001; Adams, 2013). Others found that an I_{10p} threshold of 75 mm h^{-1} was the most significant rainfall intensity for producing erosion and soil disturbance (Spiegel and Robichaud, 2007).

Because of the anomalies in saturated hydraulic conductivity (K_{sat}) reductions and the discovery of reduced K_{sat} in undisturbed forest soils, it is difficult to state that there is a common infiltration reduction threshold related to fire (Neary, 2011). High severity wildfire is commonly associated with the development of water repellency and aggravated surface runoff. This may also occur in prescribed fires due to natural repellency processes (DeBano, 2000). Water repellency is an inciting runoff factor after fire until erosion, mechanical disturbance, or climate-related disturbances break up the hydrophobic layer and erosion cuts into deeper soil horizons.

The wildfire nexus cannot be predicted because of uncertainty in the occurrence and magnitude of nexus factors, but the relative risk can be understood if the important components are considered (Miller and Ager, 2012). Steep slopes and high fuel loads are key indicators that significant soil disturbance could occur if intense rainfall follows fire events. This risk is certainly higher during the first year after fire events but could be delayed by years or even a decade.

Wildfires and prescribed fires cause a range of impacts on forest soils depending on the interactions of a nexus of fire severity, scale of fire, slope, infiltration rates, and postfire rainfall. These factors determine the degree of impact on forest soils and subsequently the need for postfire soil management. Fire is a useful tool in landscape management that can be benign or can set off serious deteriorations in soil quality that lead to long-term desertification. If parts of the nexus are absent or not inherently risky, forest soil impacts can be relatively minor or nonexistent. A low severity prescribed fire on a small landscape unit with minimal fuel loading, slopes less than 10%, and no water repellency is unlikely to damage soil or watershed functions under any but very heavy rainfall conditions. On the other hand, a high severity wildfire in a substantial area of heavy fuels, with slopes >100%, and development of water repellency may undergo serious soil damage with even moderate rainfall. Soil management is not likely to be needed in the former case but may be virtually impossible in the latter scenario. Of course, these scenarios represent opposite ends of a spectrum, and nearly every combination of the components involved in the fire nexus may be encountered in the field. Understanding the interactions of these components will improve the responses of managers as they make decisions about restoration of burned forest soils.

12.3 Trends

12.3.1 Prescribed fire use

Prescribed fire is an important management tool and has been growing in frequency in the past several decades as land manager understand more clearly its role in the ecology of forest ecosystems. The general conditions of prescribed fire conditions in relation to soil and water resource responses are depicted in

Fig. 12.5 on the lower portion of the fire severity and resource response curve (Neary et al., 2005). These conditions are typically characterized by lower air temperature, higher relative humidity, and higher soil moisture burning conditions, where fuel loading is low and fuel moisture can be high. These conditions produce lower fire intensities and, as a consequence, lower fire severity leading to reduced potential for subsequent damage to soil and water resources. Prescribed fire, by its design, usually has minor impacts on soil and water resources.

In the past, State and Federal agencies have been slow to respond to changing fire management needs. But now communities which were once more concerned about smoke issues have increased awareness of the threats posed by large wildfires. Stakeholders are urging land management agencies to accelerate the use of prescribed fire as part of a package of treatments to reduce fuels and fire risks. In addition, there is pressure on agencies to reduce fire suppression costs, which can often consume 50% or more of annual budgets. More recently, the first eight of US National Forest revised management plans in the southern Sierra Nevada region of California proposed placing 50% of their management areas in prescribed and managed natural fire zones (North et al., 2015). Another 155 National Forests are working on revised land-management plans that incorporate increased use of prescribed fire.

The number of prescribed fires in the United States over the past 20 years has increased from 279 in 1998 to 450,335 in 2018 with the largest increase in state and other land-management entities (National Interagency Fire Center, 2019). Areas burned in the same period have increased from 359,086 to 2,574,286 ha. The number of wildfires declined over the 1998–2018 time period but the area burned increased by a factor of 6.6. There were 196 wildfires over 40,470 ha with the largest reaching a size of 528,374 ha. Clearly, use of prescribed fires has increased to attempt to reduce the numbers and sizes of wildfires.

12.3.2 Fire size and severity

The trend of a growing occurrence of fire around the world brings with it many of the consequences both direct and indirect (Liu et al., 2018). This analysis indicated that the future for potential wildfire increases significantly in fire prone regions of North America, South America, central Asia, southern Europe, southern Africa, and Australia. Fire potential is projected to increase in these regions, from currently low-to-future moderate potential or from moderate-to-high potential. The increased fire risk is driven by climate warming in North and South America and Australia, and by the combination of temperature increases and desertification in the other regions. The analysis in Liu et al. (2018) indicates that future increases in wildfire trends will require substantial investment of financial resources and management actions for wildfire disaster prevention and recovery.

In a discussion to the contrary the argument is made that there is evidence of reduced fire worldwide today than centuries ago (Doerr and Santin, 2016). Regarding fire severity, limited data are available. They indicate evidence of little change in the western United States and declines in the area of high severity fire compared to 18th and 19th century conditions. The authors argue that direct fatalities from fire and economic losses also show no clear trends over the past 30 years (Doerr and Santin, 2016). Trends in indirect impacts are insufficiently quantified to be examined in any significant degree.

On the other hand, an analysis of large wildfire trends in the western United States reported a significant increase in fire numbers and area burned (Dennison et al., 2014). This was particularly true in southern mountain regions with drought. The reported increase of wildland fires in these areas has amounted to 355 km^2 year^{-1}. An analysis of wildfire in Russia demonstrated and acceleration of wildfire in the 21st Century as a result of climate change (Shvidenko and Shchepashchenko, 2011). Trends in wildfire on US Forest Service lands from 1970 to 2002 were examined in a 2005 paper in the Journal of Forestry (Calkin et al., 2005). The authors reported that the number of large fires has more than doubled over this period and the area burned has increased fourfold. The number of fires and area burned by wildfires in eastern Spain from 1941 to 1994 documented increasing fire activity in southern Europe (Piñol et al., 1998). During this time period the areas and numbers of fires were increasing significantly and were associated with high fire hazard indices.

Wildfire appears to be on the increase globally but not uniformly. Drought and elevated temperatures are major factors contributing to wildfires and the hazards they pose to natural ecosystems and humans. Wildfire sizes and severity thus have the potential to present significant hazards to human health and safety and infrastructure in the 21st Century (DeBano et al., 1998; Brown and Smith, 2000).

12.4 Desertification

Desertification is an ecosystem disturbance of considerable concern in many regions of the world. It is a human-induced or natural process which negatively affects ecosystem function and which results in disturbance to the ability of an ecosystem to accept, store, recycle water, nutrients, and energy (Glantz and Orlovsky, 1983). It is not necessarily the immediate creation of classical deserts such as the Sahara, Gobi, Sonoran, or Atacama deserts (Dregne, 1986; Walker, 1997). These types of landscapes are more one type of "end point" of the desertification process. Although desertification is commonly thought of as land degradation that is a problem of arid, semiarid, and dry subhumid regions of the world, humid regions such as Brazil and Indonesia are now experiencing desertification because of wide-scale deforestation and fire use.

Three salient features of desertification are soil erosion, reduced biodiversity, and the loss of productive capacity, such as the transition from forests to shrublands or from grassland dominated by perennial grasses to one dominated by perennial shrubs (Aubreville, 1949). It is a common misunderstanding that droughts cause desertification since dry periods are common in arid and semiarid lands and are part of the ebb and flow of climate in these regions (Walker, 1997). Well-managed lands can recover from the dry segments of climate cycles when the rainfall increases. However, continued land abuse from wildfire during droughts certainly increases the potential for permanent land degradation (MacDonald, 2000). Desertification results in the loss of the land's proper hydrologic function, biological productivity, and other ecosystem services as a result of human activities and climate change (Walker, 1997; Black, 1997). It affects one-third of the earth's surface and over a billion people (MacDonald, 2000). The Amazon region is an example of where forest harvesting, shifting cut and burn agriculture, and large-scale grazing are producing desertification of a tropical rain forest on a large scale (Malhi et al., 2008; Schiffman, 2015). Indonesia is also having problems where large areas of tropical forest have been converted to oil palm cultivation using fire as a clearing technique (Margono et al., 2014).

Desertification of formerly productive land is a complex process. It involves multiple causes, and it proceeds at varying rates in different climates. It may intensify a general climatic trend toward greater aridity, or it may initiate a change in local climate. Desertification does not occur in straightforward, easily predictable patterns that can be rigorously mapped. Deserts advance erratically, forming patches on their borders. Areas far from natural deserts can degrade quickly to barren soil or rock through poor land management. The proximity of a nearby desert has no direct relationship to desertification. Fire-induced desertification has many of the same consequences as other causes of desertification with the addition of substantial impacts in the short term.

Severe wildfires produce disturbances in the natural dynamics of wildland ecosystems that cause them to lose integrity of various components and degrade to a lower system state. Instead of recovery to their original state, they decline to a new but disturbed, stable state where recovery to the original natural dynamic could take centuries (Yackinous, 2015). Repeated fires aggravate the initial desertification and could prolong recovery or prevent it from ever occurring. There is growing evidence that new, severe fires are burning into older fire scars and aggravating desertification by accelerating vegetation type conversion, species diversity loss, and erosion (McGuire and Youberg, 2019). Other environmental effects add to the degree of desertification (Geist and Lambin, 2004). Low severity prescribed fires do not contribute to desertification in any appreciable manner so they can be used in environmentally sound fire management programs.

12.5 **Summary and conclusion**

This chapter examined the effects on soils of restoring prescribed fire to forests that have been managed under the paradigm that all fire is deleterious to ecosystem health. More specifically it contrasted the effects on soils of desired prescribed fire and undesirable wildfire. Prescribed fire is being used with increasing frequency to prevent catastrophic wildfires. In situations where forests have become overstocked due to fire exclusion, mechanical thinning, and removal of excess biomass by harvesting and prescribed fire is necessary to put these forests in ecological conditions whereby accidental or natural ignitions of wildfire do not explode into high-severity wildfires. Well-planned and controlled prescribed fires do not reach the severities or sizes that result in significant soil and other resource damage. Wildfires pose a definite risk to soil health and conditions. Prescribed fires do not. Ecologically sustainable forest management can certainly utilize prescribed fire as a tool to achieve objectives and reduce the risks of catastrophic wildfires. Once restored, forest stands can be maintained in a low wildfire risk category by periodic prescribed fires.

References

Adams, M.A., 2013. Mega-fires, tipping points and ecosystem services: managing forests and woodlands in an uncertain future. For. Ecol. Manage. 294, 250–261.

Agee, J.K., 1993. Fire Ecology of Pacific Northwest forests. Island Press, Washington, DC, p. 493.

Aubreville, A., 1949. Climats, forêts et désertification de l'Afrique tropicale. Société d'Editions Géographiques, Maritimes et Coloniales, Paris.

Bento-Gonçalves, A., Vieira, A., Úbeda, X., Martin, D., 2012. Fire and soils: key concepts and recent advances. Geoderma 191, 3–13.

Black, P.E., 1997. Watershed functions. J. Am. Water Resour. Assoc. 33, 1–11.

Brown, J.K., 2000. Introduction and fire regimes. In: Brown, J.K., Smith, J.K. (Eds.), Wildland Fire in Ecosystems—Effects of Fire on Flora. USDA Forest Service, Rocky Mountain Research Station, Ogden, UT, pp. 1–8. General Technical Report GTR-RMRS-42-Vol. 2, 257 p.

Brown, J.K., Smith, J.K., 2000. Wildland Fire in Ecosystems: Effects on Flora. USDA Forest Service, Rocky Mountain Research Station, Fort Collins, CO, General Technical Report RMRS-GTR-42, Volume 2, 257 p.

Calkin, D.E., Gebort, K.M., Jones, J.G., Neilson, R.P., 2005. Forest service large fire area burned and suppression expenditure trends 1970-2002. J. For. 103, 179–183.

Callaham Jr., M.A., Scott, D.A., O'Brien, J.J., Stanturf, J.A., 2012. Cumulative effects of fuel management on the soils of eastern U.S. In: LaFayette, R., Brooks, M.T., Potyondy, J.P., Audin, L., Krieger, S.L., Trettin, C.C. (Eds.), Cumulative Watershed Effects of Fuel Management in the Eastern United States. U.S. Department of Agriculture Forest Service, Southern Research Station, Asheville, NC, pp. 202–228. General Technical Report SRS-161.

DeBano, L.F., 1974. Chaparral soils. Proceedings of a Symposium on Living With the Chaparral. Sierra Club, San Francisco, CA, pp. 19–26.

DeBano, L.F., 1981. Water Repellent Soils: A State-of-the-Art. USDA Forest Service, Pacific Southwest Forest and Range Experiment Station, Berkeley, CA, p. 21, General Technical Report PSW 46.

DeBano, L.F., 2000. The role of fire and soil heating on water repellency in wildland environments: a review. J. Hydrol. 231–232, 195–206.

DeBano, L.F., Neary, D.G., Ffolliott, P.F., 1998. Fire's Effects on Ecosystems. John Wiley & Sons, New York, p. 333.

Dennison, P.E., Brower, S.C., Arnold, J.D., Moritz, M.A., 2014. Large wildfire trends in the western USA 1984-2011. Geophys. Res. Lett. 41, 2928–2933.

Dicosty, R.J., Callaham Jr., M.A., Stanturf, J.A., 2006. Atmospheric deposition and re-emission of mercury estimated in a prescribed forest-fire experiment in Florida, USA. Water Air Soil. Pollut. 176, 77–91.

Doerr, S.H., Santin, C., 2016. Global trends in wildfire and its impacts: perceptions versus realities in a changing world. Philos. Trans. R. Soc. B 371, 1471–2970.

Doerr, S.H., Shakesby, R.A., Walsh, R.P.D., 2000. Soil water repellency: its causes, characteristics and hydro-geomorphological significance. Earth Sci. Rev. 51, 33–65.

Dregne, H.E., 1986. Desertification of arid lands. In: El-Baz, F., Hassan, M.H.A. (Eds.), Physics of Desertification. pp. 4–34. Martinus Nijhoff, Dordrecht, The Netherlands, 473 p.

Frost, C.C., 1998. Presettlement fire frequency regimes of the United States: a first approximation. In: Pruden, T.L., Brennan, L. (Eds.), Fire in Ecosystem Management: Shifting Paradigm From Suppression to Prescription, 20. Tall Timbers Research Station, Tallahassee, FL, pp. 70–81. Proceedings, Tall Timbers Fire Ecology Conference; 1996 May 7–10.

Geist, H.J., Lambin, E.F., 2004. Dynamic causal patterns of desertification. BioScience 54, 817–829.

Glantz, M.H., Orlovsky, N.S., 1983. Desertification: a review of the concept. Desertification Control. Bull. 9, 15–22.

Hardy, C.C., Menakis, J.P., Long, D.G., Brown, J.K., 1998. Mapping historic fire regimes for the Western United States: Integrating remote sensing and biophysical data. In: Greer, J.D. (Ed.), Proceedings of the Seventh Forest Service Remote Sensing Applications Conference, 1998 April 6-10. Nassau Bay, TX. (pp. 288–300). Bethesda, MD: American Society for Photogrammetry and Remote Sensing.

Hardy, C.C., Schmidt, K.M., Menakis, J.P., Sampson, R.N., 2001. Spatial data for national fire planning and fuel management. Int. J. Wildland Fire 10, 353–372.

Hartford, R.A., Frandsen, W.H., 1992. When it's hot, it's hot … or maybe it's not (surface flaming may not portend extensive soil heating). Int. J. Wildland Fire 2, 139–144.

Heinselman, M.L., 1978. Fire in wilderness ecosystems. In: Hendee, J.C., Stankey, G.H., Lucas, R.C. (Eds.), Wilderness Management. Miscellaneous Publication No. 1365, Washington, DC, pp. 249–278. USDA Forest Service, 546 p.

Jones, G.M., Gutierrez, R.J., Tempel, D.J., Whitmore, S.A., Berigan, W.J., Peemy, M.Z., 2016. Megafires: an emerging threat to old-forest species. Front. Ecol. Environ. 14, 300–306.

Kauffman, J.B., Steele, M.D., Cummings, D., Jaramillo, V.J., 2003. Biomass dynamics associated with deforestation, fire, and conversion to cattle pasture in a Mexican tropical dry forest. For. Ecol. Manag. 176, 1–12.

Liu, Y., Stanturf, J., Goodrick, S., 2018. Trends in global wildfire potential in a changing climate. For. Ecol. Manag. 259, 685–697.

MacDonald, L.H., 2000. Evaluating and managing cumulative effects: process and constraints. Environ. Manag. 26, 299–315.

Malhi, Y., Roberts, J., Betts, R., Killeen, T., Li, W., Nobre, C., 2008. Climate change, deforestation, and the fate of the Amazon. Science 319, 169–172.

Margono, B.A., Potapov, P.V., Turubanova, S., Stolle, F., Hansen, M.C., 2014. Primary forest cover loss in Indonesia over 2000-2012. Nat. Clim. Change 4, 730–735.

McGuire, L.A., Youberg, A.M., 2019. Impacts of successive wildfire on soil hydraulic properties: Implications for debris flow hazards and system resilience. Earth Surf. Process. Landf. 44, 2236–2250.

Miller, C., Ager, A.A., 2012. Review of recent advances in risk analysis for wildfire management. Int. J. Wildland Fire 22, 1–14.

Moody, J.A., Martin, D.A., 2001. Post-fire, rainfall intensity—peak discharge relations for three mountainous watersheds in the western USA. Hydrol. Process. 15, 2981–2993.

National Interagency Fire Center, 2019. Prescribed Fires and Acres by Agency 1998 to 2018 <https://www.nifc.gov/fireInfo/fireInfo_stats_prescribed.html> (accessed 01.05.19.).

Neary, D.G., 2011. Impacts of wildfire severity on hydraulic conductivity in forest, woodland, and grassland soils. In: Elango, L. (Ed.), Hydraulic Conductivity-Issues, Determination and Applications. INTECH, Rijeka, Croatia, pp. 123–142. ISBN 978-953-307-288-3, 434 p.

Neary, D.G., Leonard, J.L., 2015. Multiple ecosystem impacts of wildfire. In: Bento, A.J.B., Vieira, A.A.B. (Eds.), Wildland Fires—A Worldwide Reality. pp. 35–112. Nova Science Publishers, Hauppauge, New York, 229 p.

Neary, D.G., Ryan, K.C., DeBano, L.F., 2005. Wildland Fire in Ecosystems: Effects of Fire on Soil and Water. USDA Forest Service, Rocky Mountain Research Station, Ogden, UT, p. 250, General Technical Report RMRS-GTR-42 Volume 4.

North, M.P., Stephens, S.L., Collins, B.M., Agee, J.K., Aplet, G., Franklin, J.F., et al., 2015. Reform forest fire management. Science 349, 1280–1281.

Piñol, J., Terradas, J., Lloret, F., 1998. Climate warming, wildfire hazard, and wildfire occurrence in coastal eastern Spain. Clim. Change 38, 345–357.

Pyne, S.J., 1982. Fire in America: A Cultural History of Wildland and Rural Fire. University of Washington Press, Seattle, p. 654.

Raison, R.J., Woods, P.V., Jakobsen, B.F., Bary, G.A.V., 1986. Soil temperatures during and following low-intensity prescribed burning in a eucalyptus pauciflora forest. Austr. J. Soil Res. 24, 33–47.

Rhoades, C.C., Fornwalt, P.J., Paschke, M.W., Shanklin, A., Jonas, J., 2015. Recovery of small pile burn scars in conifer forests of the Colorado Front Range. For. Ecol. Manag. 347, 180–187.

Rhoades, C.C., Meier, A.J., Rebertus, A.J., 2004. Soil properties in fire-consumed log burn-out openings in a Missouri oak savanna. For. Ecol. Manag. 192, 277–284.

Robichaud, P.R., Lewis, S.A., Brown, R.E., Ashmun, L.E., 2009. Emergency post-fire rehabilitation treatment effects on burned area ecology and long-term restoration. Fire Ecol. 5, 115–128.

Ryan, K.C., 2002. Dynamic interactions between forest structure and fire behavior in boreal ecosystems. Silva Fennica 36, 13–39.

Ryan, K.C., Jones, A.T., Koerner, C.L., Lee, K.M., 2012. (Eds.) Wildland Fire in Ecosystems: Effects of Fire on Cultural Resources and Archeology. USDA Forest Service, Rocky Mountain Research Station, Ogden, UT, p. 224, General Technical Report RMRS-GTR-42 Volume 3.

Sandberg, D.V., Ottmar, R.D., Peterson, J.L., 2002. (Eds.) Wildland Fire in Ecosystems: Effects of Fire on Air. USDA Forest Service, Rocky Mountain Research Station, Ogden, UT, p. 79, General Technical Report RMRS-GTR-42, Volume 5.

Schiffman, R., 2015. Rain-forest threats resume. Sci. Am. 312, 24.

Schmidt, K.M., Menakis, J.P., Hardy, C.C., Hann, W.J., Bunnell, D.L., 2002. Development of Coarse-Scale Spatial Data for Wildland Fire and Fuel Management. USDA Forest Service, Rocky Mountain Research Station, Fort Collins, CO, p. 41, General Technical Report RMRS-87.

Sc&nott, A.C., 2000. The pre-quaternary history of fire. Paleogeogr. Paleoclimatol. 164, 281–329.

Shvidenko, A.Z., Shchepashchenko, D.G., 2011. Impact of wildfire in Russia between 1998-2010 on ecosystems and the global carbon budget. Doklady Earth Sci. 441, 1678–1682.

Simard, A.J., 1991. Fire severity, changing scales, and how things hang together. Int. J. Wildland Fire 1, 23–34.

Smith, J.K., 2000. (Ed.) Wildland Fire in Ecosystems: Effects on Fauna. USDA Forest Service, Rocky Mountain Research Station, Fort Collins, CO, p. 83, General Technical Report RMRS-GTR-42, Volume 1.

Spiegel, K.M., Robichaud, P.R., 2007. First-year post-fire erosion rates in Bitterroot National Forest Montana. Hydrol. Process. 21, 998–1005.

Stephens, S.L., Burrows, N., Buyantuyev, A., Gray, R.W., Keane, R.E., Kubian, R., et al., 2014. Temperate and boreal forest mega-fires: characteristics and challenges. Front. Ecol. Environ. 12 (2), 115–122.

Walker, A.S., 1997. Deserts: Geology and Resources. <http://pubs.usgs.gov/gip/deserts/>.

White, C.R., Harley, G.L., 2016. Historical fire in longleaf pine (*Pinus palustris*) forests of south Mississippi and its relation to land use and climate. Ecosphere 7. Available from: https://doi.org/10.1002/ecs2.1458.

Wright, H.A., Bailey, A.W., 1982. Fire ecology-United States and Southern Canada. John Wiley & Sons Inc, New York, p. 501.

Yackinous, W.S., 2015. (Ed.) Understanding Complex Ecosystem Dynamics. Elsevier, Amsterdam, p. 416.

Zouhar, K., Smith, J.K., Sutherland, S., Brooks, M.L., 2008. Wildland Fire in Ecosystems: Fire and Nonnative Invasive Plants. USDA Forest Service, Rocky Mountain Research Station, Ogden, UT, p. 355, General Technical Report RMRS-GTR-42 Volume 6.

Further reading

Boisramé, G.F.S., Thompson, S.E., Kelly, M., Cavalli, J., Wilkin, K.M., Stephens, S.L., 2017. Vegetation change during 40 years of repeated managed wildfires in the Sierra Nevada, California. For. Ecol. Manag. 402, 241–252.

Cannon, S.H., Boldt, E.M., Laber, J.L., Kean, J.W., Staley, D.M., 2011. Rainfall intensity–duration thresholds for postfire debris-flow emergency-response planning. Nat. Hazards 59, 209–236.

Smith, J.K., Zouhar, K., Sutherland, S., Brooks, M.L., 2008. Chapter 1: Fire and nonnative invasive plants—introduction. In: Zouhar, K., Smith, J.K., Sutherland, S., Brooks, M.L. (Eds.), Wildland Fire in Ecosystems: Fire and Nonnative Invasive Plants. USDA Forest Service, Rocky Mountain Research Station, Ogden, UT, pp. 1–31. General Technical Report RMRS-GTR-42 Volume 6, 355 p.

Converting agricultural lands into heathlands: the relevance of soil processes

13

Rudy van Diggelen[1], Roland Bobbink[2], Jan Frouz[3], Jim Harris[4] and Erik Verbruggen[5]

[1]*Ecosystem Management Research Group, University of Antwerp, Antwerp, Belgium*
[2]*B-WARE Research Centre, Radboud University, Nijmegen, The Netherlands*
[3]*Institute for Environmental Studies, Charles University, Prague, Czech Republic*
[4]*School of Energy, Environment and Agrifood, Cranfield University, Bedfordshire, United Kingdom*
[5]*Plants and Ecosystems, University of Antwerp, Antwerp, Belgium*

13.1 Introduction

Heathlands in the Atlantic coastal region of Europe are in decline, and efforts are underway to restore them by recreating low fertility conditions. To speed up such impoverishment processes, nature managers have used alternative, rather drastic, techniques and stripped away the entire topsoil. Not only nutrients but also above- and belowground vegetation and the complete soil community are removed with such a technique, and recovery has to start from scratch.

Until the end of the 19th century, agricultural productivity in Europe was limited by natural soil fertility and locally produced fertilizers, mainly manure from animal husbandry. In regions with infertile soils, this led to a so-called infield—outfield system where organic material and animal dung were collected from large grazing areas and used to fertilize much smaller arable fields close to the settlements. In Western Europe, outfields were normally covered with heathlands, a vegetation type unique to the Atlantic coastal region that can survive on extremely impoverished soils and consists mainly of small dwarf shrubs from the *Ericaceae* (Loidi et al., 2010).

This practice changed radically at the beginning of the 20th century with the onset and massive expansion of industrial production of artificial fertilizers. Agricultural yields increased enormously because outfields could now be used for crop production, and existing farmland could produce much more than before. With the famines of the 19th century still in mind, politicians enthusiastically supported agricultural intensification programs, first at the national level and later via the Common Agricultural Policy at the European level.

Soils and Landscape Restoration. DOI: https://doi.org/10.1016/B978-0-12-813193-0.00013-8

357

From an agricultural point of view the intensification programs have been highly successful, but the ecological drawback is a large decline in surface area and quality of outfields. Heathlands are now classified as "vulnerable" at a European scale (Janssen et al., 2016) and have declined in surface between 30% and 50% during the last half century. Regionally, the decline can be even much larger (Pywell et al., 1994; Bakker et al., 2012). To counteract these unwanted developments, programs have been set up to increase heathland surface through restoration.

Starting from fertilized grasslands, a first and essential step is to lower site fertility, often through mowing and removal of the cuttings afterward but without adding fertilizer (Bakker, 1989). Unfortunately, it may take a very long time before such management results in nutrient-poor, yet species-rich grasslands (Bakker and Olff, 1995; Pywell et al., 2011; Redhead et al., 2014), and a further development to heathlands has yet to be observed. A major reason may be that nutrient availability is still too high. Nutrient balance calculations suggest that in regions with high external nutrient input, for example, via atmospheric N deposition, net nutrient export via mowing is very low (Jones et al., 2017) (Fig. 13.1).

13.2 Soil chemistry

Soil N pools are generally larger than P pools (Verhagen, 2007), but accumulation of P may nevertheless be a serious bottleneck for the restoration of oligotrophic ecosystems. Decades of intensive agriculture may result in P saturation as deep as 20−100 cm of the profile (Smolders et al., 2008). Under dry conditions, P is highly immobile and P losses consist mainly of removal by mowing while accumulated N can be lost either by denitrification, leaching of nitrate or mowing. Moreover, plants contain on average 10−20 times more N than P (Larcher, 2001; Güsewell, 2004), and mowing therefore results in much larger removal of N than of P. Nevertheless, when the P pool is small, mowing may be an effective technique to remove so much P from the soil that it becomes a limiting element for vegetation productivity (Härdtle et al., 2009). Unfortunately, this is an exception to the general rule. Normally soil P pools in formerly fertilized fields are very large, and the fraction of P that is removed via mowing is negligible. Consequently, the productivity there is limited by N (Verhagen, 2007). The sites are highly susceptible to enhanced N availability through atmospheric deposition (Bobbink et al., 2010) and require constant on-site management to keep productivity low (Jones et al., 2017). Moreover, high levels of P may inhibit mycorrhizal association establishment.

Topsoil removal on former agricultural fields is a drastic measure that removes most of the N and P (Verhagen, 2007; Klimkowska et al., 2007), but recovery of the typical heathland vegetation is in many cases poor. It has been hypothesized that suboptimal soil conditions are a major cause of this limited

FIGURE 13.1

A yearly mown grassland that has not been fertilized for over 10 years that is still rather productive and poor in species.

success, more in particular a soil pH that lies outside the optimal range for heathland (Pywell et al., 1994) (Fig. 13.2).

Decomposition of organic material and mineralization of N and P are enhanced at higher pH, and the resulting nutrient availability may become so high that heathland species are outcompeted by fast(er) growing grasses and ruderals (Pywell et al., 1994; Pywell et al., 2011). The application of elemental S (S^0) is sometimes proposed as a suitable measure after topsoil removal to lower soil pH toward the values appropriate for heathlands. Until now, the effects of S^0 are little studied; only short-term effects on soil pH and *Calluna vulgaris* establishment have been shown in the United Kingdom (Owen et al., 1999; Owen and Marrs, 2000; Lawson et al., 2004; Tibbett and Diaz, 2005).

These relationships were further investigated in a heathland restoration study in the Northern Netherlands (see case study for more details) where topsoil was removed in 2011 on c. 160 ha former agricultural land. We analyzed the effects of soil pH manipulation and inoculation with biota via herbage or crumbled sods, in all possible combinations for 7 years (Weijters et al., 2015; van der Bij et al., 2017). The addition of S^0 lowered soil pH and base cation concentrations and increased dissolved Al. We found a clear shift in soil abiotic conditions from those typical of a common grassland to conditions that lie within the range of

FIGURE 13.2

The situation 1 year after topsoil removal.

well-developed heathlands (De Graaf et al., 2009). Concentrations of available P, ammonium, and nitrate remained very low during the whole period, showing an effective reduction of site fertility after topsoil removal. Moreover, the addition of biota did not have any effect on the soil chemical parameters that we measured, implying that chemical alterations were not caused by the addition of herbage or sods. In conclusion, topsoil removal in combination with manipulation of the pH was highly successful in shifting soil abiotic conditions toward values suitable for heathlands.

13.3 Vegetation

Depleting soil nutrients by regular mowing is used widely by nature managers. Over time, the vegetation becomes less productive and more open; competition for light decreases (Kotowski and van Diggelen, 2004; Hautier et al., 2009) and less-competitive species establish. However, even when the productivity lies within the target range, the vegetation may still remain highly undersaturated in species (Klimkowska et al., 2007). Apart from abiotic constraints, biotic bottlenecks such as high resistance of an existing canopy against invasion (Tilman, 1997;

Chytrý et al., 2008) in combination with a low seed pressure of uncommon species (Bekker et al., 1997; Ozinga et al., 2009) are major reasons for low diversity.

Topsoil removal changes these conditions, making them suitable for the establishment of species from low-productive environments. Competition with existing vegetation stays low for years, and establishment gaps are widely available. However, the problem of low seed pressure of less-common species remains, and the emerging vegetation mainly reflects the species palette of the neighborhood (Verhagen et al., 2001). A well-established technique to speed up establishment is the transfer of fresh herbage with ripe seeds of well-developed communities to an abiotically restored area (Kiehl et al., 2010).

In heathland restoration, there is only limited experience with soil transfer (Pywell et al., 2011; van der Bij et al., 2018), but the available evidence suggests additional beneficial effects, at least in the first years after application. Heathland species establish fast and reach a higher cover than in treatments without soil addition. To what degree this is to be attributed to the transfer of seeds with the added soil is unknown but seems only likely for uncommon species. Differences in establishment of dominant heathland species between herbage-only and soil addition treatments therefore suggest additional beneficial effects of the soil community (Fig. 13.3).

FIGURE 13.3

The situation three years after topsoil removal and addition of soil from a well-developed heathland.

13.4 Soil microbial communities

Determining characteristics of the soil microbial community to inform progress toward a desired target condition has been well established (Harris, 2003). However, recognition of the role of the microbial community in facilitating or even determining the composition of the vegetation assembly is a more recent development (Harris, 2009; Wubs et al., 2016). Certainly, the composition of the microbial community can be changed by reducing the freely available nutrient pool to levels found in reference systems, increasing the size, and changing the composition, of the soil organic component but this is a very long-term process.

Reducing biomass through the removal of the top layers of topsoil in order to effect rapid change has shown to be effective—potentially leading to a change from the bacterial to the fungal channel of nutrient cycling. This is important as it has been suggested that bacterially dominated systems enhance rates of nutrient cycling, favoring ruderal species—particularly grasses—whereas fungal energy channels promote slow and conservative nutrient cycling (Wardle, 2002). Topsoil removal may shift the system to a different state and enable the start of heathland community development (Harris, 2013).

Van der Bij et al. (2017, 2018) have demonstrated that removal of topsoil reduces the size of the microbial biomass significantly, as might be expected. But this also shifts the F:B (fungal to bacterial) ratio from a bacterially to a fungally dominated system, which coincides with switching to a heather-dominated vegetation. Phospholipid fatty acid profiling also demonstrated that adding a soil inoculum of the target system rapidly establishes a microbial community similar to that of the target system but with a greatly reduced biomass (Van der Bij et al., 2017, 2018).

Conversion of former agricultural land into heathland not only requires changing from a bacteria-dominated microbial community into one more dominated by fungi but also changing the dominant mycorrhizal type. While most agricultural crops and temperate meadows form arbuscular mycorrhizae (AM), heather and most other *Ericaceae* species form ericoid mycorrhizae (ErM). These are formed with a select group of fungi, predominantly from the phylum Ascomycota (Read et al., 2004), and are composed of strongly inflated epidermal cells colonized by ErM fungi.

Heathland plants and their mycorrhizal fungi are a very effective symbiosis. Within higher plants, *Ericaceae* have among the lowest nutrient content and decomposition rates (Cornelissen et al., 2001). Therefore in the typically infertile systems where they occur, most of the nutrients required for plant growth have to come from slowly cycling biomass. Mycorrhizal fungi help ericaceous plants take up organic N from their litter (Kerley and Read, 1997), which is largely inaccessible to other plants due to its high recalcitrance (Jalal et al., 1982). The same is true for uptake of N from fungal necromass (Kerley and Read, 1997), both of which lead to tight recycling of scarce nutrients. In contrast, the AM fungal symbionts of most grasses and forbs are especially effective at increasing uptake of inorganic P and N but only have low capacity to release N from organic matter (Smith and Read, 2008), which means AM plants have a low competitive capacity when the recalcitrant heather litter is the main source of nutrients. Furthermore, the relatively fast turnover of AM mycelia compared with nutrient-conservative ErM mycelia (Olsson and Johnson, 2005; Verbruggen et al., 2017) leads to a high requirement of N in the former that is a poor match for systems of low fertility.

Indeed, ErM are associated with sites of lower productivity than AM systems, as well as a lower pH (Bueno et al., 2017). However, while there certainly is a favorable range of abiotic conditions that differ for these mycorrhizal types, these ranges have a high overlap (Soudzilovskaia et al., 2015; Bueno et al., 2017). Likewise, there is a large overlap of climatic and abiotic properties (nutrients, pH), under which both heather and grassland can come to dominate (Ransijn et al., 2015). The alignment of plant traits (nutrient content, growth rates), their mycorrhizal specificity, and the traits of these mycorrhizal fungi may therefore lead to a positive feedback with the power to stabilize one vegetation type when abundant.

Mono-dominant heathlands and grasslands are thought to represent two "alternative stable states" (Berendse and Scheffer, 2009). For this reason, restoration of an ecosystem of a different mycorrhizal type than the current undesired one is not just a matter of introducing appropriate fungi. In contrast, through the positive feedbacks described earlier, these fungi can be a pertinent source of soil legacy effects (Dickie et al., 2017) that may hamper the establishment of the desired vegetation. When considering restoration of heather vegetation, a situation must be put in place where the plant—mycorrhizal complex engages in a positive feedback—so for ErM plants low inorganic nutrient availability but potentially also a low abundance of AM hosts and AM propagules in soil.

13.5 Soil faunal communities

The composition of soil fauna communities of productive grasslands is entirely different from those of oligotrophic heathlands, both in size and quality (Frouz et al., 2009). On the one hand, macrofauna communities of heathlands are small and to a large extent dominated by predators and omnivores such as centipedes, spiders, and carabid beetles. Grassland macrofaunal communities, on the other hand, are large and consist mainly of earthworms and macrosaprophagous groups involved in litter fragmentation, such as some families of Diptera and millipedes. In other words, grassland communities are dominated by species that process the litter and contribute to bioturbation (i.e., mixing of organic matter into the mineral soil). Decomposition studies have emphasized the important role of macrofauna in grasslands and shown that raw organic matter is converted much faster into humus when it can be accessed by macrofauna (Frouz et al., 2009). Also, the composition of earthworm communities differs. Anecic and endogeic species make burrows in the soil and dominate grassland communities while litter-dwelling epigeic species dominate in heathland (Frouz et al., 2009).

Heathlands on the contrary harbor much larger populations of soil mesofauna of especially oribatid mites. These do not live in the soil but instead on top of it in the fermentation (Oe) layer that consists of partly decomposed plant material (Frouz et al., 2009). The mesofauna not only use this layer as habitat but also actively contribute to its formation by shredding raw organic remains into small, partly decomposed litter fragments (Frouz et al., 2009).

The distribution of organic matter in the soil profiles differs between habitats and is related both to the composition of the soil fauna and that of the vegetation. A high level of bioturbation is associated with high nutrient availability and fast-growing plants that produce large quantities of easily decomposable litter (Frouz, 2018). The microbial community and indeed the whole soil food web shift in the direction of a bacteria-driven system with fast decomposition and nutrient release. Bioturbation promotes the formation of an A horizon rich in organic matter and with high sorption capacity. This implies that an important part of the released nutrients can be stored in an easy accessible form in the soil, thus providing good conditions for fast-growing competitive plant species that produce highly decomposable organic matter. Conversely, accumulation of litter on the surface is characteristic of a food web dominated by fungi and characterized by low nutrient turnover rates. Stress-tolerant plant species with a conservative growth strategy dominate and produce recalcitrant organic matter (Frouz, 2018).

Converting productive agricultural grasslands into oligotrophic heathlands requires replacing macrofauna living in an organic layer in the soil with mesofauna living in an organic layer on top of the soil. Simply substituting the dominant grass vegetation by heather and/or other oligotrophic species, for example, by sowing with a high seed density, does not automatically solve this problem. Saprophagous macrofauna live in the A horizon or under the litter on the soil surface, and these habitats remain there even if we replace the grassland vegetation. Conversely, a typical heathland mesofauna community is adapted to the litter layer on the surface, and this needs a certain time to develop. All in all, this implies that soil community processes may significantly retard ecosystem changes (Foster et al., 2003; Kardol et al., 2007; Brudvig et al., 2013; Hahn and Orrock, 2015).

Previous research showed that complete topsoil removal is highly effective in removing the saprophagous macrofauna community (Frouz et al., 2009; van der Bij et al., 2018), but the formation of a litter layer on the soil surface may take a long time (Frouz et al., 2009). The large-scale field experiment in the Dwingelderveld (see case study) showed a significantly increased recovery rate of the mesofaunal community after inoculation topsoil removal sites with fresh herbage, or even better sods, from undisturbed heathlands. The increased recovery of oribatid mites is especially relevant because this group is typically quite slow to colonize (Frouz et al., 2009), due to their low reproduction rate and poor dispersal capacity. Sod transfer is considered particularly effective because it reduces several constraints at the same time: sods transfer both soil fauna and part of the habitat that is needed by this fauna, thus facilitating its establishment (van der Bij et al., 2018).

13.6 **Effects of alternative restoration strategies**

Decades of mowing of existing grasslands have resulted in less-productive and more open vegetation but unfortunately so far have not led to heathland development. Often this is attributed to depleted seed banks and limited dispersal rates (Kardol et al., 2008; Ozinga et al., 2009), but more inclusive views see a more active role for the soil community. In this view, microbial communities can keep the system locked in a previous stage by enhancing the performance of adapted species (Grime et al., 1987; van der Heijden, 2004; Dudenhoeffer et al., 2018). Since grasslands are dominated by bacteria and AM and heathlands by fungi and ErM, this suggests that, as long as the soil community has not been replaced, the performance of grasses is likely to be better than that of heather.

Such shift in soil community composition is not self-evident in existing vegetation. The close match between mycorrhizal type and vegetation composition implies that heather and ErM can only establish when they arrive at the same time in large enough numbers to be able to outcompete both the existing vegetation and the current AM. In addition, the legacy of the organic matter distribution in the soil profile can be a barrier (Schmidt et al., 2011). As long as organic matter stays well mixed with the upper soil layers and the soil fauna consists of bioturbators, decomposition rates will remain high (Frouz, 2018), and site fertility might still be high enough with a decreased nutrient pool size to enable competitive grassland species to outperform heathland vegetation.

Topsoil removal eliminates several of the constraints mentioned but adds others. Potentially superior competitors are no longer present, and soil legacies in the form of accessible and easily degradable organic matter have disappeared. At the same time the soil seed bank and the soil food web are gone, and species have to reach the site via dispersal. Dispersal probability decreases with distance (Bullock et al., 2017), and both the vegetation and the soil community are therefore mainly a reflection of the ecosystems in the neighborhood. On the one hand, mycorrhizal fungi are generally less dispersal limited than plants (Peay et al., 2010; Honnay et al., 2017), and this also is likely true for other microorganisms. Plant species on the other hand have more problems with dispersal, and the species composition of restored sites often consists predominantly of good dispersers (Kirmer et al., 2008). The same is true for certain soil faunal groups such as oribatid mites (Frouz et al., 2009). The soil community suffers from the additional constraint that most organic matter has been removed which is exactly the substrate they live on. Consequently, topsoil removal sites close to or surrounded by undisturbed heathlands can develop slowly in the desired direction but with a limited number of species (Verhagen et al., 2001; Frouz et al., 2009) and a lower microbial biomass (van der Bij et al., 2017). Isolated topsoil removal sites often develop in a totally different direction (van der Bij et al., 2017).

Inoculation of topsoil removal sites with sods solves many of the problems mentioned. Both the vegetation and the soil community are added, together with some organic matter of the right composition, and more or less in the right place in the profile, namely, on the soil surface. Heathland development goes much faster with inoculation than in noninoculated topsoil removal sites, but there is still ongoing community change (see case study), suggesting that developments have not finished yet. Moreover, many characteristics of the soil community still differ considerably from those in undisturbed heathlands. Nevertheless, sites treated this way are closest to undisturbed reference heathlands.

Supplying topsoil removal sites with only herbage brings the system into an undefined and, potentially static, intermediate state. Together with the herbage, some of the belowground community is introduced in small numbers in the restoration site, but the site remains open to dispersal from the surroundings. Organic matter availability is limited, and first needs to be built up by the establishing vegetation. The outcome of such a process is highly unpredictable, but a mismatch between above- and belowground communities is likely (van der Bij et al., 2018). This could lead to a lowered performance of heather as compared to competing species and might ultimately lead to a failure of restoration attempts that seem successful during early stages but then develop back into species-poor grasslands of low conservation value.

13.7 The Noordenveld experiment

The Noordenveld is a former heathland of about 160 ha in the center of the National Park Dwingelderveld in the Northern part of the Netherlands. It was converted into arable fields and grasslands in the 1930s and restored back in 2011 when c. 30 cm of topsoil was removed to quickly reach the low nutrient levels required for heathlands. This restoration was accompanied by a large-scale field experiment with plots of 15 m × 15 m. Different restoration measures were compared: topsoil removal only, topsoil removal + addition of freshly mown herbage from nearby heathlands with ripe seeds to eliminate possible plant dispersal barriers, and topsoil removal + addition of sod cut soil from nearby heathlands to eliminate possible dispersal barriers for both plants and soil organisms. In addition, soil pH was manipulated in a full factorial treatment independent of biota addition: (1) pH was lowered by addition of elemental sulfur (S^0), (2) pH unaltered, and (3) pH increased by addition of dolomite. Soil chemistry and vegetation were monitored each year. Soil fauna and soil microbes were measured immediately after topsoil removal, in the third year after the start of the experiment and will be again in the seventh year. The results presented here (Fig. 19.1) are those in the third year after topsoil removal and based upon Weijters et al. (2015) and van der Bij et al. (2018). By that stage, there were no significant effects of pH manipulation on the performance of biota.

The composition of the soil microorganism community in terms of the F:B ratio clearly differed between treatments with the sod addition treatment being closest to the reference. The control treatment with topsoil removal only is still close to the starting situation, and the treatment with herbage addition shows

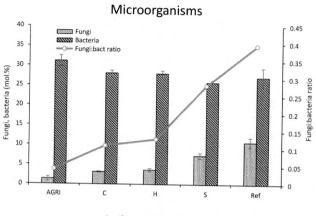

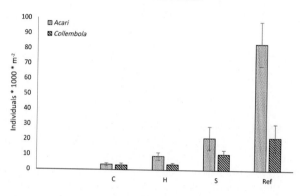

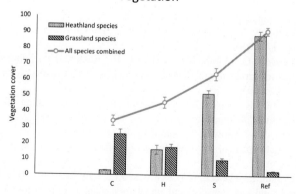

FIGURE 13.4

Response of several biotic parameters on alternative restoration treatments. AGRI, starting situation before topsoil removal; C, control; H, herbage; S, sods; and Ref, reference.

intermediate values. The results also demonstrate that the changes in ratio are not caused by differences in bacterial content but instead by increases in the amount of fungal material when going from control to sod addition and reference.

The soil mesofauna show a similar trend, but the absolute numbers are still much lower than those in the reference site. Moreover, the relative abundance of springtails (*Collembola*) as compared to soil mites (*Acari*) was much closer to one in all treatments than in the reference site, where the drought-resistant *Acari* are much more common than the drought-avoiding *Collembola*. The absolute numbers in all experiments are much lower than in the reference site, possibly because the organic matter content is much lower.

Finally, the vegetation composition shows two clearly diverging trends: from control with only topsoil removal to reference, there is a clear increase in both total vegetation cover and percentage of heathland plant species; whereas the opposite is true for grassland species. In other words, after 3 years the control sites are still a grassland—admittedly of much lower productivity—and the sites with crushed sod addition are already close to a heathland, though with lower standing biomass (Fig. 13.4).

References

Bakker, J.P. (Ed.), 1989. Nature Management by Grazing and Cutting: On the Ecological Significance of Grazing and Cutting Regimes Applied to Restore Former Species-Rich Grassland Communities in the Netherlands. Kluwer, Dordrecht.

Bakker, J.P., Olff, H., 1995. Nutrient dynamics during restoration of fen meadows by hay making without fertilizer application. In: Wheeler, B.D., Shaw, S.C., Fojt, W.J., Robertson, A. (Eds.), Restoration of Temperate Wetlands. Wiley, Chichester, pp. 143–166.

Bakker, J.P., Van Diggelen, R., Bekker, R.M., Marrs, R.H., 2012. Restoration of dry grasslands and heathlands. In: van Andel, J., Aronson, J. (Eds.), Restoration Ecology. Wiley, Chichester, pp. 173–188.

Bekker, R.M., Verweij, G., Smith, R.E.N., Reiné, R., Bakker, J.P., Schneider, S., 1997. Soil seed banks in European grasslands: does land use affect regeneration perspectives? J. Appl. Ecol. 34, 1293–1310.

Berendse, F., Scheffer, M., 2009. The angiosperm radiation revisited, an ecological explanation for Darwin's "abominable mystery". Ecol. Lett. 12, 865–872.

Bobbink, R., Hicks, K., Galloway, J., Spranger, T., Alkemade, R., Ashmore, M., et al., 2010. Global assessment of nitrogen deposition effects on terrestrial plant diversity: a synthesis. Ecol. Appl. 20, 30–59.

Brudvig, L.A., Grman, E., Habeck, C.W., Orrock, J.L., Ledvina, J.A., 2013. Strong legacy of agricultural land use on soils and understory plant communities in longleaf pine woodlands. For. Ecol. Manage. 310, 944–955.

Bueno, C.G., Moora, M., Gerz, M., Davison, J., Öpik, M., Pärtel, M., et al., 2017. Plant mycorrhizal status, but not type, shifts with latitude and elevation in Europe. Global Ecol. Biogeogr. 26, 690–699.

Bullock, J.M., Mallada González, L., Tamme, R., Götzenberger, L., White, S.M., Pärtel, M., et al., 2017. A synthesis of empirical plant dispersal kernels. J. Ecol. 105, 6−19.

Chytrý, M., Jarošík, V., Pyšek, P., Hájek, O., Knollová, I., Tichý, L., et al., 2008. Separating habitat invasibility by alien plants from the actual level of invasion. Ecology 89, 1541−1553.

Cornelissen, J., Aerts, R., Cerabolini, B., Werger, M., Van der Heijden, M., 2001. Carbon cycling traits of plant species are linked with mycorrhizal strategy. Oecologia 129, 611−619.

De Graaf, M.C.C., Bobbink, R., Smits, N.A.C., Van Diggelen, R., Roelofs, J.G.M., 2009. Biodiversity, vegetation gradients and key biogeochemical processes in the heathland landscape. Biol. Conserv. 142, 2191−2201.

Dickie, I.A., Bufford, J.L., Cobb, R.C., Grelet, G., Hulme, P.E., Klironomos, J., et al., 2017. The emerging science of linked plant − fungal invasions. N. Phytol. 215, 1314−1332.

Dudenhoeffer, J.-H., Ebeling, A., Klein, A.-M., Wagg, C., 2018. Beyond biomass: soil feedbacks are transient over plant life stages and alter fitness. J. Ecol. 106, 230−241.

Foster, D., Swanson, F., Aber, J., Burke, I., Brokaw, N., Tilman, D., et al., 2003. The importance of land-use legacies to ecology and conservation. Bioscience 53, 77−88.

Frouz, J., 2018. Effects of soil macro- and mesofauna on litter decomposition and soil organic matter stabilization. Geoderma 332, 161−172.

Frouz, J., Van Diggelen, R., Pižl, V., Starý, J., Háněl, L., Tajovský, K., et al., 2009. The effect of topsoil removal in restored heathland on soil fauna, topsoil microstructure, and cellulose decomposition: implications for ecosystem restoration. Biodivers. Conserv. 18, 3963−3978.

Grime, J.P., Mackey, J.M.L., Hillier, S.H., Read, D.J., 1987. Floristic diversity in a model system using experimental microcosms. Nature 328, 420−422.

Güsewell, S., 2004. N:P ratios in terrestrial plants: variation and functional significance: Tansley review. N. Phytol. 164, 243−266.

Hahn, P.G., Orrock, J.L., 2015. Land-use history alters contemporary insect herbivore community composition and decouples plant-herbivore relationships. J. Anim. Ecol. 84, 745−754.

Härdtle, W., von Oheimb, G., Gerke, A.-K., Niemeyer, M., Niemeyer, T., Assmann, T., et al., 2009. Shifts in N and P budgets of heathland ecosystems: effects of management and atmospheric inputs. Ecosystems 12, 298−310.

Harris, J., 2009. Soil microbial communities and restoration ecology: facilitators or followers? Science 325, 573−574.

Harris, J.A., 2003. Measurements of the soil microbial community for estimating the success of restoration. Eur. J. Soil Sci. 54, 801−808.

Harris, J.A., 2013. Case study: do feedbacks from the soil biota secure novelty in ecosystems? Novel Ecosystems. John Wiley & Sons, Ltd, pp. 124−126.

Hautier, Y., Niklaus, P.A., Hector, A., 2009. Competition for light causes plant biodiversity loss after eutrophication. Science 324, 636−638.

Honnay, O., Helsen, K., Van Geel, M., 2017. Plant community reassembly on restored semi-natural grasslands lags behind the assembly of the arbuscular mycorrhizal fungal communities. Biol. Conserv. 212, 196−208.

Jalal, M.A., Read, D.J., Haslam, E., 1982. Phenolic composition and its seasonal variation in *Calluna vulgaris*. Phytochemistry 21, 1397−1401.

Janssen, J.A.M., Rodwell, J.S., García Criado, M., Gubbay, S., Haynes, T., Nieto, A., et al., 2016. European Red List of Habitats. Part 2. Terrestrial and Freshwater Habitats. Publications Office of the European Union, Luxembourg.

Jones, L., Stevens, C., Rowe, E.C., Payne, R., Caporn, S.J.M., Evans, C.D., et al., 2017. Can on-site management mitigate nitrogen deposition impacts in non-wooded habitats? Biol. Conserv. 212, 464–475.

Kardol, P., Cornips, N.J., van Kempen, M.M.L., Bakx-Schotman, J.M.T., van der Putten, W.H., 2007. Microbe-mediated plant-soil feedback causes historical contingency effects in plant community assembly. Ecol. Monogr. 77, 147–162.

Kardol, P., Wal, A.V., der, Bezemer, T.M., Boer, W., de, Duyts, H., Holtkamp, R., et al., 2008. Restoration of species-rich grasslands on ex-arable land: seed addition outweighs soil fertility reduction. Biol. Conserv. 141, 2208–2217.

Kerley, S.J., Read, D.J., 1997. The biology of mycorrhiza in the *Ericaceae*. XIX. Fungal mycelium as a nitrogen source for the ericoid mycorrhizal fungus Hymenoscyphus ericae and its host plants. N. Phytol. 136, 691–701.

Kiehl, K., Kirmer, A., Donath, T.W., Rasran, L., Hölzel, N., 2010. Species introduction in restoration projects – evaluation of different techniques for the establishment of semi-natural grasslands in Central and Northwestern Europe. Basic. Appl. Ecol. 11, 285–299.

Kirmer, A., Tischew, S., Ozinga, W.A., von Lampe, M., Baasch, A., van Groenendael, J.M., 2008. Importance of regional species pools and functional traits in colonization processes: predicting re-colonization after large-scale destruction of ecosystems. J. Appl. Ecol. 45, 1523–1530.

Klimkowska, A., Van Diggelen, R., Bakker, J.P., Grootjans, A.P., 2007. Wet meadow restoration in Western Europe: a quantitative assessment of the effectiveness of several techniques. Biol. Conserv. 140, 318–328.

Kotowski, W., van Diggelen, R., 2004. Light as an environmental filter in fen vegetation. J. Veg. Sci. 15, 583–594.

Larcher, W., 2001. Ökophysiologie der Pflanzen: Leben, Leistung und Streßbewältigung der Pflanzen in ihrer Umwelt. Ulmer, Stuttgart.

Lawson, C.S., Ford, M.A., Mitchley, J., Warren, J.M., 2004. The establishment of heathland vegetation on ex-arable land: the response of *Calluna vulgaris* to soil acidification. Biol. Conserv. 116, 409–416.

Loidi, J., Biurrun, I., Campos, J.A., García-Mijangos, I., Herrera, M., 2010. A biogeographical analysis of the European Atlantic lowland heathlands: biogeographical analysis of European Atlantic lowland heathlands. J. Veg. Sci. 21, 832–842.

Olsson, P.A., Johnson, N.C., 2005. Tracking carbon from the atmosphere to the rhizosphere. Ecol. Lett. 8, 1264–1270.

Owen, K.M., Marrs, R.H., 2000. Creation of heathland on former arable land at Minsmere, Suffolk, UK: the effects of soil acidification on the establishment of *Calluna* and ruderal species. Biol. Conserv. 93, 9–18.

Owen, K.M., Marrs, R.H., Snow, C.S.R., Evans, C.E., 1999. Soil acidification—the use of sulphur and acidic plant materials to acidify arable soils for the recreation of heathland and acidic grassland at Minsmere, UK. Biol. Conserv. 87, 105–121.

Ozinga, W.A., Römermann, C., Bekker, R.M., Prinzing, A., Tamis, W.L.M., Schaminée, J.H.J., et al., 2009. Dispersal failure contributes to plant losses in NW Europe. Ecol. Lett. 12, 66–74.

Peay, K.G., Garbelotto, M., Bruns, T.D., 2010. Evidence of dispersal limitation in soil microorganisms: isolation reduces species richness on mycorrhizal tree islands. Ecology 91, 3631–3640.

Pywell, R.F., Webb, N.R., Putwain, P.D., 1994. Soil fertility and its implications for the restoration of heathland on farmland in southern Britain. Biol. Conserv. 70, 169–181.

Pywell, R.F., Meek, W.R., Webb, N.R., Putwain, P.D., Bullock, J.M., 2011. Long-term heathland restoration on former grassland: the results of a 17-year experiment. Biol. Conserv. 144, 1602–1609.

Ransijn, J., Kepfer-Rojas, S., Verheyen, K., Riis-Nielsen, T., Schmidt, I.K., 2015. Hints for alternative stable states from long-term vegetation dynamics in an unmanaged heathland. J. Veg. Sci. 26, 254–266.

Read, D.J., Leake, J.R., Perez-Moreno, J., 2004. Mycorrhizal fungi as drivers of ecosystem processes in heathland and boreal forest biomes. Can. J. Bot. 82, 1243–1263.

Redhead, J.W., Sheail, J., Bullock, J.M., Ferreruela, A., Walker, K.J., Pywell, R.F., 2014. The natural regeneration of calcareous grassland at a landscape scale: 150 years of plant community re-assembly on Salisbury Plain, UK (L. Fraser, Ed.), Appl. Veg. Sci., 17. pp. 408–418.

Schmidt, M.W.I., Torn, M.S., Abiven, S., Dittmar, T., Guggenberger, G., Janssens, I.A., et al., 2011. Persistence of soil organic matter as an ecosystem property. Nature 478, 49–56.

Smith, S.E., Read, D.J., 2008. Mycorrhizal Symbiosis, third ed. Academic Press, London.

Smolders, A.J.P., Lucassen, E.C.H.E.T., van der Aalst, M., Lamers, L.P.M., Roelofs, J.G. M., 2008. Decreasing the abundance of *Juncus effusus* on former agricultural lands with noncalcareous sandy soils: possible effects of liming and soil removal. Restor. Ecol. 16, 240–248.

Soudzilovskaia, Na, Douma, J.C., Akhmetzhanova, Aa, van Bodegom, P.M., Cornwell, W.K., Moens, E.J., et al., 2015. Global patterns of plant root colonization intensity by mycorrhizal fungi explained by climate and soil chemistry. Global Ecol. Biogeogr. 24, 371–382.

Tibbett, M., Diaz, A., 2005. Are sulfurous soil amendments (S^0, Fe(II)SO$_4$, Fe(III)SO$_4$) an effective tool in the restoration of heathland and acidic grassland after four decades of rock phosphate fertilization? Restor. Ecol. 13, 83–91.

Tilman, D., 1997. Community invasibility, recruitment limitation, and grassland biodiversity. Ecology 78, 81–92.

van der Bij, A.U., Pawlett, M., Harris, J.A., Ritz, K., van Diggelen, R., 2017. Soil microbial community assembly precedes vegetation development after drastic techniques to mitigate effects of nitrogen deposition. Biol. Conserv. 212, 476–483.

van der Bij, A.U., Weijters, M.J., Bobbink, R., Harris, J.A., Pawlett, M., Ritz, K., et al., 2018. Facilitating ecosystem assembly: plant-soil interactions as a restoration tool. Biol. Conserv. 220, 272–279.

van der Heijden, M.G.A., 2004. Arbuscular mycorrhizal fungi as support systems for seedling establishment in grassland. Ecol. Lett. 7, 293–303.

Verbruggen, E., Pena, R., Fernandez, C.W., Soong, J.L., 2017. Mycorrhizal interactions with saprotrophs and impact on soil carbon storage. In: Johnson, N.C., Gehring, C., Jansa, J. (Eds.), Mycorrhizal Mediation of Soil. Fertility, Structure, and Carbon Storage. Elsevier, Amsterdam, pp. 441–460.

Verhagen, H.M.C., 2007. Changing Land Use: Restoration Perspectives of Low Production Communities on Agricultural Fields After Top Soil Removal (Ph.D. thesis). University of Groningen.

Verhagen, R., Klooker, J., Bakker, J.P., Diggelen, Rvan, 2001. Restoration success of low-production plant communities on former agricultural soils after top-soil removal. Appl. Veg. Sci. 4, 75–82.

Wardle, D.A., 2002. Communities and Ecosystems. Princeton University Press, Princeton.

Weijters, M.J., van der Bij, A.U., Bobbink, R., Van Diggelen, R., Harris, J.A., Pawlett, M., et al., 2015. Praktijkproef heideontwikkeling op voormalige landbouwgrond in het Noordenveld. Resultaten 2011-2014. Report B-Ware. Nijmegen.

Wubs, E.R.J., van der Putten, W.H., Bosch, M., Bezemer, T.M., 2016. Soil inoculation steers restoration of terrestrial ecosystems. Nat. Plants 2, 16107.

Socioecological soil restoration in urban cultural landscapes

Loren B. Byrne

Department of Biology, Marine Biology and Environmental Science, Sustainability Studies Program, Roger Williams University, Bristol, RI, United States

14.1 Introduction

Around the world, agricultural and undisturbed native soils continue to be transformed by urbanization (Seto et al., 2011; FAO and ITPS, 2015; Li et al., 2018b). In the United States, states have lost, on average, more than 6% and 48% of their highly and moderately highly food-producing soils, respectively, to urban land uses, with losses in some states—including food-producing "breadbasket" ones—of more than 30% and 70%, respectively (Nizeyimana et al., 2001). In Sub-Saharan Africa, Muchelo (2017) estimated that urban landscapes replaced >74% of high-quality agricultural land between 1989 and 2015. Such soil-transformation trends are predicted to continue for the foreseeable future, raising concerns—if not alarm—about the future of society's "soil security" (Koch et al., 2013), and the abilities of scientists, land managers, and policy makers to respond to such threats (Lawler et al., 2014; Amundson et al., 2015; FAO and ITPS, 2015; Smith et al., 2016; Field et al., 2016). In addition, urbanization, alongside continued agriculturally-driven conversion of native ecosystems, has been found to threaten unique soil types with "extinction." This has caused soil scientists to call for conserving the Earth's pedodiversity, including 4540 rare and 508 endangered soil series in the United States (where 31 soil series have already become extinct) (Amundson et al., 2003; Tennesen, 2014).

In addition to decreasing pedodiversity, historical and current human activities interact to alter urban "brown infrastructure" (Szlavecz et al., 2018) in ways that affect its ability to provide the ecosystem services (e.g., nutrient cycling, water regulation, supporting plants) that enhance the well-being of urban residents (Elmqvist et al., 2015; Pavao-Zuckerman and Pouyat, 2017). Because many human activities negatively impact urban soils, these soils are often stereotyped as "degraded" due to reductions in their ecosystem services and increased generation of ecosystem disservices (Szlavecz et al., 2018). Though there is no doubt that urbanization harms soils in significant ways, many studies have revealed that degradation is not universal and that some urban soils *do* provide ecosystem services to varying degrees. As such, it is better to view urban soils in a spatially explicit, site-specific way with soil patches existing along a gradient, ranging from no or low (as in native ecosystem remnants) to severe degradation (as in industrially polluted sites) (sensu Baer et al., 2012). Given this tension between possible degradation and potential to contribute needed ecosystem services, the ecology of urban soils and their service-providing relationships to humans are "growing concerns" for science and society in a continuously urbanizing Anthropocene (De Kimpe and Morel, 2000).

These concerns related to urbanization of soils—loss of prime agricultural land, soil endangerment and extinction, reduced ecosystem services—motivate examination of ways to conserve and manage urban soils to improve the quality and sustainability of urban social-ecosystems and, by extension, surrounding regions and the earth system. To this end, a focal question is, how can degraded urban soils be restored (from any level of degradation to any degree of improvement) to help preserve Earth's pedo- and soil-biological diversity and improve the ecosystem services that support the well-being and security of urban residents? Advances toward an answer to this question have begun but remain embryonic (Crossman et al., 2007; Pavao-Zuckerman, 2008; Sloan et al., 2012; da Silva et al., 2018), despite increased study of urban ecosystems, urban soils, and their ecosystem services over the past several decades (e.g., Pavao-Zuckerman and Byrne, 2009; Elmqvist et al., 2015; Pickett et al., 2016; Setälä et al., 2014; Herrmann et al., 2017; Lepczyk et al., 2017; Anne et al., 2018). The lack of urban soil restoration knowledge is due, in part, to the fact that restoration ecologists have only recently begun placing formal, systematic emphasis on the need to consider soils (Heneghan et al., 2008; Baer et al., 2012; Moorhead, 2015; Perring et al., 2015) and urban systems as part of their research and practice (Lindig-Cisneros and Zedler, 2000; Gobster, 2001; Handel et al., 2013; Sack, 2013; Standish et al., 2013; Perring et al., 2015; Norris et al., 2017). (Reflecting this, a search at https://scholar.google.com/ for "urban soil restoration" with quotation marks only returned eight references in July 2020. In contrast, similar searches revealed that substantially more work has been done about urban wetland and aquatic ecosystem restoration (e.g., Windham et al. 2004, Bernhardt and Palmer 2007). Awareness is also growing about the need to consider sociocultural and landscape-scale variables as part of effective restoration work (Holl et al., 2003; Moreira et al., 2006; Menz et al., 2013; Hobbs et al., 2014; Perring et al., 2015; Aronson et al., 2017; Jellinek et al., 2019). These should also inform urban soil restoration, given that urban soils are impacted by socioculturally-driven decisions and are embedded in complex mosaics of highly heterogenous land uses and covers (Byrne et al., 2008; Pickett and Cadenasso, 2009; Ossola and Livesley, 2016; Lepczyk et al., 2017).

The goal of this chapter is to provide a summative overview of topics (derived from a primary literature review) that frame urban soil restoration, especially those pertaining to managing and improving urban ecosystem services from a landscape perspective. By necessity, a landscape perspective includes examining patch-level management (e.g., in parcels of <1 ha), since the characteristics of a landscape mosaic arise from the emergent, interacting characteristics of its patches and, in turn, the landscape mosaic affects site-specific conditions and management needs. Given this feedback relationship, restoring individual urban soil patches is integral to landscape-scale restoration. For the patch-level, topics to be examined include causes and consequences of urban soil degradation; key methods for improving soil conditions (focusing on compaction, organic matter, and pollutants); the creation of manufactured soils; and other biotic and ecological community factors that affect restoration outcomes. Regarding landscape and sociocultural contexts of urban restoration projects, two concluding sections discuss the complex challenges of planning and managing cultural landscapes and the need for stakeholder engagement and education efforts to help ensure successful outcomes.

Before examining those core topics, it is useful, and probably necessary, to first briefly address a question that arose during my literature review: what exactly does soil restoration "mean" in the context of urbanized landscapes? This question was inspired by current debates about the definition and standards of ecological restoration more generally. Though largely a philosophical and semantic concern, these debates relate to restoration ecology's formal scholarship pertaining to human-dominated landscapes and have direct implications for a second critical question: To what degree is urban soil restoration possible and practical? By addressing these issues explicitly, I hope to illustrate how urban soil restoration fits easily—and perhaps should even be central to—restoration science and practice in the Anthropocene.

14.2 What is urban soil restoration and is it possible?

Traditionally, a central goal of ecological restoration has been to return ecosystems to a historical baseline state that existed before degradation (i.e., restoration sensu stricto; Sack, 2013; Balaguer et al., 2014). Though a worthy goal that has guided successful projects, its practicality, utility, and relevance—including for soil (Baer et al., 2012) and urban restoration (Handel et al., 2013; Zeunert, 2013; Norris et al., 2017)—have been increasingly called into question. Main concerns, among others, include the impossibility in many cases of defining a relevant historical system, the severity and scale of degradation (e.g., due to establishment of invasive species), and the rapid pace of anthropogenically driven environmental changes, all of which can prevent the reestablishment of historical conditions (Seastadt et al, 2008; Suding, 2011; Hobbs et al., 2014). These concerns have led to calls for expanding the scope, flexibility, and goals of restoration (especially in urban landscapes) to include novel and engineered ecosystems, ecosystem services and other socioculturally-focused outcomes (such as cultural landscapes, education, and aesthetics), an approach I will call integrated socioecological restoration to connote this broader, transdisciplinary approach (Suding, 2011; Handel et al., 2013; Sack, 2013; Standish et al., 2013; Hobbs et al., 2013, 2014; Balaguer et al., 2014; Zeunert, 2013; Perring et al., 2015; Higgs et al., 2018; Jellinek et al., 2019). Replying to such calls, Aronson et al. (2018) countered that ecological restoration should remain defined sensu stricto (reflecting the Society for Ecological Restoration's 2016 standards which they cite), while other "restorative activities" for "creating or repairing human-made systems designed to meet short-term human needs and desires" (p. 916) should be called rehabilitation, reclamation, ecological engineering, or landscape design. Though I am sympathetic to sensu stricto restoration and support its goals in certain contexts, I propose (as have others; see Stanturf et al. Chapter 1, this volume) that an integrated social-ecological systems view is more realistic and appropriate for examining urban landscapes. As such, in this chapter "restoration" is used in a holistic sense to encompass all approaches within the "family of restorative activities" that are used to achieve diverse sociocultural and ecological outcomes (sensu Aronson et al., 2017).

Nonetheless, briefly thinking about urban soil restoration in the sensu stricto sense is innovative and exciting, perhaps a soil restorationist's dream. It could be a way to reestablish soils that have become extinct or endangered by urbanization, thereby conserving pedodiversity and soil biodiversity (Parker, 2010; Ibanez et al., 2012; Tennesen, 2014). If resources could be secured to attempt this, it would be fun, educational, and newsworthy to try, even if using a common soil type as the reference. Such a project, especially "resurrecting" an extinct soil (Amundson et al., 2003), would be an excellent "proof of concept" demonstration project and novel way to test ecological theories through restoration (i.e., Bradshaw's (1987) infamous acid test; also see Pavao-Zuckerman and Byrne, 2009). On the other hand, this prospect is nightmarish, perhaps a fool's errand. In extremely altered urban soils and landscapes, the challenges of sensu stricto restoration might be insurmountable in nearly all situations (Pavao-Zuckerman, 2008). The duration of many urban settlements (hundreds to thousands of years) complicates the notion of what a historical reference for some urban soils would even be. For example, in a small Scottish town Davidson et al. (2006) found that a 1-m-deep layer of human-deposited organic wastes from ∼200 to 400 years later was still present in the contemporary topsoil. Such severe alterations create legacies (or "memories" sensu Moreira et al., 2006) that are virtually unrestorable, including long-established non-native earthworm populations (Boyer et al., 2016) and elevated nutrient loads (Lewis et al., 2006). (Rather than be restored, perhaps these anthrosols should be valued and conserved as contributions to the Earth's pedodiversity.) In addition, evidence-based restoration practices to reverse degradation (e.g., adding organic matter; see Section 14.4) can place soils on a trajectory to become more dissimilar, rather than more similar, to historical counterparts. The urban sociocultural context may also preclude the goal of soil restoration sensu stricto because historical soils may not provide desired services (which could be why they have been altered; Moreira et al., 2006). Finally, given that soils (especially their profiles) form over a long time through complex interactions among soil formation factors—for which the "ecological past is extinct" (Handel et al., 2013: 694)—it seems impossible to think that attempts at urban soil restoration sensu stricto would be successful, at least on human timescales. These challenges may explain why I did not find evidence of any projects that explicitly identified this goal (including outside the primary literature) and why it will not be considered as an outcome for the restoration methods discussed later.

In contrast, a more meaningful, impactful—and possible—future for urban soil restoration, especially at landscape scales, is to be found through the lens of integrated socioecological restoration. Using this approach, urban soil restoration is highly desirable for a range of pragmatic, landscape-level sociocultural goals (including the traditional goal of reestablishing native biodiversity) (Pavao-Zuckerman, 2008; Handel et al., 2013; Perring et al., 2013; Standish et al., 2013). This broader perspective applies well to "novel urban soil ecosystems" which need novel management goals that may preclude recreating historical, preurban soil conditions (Pavao-Zuckerman, 2008; Seastadt et al., 2008; Perring et al., 2013; Standish et al., 2013; Hobbs et al., 2014; Zeunert, 2013; Egerer et al., 2018). Instead, restoration in urban contexts will often lead to different, but preferred, systems that include nonnative species, are sustainable and resilient (i.e., require no or fewer management inputs, especially after disturbances) and achieve desired social-ecological goals (e.g., enhanced aesthetics, carbon sequestration, and water, temperature and pest regulation; Elmqvist et al., 2015; Zeunert, 2013). One such goal that integrated socioecological urban soil restoration can embrace is that of reconciliation ecology, which seeks to manage human-dominated landscapes in ways that allow desirable species to co-inhabit them (Lundholm and Richardson, 2010). To such ends, a broader restoration framework allows for choosing plant species that can survive and reproduce in urban soils, even if they are non-native, rather than trying to restore soils to support historical plant communities (which the current urban context may prevent anyway; e.g., Hitchmough, 2008; Perring et al., 2015; Fischer et al., 2013). Using this broader definition, restored urban soils are a means to other ends (including ecosystem services and human well-being), rather than the restoration ends themselves (as suggested by the sensu stricto definition).

In this holistic, transdisciplinary spirit, I suggest the following definition for "urban soil restoration": the science and practice of managing urban soils to achieve integrated social-ecological goals for improving the sustainability of urban cultural landscapes and well-being of urban citizens and desired biodiversity. Though this definition does not align directly with restoration sensu stricto, it does not preclude such goals nor ignore the consideration of restoring soils and their biodiversity simply for their own sakes. Instead, it provides valuable flexibility (Higgs et al., 2018) that allows for using the word in ways that create more options for diverse social-ecological goals, thereby fostering productive collaborations among urban ecologists, planners, policy makers, and the public (Standish et al., 2013). It achieves this by accepting—even enthusiastically embracing-the reality that many urban habitats are permanently novel, some intentionally so; that people have diverse motivations and desires for managing urban landscapes; and that restorationists have a responsibility to help improve the well-being of urban societies. The following six sections highlight concerns and methods that inform the science and practice of integrated socioecological urban soil restoration.

14.3 Causes and consequences of urban soil degradation

Urbanization alters soil characteristics in many direct and indirect ways (Byrne, 2007; Byrne et al., 2008; Pavao-Zuckerman, 2008; Pickett and Cadenasso, 2009; Sloan et al., 2012). For instance, Herrmann et al. (2018) found that urbanization substantially modified soil profiles by deepening A horizons and eliminating B horizons. Such structural alterations are attributable to a suite of first-order causes of soil degradation associated with creating and managing urban infrastructure (including gray, green and blue varieties; see Li et al., 2017): scraping, removing, mixing, transporting, storing, replacing, and grading soils with heavy machinery (Fig. 14.1). Such activities degrade soil in many ways but especially through reduced OM (due to disturbance-induced decomposition; Chen et al., 2013, 2014) and compaction (but not always: see Edmondson et al., 2011). Often, subsoil is exposed or excavated and used to replace (or cover) original topsoil which reduces the quality of the new surface soil (Cheng and Grewal, 2009; Fig. 14.2A). Erosion caused by construction (Fig. 14.1D) and sealing with impervious surfaces also contribute to the physicochemical degradation of soils (Scalenghe and Marsan, 2009; Szlavecz et al., 2018), especially through loss of OM and nutrients (e.g., Jimenez et al., 2013; Raciti et al., 2012). These changes, alongside the addition of materials (e.g., intentionally as fill, via littering, left-behind byproducts of construction) such as bricks, gravel, broken-up concrete (Fig. 14.1C), dredged waterway sediments and other wastes (which may be challenging or impossible to remove during restoration work), create unique anthrosols that are often assessed as degraded compared to preurban soils due to properties that reduce their service-providing abilities (Meuser, 2010; Brose et al., 2016).

Second-order soil degradation occurs within urbanized landscapes via human activities that generate a variety of pollutants (e.g., heavy metals, fossil fuels, organic molecules such as PAHs and PCBs, excess nitrogen, salts). Common causes include industrial manufacturing (spills and wastes), transportation (spills and combustion of gasoline), and creation, management, deterioration, and disposal of infrastructure and other products (e.g., flaking lead-based paint; concrete erosion; treated wood; applying deicing salts to paved surfaces; waste dump sites; Figs. 14.1F and 14.3) (reviewed by Marcotullio et al., 2008; Meuser, 2010; Menefee and Hettiarachchi, 2017; Duarte et al., 2018; Li et al., 2018a). Heavy metals can also come from natural parent materials (bedrock) which creates a background level that is important to quantify because it can contribute to exceeding health risk thresholds when anthropogenic sources are added (Thomas and Lavkulich, 2015).

At smaller scales, management and recreation activities can degrade soil properties through fertilizer and pesticide applications, repeated lawn mowing, excessive human foot traffic (Fig. 14.1F) and removing detritus (e.g., lawn clippings, fallen tree leaves), all of which can also disrupt food webs and nutrient cycles. To counter these ecological changes, composts, biosolids, and mulches are often used as soil amendments (Fig. 14.2B; see next section). However, depending on their source, application rate and how they were processed, they may cause problems

FIGURE 14.1

Many activities in urbanized landscapes degrade soils including small-scale construction projects in which machinery (A) digs up and (B) scrapes soils, and (C) compacts and covers soils with rocks and concrete. (D) Such construction activities can lead to soil erosion even when steps (like barriers) are taken to try to protect soils. (E) Soil excavated from construction sites is often saved for later use (as in this 10-m-high pile). (F) After construction, human activities such as walking and applying deicing salts to sidewalks can interact to further degrade soils.

Courtesy of the author except (D) which is courtesy of Mac Callaham.

such as hydrophobicity, excessive nutrient loads, and introduced contaminants such as weed seeds, pathogens, salts (from food wastes), pharmaceuticals, and heavy metals (e.g., from sewage pipes) (De Miguel et al., 1998; Meuser, 2010; Sloan et al., 2012; Basta et al., 2016; Egendorf et al., 2018).

Urbanization-driven soil degradation has many consequences for biodiversity, ecosystem services, human well-being, and urban sustainability (reviewed in

(C)

INGREDIENTS:
Formulated from high-quality
organic and inorganic materials
derived from one or more of the
following: composted forest
products, peat, and sand
(sand not exceeding 10%)

FIGURE 14.2

Urban soil profiles and properties are often altered by human activities. (A) Subsoils which have more clay content and less organic matter (and are lighter in color) spread over topsoil (which is darker due to more organic matter) create unique urban profile discontinuities (*arrow*) and degraded surface soils that need restoration. (B) Restoration goals can be pursued through inputs of organic matter such as bagged soil blendsand manure, and wood mulch. (C) The ingredients listed on a bag of "topsoil" indicate its human-derived origin and reflect the need to consider the origins of restoration amendments to ensure desired outcomes. (D) Vegetable and ornamental gardens are places where landscape management activities such as organic matter inputs can help restore the urban soil landscape.

Courtesy of the author.

Pavao-Zuckerman and Pouyat, 2017; Szlavecz et al., 2018). For example, compaction can prevent certain organisms from surviving (e.g., cicadas, Moriyama and Numata, 2015) and reduces stormwater infiltration (e.g., Mohammadshirazi et al., 2016). Chemical contaminants can alter soil communities, affect plant health, and raise concerns for consuming food from urban gardens (though this risk, especially for lead, is generally perceived to be low and can be remediated, see Section 14.5; Henry et al., 2015). The multivariate causes and consequences of urbanizing soils lead to the general perception that urban soils are "degraded" and need management to improve them. Though often true, it is essential to note that

FIGURE 14.3

Knowing the history of land uses in a location can inform restoration projects. For example, soils under (A) scrap yards and (B) abandoned gas stations may need additional analyses for specific contaminants to determine whether remediation efforts are needed to make the soils suitable for other uses.

Courtesy of the author.

this should not be assumed for all urban soils, which are highly heterogenous; many, including unmanaged ones, have good quality and are able to support high biodiversity and provide valuable ecosystem services (Herrmann et al., 2017; Joimel et al., 2017; Szlavecz et al., 2018; Singh et al., 2019), including protecting waterways and humans from pollutants (e.g., Setälä et al., 2017). As such, site-specific evaluation is essential (though challenging because of resources needed for detailed analyses) to determine if urban soils in a location need restoration interventions (Pavao-Zuckerman, 2008; Montgomery et al., 2016; Anne et al., 2018). When they do, some of the approaches reviewed in the following sections may be appropriate.

14.4 Decompacting and adding organic matter

In many situations the most likely first steps to restore degraded urban soils are decompaction to decrease bulk density and replace lost nutrients by adding organic matter (OM), both of which will increase plant success. Often, these two treatments have been applied and studied in combination because decompaction methods can simultaneously integrate OM into the soil and OM helps improve soil aggregation that can maintain lowered bulk density and prevent re-compaction. Though practical, this makes it difficult to separate the effects of the two factors on restoration outcomes from each other. Further, the vast literature about these methods (including studies about agricultural systems that can inform urban projects, e.g., Tejada et al., 2009) is complicated by large variation in study systems (e.g., gardens, industrial sites, vacant lots, forests), experimental methods, and dependent variables. As such, the goal here is not a detailed review but rather to briefly synthesize previous studies that have examined whether these two steps can improve urban soils.

Decompaction methods break apart volumes of soil to create pore spaces for water, air, and organism movement, including root growth. Methods range from shallow rototilling to deeper "ripping" or fracturing which involves digging up and dropping soil in large volumes which is part of the "scoop and dump" and "profile rebuilding" methods (Sax et al., 2017). Mohammadshirazi et al. (2016) found that 15-cm deep tilling reduced bulk density and by itself increased water infiltration by 19–33 times relative to compacted soil and reduced erosion by 60%–82%; adding compost did not change these values. Other studies have similarly observed bulk density reductions with decompaction methods alone but also found that adding OM reduced it further with better outcomes for water infiltration and tree growth; however, results varied across tillage depth, soil types, locations, and species (e.g., Olson et al., 2013; Chen et al., 2014; Layman et al., 2016; McGrath and Henry, 2016; Somerville et al., 2018). In contrast, Loper et al. (2010) concluded from their experiment that tillage was not needed because it did not enhance restoration benefits compared to only adding OM. As such, the benefits of tilling as a standalone method appear to be useful only in some conditions (Somerville et al., 2018). One of those may be repairing severely compacted soils exposed after the removal of impervious surfaces (using a deeper tilling method named "suburban subsoiling" by Schwartz and Smith, 2016). Additional research to elucidate this context-dependency and isolate the effects of decompaction apart from OM inputs are needed to guide efficient and effective urban soil restoration.

Adding OM to degraded soils provides food for organisms (especially microbes) who can help create favorable soil conditions through their contributions to aggregate formation and soil fertility (e.g., via nitrogen mineralization). Studies have examined a diversity of OM amendments including compost created from dairy cow manure and municipal food and yard wastes; biosolids made from human sewage wastes; biochar (burned OM of all types); and blends that incorporate any combination of those materials, sometimes with chipped wood (e.g., sawdust), sand, or sediments dredged from water bodies (Fig. 14.2B and C). Regardless of OM type, nearly all studies have found consistently significant and positive results of OM additions on physicochemical and biotic variables, including fertility levels, microbes, earthworms, and plant health and growth (Loschinkohl and Boehm, 2001; McIvor et al., 2012; Chen et al., 2013; Scharenbroch et al., 2013; Oldfield et al., 2014, 2015; Carlson et al., 2015; Basta et al., 2016; Beniston et al., 2016; Badzmierowski et al., 2019; also see relevant studies cited above). Though most studies incorporate OM into the soil, leaving it on the surface (as mulch) is a lower cost strategy that can also have positive effects (Sæbø and Ferrini, 2006; Byrne et al., 2008; Scharenbroch, 2009). This includes leaving lawn clippings and fallen tree leaves on the soil surface as in situ OM additions which can also help protect and improve the sustainability of urban soils and landscapes through reduced management inputs (e.g., Knot et al., 2017).

The above-cited studies have revealed variation among OM types in the degree and duration of their effects and contextual factors (e.g., soil texture, plant species) that impact results (also see Weindorf et al., 2006; Alvarez-Campos and

Evanylo, 2019). For example, Iannone et al. (2013) tried using shredded OM from an invasive shrub to prevent its continued invasion in urban forests but it did not work. Thus any and all OM amendments cannot be assumed to provide universal solutions as some may work better than others in certain situations (Sæbø and Ferrini, 2006; Larney and Angers, 2012; Scharenbroch et al., 2013; Basta et al., 2016). Efficacy of OM inputs is also related to their quantity and quality (i.e., chemical composition) which may have negative impacts on soil fertility (e.g., high C:N causing nitrogen immobilization; Cogger, 2005), water quality (Sloan et al., 2012), and accumulation of introduced contaminants (see previous section). Therefore some caution is warranted about type of OM to use, how much to apply and the method of application, especially to reduce landscape-scale risks of large-scale applications (e.g., Egendorf et al., 2018). Nonetheless, in many, if not most, situations, available evidence indicates that the benefits of OM additions outweigh the risks and that adding OM improves degraded urban soils in many ways (a conclusion supported by the previous reviews of Cogger, 2005; Scharenbroch, 2009; Larney and Angers, 2012). However, as Oldfield et al. (2015) found, the benefits of OM amendments can take several years to appear, so, as with any restoration project, patience may be needed to see results. Research frontiers include application rates needed to achieve desired outcomes (Alvarez-Campos and Evanylo, 2019); how long benefits persist; how soil biodiversity and ecosystem services (e.g., carbon sequestration) are impacted over the short and long term; and whether repeated OM amendments can further enhance restoration improvements (Scharenbroch, 2009; Sloan et al., 2012; Doroski et al., 2018). An unexplored question is whether there are situations in which OM amendments would create ecosystem disservices, as in soils already prone to prolonged water saturation (e.g., clayey, low-lying) where OM may interfere with favorable drainage.

14.5 Remediating pollution

Alongside compaction and OM levels, a third major concern about urban soils is the contaminants they may contain. This issue has been studied extensively in cities and regions around the world because of its relevance for human health and water quality; perhaps more is known about pollutants than any other urban soil variable. However, making generalizations about patterns of urban soil pollution is challenging (if not impossible) given the high levels of (often fine-scale) spatial variation that has been documented (e.g., Delbecque and Verdoodt, 2016; Paltseva et al., 2018). Site-specific assessment is needed to determine what remediation, if any, is needed at a site, which can be frustrating given the resources needed for detailed analyses. Knowing as much as possible about the historical land use and cover is useful to guide decisions about which pollutants to focus on (Fig. 14.3; Delbecque and Verdoodt, 2016). Such decisions may also be informed by information from previous research that is briefly reviewed in this section (also see Meuser, 2013 and references in Section 14.3).

Remediating soil pollutants is often a necessary prerequisite to establishing plant communities. However, some polluted sites can support spontaneous and installed vegetation without remediation; such flora may help advance restoration and reduce the need for other interventions by improving soil conditions, in part through facilitating development of microbial communities that can degrade organic pollutants (e.g., PAHs) and improve soils in other ways, thereby facilitating further establishment of vegetation (Krumins et al., 2015; Pregitzer et al., 2016; Menefee and Hettiarachchi, 2017; Singh et al., 2019). Such phyto- and bioremediation approaches (Song et al., 2019, see Section 14.7) may be efficient ways to detoxify or immobilize certain pollutants as the ecosystem undergoes successional changes such as accumulating soil OM, which can increase the efficacy of phytoremediation (e.g., Chirakkara and Reddy, 2015).

Unfortunately, some soil pollutants cannot be bioremediated (Menefee and Hettiarachchi, 2017), which necessitates other intervention approaches. Ex situ remediation—removing soils from the original site to clean or dispose of them elsewhere—is a well-established method, as is the in situ approach of covering (capping) contaminated soils with layers of clean soil (Laidlaw et al., 2017). In many situations these are likely to be impractical and cost-prohibitive and therefore only used in extreme cases so they will not be reviewed further (for more see Liu et al., 2018). After soil removal or capping, the installation of constructed (or engineered) soils may be an effective way to restore parts of urban landscapes, an emerging approach that will be discussed in Section 14.6.

In contrast to ex situ, in situ remediation methods are more amenable to many restoration efforts as might be desired for urban community gardens, vacant lots, parks, and small parcels impacted by industrial activities or gray infrastructure (Fig. 14.3; Liu et al., 2018). Research to refine methods and develop new ones (e.g., using electrical currents, nanoparticles, and surfactants) show promise for enhancing future success of pollution remediation (Caliman et al., 2011; Mao et al., 2015; Liu et al., 2018; Xu et al., 2019; Song et al., 2019). Currently, substantial research indicates that organic and inorganic amendments—including composts, biosolids, biochar, lime, clay, and materials containing phosphorus and iron—work well to remediate many (but not all) key pollutants, including lead, cadmium, PAHs, and pesticides (Meuser, 2013; Henry et al., 2015; Kargar et al., 2015; Kästner and Miltner, 2016; Morillo and Villaverde, 2017; Obrycki et al., 2017; Lwin et al., 2018; Menefee and Hettiarachchi, 2017). Amendments work in several ways, including dilution which reduces the per-volume concentration and surface binding (immobilization) which makes pollutants biologically unavailable, both of which reduce plant uptake and human health risks (Liu et al., 2018; Menefee and Hettiarachchi, 2017; Song et al., 2019). Many authors have concluded that, especially with such remediation, the environmental and health risks of lead, for example, in urban soils is very low (e.g., Henry et al., 2015; Brown et al., 2016; Menefee and Hettiarachchi, 2017). As such, amendments, especially OM and phosphates, are being used to remediate pollutants in urban vegetable garden soils (e.g., Henry et al., 2015; Paltseva et al., 2018).

Though scientific understanding has progressed rapidly, challenges and concerns with remediation via amendments remain. The efficacy of amendments varies from soil to soil because texture, pH, biota, interactions with co-contaminants, and properties of organic amendments (e.g., compost age) can affect the immobilization of manypollutants, including through interactions among such variables (Henry et al., 2015; Ye et al., 2017; Basta et al., 2016 Huang et al., 2016; Song et al., 2019). Further, since immobilization does not actually remove pollutants from the site, they could become problematic in the future (Lwin et al., 2018; Song et al., 2019). For human health, contaminants remaining in the soil create some risks because of possible soil ingestion via hand-to-mouth transfer (especially by children) and from dust that is inhaled or deposited onto food crops and in buildings (Henry et al., 2015; Brown et al., 2016; Laidlaw et al., 2017; Paltseva et al., 2018). [This may relate to the finding by Laidlaw et al. (2017) that a reduction in soil lead levels, due to flooding, was correlated with reductions in children's blood levels in New Orleans, LA.] Finally, as noted, the need for site-specific data about soil conditions and resources for such analyses (including expert consultation) present obstacles to evidence-based decision-making about soil remediation. Even when sufficient data are available, urban restoration projects will be challenged by the many unknowns about how the complex, often unique, combinations of soil conditions and contaminants across urban landscapes will affect remediation results.

14.6 Manufacturing soils and greening roofs

In some urban locations, high-quality soil is wanted where it was removed (e.g., to deal with contamination or for construction of basements, swimming pools, pavement, gardens, and other infrastructure) and/or could not form naturally on human-timescales (e.g., over trash dumps, within densely paved areas, on roofs) (Fig. 14.4A−E). Creating soils de novo for such places can be viewed as a form of urban soil restoration because doing so will improve the landscape's abilities to provide associated ecosystem services, especially by increasing permeable surface area. These manufactured soils—also called Technosols and constructed, designed, or engineered soils—have unique properties that are usually dictated by specific "recipes" comprising varied mixtures of "growing material" (OM) and "structural material" (inorganic sediments such as rocks and crushed bricks) for management goals such as supporting trees (Fig. 14.4A) and rapid stormwater infiltration as in rain gardens (Fig. 14.4B) (Sloan et al., 2012; Yilmaz et al., 2018). Though knowledge about their development and function remain in its infancy, foundational research has revealed that their pedogenesis processes (e.g., aggregate formation, clay illuviation) and properties are affected by interactions among the materials used to create them and other soil formation factors, including precipitation and the biota that colonize them (Handel et al., 1997; Badin et al., 2009; Scharenbroch and Johnston, 2011; Hafeez et al., 2012; Huot et al.,

FIGURE 14.4

Restored soil patches in urbanized landscapes include those created for purposes such as (A) supporting street tree, (B) creating rain gardens in parking lots, (C) establishing green roofs, (D) filling in swimming pools, and (E) creating raised beds in community vegetable gardens. (F) All of these places can serve as hands-on educational sites to engage students and residents in learning about soils (e.g., examining soil texture as shown here) and their ecosystem services.

Courtesy (A, B and F) author; (C) TonyTheTiger used via CC BY-SA 3.0; (D) Gerald Hess; (E) Jeff Schuler via CC-BY-2.0.

2015; Deeb et al., 2017; Vergnes et al., 2017; Scharenbroch et al., 2018). Here, two specific areas of research are reviewed, focusing on soils "restored" to support street trees and green roofs, both of which provide valuable ecosystem services and represent an increasing proportion of urban cultural landscapes (Oberndorfer et al., 2007; Mullaney et al., 2015). [Less research has been done on soils created for rain gardens but see Mehring et al. (2016), Shuster et al. (2017), and Funai and Kupec (2019).]

Keeping street trees alive and healthy is challenging, especially regarding root growth and access to water and nutrients. Research has shown that trees growing in manufactured soils in tree pits or under pavement (specifically called structural soils; Fig. 14.4A) grow better and survive longer, and are healthier than those growing in compacted, urbanized soils (reviewed in Mullaney et al., 2015; Ow and Ghosh, 2017). In large part, this is due to the availability of pore spaces for roots and water to move into. However, variation in structural soil properties (e.g., chemistry and water retention) can impact trees negatively, including rooting depth, so care must be taken to design them properly for specific contexts and avoid mistakes during installation (Bartens et al., 2009; Bühler et al., 2017; Cannavo et al., 2018). This is especially important to prevent trees from falling over, which can be effectively achieved with structural soils (though with variation among tree species; Bartens et al., 2010; Ow and Mohd. Yusof, 2018). As for all urban soils, structural soils can be degraded by pollutants in stormwater run-off such as deicing salts which can increase tree mortality (Ordóñez-Barona et al., 2018). Given this and pedogenic changes that may be undesirable (Scharenbroch et al., 2018), it seems plausible that street tree soils and other urban Technosols may themselves eventually need restoration to improve their conditions and services, a wholly unexplored area for future studies.

Technosols constructed for green roofs (Fig. 14.4C) are also subject to pedogenic processes that change their characteristics significantly over time (even within 4 years: Bouzouidja et al., 2018), including successional-like development of their communities (Schrader and Böning, 2006; Ksiazek-Mikenas et al., 2018). In the short term, substrates used in green roof soil construction directly and indirectly affect plant performance (reviewed in Kazemi and Mohorko, 2017) and colonization by soil biota, including a diverse suite of bacteria, fungi, and arthropods like mites, collembola, spiders, beetles, ants, and bees (e.g., McGuire et al., 2013; John et al., 2014; MacIvor and Lundholm, 2011; Rumble and Gange, 2013; Molineux et al., 2015). Studies have found that the site-specific characteristics of a green roof (including the plants) affect its biota more than the surrounding landscape, leading to different communities among roofs (Madre et al., 2013; McGuire et al., 2013; Kyrö et al., 2018; Ksiazek-Mikenas et al., 2018; Schindler et al., 2019). Though this makes sense given that roofs are fragmented from each other and ground-level soil, Braaker et al. (2014, 2017) found evidence for biotic connectivity among roofs and surrounding habitats for more mobile arthropods (e.g., bees, weevils, running spiders). Nevertheless, the overall uniqueness of a roof's soil biodiversity is also enhanced by biota introduced via compost (Joimel et al., 2018), the soil of installed plants (Rumble et al., 2018) and microbial inoculations (Molineux et al., 2014; Young et al., 2015; Rumble and Gange, 2017).

Though green roofs provide habitats for diverse soil communities, in most cases this is a secondary goal—or probably an unappreciated byproduct—relative to primary goals for other ecosystem services (including providing habitat for aboveground biota). Nonetheless, creating green roof soils in ways that will specifically restore and conserve targeted soil populations is worthy of further exploration (see Williams et al., 2014). For instance, Kadas (2006) found that 10% of ground-dwelling arthropod species collected on green roofs in London were of conservation interest and could benefit from additional green roof habitat. From a landscape perspective, green roofs, street tree pits, rain gardens and other Technosol patches (Fig. 14.4A−E) could be created using a variety of "parent materials" and in strategic places (i.e., to increase connectivity and variety of environmental context) to maximize the restoration of soil biodiversity (i.e., increase beta and gamma diversity) and ecosystem services across urbanized landscapes (Ksiazek-Mikenas et al., 2018; Schindler et al., 2019).

14.7 Working with biota as restoration partners and foes

That physicochemical soil degradation affects soil organisms is well-established. In contrast, how plants and soil organisms affect urban soil properties and restoration outcomes has been investigated much less. As noted in Section 14.5, plants and microbes can provide "nature-based" pollution remediation in urban soils (Song et al., 2019), and studies identifying which species are more effective bioremediators generate insights for successful use of this approach (e.g., Jensen et al., 2009; Dadea et al., 2017; Liu et al., 2018; Shuttleworth et al., 2018). Favorable restoration outcomes can be achieved with such restoration partners but "restoration foes" can interfere with restoration goals.

Simply letting nature take its course (naturalization) is one possible management decision for degraded urban lands, especially in the context of financial and expertise constraints, and possibilities that interventions will negatively affect the functions of soil biota (for instance, liming was found to reduce the remediation abilities of bacterial communities: Hesse et al., 2019). Even in polluted and constructed soils (e.g., covering landfills), plants, including native ones, and soil biota, including mycorrhizal fungi, can establish and survive well (e.g., Handel et al., 1997; Fischer et al., 2013; Pregitzer et al., 2016; Everingham et al., 2019; Singh et al., 2019). Spontaneous, unmanaged vegetation can stabilize the soil, preventing erosion and movement of pollutants (Song et al., 2019), though Setälä et al. (2017) observed differences between deciduous and coniferous tree communities in their levels of water and heavy metal retention. Naturalization may be a good option for other degradation such as compaction from human traffic; for example, in an urban forested park Millward et al. (2011) found that allowing understory plants to grow without management (and keeping humans out) for

6 years reduced soil bulk density and increased water infiltration. As such, an important lesson for ecological restoration is that, when humans do nothing, wild and weedy restoration partners can sometimes provide ecosystem services that advance restoration goals.

When restoration goals necessitate intentional plant installation, existing urban soil properties can affect successful establishment, leading to the recommendations that plants should be chosen that match the local conditions—an important aspect of identifying restoration partners (Hitchmough, 2008; Haan et al., 2012). After installation, plants can improve soil conditions (in addition to that caused by initial soil preparation) by adding OM to the soil via roots and detritus (e.g., Vannucchi et al., 2015) and reducing bulk density and creating channels for water flow via root growth (e.g., Bartens et al., 2008). In restored prairies created in urban lawns, Johnston et al. (2016) found trends of improved soil conditions after 15 years (but see Yost et al. (2016) for more complex outcomes). Installed plants also influence restoration of soil microbial communities, such that they can become more like reference ecosystems (Gellie et al., 2017).

In addition to plants, soil microbes and animals, should be considered as potential restoration friends. For example, reestablishing microbial root mutualists (i.e., nitrogen-fixing bacteria, mycorrhizal fungi) that can improve host—plant establishment and health may be aided by inoculations to plants before installation (Fini et al., 2011; Bashan et al., 1999). Animals, such as arthropods and ecosystem engineers such as mammals and earthworms (even if they are not native), may also help improve soil conditions; thus, management to conserve or restore them could be an effective strategy for urban soil restoration (Byers et al., 2006; Snyder and Hendrix, 2008). An example is from Australia where bandicoots inhabiting urban landscapes dig into soil for food such as mycorrhizal fungi fruiting bodies; through their bioturbation and dispersal of fungal spores in their scat, bandicoots might aid in the restoration of a declining *Eucalyptus* species in degraded urban forest remnants (Tay et al., 2018). Though fascinating, such mutualistic animal—plant—soil—restoration relationships have rarely been examined in urban contexts and deserve much more attention.

Certain organisms can also create ecosystem disservices that impact urban restoration efforts. For instance, nonnative invasive species (e.g., plants, insects, worms), which are often especially successful in urbanized landscapes, are a form of "biological pollution" that can degrade soil conditions including bulk density, biodiversity and biogeochemical cycles (Szlavecz et al., 2006; Heneghan et al., 2009; Vilà et al., 2011; Ferlian et al., 2018). In turn, invasive-degraded soils may facilitate continued invasion and persistence of invasive species, with soil legacies (e.g., seed banks, altered structure and nutrient levels) remaining even after invasives are removed (e.g., Heneghan et al., 2009; Corbin and D'Antonio, 2012; Overdyck and Clarkson, 2012). Along these lines, studies have found that restoration interventions do not prevent regrowth of nonnative weeds (e.g., due to dispersal via birds from surrounding landscapes; Sullivan et al., 2009) and might actually promote them in some contexts (e.g., compost amendments can increase nonnative seedling recruitment: Doroski et al., 2018). Aboveground native animals may also interfere with achieving soil restoration goals because they can alter soil conditions alone and in combination with invasive species. For example, exclosure experiments revealed many direct and indirect negative effects of white-tailed deer (which are overpopulated in many U.S. urbanized landscapes) on variables relevant to urban soil restoration, including compaction, native plant survival (Shelton et al., 2014) and earthworm densities and biomass (Dávalos et al., 2015; Mahon and Crist, 2019). On the positive side, results from these studies suggest that when deer are excluded, soil conditions can "restore" themselves with potentially positive outcomes for native plants. This reflects the need for soil restorationists to consider a wide range of variables and management practices (not just soil and plant ones) to facilitate restoration success in urban landscapes.

14.8 Urban soil patches in cultural landscapes

More so than for other systems, urban social-ecosystems have unique characteristics that affect (and may limit) restoration work within them, especially regarding soils (Pavao-Zuckerman, 2008; Handel et al., 2013; Norris et al., 2017). In particular, urban places are best viewed as cultural landscapes (Moreira et al., 2006; Zeunert, 2013) in which extreme social-ecological spatial heterogeneity (patchiness) arises from parcels that have diverse sizes, owners, land uses and covers, management inputs and histories. Further, because patches do not exist in isolation, their characteristics—and restoration outcomes within them—will be influenced by their spatial context, especially due to cross-patch movement of ecological and sociocultural variables (e.g., water, organisms, information from neighbors and advertisements: Holl et al., 2003; Byrne and Grewal, 2008; Jellinek et al., 2019). Emergent aboveground landscape patterns—which are arguably what humans (mostly) respond to and manage—directly, indirectly (through habitat structure effects; Byrne, 2007), and collectively create heterogeneous "urban soil mosaics" (jointly with preurban spatial soil variation; Pickett and Cadenasso, 2009; Ossola and Livesley, 2016). To date, the few pertinent studies about these mosaics indicate that spatial patterns of urban soils affect landscape-level ecosystem service patterns and therefore overall urban sustainability (Edmondson et al., 2014a; Grafius et al., 2018; Ziter and Turner, 2018; Steele and Wolz, 2019). This suggests that urban soil restoration aimed at improving these outcomes should be done at the landscape-scale, even more so than in other systems (Crossman et al., 2007; Setälä et al., 2014).

Unfortunately, high spatial variation of social-ecological variables—including soil degradation, desired ecosystem services, and human communities—within urban cultural landscapes makes a larger scale approach for (both above- and belowground) restoration complicated and challenging. The spatial variation is compounded by diverse, interacting ecological and sociocultural variables within and between patches that give rise to novel ecosystem conditions at different locations; in turn, a patch's unique combinations of soil (e.g., OM, texture) and social characteristics (e.g., ownership, management resources, surrounding neighborhood, desired services) may dictate restoration possibilities (Pavao-Zuckerman, 2008). Further, conditions are temporally dynamic so that what is desired and possible for restoration at a location can change quickly over time (e.g., due to changing political and economic contexts: Xin et al., 2018). The emergent spatiotemporal complexity that emerges in urban cultural landscapes raises many questions for urban soil restoration efforts. None is more central than, which patches should be prioritized for restoration to achieve the best social-ecological outcomes for a given landscape?

In addition to knowing a lot about patch-specific soil, environmental and sociocultural variables across a landscape, comprehensively answering that question requires consideration of additional concepts, context, and methods, especially transdisciplinary ones, beyond those discussed above. Though these add to the complexity of pursuing landscape-level urban soil restoration, success can be advanced through integration of knowledge and methods from diverse fields that examine the spatiotemporal social-ecological complexity of urban cultural landscapes. In particular, urban and ecological planning (or design) are rich multidimensional fields with long histories and well-developed approaches (a detailed review of which is outside the scope here; for overviews see Levy, 2016; Heymans et al., 2019). A brief review of urban/ecological planning literature that specifically addresses soils and ecosystem services revealed oft-mentioned points, discussed in this section, that are needed to develop a fully integrated landscape-scale socioecological view of urban soil restoration.

The need to integrate soils and ecosystem services into urban planning is increasingly appreciated but needs further advances in both research and practice to guide improvements in urban landscapes' resilience and sustainability (Sack, 2013; Ahern et al., 2014; da Silva et al., 2018; Anne et al., 2018; Wilkerson et al., 2018; Heymans et al., 2019). Progress has been made in mapping urban ecosystem services and their economic values which can inform spatial planning decisions about restoration priorities (e.g., Crossman et al., 2013; Xin et al., 2018; Terzi et al., 2019), including incremental approaches in which patches are restored in sequence (Hobbs et al., 2014; Perring et al., 2015). However, planning is inherently complicated because of interpatch variation of many sociocultural and ecological variables which brings forth the concepts of landscape multifunctionality, complementarity, and trade-offs. Briefly, these define the reality that many goals exist for a landscape, but that no patches can contribute to all the goals and some patches might interfere with them. As such, decisions must be made about how to restore individual patches to maximize their contributions to different goals in complementary ways so that together they collectively maximize overall benefits (while minimizing risks and harms) across the landscape (Hobbs et al., 2014; Setälä et al., 2014; Perring et al., 2015; Zeunert, 2013; Xin et al., 2018). Therefore, thoughtful, spatially coordinated planning is needed to determine what restoration methods should be prioritized in what locations using optimization approaches, including spatial and economic modeling (Crossman et al., 2007; Menz et al., 2013; Setälä et al., 2014; da Silva et al., 2018). For instance, Dusza et al. (2017) concluded that differently designed Technosols exhibit trade-offs among various functions and benefits. Thus placement of new Technosol patches of different designs (i.e., material mixes) (see Section 14.6) could be determined by identifying prioritized needs for different locations (e.g., one type of Technosol to improve stormwater drainage in a site with high impervious surface cover (Fig. 14.4B and C) and another type in locations where ensuring tree stability (Fig. 14.4A) is a goal (e.g., Bartens et al., 2010). Other patches (e.g., vacant lots) could be identified as priorities for "sociocultural restoration" in which soils are managed primarily to improve socioeconomic goals such as food production (Fig. 14.4E; Egerer et al., 2018). The prospect of helping create and implement unified "multifunctional landscape soil restoration plans" provides inspiration for future developments and collaborations at the intersection of urban and restoration ecology, along with other fields.

Urban planning for landscape-scale soil restoration also requires working with many stakeholders (including scholars, practitioners, policy makers, organizations, and the public) who have diverse beliefs, interests, perceptions, values, and goals (Moreira et al., 2006; Standish et al., 2013; Ahern et al., 2014; Aronson et al., 2017). In this context, a variety of needs and motivations should be considered when planning urban soil restoration, including economic development, human health and safety, food production, clean water and air, aesthetics, recreation, and biodiversity conservation (e.g., Mills et al., 2017; McHale et al., 2018; Jellinek et al., 2019; Wilkerson et al., 2018). Many of these motivations may be linked to management of green and blue infrastructure so that soil restorationists need to remember to be open-minded and flexible when justifying soil restoration in discussions with diverse stakeholders; soil restoration may be best framed as the means to achieve other ends (see plans reviewed in da Silva et al., 2018). In this vein, effective communication, cooperation, and collaboration have been repeatedly emphasized as essential for successful urban restoration (e.g., Holl et al., 2003; Crossman et al., 2007; Standish et al., 2013; Hobbs et al., 2014; Perring et al., 2015; Anne et al., 2018; Jellinek et al., 2019). To facilitate these, decision support systems have been advocated for identifying priorities and building consensus around goals and management plans (Crossman et al., 2007; Hobbs et al., 2014; Aronson et al., 2017; Anne et al., 2018). For example, in a landscape-level urban park restoration project, Gobster (2001) found that a collaborative community−based process emphasizing shared goals and compatibility among outcomes resulted in satisfying outcomes. Indeed, ensuring that community members are included in the planning process and that their values and needs are addressed (i.e., the "democratization of ecosystem services" sensu McHale et al., 2018) may be core requirements to generate local support and improve a project's chances of short- and long-term success (Hychka and Druschke, 2017; Jellinek et al., 2019).

Even in otherwise-ideal situations, inherent features of urban cultural landscapes will challenge larger scale restoration planning and implementation. Conflicts within the planning process are probably inevitable given diverse stakeholders' worldviews and backgrounds; thus difficult compromises and trade-offs will have to be part of a negotiation process (Gobster, 2001; Jellinek et al., 2019; Heymans et al., 2019). Changing political climates and administrations; short funding cycles and lack of financial resources; and turnover in project staff and parcel ownership create additional constraints for urban restoration success (Hychka and Druschke, 2017; Norris et al., 2017). Even desired environmental outcomes of restoration efforts may create dissatisfaction andbacklash within neighborhoods which can overshadow benefits or prevent further progress. For instance, spontaneous plant communities that might advance soil restoration, as in vacant lots, have potential social downsides (i.e., disservices such as undesired appearance, production of allergens) that need to be considered within the planning process (Riley et al., 2018). Finally, a key constraint in many places is likely to be a lack of local urban and soil restoration expertise to carry out projects; in such cases, it may be necessary to seek out experienced professionals from outside the community (Norris et al., 2017). In the future, demand for such professionals will hopefully surge due to increasing inclusion of brown infrastructure and associated ecosystem services in urban ecological planning efforts.

14.9 The future of urban soil restoration

Assuming that demand for urban soil ecological knowledge (sensu Heneghan et al., 2008; Pavao-Zuckerman, 2008; Baer et al., 2012) for restoration projects will increase, the world will need more urban soil restoration scientists, practitioners, and educators. What should they be prepared to do? In other words, what might future research, practice, and education in urban soil restoration look like? Foundations of answers to these questions are provided by the topics discussed previously. This concluding section presents additional ideas about the future of these three dimensions of urban soil restoration. For all three, an integrated social-ecological, landscape-scale approach is needed, one that recognizes the legitimacy of diverse motivations, goals, and outcomes (especially enhancing urban ecosystem services) for improving urban soils.

14.9.1 Research

Overall, though knowledge about how to improve the "urban soil landscape" has been increasing (especially regarding the multiple benefits of OM additions), gaps in the literature point to opportunities for conducting foundational research needed to inform urban soil restoration. Frontiers include methods for and benefits of reestablishing and conserving desirable (even endangered) soil populations and communities in degraded soils (i.e., what are the "best" abundances, species and biodiversity for various goals?); diverse, reciprocal relationships between brown and other colors of infrastructure (Li et al., 2017); effects of climate change on urban soils, their services and restoration outcomes (e.g., Rawlins et al., 2015); the process of "anthropedogenesis" in manufactured and intensively managed soils (e.g., Bouzouidja et al., 2018; Scharenbroch et al., 2018); and the potential for nature-based solutions (Song et al., 2019) to provide inexpensive, widely adoptable restoration outcomes in urban locations where other methods cannot be used. More transdisciplinary research is also needed that links sociocultural with soil variables, such as how income and knowledge affect gardening activities that affect soils (Fig. 14.2), which can impact landscape-scale patterns of ecosystem services and therefore restoration plans (Byrne and Grewal, 2008; Egerer et al., 2018; Wilkerson et al., 2018; Ziter and Turner, 2018). For these and many other topics, the need for long-term studies (>5 years) and site-specific information presents major challenges for the research community (Pavao-Zuckerman, 2008). Another relatively unexplored frontier is cross-system transfer of restoration insights: urban soil restoration can learn from work done in other (social-)ecosystems (e.g., mines and agroecosystems) and vice versa. Advancing such knowledge transfer depends on developing common vocabulary and theories which is facilitated by using the word "restoration" in an open, flexible, and integrated way—in contrast to the restrictive sensu stricto meaning—as discussed in Section 14.2.

Advances in some frontiers of basic urban soil research will provide direct insights for restoration practice. In particular, little is known about strategies for effective and efficient monitoring and assessment of urban soils, their ecology and quality for sustaining human well-being; future studies about these are needed to inform the development of indicators of successful restoration (Holl et al., 2003; Pavao-Zuckerman, 2008; da Silva et al., 2018; Anne et al., 2018; Callaghan et al., 2019). Two promising directions for such work are the Cornell Soil Health Test that was revised for use in urban landscapes (Schindelbeck et al., 2008) and the Urban Soil Quality Index (Scharenbroch and Catania, 2012). Data collected from monitoring can be used to adjust management activities and set new goals as part of an adaptive ecosystem management approach (e.g., Holl et al., 2003; Pavao-Zuckerman, 2008; Ahern et al., 2014; Hychka and Druschke, 2017). Further research about this approach, in general and specifically for soils, will be of enormous benefit to urban restoration practitioners, including through case studies that provide widely-applicable, transferable insights.

14.9.2 Practice

What might an idealized "urban soil restorationist" job advertisement say? In many ways, it would be the same as for restorationists working in other systems: knowledge and skills relating to analyzing and managing soil conditions and biodiversity (especially controlling invasive species and introducing beneficial biota), remediating pollution, communicating effectively, and working on teams. However, soil restoration practice in urban landscapes requires different kinds and scopes of soil and social-ecological knowledge and abilities, especially pertaining to urban- and landscape-centric concepts (e.g., Technosols, infrastructure, planning, multifunctionality, trade-offs, GIS); working with very diverse land covers, stakeholders and goals; and, of course, integrated, transdisciplinary understanding of the complex social-ecological dynamics of urban systems. As such, urban soil restorationists should have broader training that includes foundations in social sciences and urban planning to help prepare them for the challenges of outreach, negotiation, and planning with many people and organizations, some of whom may be hostile to restoration projects. In this context a job ad for urban restorationists might include the role of central coordinator (or liaison) who oversees individualized management plans for patches and advances a "metaplan" for the landscape-level. This coordinator might also write grants to obtain funding, especially for crucial site-specific soil analyses and long-term monitoring; such funding could be in the form of "payments for ecosystem services" if data can be collected to show that restoring urban soils will provide financially valuable outcomes (Perring et al., 2015; Jellinek et al., 2019). Alongside broad urban ecological knowledge (including about soils), such a person would need visionary leadership and excellent "people skills" (Handel et al., 2013; Hychka and Druschke, 2017). Because it is not reasonable to expect any one person to possess all the requirements of such an "idealized" job ad, all urban restorationists should be able to learn from and collaborate effectively with people from diverse disciplinary and sociocultural backgrounds (e.g., Crossman et al., 2007; Standish et al., 2013; Anne et al., 2018; Jellineck et al., 2018). Though perhaps Quixotic, this broad vision for a future urban soil restorationist's career suggests an exciting opportunity to design educational programs in ways that would give students the integrated, transdisciplinary background to succeed in many related positions.

14.9.3 Education

In addition to educating practitioners, a vibrant future for urban soil restoration depends on educating all urban residents about urban soils and the value of improving their abilities to support human well-being. To advance this goal—and overcome societal underappreciation for soil more generally—creative, experiential education programs are needed to reach people of all ages and backgrounds. This can be achieved by using soils and their restoration in citizen (civic) science projects and urban designed experiments that invite people to help generate data or participate in planning processes (Felson et al., 2013; Ahern et al., 2014; Callaghan et al., 2019). Such engagement can help reconnect people with nature (Fig. 14.4F), thereby increasing their general environmental knowledge and stewardship (Pavao-Zuckerman, 2008; Felson et al., 2013) and, hopefully, support for restoration (Standish et al., 2013; Jellinek et al., 2019). In formal education settings (classrooms), urban soils, landscapes, and their restoration have a valuable place as part of "pedagogy for the pedosphere" (Byrne et al., 2016) that advocates using dynamic, learner-centered teaching methods to help students connect soils with their lives (e.g., see example lessons in Harms et al., 2014; Dooling, 2015). One way to do this is to exploit everyday urban places, for example, lawns and gardens, as focal places for soil education (Fig. 14.2D). Though not thought of as restoration per se, typical gardening activities (e.g., composting; Fig. 14.2) aimed at improving diverse soil services (e.g., supporting beautiful plants, water infiltration, and favorable nutrient cycling: Edmondson et al., 2014b) can be important members of the "family of restoration activities" (Aronson et al., 2017). It may be that optimized patterns of soil restoration and ecosystem services across urban landscapes cannot be achieved without working with managers (especially homeowners) of lawns and gardens in individual parcels (Standish et al., 2013; Ziter and Turner, 2018). To help people see their yards (and even roofs) as places for urban soil restoration that enhances landscape-scale ecosystem services, the framework of ecological landscaping is a useful concept because it emphasizes an across-patch, systems views of urban landscapes, and the responsibility that individuals have for contributing to larger scale environmental outcomes (Byrne and Grewal, 2008). In addition to professional teachers of all types, such concepts could be conveyed to people through new career paths in urban soil outreach and education (Pavao-Zuckerman and Byrne, 2009). To these ends, an example of an organization leading the way is The Urban Soils Institute in New York City (https://projectsoils.org/), which is using diverse and creative approaches, including art-science projects, to help educate the public about the beauty and social-ecological value of urban soils (e.g., https://usi.nyc/divisions/education-outreach/).

14.9.4 Final thoughts

Though often unacknowledged and underappreciated by urban societies—and even many scientists—urban soils are the essential brown infrastructure that supports gray and green infrastructure and helps regulate blue infrastructure (Fig. 14.4A−C; Li et al., 2017). Soil degradation negatively affects the other infrastructures and diverse ecosystem services in ways that reduce the well-being and security of urban social-ecological systems. More so for urban soils than soils in other places, reversing this degradation through "ecological restoration is an investment, not an expense" (Handel et al., 2013, p. 667). This investment, in both science and practice, can create needed ecosystem services, jobs, education and recreation opportunities, and, ultimately, compelling examples of sustainable, thriving urban cultural landscapes (Elmqvist et al., 2015; Norris et al., 2017). One of the main challenges for the future of urban soil restoration is convincing many people (especially policy makers and planners) that urban soils are not just "dirt" and that, even if severely degraded, they can be improved to benefit the health of urban people, economies and biodiversity, as exemplified by examples of successful projects (see Lindig-Cisneros and Zedler, 2000; Norris et al., 2017).

As interest in urban restoration increases further, more knowledge will be needed about how to integrate soils and their associated ecosystem services into landscape-level restoration projects and urban planning so that their benefits are maximized (Setälä et al., 2014; Pavao-Zuckerman and Pouyat, 2017; Perring et al., 2015; da Silva et al., 2018; Anne et al., 2018). Generating and disseminating the (often site-specific) information and methods required for effective urban soil restoration, especially in the context of landscape-level planning and management, will be an ongoing challenge. Based on literature reviewed for this chapter, foundations of the science and practice of urban soil restoration are well-established but remain nascent and require substantial development to support the critical outcomes needed to create sustainable urban cultural landscapes. To borrow phrasing from Pavao-Zuckerman and Byrne (2009), urban soil restoration scientists, practitioners, and educators have barely "scratched the surface" of their subject. Given this and trends of continuously expanding urbanized landscapes, endless opportunities exist for them—and many future generations—to "dig deeper" into urban soils and help restore them long into the urbanizing Anthropocene.

Acknowledgment

I thank Mac Callaham for the invitation to contribute this chapter, his comments that improved the final version (including the insightful question about whether anthrosols contribute to pedodiversity), and for being an all-around great guy.

References

Ahern, J., Cilliers, S., Niemelä, J., 2014. The concept of ecosystem services in adaptive urban planning and design: a framework for supporting innovation. Landsc. Urban Plann. 125, 254–259.

Alvarez-Campos, O., Evanylo, G.K., 2019. Plant available nitrogen estimation tools for a biosolids-amended, clayey urban soil. Soil Sci. Soc. Am. J. 83, 808–816.

Amundson, R., Guo, Y., Gong, P., 2003. Soil diversity and land use in the United States. Ecosystems 6 (5), 470–482.

Amundson, R., Berhe, A.A., Hopmans, J.W., Olson, C., Sztein, A.E., Sparks, D.L., 2015. Soil and human security in the 21st century. Science 348 (6235), 647.

Anne, B., Geoffroy, S., Cherel, J., Warot, G., Marie, S., Noël, C.J., et al., 2018. Towards an operational methodology to optimize ecosystem services provided by urban soils. Landsc. Urban Plann. 176, 1–9.

Aronson, J., Blignaut, J.N., Aronson, T.B., 2017. Conceptual frameworks and references for landscape-scale restoration: reflecting back and looking forward. Ann. Mo. Bot. Garden 102 (2), 188–201.

Aronson, J.C., Simberloff, D., Ricciardi, A., Goodwin, N., 2018. Restoration science does not need redefinition. Nat. Ecol. Evol. 2 (6), 916.

Badin, A.L., Bedell, J.P., Delolme, C., 2009. Effect of water content on aggregation and contaminant leaching: the study of an urban Technosol. J. Soils Sediments 9 (6), 653–663.

Badzmierowski, M.J., Evanylo, G.K., Ervin, E.H., Boyd, A., Brewster, C., 2019. Biosolids-based amendments improve tall fescue establishment and urban soils. Crop Sci. 59, 1273–1284.

Baer, S.G., Heneghan, L., Eviner, V., 2012. Applying soil ecological knowledge to restore ecosystem services. In: Wall, D.H. (Ed.), Soil Ecology and Ecosystem Services. Oxford University Press, Oxford, pp. 377–393.

Balaguer, L., Escudero, A., Martin-Duque, J.F., Mola, I., Aronson, J., 2014. The historical reference in restoration ecology: re-defining a cornerstone concept. Biol. Conserv. 176, 12–20.

Bartens, J., Day, S.D., Harris, J.R., Dove, J.E., Wynn, T.M., 2008. Can urban tree roots improve infiltration through compacted subsoils for stormwater management? J. Environ. Qual. 37 (6), 2048–2057.

Bartens, J., Day, S.D., Harris, J.R., Wynn, T.M., Dove, J.E., 2009. Transpiration and root development of urban trees in structural soil stormwater reservoirs. Environ. Manage. 44 (4), 646–657.

Bartens, J., Wiseman, P.E., Smiley, E.T., 2010. Stability of landscape trees in engineered and conventional urban soil mixes. Urban For. Urban Green. 9 (4), 333–338.

Bashan, Y., Rojas, A., Puente, M.E., 1999. Improved establishment and development of three cactus species inoculated with *Azospirillum brasilense* transplanted into disturbed urban desert soil. Can. J. Microbiol. 45 (6), 441–451.

Basta, N.T., Busalacchi, D.M., Hundal, L.S., Kumar, K., Dick, R.P., Lanno, R.P., et al., 2016. Restoring ecosystem function in degraded urban soil using biosolids, biosolids blend, and compost. J. Environ. Qual. 45 (1), 74–83.

Beniston, J.W., Lal, R., Mercer, K.L., 2016. Assessing and managing soil quality for urban agriculture in a degraded vacant lot soil. Land Degrad. Dev. 27 (4), 996–1006.

Bernhardt, E.S., Palmer, M.A., 2007. Restoring streams in an urbanizing world. Freshwater Biol. 52 (4), 738–751.

Bouzouidja, R., Rousseau, G., Galzin, V., Claverie, R., Lacroix, D., Séré, G., 2018. Green roof ageing or Isolatic Technosol's pedogenesis? J. Soils Sediments 18 (2), 418–425.

Boyer, S., Kim, Y.N., Bowie, M.H., Lefort, M.C., Dickinson, N.M., 2016. Response of endemic and exotic earthworm communities to ecological restoration. Restor. Ecol. 24 (6), 717–721.

Bradshaw, A.D., 1987. Restoration: an acid test for ecology. In: Jordan, W.R., Giplin, M. E., Aber, J.D. (Eds.), Restoration Ecology: A Synthetic Approach to Ecological Research. Cambridge University Press, Cambridge.

Braaker, S., Ghazoul, J., Obrist, M.K., Moretti, M., 2014. Habitat connectivity shapes urban arthropod communities: the key role of green roofs. Ecology 95 (4), 1010–1021.

Braaker, S., Obrist, M.K., Ghazoul, J., Moretti, M., 2017. Habitat connectivity and local conditions shape taxonomic and functional diversity of arthropods on green roofs. J. Anim. Ecol. 86 (3), 521–531.

Brose, D.A., Hundal, L.S., Oladeji, O.O., Kumar, K., Granato, T.C., Cox, A., et al., 2016. Greening a steel mill slag brownfield with biosolids and sediments: a case study. J. Environ. Qual. 45 (1), 53–61.

Brown, S.L., Chaney, R.L., Hettiarachchi, G.M., 2016. Lead in urban soils: a real or perceived concern for urban agriculture? J. Environ. Qual. 45 (1), 26–36.

Bühler, O., Ingerslev, M., Skov, S., Schou, E., Thomsen, I.M., Nielsen, C.N., et al., 2017. Tree development in structural soil—an empirical below-ground in-situ study of urban trees in Copenhagen, Denmark. Plant Soil 413 (1–2), 29–44.

Byers, J.E., Cuddington, K., Jones, C.G., Talley, T.S., Hastings, A., Lambrinos, J.G., et al., 2006. Using ecosystem engineers to restore ecological systems. Trends Ecol. Evol. 21 (9), 493–500.

Byrne, L.B., 2007. Habitat structure: a fundamental concept and framework for urban soil ecology. Urban Ecosyst. 10 (3), 255–274.

Byrne, L.B., Grewal, P., 2008. Introduction to ecological landscaping: a holistic description and framework to guide the study and management of urban landscape parcels. Cities Environ. 1 (2), article 3, 20 pp. Available from: https://digitalcommons.lmu.edu/cate/vol1/iss2/3/.

Byrne, L.B., Bruns, M.A., Kim, K.C., 2008. Ecosystem properties of urban land covers at the aboveground–belowground interface. Ecosystems 11 (7), 1065–1077.

Byrne, L.B., Thiet, R.K., Chaudhary, V.B., 2016. Pedagogy for the pedosphere. Front. Ecol. Environ. 14 (5), 238–240.

Caliman, F.A., Robu, B.M., Smaranda, C., Pavel, V.L., Gavrilescu, M., 2011. Soil and groundwater cleanup: benefits and limits of emerging technologies. Clean. Technol. Environ. Policy 13 (2), 241–268.

Cannavo, P., Guénon, R., Galopin, G., Vidal-Beaudet, L., 2018. Technosols made with various urban wastes showed contrasted performance for tree development during a 3-year experiment. Environ. Earth Sci. 77 (18), 650.

Callaghan, C.T., Major, R.E., Lyons, M.B., Martin, J.M., Wilshire, J.H., Kingsford, R.T., et al., 2019. Using citizen science data to define and track restoration targets in urban areas. J. Appl. Ecol. 56, 1998–2006.

Carlson, J., Saxena, J., Basta, N., Hundal, L., Busalacchi, D., Dick, R.P., 2015. Application of organic amendments to restore degraded soil: effects on soil microbial properties. Environ. Monit. Assess. 187 (3), 109.

Chen, Y., Day, S.D., Wick, A.F., Strahm, B.D., Wiseman, P.E., Daniels, W.L., 2013. Changes in soil carbon pools and microbial biomass from urban land development and subsequent post-development soil rehabilitation. Soil Biol. Biochem. 66, 38–44.

Chen, Y., Day, S.D., Wick, A.F., McGuire, K.J., 2014. Influence of urban land development and subsequent soil rehabilitation on soil aggregates, carbon, and hydraulic conductivity. Sci. Total Environ. 494, 329–336.

Cheng, Z., Grewal, P.S., 2009. Dynamics of the soil nematode food web and nutrient pools under tall fescue lawns established on soil matrices resulting from common urban development activities. Appl. Soil Ecol. 42 (2), 107–117.

Chirakkara, R.A., Reddy, K.R., 2015. Biomass and chemical amendments for enhanced phytoremediation of mixed contaminated soils. Ecol. Eng. 85, 265–274.

Cogger, C.G., 2005. Potential compost benefits for restoration of soils disturbed by urban development. Compost. Sci. Util. 13 (4), 243–251.

Corbin, J., D'Antonio, C., 2012. Gone but not forgotten? Invasive plants' legacies on community and ecosystem properties. Invasive Plant Sci. Manage. 5 (1), 117–124.

Crossman, N.D., Bryan, B.A., Ostendorf, B., Collins, S., 2007. Systematic landscape restoration in the rural–urban fringe: meeting conservation planning and policy goals. Biodivers. Conserv. 16 (13), 3781–3802.

Crossman, N.D., Burkhard, B., Nedkov, S., Willemen, L., Petz, K., Palomo, I., et al., 2013. A blueprint for mapping and modelling ecosystem services. Ecosyst. Serv. 4, 4–14.

Dadea, C., Russo, A., Tagliavini, M., Mimmo, T., Zerbe, S., 2017. Tree species as tools for biomonitoring and phytoremediation in urban environments: a review with special regard to heavy metals. Arboricult. Urban For. 43 (4), 155–167.

Dávalos, A., Simpson, E., Nuzzo, V., Blossey, B., 2015. Non-consumptive effects of native deer on introduced earthworm abundance. Ecosystems 18 (6), 1029–1042.

Davidson, D.A., Dercon, G., Stewart, M., Watson, F., 2006. The legacy of past urban waste disposal on local soils. J. Archaeol. Sci. 33 (6), 778–783.

da Silva, R.T., Fleskens, L., van Delden, H., van der Ploeg, M., 2018. Incorporating soil ecosystem services into urban planning: status, challenges and opportunities. Landsc. Ecol. 33 (7), 1087–1102.

De Kimpe, C.R., Morel, J.L., 2000. Urban soil management: a growing concern. Soil Sci. 165 (1), 31–40.

De Miguel, E., De Grado, M.J., Llamas, J.F., Martin-Dorado, A., Mazadiego, L.F., 1998. The overlooked contribution of compost application to the trace element load in the urban soil of Madrid (Spain). Sci. Total Environ. 215 (1–2), 113–122.

Deeb, M., Desjardins, T., Podwojewski, P., Pando, A., Blouin, M., Lerch, T.Z., 2017. Interactive effects of compost, plants and earthworms on the aggregations of constructed Technosols. Geoderma 305, 305–313.

Delbecque, N., Verdoodt, A., 2016. Spatial patterns of heavy metal contamination by urbanization. J. Environ. Qual. 45 (1), 9–17.

Dooling, S.E., 2015. Novel landscapes: challenges and opportunities for educating future ecological designers and restoration practitioners. Ecol. Restor. 33 (1), 96–110.

Doroski, D.A., Felson, A.J., Bradford, M.A., Ashton, M.P., Oldfield, E.E., Hallett, R.A., et al., 2018. Factors driving natural regeneration beneath a planted urban forest. Urban For. Urban Green. 29, 238–247.

Dusza, Y., Barot, S., Kraepiel, Y., Lata, J.-C., Abbadie, L., Raynaud, X., 2017. Multifunctionality is affected by interactions between green roof plant species, substrate depth, and substrate type. Ecology and Evolution 7 (7), 2357–2369.

Duarte, A.C., Cachada, A., Rocha-Santos, T.A.P. (Eds.), 2018. Soil Pollution: From Monitoring to Remediation. Academic Press, London.

Edmondson, J.L., Davies, Z.G., McCormack, S.A., Gaston, K.J., Leake, J.R., 2011. Are soils in urban ecosystems compacted? A citywide analysis. Biol. Lett. 7 (5), 771–774.

Edmondson, J.L., Davies, Z.G., McCormack, S.A., Gaston, K.J., Leake, J.R., 2014a. Land-cover effects on soil organic carbon stocks in a European city. Sci. Total Environ. 472, 444–453.

Edmondson, J.L., Davies, Z.G., Gaston, K.J., Leake, J.R., 2014b. Urban cultivation in allotments maintains soil qualities adversely affected by conventional agriculture. J. Appl. Ecol. 51 (4), 880–889.

Egendorf, S.P., Cheng, Z., Deeb, M., Flores, V., Paltseva, A., Walsh, D., Groffman, P., Howard, W.M., 2018. Constructed soils for mitigating lead (Pb) exposure and promoting urban community gardening: The New York City Clean Soil Bank pilot study. Landscape and Urban Planning 175, 184–194.

Egerer, M., Ossola, A., Lin, B.B., 2018. Creating socioecological novelty in urban agroecosystems from the ground up. BioScience 68 (1), 25–34.

Elmqvist, T., Setälä, H., Handel, S.N., Van Der Ploeg, S., Aronson, J., Blignaut, J.N., et al., 2015. Benefits of restoring ecosystem services in urban areas. Curr. Opin. Environ. Sustain. 14, 101–108.

Everingham, S.E., Hemmings, F., Moles, A.T., 2019. Inverted invasions: native plants can frequently colonise urban and highly disturbed habitats. Austral Ecol. 44, 702–712.

FAO and ITPS. 2015. Status of the World's Soil Resources (SWSR) – Technical Summary. Food and Agriculture Organization of the United Nations and Intergovernmental Technical Panel on Soils, Rome.

Felson, A.J., Bradford, M.A., Terway, T.M., 2013. Promoting earth stewardship through urban design experiments. Front. Ecol. Environ. 11 (7), 362–367.

Ferlian, O., Eisenhauer, N., Aguirrebengoa, M., Camara, M., Ramirez-Rojas, I., Santos, F., et al., 2018. Invasive earthworms erode soil biodiversity: A meta-analysis. J. Animal Eco 87 (1), 162–172.

Field, D.J., Morgan, C.L., McBratney, A.B. (Eds.), 2016. Global Soil Security. Springer, New York.

Fini, A., Frangi, P., Amoroso, G., Piatti, R., Faoro, M., Bellasio, C., et al., 2011. Effect of controlled inoculation with specific mycorrhizal fungi from the urban environment on growth and physiology of containerized shade tree species growing under different water regimes. Mycorrhiza 21 (8), 703–719.

Fischer, L.K., von der Lippe, M., Rillig, M.C., Kowarik, I., 2013. Creating novel urban grasslands by reintroducing native species in wasteland vegetation. Biol. Conserv. 159, 119–126.

Funai, J.T., Kupec, P., 2019. Evaluation of three soil blends to improve ornamental plant performance and maintain engineering metrics in bioremediating rain gardens. Water Air Soil Pollut. 230 (1), 3.

Gellie, N.J., Mills, J.G., Breed, M.F., Lowe, A.J., 2017. Revegetation rewilds the soil bacterial microbiome of an old field. Mol. Ecol. 26 (11), 2895–2904.

Gobster, P.H., 2001. Visions of nature: conflict and compatibility in urban park restoration. Landsc. Urban Plann. 56 (1–2), 35–51.

Grafius, D.R., Corstanje, R., Harris, J.A., 2018. Linking ecosystem services, urban form and green space configuration using multivariate landscape metric analysis. Landsc. Ecol. 33 (4), 557–573.

Gross, M., Hoffmann-Riem, H., 2005. Ecological restoration as a real-world experiment: designing robust implementation strategies in an urban environment. Public Understanding Sci. 14 (3), 269–284.

Haan, N.L., Hunter, M.R., Hunter, M.D., 2012. Investigating predictors of plant establishment during roadside restoration. Restor. Ecol. 20 (3), 315–321.

Hafeez, F., Martin-Laurent, F., Béguet, J., Bru, D., Cortet, J., Schwartz, C., et al., 2012. Taxonomic and functional characterization of microbial communities in Technosols constructed for remediation of a contaminated industrial wasteland. J. Soils Sediments 12 (9), 1396–1406.

Handel, S.N., Robinson, G.R., Parsons, W.F., Mattei, J.H., 1997. Restoration of woody plants to capped landfills: root dynamics in an engineered soil. Restor. Ecol. 5 (2), 178–186.

Handel, S.N., Saito, O., Takeuchi, K., 2013. Restoration ecology in an urbanizing world. In: Elmqvist, T., et al., (Eds.), Urbanization, Biodiversity and Ecosystem Services: Challenges and Opportunities. Springer, Dordrecht, pp. 665–698.

Harms, A.M.R., DeAnn, R.P., Hettiarachchi, G.M., Attanayake, C., Martin, S., Thien, S.J., 2014. Harmony park: a decision case on gardening on a brownfield site. Nat. Sci. Educ. 43, 33–41.

Heneghan, L., Miller, S.P., Baer, S., Callaham Jr, M.A., Montgomery, J., Pavao-Zuckerman, M., et al., 2008. Integrating soil ecological knowledge into restoration management. Restor. Ecol. 16 (4), 608–617.

Heneghan, L., Umek, L., Bernau, B., Grady, K., Iatropulos, J., Jabon, D., et al., 2009. Ecological research can augment restoration practice in urban areas degraded by invasive species—examples from Chicago Wilderness. Urban Ecosyst. 12 (1), 63–77.

Henry, H., Naujokas, M.F., Attanayake, C., Basta, N.T., Cheng, Z., Hettiarachchi, G.M., et al., 2015. Bioavailability-based in situ remediation to meet future lead (Pb) standards in urban soils and gardens. Environ. Sci. Technol. 49 (15), 8948–8958.

Herrmann, D.L., Shuster, W.D., Garmestani, A.S., 2017. Vacant urban lot soils and their potential to support ecosystem services. Plant Soil 413 (1–2), 45–57.

Herrmann, D.L., Schifman, L.A., Shuster, W.D., 2018. Widespread loss of intermediate soil horizons in urban landscapes. Proc. Natl. Acad. Sci. U.S.A. 115 (26), 6751–6755.

Hesse, E., Padfield, D., Bayer, F., Van Veen, E.M., Bryan, C.G., Buckling, A., 2019. Anthropogenic remediation of heavy metals selects against natural microbial remediation. Proc. R. Soc. B 286, <https://doi.org/10.1098/rspb.2019.0804>.

Heymans, A., Breadsell, J., Morrison, G.M., Byrne, J.J., Eon, C., 2019. Ecological urban planning and design: a systematic literature review. Sustainability 11 (13), 3723.

Higgs, E.S., Harris, J.A., Heger, T., Hobbs, R.J., Murphy, S.D., Suding, K.N., 2018. Keep ecological restoration open and flexible. Nat. Ecol. Evol. 2 (4), 580.

Hitchmough, J.D., 2008. New approaches to ecologically based, designed urban plant communities in Britain: do these have any relevance in the United States? Cities Environ. 1 (2), article 10, 15 pp. <https://digitalcommons.lmu.edu/cate/vol1/iss2/10/>.

Hobbs, R.J., Higgs, E., Hall, C.M., Bridgewater, P., Chapin III, F.S., Ellis, E.C., et al., 2014. Managing the whole landscape: historical, hybrid, and novel ecosystems. Front. Ecol. Environ. 12 (10), 557–564.

Holl, K.D., Crone, E.E., Schultz, C.B., 2003. Landscape restoration: moving from generalities to methodologies. BioScience 53 (5), 491–502.

Huang, M., Zhu, Y., Li, Z., Huang, B., Luo, N., Liu, C., et al., 2016. Compost as a soil amendment to remediate heavy metal-contaminated agricultural soil: mechanisms, efficacy, problems, and strategies. Water Air Soil Pollut. 227 (10), 359. <https://doi.org/10.1007/s11270-016-3068-8>.

Huot, H., Simonnot, M.O., Morel, J.L., 2015. Pedogenetic trends in soils formed in technogenic parent materials. Soil Sci. 180 (4/5), 182–192.

Hychka, K., Druschke, C.G., 2017. Adaptive management of urban ecosystem restoration: learning from restoration managers in Rhode Island, USA. Soc. Nat. Resour. 30 (11), 1358–1373.

Iannone, B.V., Umek, L.G., Heneghan, L., Wise, D.H., 2013. Amending soil with mulched European buckthorn (*Rhamnus cathartica*) does not reduce reinvasion. Ecol. Restor. 31 (3), 264–273.

Ibanez, J.J., Krasilnikov, P.V., Saldana, A., 2012. Archive and refugia of soil organisms: applying a pedodiversity framework for the conservation of biological and non-biological heritages. J. Appl. Ecol. 49 (6), 1267–1277.

Jellinek, S., Wilson, K.A., Hagger, V., Mumaw, L., Cooke, B., Guerrero, A.M., et al., 2019. Integrating diverse social and ecological motivations to achieve landscape restoration. J. Appl. Ecol. 56 (1), 246–252.

Jensen, J.K., Holm, P.E., Nejrup, J., Larsen, M.B., Borggaard, O.K., 2009. The potential of willow for remediation of heavy metal polluted calcareous urban soils. Environ. Pollut. 157 (3), 931–937.

Jimenez, M.D., Ruiz-Capillas, P., Mola, I., Pérez-Corona, E., Casado, M.A., Balaguer, L., 2013. Soil development at the roadside: a case study of a novel ecosystem. Land Degrad. Dev. 24 (6), 564–574.

John, J., Lundholm, J., Kernaghan, G., 2014. Colonization of green roof plants by mycorrhizal and root endophytic fungi. Ecol. Eng. 71, 651–659.

Johnston, M.R., Balster, N.J., Zhu, J., 2016. Impact of residential prairie gardens on the physical properties of urban soil in Madison, Wisconsin. J. Environ. Qual. 45 (1), 45–52.

Joimel, S., Schwartz, C., Hedde, M., Kiyota, S., Krogh, P.H., Nahmani, J., et al., 2017. Urban and industrial land uses have a higher soil biological quality than expected from physicochemical quality. Sci. Total Environ. 584, 614–621.

Joimel, S., Grard, B., Auclerc, A., Hedde, M., Le Doaré, N., Salmon, S., et al., 2018. Are Collembola "flying" onto green roofs? Ecol. Eng. 111, 117–124.

Kargar, M., Clark, O.G., Hendershot, W.H., Jutras, P., Prasher, S.O., 2015. Immobilization of trace metals in contaminated urban soil amended with compost and biochar. Water Air Soil Pollut. 226 (6), 191. <https://doi.org/10.1007/s11270-015-2450-2>.

Kästner, M., Miltner, A., 2016. Application of compost for effective bioremediation of organic contaminants and pollutants in soil. Appl. Microbiol. Biotechnol. 100, 3433–3449.

Kadas, G., 2006. Rare invertebrates colonizing green roofs in London. Urban Habitats 4 (1), 66–86.

Kazemi, F., Mohorko, R., 2017. Review on the roles and effects of growing media on plant performance in green roofs in world climates. Urban For. Urban Green. 23, 13–26.

Knot, P., Hrabe, F., Hejduk, S., Skladanka, J., Kvasnovsky, M., Hodulikova, L., et al., 2017. The impacts of different management practices on botanical composition, quality, colour and growth of urban lawns. Urban For. Urban Green. 26, 178–183.

Koch, A., McBratney, A., Adams, M., Field, D., Hill, R., Crawford, J., et al., 2013. Soil security: solving the global soil crisis. Global Policy 4, 434–441.

Krumins, J.A., Goodey, N.M., Gallagher, F., 2015. Plant–soil interactions in metal contaminated soils. Soil Biol. Biochem. 80, 224–231.

Ksiazek-Mikenas, K., Herrmann, J., Menke, S.B., Köhler, M., 2018. If you build it, will they come? Plant and arthropod diversity on urban green roofs over time. Urban Nat. 1, 52–72.

Kyrö, K., Brenneisen, S., Kotze, D.J., Szallies, A., Gerner, M., Lehvävirta, S., 2018. Local habitat characteristics have a stronger effect than the surrounding urban landscape on beetle communities on green roofs. Urban For. Urban Green. 29, 122–130.

Laidlaw, M.A., Filippelli, G.M., Brown, S., Paz-Ferreiro, J., Reichman, S.M., Netherway, P., et al., 2017. Case studies and evidence-based approaches to addressing urban soil lead contamination. Appl. Geochem. 83, 14–30.

Larney, F.J., Angers, D.A., 2012. The role of organic amendments in soil reclamation: a review. Can. J. Soil Sci. 92 (1), 19–38.

Lawler, J.J., Lewis, D.J., Nelson, E., Plantinga, A.J., Polasky, S., Withey, J.C., et al., 2014. Projected land-use change impacts on ecosystem services in the United States. Proc. Natl. Acad. Sci. U.S.A. 111, 7492–7497.

Layman, R.M., Day, S.D., Mitchell, D.K., Chen, Y., Harris, J.R., Daniels, W.L., 2016. Below ground matters: urban soil rehabilitation increases tree canopy and speeds establishment. Urban For. Urban Green. 16, 25–35.

Lepczyk, C.A., Aronson, M.F., Evans, K.L., Goddard, M.A., Lerman, S.B., MacIvor, J.S., 2017. Biodiversity in the city: fundamental questions for understanding the ecology of urban green spaces for biodiversity conservation. BioScience 67 (9), 799–807.

Lewis, D.B., Kaye, J.P., Gries, C., Kinzig, A.P., Redman, C.L., 2006. Agrarian legacy in soil nutrient pools of urbanizing arid lands. Global Change Biol. 12 (4), 703—709.

Levy, J.M., 2016. Contemporary Urban Planning. Routledge, New York.

Li, F., Liu, X., Zhang, X., Zhao, D., Liu, H., Zhou, C., et al., 2017. Urban ecological infrastructure: an integrated network for ecosystem services and sustainable urban systems. J. Cleaner Prod. 163, S12—S18.

Li, G., Sun, G.X., Ren, Y., Luo, X.S., Zhu, Y.G., 2018a. Urban soil and human health: a review. Eur. J. Soil Sci. 69 (1), 196—215.

Li, X., Yang, L., Ren, Y., Li, H., Wang, Z., 2018b. Impacts of urban sprawl on soil resources in the Changchun—Jilin Economic Zone, China, 2000—2015. Int. J. Environ. Res. Public Health 15, 1186.

Lindig-Cisneros, R., Zedler, J.B., 2000. Restoring urban habitats: a comparative study. Ecol. Restor. 18 (3), 184—192.

Liu, L., Li, W., Song, W., Guo, M., 2018. Remediation techniques for heavy metal-contaminated soils: principles and applicability. Sci. Total Environ. 633, 206—219.

Lundholm, J.T., Richardson, P.J., 2010. Habitat analogues for reconciliation ecology in urban and industrial environments. J. Appl. Ecol. 47 (5), 966—975.

Loper, S., Shober, A.L., Wiese, C., Denny, G.C., Stanley, C.D., Gilman, E.F., 2010. Organic soil amendment and tillage affect soil quality and plant performance in simulated residential landscapes. HortScience 45 (10), 1522—1528.

Loschinkohl, C., Boehm, M.J., 2001. Composted biosolids incorporation improves turfgrass establishment on disturbed urban soil and reduces leaf rust severity. HortScience 36 (4), 790—794.

Lwin, C.S., Seo, B.H., Kim, H.U., Owens, G., Kim, K.R., 2018. Application of soil amendments to contaminated soils for heavy metal immobilization and improved soil quality—a critical review. Soil Sci. Plant Nutr. 64 (2), 156—167.

MacIvor, J.S., Lundholm, J., 2011. Insect species composition and diversity on intensive green roofs and adjacent level-ground habitats. Urban Ecosyst. 14 (2), 225—241.

Madre, F., Vergnes, A., Machon, N., Clergeau, P., 2013. A comparison of 3 types of green roof as habitats for arthropods. Ecol. Eng. 57, 109—117.

Mahon, M.B., Crist, T.O., 2019. Invasive earthworm and soil litter response to the experimental removal of white-tailed deer and an invasive shrub. Ecology 100 (5), e02688. <https://doi.org/10.1002/ecy.2688>.

Mao, X., Jiang, R., Xiao, W., Yu, J., 2015. Use of surfactants for the remediation of contaminated soils: a review. J. Hazard. Mater. 285, 419—435.

Marcotullio, P.J., Braimoh, A.K., Onishi, T., 2008. The impact of urbanization on soils. In: Braimoh, A.K., Vlek, P.L.G. (Eds.), Land Use and Soil Resources. Springer, Stockholm, pp. 201—250.

McIvor, K., Cogger, C., Brown, S., 2012. Effects of biosolids based soil products on soil physical and chemical properties in urban gardens. Compost. Sci. Util. 20 (4), 199—206.

McGrath, D., Henry, J., 2016. Organic amendments decrease bulk density and improve tree establishment and growth in roadside plantings. Urban For. Urban Green. 20, 120—127.

McGuire, K.L., Payne, S.G., Palmer, M.I., Gillikin, C.M., Keefe, D., Kim, S.J., et al., 2013. Digging the New York City skyline: soil fungal communities in green roofs and city parks. PLoS One 8 (3), e58020. <https://doi.org/10.1371/journal.pone.0058020>.

McHale, M.R., Beck, S.M., Pickett, S.T., Childers, D.L., Cadenasso, M.L., Rivers III, L., et al., 2018. Democratization of ecosystem services—a radical approach for assessing nature's benefits in the face of urbanization. Ecosyst. Health Sustain. 4 (5), 115–131.

Mehring, A.S., Hatt, B.E., Kraikittikun, D., Orelo, B.D., Rippy, M.A., Grant, S.B., et al., 2016. Soil invertebrates in Australian rain gardens and their potential roles in storage and processing of nitrogen. Ecol. Eng. 97, 138–143.

Menefee, D.S., Hettiarachchi, G.M., 2017. Contaminants in urban soils: bioavailability and transfer. In: Lal, R., Stewart, B.A. (Eds.), Urban Soils. CRC Press, Boca Raton, FL, pp. 175–198.

Menz, M.H., Dixon, K.W., Hobbs, R.J., 2013. Hurdles and opportunities for landscape-scale restoration. Science 339 (6119), 526–527.

Meuser, H., 2010. Contaminated Urban Soils. Springer, New York.

Meuser, H., 2013. Soil Remediation and Rehabilitation: Treatment of Contaminated and Disturbed Land. Springer, Dordrecht.

Mills, J.G., Weinstein, P., Gellie, N.J., Weyrich, L.S., Lowe, A.J., Breed, M.F., 2017. Urban habitat restoration provides a human health benefit through microbiome rewilding: the microbiome rewilding hypothesis. Restor. Ecol. 25 (6), 866–872.

Millward, A.A., Paudel, K., Briggs, S.E., 2011. Naturalization as a strategy for improving soil physical characteristics in a forested urban park. Urban Ecosyst. 14 (2), 261–278.

Mohammadshirazi, F., Brown, V.K., Heitman, J.L., McLaughlin, R.A., 2016. Effects of tillage and compost amendment on infiltration in compacted soils. J. Soil Water Conserv. 71 (6), 443–449.

Molineux, C.J., Connop, S.P., Gange, A.C., 2014. Manipulating soil microbial communities in extensive green roof substrates. Sci. Total Environ. 493, 632–638.

Molineux, C.J., Gange, A.C., Connop, S.P., Newport, D.J., 2015. Are microbial communities in green roof substrates comparable to those in post-industrial sitesA preliminary study. Urban Ecosyst. 18 (4), 1245–1260.

Montgomery, J.A., Klimas, C.A., Arcus, J., DeKnock, C., Rico, K., Rodriguez, Y., et al., 2016. Soil quality assessment is a necessary first step for designing urban green infrastructure. J. Environ. Qual. 45 (1), 18–25.

Morillo, E., Villaverde, J., 2017. Advanced technologies for the remediation of pesticide-contaminated soils. Sci. Total Environ. 586, 576–597.

Moriyama, M., Numata, H., 2015. Urban soil compaction reduces cicada diversity. Zool. Lett. 1 (1), 19.

Moorhead, K.K., 2015. A pedogenic view of ecosystem restoration. Ecol. Restor. 33 (4), 341–351.

Moreira, F., Queiroz, A.I., Aronson, J., 2006. Restoration principles applied to cultural landscapes. J. Nat. Conserv. 14 (3–4), 217–224.

Muchelo, R.O., 2017. Urban Expansion and Loss of Prime Agricultural Land in Sub-Saharan Africa: A Challenge to Soil Conservation and Food Security (Ph.D. thesis). University of Sydney. <http://hdl.handle.net/2123/18116>.

Mullaney, J., Lucke, T., Trueman, S.J., 2015. A review of benefits and challenges in growing street trees in paved urban environments. Landsc. Urban Plann. 134, 157–166.

Nizeyimana, E.L., Petersen, G.W., Imhoff, M.L., Sinclair, H.R., Waltman, S.W., Reed-Margetan, D.S., et al., 2001. Assessing the impact of land conversion to urban use on soils with different productivity levels in the USA. Soil Sci. Soc. Am. J. 65 (2), 391–402.

Norris, J.H., Bowers, K., Murphy, S.D., 2017. Ecological restoration in an urban context. In: Allison, S.K., Murphy, S.D. (Eds.), Routledge Handbook of Ecological and Environmental Restoration. Taylor & Francis, London, pp. 772-385.

Oberndorfer, E., Lundholm, J., Bass, B., Coffman, R.R., Doshi, H., Dunnett, N., et al., 2007. Green roofs as urban ecosystems: ecological structures, functions, and services. BioScience 57 (10), 823−833.

Obrycki, J.F., Basta, N.T., Culman, S.W., 2017. Management options for contaminated urban soils to reduce public exposure and maintain soil health. J. Environ. Qual. 46 (2), 420−430.

Oldfield, E.E., Felson, A.J., Wood, S.A., Hallett, R.A., Strickland, M.S., Bradford, M.A., 2014. Positive effects of afforestation efforts on the health of urban soils. For. Ecol. Manage. 313, 266−273.

Oldfield, E.E., Felson, A.J., Auyeung, D.N., Crowther, T.W., Sonti, N.F., Harada, Y., et al., 2015. Growing the urban forest: tree performance in response to biotic and abiotic land management. Restor. Ecol. 23 (5), 707−718.

Olson, N.C., Gulliver, J.S., Nieber, J.L., Kayhanian, M., 2013. Remediation to improve infiltration into compact soils. J. Environ. Manage. 117, 85−95.

Ordóñez-Barona, C., Sabetski, V., Millward, A.A., Steenberg, J., 2018. De-icing salt contamination reduces urban tree performance in structural soil cells. Environ. Pollut. 234, 562−571.

Ossola, A., Livesley, S.J., 2016. Drivers of soil heterogeneity in the urban landscape. In: Francis, R.A., Millington, J.D.A., Chadwick, M.A. (Eds.), Urban Landscape Ecology. Routledge, London, pp. 37−59.

Overdyck, E., Clarkson, B.D., 2012. Seed rain and soil seed banks limit native regeneration within urban forest restoration plantings in Hamilton City, New Zealand. N.Z. J. Ecol. 36, 177−190.

Ow, L.F., Ghosh, S., 2017. Growth of street trees in urban ecosystems: structural cells and structural soil. J. Urban Ecol. 3 (1), <https://doi.org/10.1093/jue/jux017>.

Ow, L.F., Mohd. Yusof, M.L., 2018. Stability of four urban trees species in engineered and regular urban soil blends. J. Urban Ecol. 4 (1), <https://doi.org/10.1093/jue/juy014>.

Paltseva, A., Cheng, Z., Deeb, M., Groffman, P.M., Maddaloni, M., 2018. Variability of bioaccessible lead in urban garden soils. Soil Sci. 183 (4), 123−131.

Parker, S.S., 2010. Buried treasure: soil biodiversity and conservation. Biodivers. Conserv. 19 (13), 3743−3756.

Pavao-Zuckerman, M.A., 2008. The nature of urban soils and their role in ecological restoration in cities. Restor. Ecol. 16 (4), 642−649.

Pavao-Zuckerman, M.A., Byrne, L.B., 2009. Scratching the surface and digging deeper: exploring ecological theories in urban soils. Urban Ecosyst. 12 (1), 9−20.

Pavao-Zuckerman, M., Pouyat, R.V., 2017. The effects of urban expansion on soil health and ecosystem services: an overview. In: Gardi, C. (Ed.), Urban Expansion, Land Cover and Soil Ecosystem Services. Routledge, London, pp. 123−145.

Perring, M.P., Manning, P., Hobbs, R.J., Lugo, A.E., Ramalho, C.E., Standish, R.J., 2013. Novel urban ecosystems and ecosystem services. In: Hobbs, R.J., Higgs, E.S., Hall, C. (Eds.), Novel Ecosystems: Intervening in the New Ecological World Order. John-Wiley & Sons, London, pp. 310−325.

Perring, M.P., Standish, R.J., Price, J.N., Craig, M.D., Erickson, T.E., Ruthrof, K.X., et al., 2015. Advances in restoration ecology: rising to the challenges of the coming decades. Ecosphere 6 (8), 1−25. <https://doi.org/10.1890/ES15-00121.1>.

Pickett, S.T., Cadenasso, M.L., 2009. Altered resources, disturbance, and heterogeneity: a framework for comparing urban and non-urban soils. Urban Ecosyst. 12 (1), 23–44.

Pickett, S.T., Cadenasso, M.L., Childers, D.L., McDonnell, M.J., Zhou, W., 2016. Evolution and future of urban ecological science: ecology in, of, and for the city. Ecosyst. Health Sustainability 2 (7), e01229. <https://doi.org/10.1002/ehs2.1229>.

Pregitzer, C.C., Sonti, N.F., Hallett, R.A., 2016. Variability in urban soils influences the health and growth of native tree seedlings. Ecol. Restor. 34 (2), 106–116.

Raciti, S.M., Hutyra, L.R., Finzi, A.C., 2012. Depleted soil carbon and nitrogen pools beneath impervious surfaces. Environ. Pollut. 164, 248–251.

Rawlins, B.G., Harris, J., Price, S., Bartlett, M., 2015. A review of climate change impacts on urban soil functions with examples and policy insights from England, UK. Soil Use Manage. 31, 46–61.

Riley, C., Perry, K., Ard, K., Gardiner, M., 2018. Asset or liability? Ecological and sociological tradeoffs of urban spontaneous vegetation on vacant land in shrinking cities. Sustainability 10 (7), 2139.

Rumble, H., Gange, A.C., 2013. Soil microarthropod community dynamics in extensive green roofs. Ecol. Eng. 57, 197–204.

Rumble, H., Gange, A.C., 2017. Microbial inoculants as a soil remediation tool for extensive green roofs. Ecol. Eng. 102, 188–198.

Rumble, H., Finch, P., Gange, A.C., 2018. Green roof soil organisms: anthropogenic assemblages or natural communities? Appl. Soil Ecol. 126, 11–20.

Sack, C., 2013. Landscape architecture and novel ecosystems: ecological restoration in an expanded field. Ecol. Process. 2 (1), 35. <https://doi.org/10.1186/2192-1709-2-35>.

Sæbø, A., Ferrini, F., 2006. The use of compost in urban green areas—a review for practical application. Urban For. Urban Green. 4 (3–4), 159–169.

Sax, M.S., Bassuk, N., van Es, H., Rakow, D., 2017. Long-term remediation of compacted urban soils by physical fracturing and incorporation of compost. Urban For. Urban Green. 24, 149–156.

Scalenghe, R., Marsan, F.A., 2009. The anthropogenic sealing of soils in urban areas. Landsc. Urban Plann. 90 (1–2), 1–10.

Scharenbroch, B.C., 2009. A meta-analysis of studies published in Arboriculture & Urban Forestry relating to organic materials and impacts on soil, tree, and environmental properties. J. Arboricult. 35 (5), 221.

Scharenbroch, B.C., Catania, M., 2012. Soil quality attributes as indicators of urban tree performance. Arboricult. Urban For. 38 (5), 214.

Scharenbroch, B.C., Johnston, D.P., 2011. A microcosm study of the common night crawler earthworm (*Lumbricus terrestris*) and physical, chemical and biological properties of a designed urban soil. Urban Ecosyst. 14 (1), 119–134.

Scharenbroch, B.C., Meza, E.N., Catania, M., Fite, K., 2013. Biochar and biosolids increase tree growth and improve soil quality for urban landscapes. J. Environ. Qual. 42 (5), 1372–1385.

Scharenbroch, B.C., Fite, K., Kramer, E., Uhlig, R., 2018. Pedogenic processes and urban tree health in engineered urban soils in Boston, Massachusetts, USA. Soil Sci. 183 (4), 159–167.

Schindelbeck, R.R., van Es, H.M., Abawi, G.S., Wolfe, D.W., Whitlow, T.L., Gugino, B.K., et al., 2008. Comprehensive assessment of soil quality for landscape and urban management. Landsc. Urban Plann. 88 (2–4), 73–80.

Schindler, B.Y., Vasl, A., Blaustein, L., Gurevich, D., Kadas, G.J., Seifan, M., 2019. Fine-scale substrate heterogeneity does not affect arthropod communities on green roofs. PeerJ 7, e6445. <https://doi.org/10.7717/peerj.6445>.

Schrader, S., Böning, M., 2006. Soil formation on green roofs and its contribution to urban biodiversity with emphasis on Collembolans. Pedobiologia 50 (4), 347−356.

Schwartz, S.S., Smith, B., 2016. Restoring hydrologic function in urban landscapes with suburban subsoiling. J. Hydrol. 543, 770−781.

Seastadt, T.R., Hobbs, R.J., Suding, K.N., 2008. Management of novel ecosystems: are novel approaches required? Front. Ecol. Environ. 6 (10), 547−553.

Setälä, H., Bardgett, R.D., Birkhofer, K., Brady, M., Byrne, L., De Ruiter, P.C., et al., 2014. Urban and agricultural soils: conflicts and trade-offs in the optimization of ecosystem services. Urban Ecosyst. 17, 239−253.

Setälä, H., Francini, G., Allen, J.A., Jumpponen, A., Hui, N., Kotze, D.J., 2017. Urban parks provide ecosystem services by retaining metals and nutrients in soils. Environ. Pollut. 231, 451−461.

Seto, K.C., Fragkias, M., Güneralp, B., Reilly, M.K., 2011. A meta-analysis of global urban land expansion. PLoS One 6 (8), e23777. <https://doi.org/10.1371/journal.pone.0023777>.

Shelton, A.L., Henning, J.A., Schultz, P., Clay, K., 2014. Effects of abundant white-tailed deer on vegetation, animals, mycorrhizal fungi, and soils. For. Ecol. Manage. 320, 39−49.

Shuster, W., Darner, R., Schifman, L., Herrmann, D., 2017.). Factors contributing to the hydrologic effectiveness of a rain garden network (Cincinnati, OH, USA). Infrastructures 2 (3), 11.

Shuttleworth, A.B., Newman, A.P., Nnadi, E.O., 2018. Bioremediation in urban pollution mitigation: applications to solid media. In: Charlesworth, S.M., Booth, C.A. (Eds.), Urban Pollution: Science and Management. John Wiley & Sons, Hoboken, pp. 277−291.

Singh, J.P., Vaidya, B.P., Goodey, N.M., Krumins, J.A., 2019. Soil microbial response to metal contamination in a vegetated and urban brownfield. J. Environ. Manage. 244, 313−319.

Sloan, J.J., Ampim, P.A., Basta, N.T., Scott, R., 2012. Addressing the need for soil blends and amendments for the highly modified urban landscape. Soil Sci. Soc. Am. J. 76 (4), 1133−1141.

Smith, P., House, J.I., Bustamante, M., Sobocká, J., Harper, R., Pan, G., et al., 2016. Global change pressures on soils from land use and management. Global Change Biol. 22 (3), 1008−1028.

Snyder, B.A., Hendrix, P.F., 2008. Current and potential roles of soil macroinvertebrates (earthworms, millipedes, and isopods) in ecological restoration. Restor. Ecol. 16 (4), 629−636.

Somerville, P.D., May, P.B., Livesley, S.J., 2018. Effects of deep tillage and municipal green waste compost amendments on soil properties and tree growth in compacted urban soils. J. Environ. Manage. 227, 365−374.

Song, Y., Kirkwood, N., Maksimović, Č., Zhen, X., O'Connor, D., Jin, Y., et al., 2019. Nature based solutions for contaminated land remediation and brownfield redevelopment in cities: a review. Sci. Total Environ. 663, 568−579.

Standish, R.J., Hobbs, R.J., Miller, J.R., 2013. Improving city life: options for ecological restoration in urban landscapes and how these might influence interactions between people and nature. Landsc. Ecol. 28 (6), 1213–1221.

Steele, M.K., Wolz, H., 2019. Heterogeneity in the land cover composition and configuration of US cities: implications for ecosystem services. Landsc. Ecol. 34, 1247–1261.

Sullivan, J.J., Meurk, C., Whaley, K.J., Simcock, R., 2009. Restoring native ecosystems in urban Auckland: urban soils, isolation, and weeds as impediments to forest establishment. N.Z. J. Ecol. 33, 60–71.

Suding, K.N., 2011. Toward an era of restoration in ecology: successes, failures, and opportunities ahead. Annu. Rev. Ecol. Evol. Syst. 42, 465–487.

Szlavecz, K., Placella, S.A., Pouyat, R.V., Groffman, P.M., Csuzdi, C., Yesilonis, I., 2006. Invasive earthworm species and nitrogen cycling in remnant forest patches. Appl. Soil Ecol. 32 (1), 54–63.

Szlavecz, K., Yesilonis, I., Pouyat, R., 2018. Soil as a foundation to urban biodiversity. In: Ossola, A., Niemelä, J. (Eds.), Urban Biodiversity. Routledge, New York.

Tay, N.E., Hopkins, A.J., Ruthrof, K.X., Burgess, T., Hardy, G.E.S., Fleming, P.A., 2018. The tripartite relationship between a bioturbator, mycorrhizal fungi, and a key Mediterranean forest tree. Austral Ecol. 43 (7), 742–751.

Tejada, M., Hernandez, M.T., Garcia, C., 2009. Soil restoration using composted plant residues: effects on soil properties. Soil Tillage Res. 102, 109–117.

Tennesen, M., 2014. Rare earth. Science 346, 692–695.

Terzi, F., Tezer, A., Turkay, Z., Uzun, O., Köylü, P., Karacor, E., et al., 2019. An ecosystem services-based approach for decision-making in urban planning. J. Environ. Plann. Manage. 63, 1–20.

Thomas, E.C., Lavkulich, L.M., 2015. Anthropogenic effects on metal content in urban soil from different parent materials and geographical locations: a Vancouver, British Columbia, Canada, study. Soil Sci. 180 (4/5), 193–197.

Vannucchi, F., Malorgio, F., Pezzarossa, B., Pini, R., Bretzel, F., 2015. Effects of compost and mowing on the productivity and density of a purpose-sown mixture of native herbaceous species to revegetate degraded soil in anthropized areas. Ecol. Eng. 74, 60–67.

Vergnes, A., Blouin, M., Muratet, A., Lerch, T.Z., Mendez-Millan, M., Rouelle-Castrec, M., et al., 2017. Initial conditions during Technosol implementation shape earthworms and ants diversity. Landsc. Urban Plann. 159, 32–41.

Vilà, M., José, L.E., Hejda, M., Hulme, P.E., Jarošík, V., Maron, J.L., et al., 2011. Ecological impacts of invasive alien plants: a meta-analysis of their effects on species, communities and ecosystems. Ecology Lett 14 (7), 702–708.

Weindorf, D.C., Zartman, R.E., Allen, B.L., 2006. Effect of compost on soil properties in Dallas, Texas. Compost. Sci. Util. 14 (1), 59–67.

Wilkerson, M.L., Mitchell, M.G., Shanahan, D., Wilson, K.A., Ives, C.D., Lovelock, C.E., et al., 2018. The role of socio-economic factors in planning and managing urban ecosystem services. Ecosyst. Serv. 31, 102–110.

Williams, N.S., Lundholm, J., Scott MacIvor, J., 2014. Do green roofs help urban biodiversity conservation? J. Appl. Ecol. 51 (6), 1643–1649.

Windham, L., Laska, M.S., Wollenberg, J., 2004. Evaluating urban wetland restorations: case studies for assessing connectivity and function. Urban Habitats 2 (1), 130–146.

Xin, Z., Li, C., Liu, H., Shang, H., Ye, L., Li, Y., Zhang, C., 2018. Evaluation of temporal and spatial ecosystem services in dalian, china: implications for urban planning. Sustainability 10 (4), 1247.

Xu, J., Liu, C., Hsu, P.C., Zhao, J., Wu, T., Tang, J., et al., 2019. Remediation of heavy metal contaminated soil by asymmetrical alternating current electrochemistry. Nat. Commun. 10 (1), 2440. <https://doi.org/10.1038/s41467-019-10472-x>.

Ye, S., Zeng, G., Wu, H., Zhang, C., Liang, J., Dai, J., et al., 2017. Co-occurrence and interactions of pollutants, and their impacts on soil remediation—a review. Crit. Rev. Environ. Sci. Technol. 47 (16), 1528—1553.

Yilmaz, D., Cannavo, P., Séré, G., Vidal-Beaudet, L., Legret, M., Damas, O., et al., 2018. Physical properties of structural soils containing waste materials to achieve urban greening. J. Soils Sediments 18 (2), 442—455.

Yost, J.L., Egerton-Warburton, L.M., Schreiner, K.M., Palmer, C.E., Hartemink, A.E., 2016. Impact of restoration and management on aggregation and organic carbon accumulation in urban grasslands. Soil Sci. Soc. Am. J. 80 (4), 992—1002.

Young, T., Cameron, D.D., Phoenix, G.K., 2015. Using AMF inoculum to improve the nutritional status of *Prunella vulgaris* plants in green roof substrate during establishment. Urban For. Urban Green. 14 (4), 959—967.

Zeunert, J., 2013. Challenging assumptions in urban restoration ecology. Landsc. J. 32 (2), 231—242.

Ziter, C., Turner, M.G., 2018. Current and historical land use influence soil-based ecosystem services in an urban landscape. Ecol. Appl. 28 (3), 643—654.

Index

Printed in the United States
By Bookmasters